Lecture Notes in Control and Information Sciences - Proceedings

Series editors

This distinguished conference proceedings series publishes the latest research developments in all areas of control and information sciences—quickly, informally and at a high level. Typically based on material presented at conferences, workshops and similar scientific meetings, volumes published in this series will constitute comprehensive state-of-the-art references on control and information science and technology studies.

More information about this series at http://www.springer.com/series/15828

Roberto Tempo · Stephen Yurkovich
Pradeep Misra
Editors

Emerging Applications of Control and Systems Theory

A Festschrift in Honor of Mathukumalli Vidyasagar

Springer

Editors
Roberto Tempo (deceased)
IEIIT—CNR
Politecnico di Torino
Turin
Italy

Pradeep Misra
Russ Engineering Center
Wright State University
Dayton, OH
USA

Stephen Yurkovich
Department of Systems Engineering
University of Texas at Dallas
Richardson, TX
USA

ISSN 2522-5383 ISSN 2522-5391 (electronic)
Lecture Notes in Control and Information Sciences - Proceedings
ISBN 978-3-319-67067-6 ISBN 978-3-319-67068-3 (eBook)
https://doi.org/10.1007/978-3-319-67068-3

Library of Congress Control Number: 2017962042

Printed on acid-free paper

This Springer imprint is published by the registered company Springer International Publishing AG part of Springer Nature
The registered company address is: Gewerbestrasse 11, 6330 Cham, Switzerland

To Prof. Mathukumalli Vidyasagar on his 70th birthday.

M. Vidyasagar—A Brilliant Intellect

Photograph Courtesy: The University of Texas at Dallas

Vidya is a Sanskrit word and it signifies knowledge and learning. The word *Sagar* translates into sea or ocean. Taken together, the word *Vidyasagar* means an ocean of knowledge or sea of learning. Those of us who have had the privilege of knowing Prof. M. Vidyasagar, and even those who know him only through his scholarly achievements, will recognize the rare and remarkable match between the person and the name.

My first exposure to Sagar's remarkable intellect was in 1980. I was a doctoral student at the University of Florida under the supervision of late Professor Kalman. I was researching topics in algebraic system theory such as linear systems over rings and matrix fraction representations of linear systems. I received a very kind and quite unexpected letter from Sagar inviting me to the University of Waterloo for a visit. As a beginning graduate student, I greatly appreciated this as a wonderful opportunity. During the week-long visit, we had many discussions along with a little bit of tennis. As I was eagerly explaining my latest results and ideas to him, I quickly realized that he knew a great deal more about these topics and their relations to other problems in control and system theory. I was deeply impressed by his formidable and penetrating mind: not only was he able to answer questions as soon as they were posed but did so in a way that demonstrated a great understanding of the technical issues. During this visit, he and his wife Shakunthala offered their warm and generous friendship at a personal level.

Over the last 37 years, my admiration and respect for Sagar as a scholar with unparalleled depth and range has grown ever more profound. And he has been a great friend and mentor, personally and professionally, for which I am most grateful.

In a later edition of the book *Feedback Systems: Input-Output Properties*, Sagar writes:

> It is difficult to believe that more than 30 years have passed since Charlie and I published Feedback Systems: Input-Output Properties. In spite of the passage of time, the book continues to get cited, primarily because there are many results in this book that are simply not found anywhere else. Mathematics (or control theory) experiences its own winds of fashion, just as any other subject does.

This passage highlights some of the salient characteristics of Sagar's contributions. They have outlasted the winds of fashion in control theory and continue to remain valuable and relevant. They are foundational in nature and address the most essential issues in the field. They are solid and authoritative. For any researcher, neophyte or seasoned, they can be relied upon to provide excellent building blocks for future work.

Sagar's book *Control Systems: A Factorization Approach* had a great influence on me during the 1980s. I used it to teach advanced graduate level courses in multivariable control at the University of Minnesota. For my graduate students, it was core material they had to master in order to progress in their studies. It was a potent combination of ideas and tools from advanced algebra as applied to essential problems in multivariable control using stable matrix factorizations for transfer functions. Moreover, it also contained the essential learnings from Sagar's prior work in stability theory, feedback systems, and input–output analysis. It also brought in concepts, tools, and techniques from analysis and topology to complement the advanced algebra.

Sagar's work has covered vast ground from circuit theory to power systems to control theory to robotics to learning theory to cancer biology. In each case, he was able to quickly master the state of the art and become a leader in the field. He has been able to renew himself while building on his prior work and knowledge base. There is a certain flow—widening and deepening—in his intellectual journey which is breathtaking and inspiring. Equally impressive is his stellar career as a leader and administrator in government, private, and academic organizations.

Sagar's insatiable curiosity, razor-sharp intellect, wide range of scientific interests, enormous base of experiences, and natural talent for focusing on the essence of any situation make him a unique person to seek perspective and advice. I have been very fortunate to be able to connect with him and seek his insights and analysis throughout my career. Invariably, these conversations and discussions have left me with a new perspective and understanding, clearing the way to progress and new avenues.

As I reflect on Sagar's remarkable career and contributions, I wonder what we and, more importantly, future generations could and should learn from him.

Above all, his demonstrated excellence in working on important problems and subjects at the leading edge of research should inspire us to periodically renew ourselves to work on the frontiers of knowledge. He is a great example of constantly learning about new areas and being open to new approaches and avenues. This insatiable curiosity has been a major pillar of his wide-ranging research career. It is trite to say that the pace of discovery and research is getting ever faster. A natural implication is that research areas and fields become mature more quickly. At the same time, research careers can last for five or more decades. Hence, the ability to renew oneself is bound to be extremely important. There is no better role model in the field of systems and control than Sagar to seek inspiration for renewal and excellence.

Working on foundational questions is of great importance for making long-lasting contributions to knowledge. Sagar exemplifies this trait and ability. It is easy to get caught up in peripheral questions and marginal issues. Only a steadfast commitment and determination to get to the roots and foundations can prevent one from getting distracted. If only we could emulate Sagar in defining and choosing foundational questions and topics, we will be well served.

Finally, Sagar has generously helped mentor a large group of fellow researchers and scholars. Research careers are full of challenges and opportunities. Each one benefits from the work and contributions of others. We can learn from Sagar how to contribute to the intellectual and professional growth of (junior) colleagues in the research community.

Sagar has made indelible imprints on all the research fields he has touched. He has the gift of a most brilliant mind and intellect. For me, he has been a role model and a great friend. I am grateful for all the benefits I have been able to derive from him. I hope to enjoy his mentorship and friendship for decades to come. And I wish him continued successes and fulfillment in all spheres of life.

List of Publications of M. Vidyasagar

Books

1. C. A. Desoer and M. Vidyasagar, *Feedback Systems: Input-Output Properties*, Academic Press, New York, 1975.
2. M. Vidyasagar, *Nonlinear Systems Analysis*, Prentice-Hall, Englewood Cliffs, NJ, 1978.

3. M. Vidyasagar, *Input-Output Analysis of Large-Scale Interconnected Systems: Decomposition, Well-Posedness and Stability*, Springer-Verlag, New York, 1981.
4. M. Vidyasagar, *Control System Synthesis: A Factorization Approach*, MIT Press, Cambridge, MA, 1985.
5. M. W. Spong and M. Vidyasagar, *Robot Dynamics and Control*, John Wiley, New York, 1989.
6. M. Vidyasagar, *Nonlinear Systems Analysis*, (Second Edition), Prentice-Hall, Englewood Cliffs, NJ, 1993.
7. M. Vidyasagar, *A Theory of Learning and Generalization: With Applications to Neural Networks and Control Systems*, Springer-Verlag, London, 1997.
8. M. Vidyasagar, *Learning and Generalization With Applications to Neural Networks*, (Second Edition), Springer-Verlag, London, 2003.
9. M. W. Spong, S. R. Hutchinson and M. Vidyasagar, *Robot Modeling and Control*, John Wiley and Sons, New York, 2006.
10. M. Vidyasagar, *Computational Cancer Biology: An Interaction Network Approach*, Springer-Verlag, London, 2013.
11. M. Vidyasagar, *Hidden Markov Processes: Theory and Applications to Biology*, Princeton University Press, 2014.
12. M. Vidyasagar, *An Introduction to Compressed Sensing*, under preparation.

Republished Books

1. M. Vidyasagar, *Nonlinear Systems Analysis*, Society of Industrial and Applied Mathematics, (SIAM Classics Series), Philadelphia, 2002. (Reissue of 1993 book published by Prentice-Hall.)
2. C. A. Desoer and M. Vidyasagar, *Feedback Systems: Input-Output Properties*, (SIAM Classics Series), Philadelphia, 2009. (Reissue of the 1975 book published by Academic Press.)
3. M. Vidyasagar, *Control System Synthesis: A Factorization Approach*, Morgan & Claypool, 2011. (Completely re-typeset version, with several typos corrected, of Book No. 4 above.)

Publications in Journals

1. M. Vidyasagar and T. J. Higgins, "A controllability criterion for stable linear systems," *IEEE Trans. Auto. Control*, **AC-15(3)**, 391–392, June 1970.
2. J. A. Heinen and M. Vidyasagar, "Lagrange stability and higher order derivatives of Lyapunov functions," *Proc. IEEE*, **58(7)**, 1174, July 1970.
3. M. Vidyasagar, "A controllability criterion for linear systems with unbounded operators," *IEEE Trans. Auto. Control*, **AC-15(4)**, 491–492, August 1970.
4. M. Vidyasagar, "On the controllability of infinite-dimensional linear systems," *J. Opt. Thy. and Appl.*, **6**, 171–173, August 1970.

5. M. Vidyasagar, “An algebraic method of finding a dual graph of a given graph,” *IEEE. Trans. Circ. Thy.*, **CT-17(4)**, 434–436, August 1970.
6. M. V. Subbarao and M. Vidyasagar, “On Watson’s quintuple product identity,” *Proc. Amer. Math. Soc.*, **26**, 23–27, September 1970.
7. M. Vidyasagar, “A novel method of evaluating e^{At} in closed form,” *IEEE Trans. Auto. Control*, **AC-15(5)**, 600–601, October 1970.
8. H. Schneider and M. Vidyasagar, “Cross-positive matrices,” *SIAM J. Num. Anal.*, **7**, 508–519, December 1970.
9. M. Vidyasagar, “An extension of bounded-input-bounded-output stability,” *IEEE Trans. Auto. Control*, **AC-15(6)**, 702–703, December 1970.
10. M. Vidyasagar, “A comment on ‘A definition and some results for distributed system observability’,” *IEEE Trans. Auto. Control*, **AC-16**, 209, February 1971.
11. M. Vidyasagar and J. A. Heinen, “Estimating the transient response and settling time of time-varying linear systems,” *Int. J. Sys. Sci.*, **1**, 257–263, March 1971.
12. M. Vidyasagar, “Optimal control by direct inversion of a positive definite operator in Hilbert space,” *J. Opt. Thy. and Appl.*, **7**, 173–177, March 1971.
13. M. Vidyasagar, “On a class of inverse optimal control problems,” *J. Opt. Thy. and Appl.*, **7**, 287–301, April 1971.
14. M. Vidyasagar, “Causal systems and feedforward loops,” *IEEE Trans. Auto. Control*, **AC-16(2)**, 209, April 1971.
15. M. Vidyasagar, “On the L_2-stability of time-varying linear systems,” *IEEE Trans. Auto. Control*, **AC-16(3)**, 268–269, June 1971.
16. M. Vidyasagar, “A characterization of e^{At} and a constructive proof of the controllability criterion,” *IEEE Trans. Auto.Control*, **AC-16(4)**, 370–371, August 1971.
17. C. A. Desoer and M. Vidyasagar, “General necessary conditions for input-output stability,” *Proc. IEEE*, **59(8)**, 1255–1256, August 1971.
18. M. Vidyasagar and J. C. Giguère, “Equivalence of the stability properties of two related classes of feedback systems,” *Proc. IEEE*, **59(12)**, 1727–1728, December 1971.
19. G. E. Andrews, M. V. Subbarao and M. Vidyasagar, “A family of combinatorial identities,” *Can. Math. Bull.*, **15(1)**, 11–18, January 1972.
20. M. Vidyasagar, “Input-output properties of a broad class of linear time-invariant multivariable systems,” *SIAM J. Control*, **10(1)**, 203–209, February 1972.
21. S. A. Gracovetsky and M. Vidyasagar, “A simple iterative method of supoptimal control of linear time-lag systems with quadratic costs,” *Int. J Control*, **16**, 997–1002, May 1972.
22. M. Vidyasagar, “An instability condition for nonlinear time-varying feedback systems,” *Proc. IEEE*, **60(6)**, 762, June 1972.
23. M. Vidyasagar, “L_p-stability of time-varying linear feedback systems,” *IEEE Trans. Auto. Control*, **AC-17(3)**, 412–414, June 1972.

24. M. Vidyasagar, "A controllability condition for nonlinear systems," *IEEE Trans. Auto. Control*, **AC-17(4)**, 569–570, August 1972.
25. J. C. Giguère, M. Vidyasagar and M. N. S. Swamy, "Input-output stability of two broad classes of lumped-distributed systems," *J. Franklin Inst.*, **294**, 203–213, September 1972.
26. M. Vidyasagar, "Some applications of the spectral radius concept to nonlinear feedback stability," *IEEE Trans. Circ. Thy.*, **CT-19(6)**, 607–615, November 1972.
27. S. A. Gracovetsky and M. Vidyasagar, "Suboptimal control of neutral systems," *Int. J. Control*, **18**, 121–128, January 1973.
28. M. Vidyasagar and T. J. Higgins, "A basic theorem on distributed control and point control," *ASME Trans., J. Dyn. Sys. Meas. and Control*, **95(1)**, 64–67, March 1973.
29. M. Vidyasagar, "An elementary proof of a factorization theorem for positive operators," *IEEE Trans. Auto. Control*, **AC-18(4)**, 403–404, August 1973.
30. M. Vidyasagar and G. S. Mueller, "Optimal control of linear systems with a class of nonquadratic convex cost functionals," *Int. J. Control*, **19**, 657–600, September 1973.
31. M. Vidyasagar, "Maximum power transfer in n-ports with passive loads," *IEEE Trans. Circ. and Sys.*, **CAS-21(5)**, 327–330, May 1974.
32. N. K. Bose and M. Vidyasagar, "Proof of Talbot's conjecture and consequences," *IEEE Trans. Circ. and Sys.*, **CAS-21(9)**, 701, September 1974.
33. M. Vidyasagar, "Simplified graphical stability criteria for distributed feedback systems," *IEEE Trans. Auto. Control*, **AC-20(3)**, 440–442, June 1975.
34. M. Vidyasagar, "Conditions for a feedback transfer matrix to be proper," *IEEE Trans. Auto. Control*, **AC-20(4)**, 570–571, August 1975.
35. S. R. K. Dutta and M. Vidyasagar, "Optimal design of nonlinear DC transistor circuits without solving network equations," *IEEE Trans. Circ. and Sys.*, **CAS-22(8)**, 661–665, August 1975.
36. M. Vidyasagar, "Coprime factorizations and stability of multivariable distributed feedback systems," *SIAM J. Control*, **13(4)**, 1144–1155, November 1975.
37. M. Vidyasagar, "On the existence of optimal controls," *J. Opt. Thy. and Appl.*, **17**, 273–278, November 1975.
38. M. Vidyasagar, "A new approach to N-person nonzero sum linear differential games," *J. Opt. Thy. and Appl.*, **18**, 171–175, January 1976.
39. M. Vidyasagar, "L_2-instability criteria for interconnected systems," *SIAM J. Control*, **15(1)**, 312–328, February 1977.
40. M. Vidyasagar and M. A. L. Thathachar, "A note on feedback stability and instability in the Marcinkiewicz space M_2," *IEEE Trans. Circ. and Sys.*, **CAS-24(3)**, 127–131, March 1977.
41. M. Vidyasagar, "On the instability of large-scale systems," *IEEE Trans. Auto. Control*, **AC-22(2)**, 267–269, April 1977.

42. S. R. K. Dutta and M. Vidyasagar, "Worst-case design of DC transistor circuits," *IEEE Trans. Circ. and Sys.*, **CAS-24(5)**, 273–274, May 1977.
43. M. K. Sundareshan and M. Vidyasagar, "L_2-stability of large-scale dynamical systems—Criteria via positive operator theory," *IEEE Trans. Auto. Control*, **AC-22(3)**, 396–399, June 1977.
44. M. Vidyasagar, "Instability of feedback systems," *IEEE Trans. Auto. Control*, **AC-22(3)**, 466–467, June 1977.
45. S. R. K. Dutta and M. Vidyasagar, "New algorithms for constrained minimax optimization," *Math. Programming*, **13**, 140–155, October 1977.
46. M. Vidyasagar, "L_2-stability of interconnected systems using a reformulation of the passivity theorem," *IEEE Trans. on Circ.and Sys.*, **CAS-24(11)**, 637–645, November 1977.
47. M. Vidyasagar, "On matrix measures and convex Liapunov functions," *J. Math. Anal. Appl.*, **62**, 90–103, January 1978.
48. R. A. El-Attar and M. Vidyasagar, "Order reduction by ℓ_1- and ℓ_∞-norm minimization," *IEEE Trans. Auto. Control*, **AC-23(4)**, 731–734, August 1978.
49. M. Vidyasagar, "On the use of right-coprime factorizations in distributed feedback systems containing unstable subsystems," *IEEE Trans. Circ. and Sys.*, **CAS-25(11)**, 916–925, November 1978.
50. M. Vidyasagar, "L_∞-stability criteria for interconnected systems using exponential weighting," *IEEE Trans. Circ. and Sys.*, **CAS-25(11)**, 946–947, November 1978.
51. R. A. El-Attar and M. Vidyasagar, "System order reduction using the induced operator norm and its application to linear regulation," *J. Franklin Inst.*, **306**, 457–474, December 1978.
52. R. A. El-Attar, M. Vidyasagar and S. R. K. Dutta, "New algorithms for ℓ_1-norm minimization with application to nonlinear ℓ_1-norm approximation," *SIAM J. Num. Anal.*, **16**, 70–86, February 1979.
53. R. A. El-Attar and M. Vidyasagar, "Subsystem simplification in large-scale system analysis," *IEEE Trans. Auto. Control*, **AC-24(2)**, 321–323, April 1979.
54. M. Vidyasagar, "New passivity-type criteria for large-scale interconnected systems," *IEEE Trans. Auto. Control*, **AC-24(4)**, 575–579, August 1979.
55. M. Vidyasagar, "On the well-posedness of large-scale interconnected systems," *IEEE Trans. Auto. Control*, **AC-25(3)**, 413–420, June 1980.
56. M. Vidyasagar, "On the stabilization of nonlinear systems using state detection," *IEEE Trans. Auto. Control*, **AC-25(3)**, 504–509, June 1980.
57. M. Vidyasagar, "Decomposition techniques for large-scale systems with nonadditive interconnections," *IEEE Trans. Auto. Control*, **AC-25(4)**, 773–779, August 1980.
58. M. Vidyasagar, "New L_2-instability criteria for large-scale interconnected systems," *IEEE Trans. Circ. and Sys.*, **CAS-27(10)**, 970–973, October 1980.
59. M. Vidyasagar, "L_∞-instability of large-scale interconnected systems using orthogonal decomposition and exponential weighting," *IEEE Trans. Circ. and Sys.*, **CAS-27(10)**, 973–976, October 1980.

60. P. J. Moylan, A. Vannelli and M. Vidyasagar, "On the stability and well-posedness of interconnected nonlinear dynamical systems," *IEEE Trans. Circ. and Sys.*, **CAS-27(11)**, 1097–1102, November 1980.
61. N. Viswanadham and M. Vidyasagar, "Stabilization of linear and nonlinear dynamical systems using an observer-controller configuration," *Sys. Control Lett.*, **1**, 87–91, August 1981.
62. W. Kotiuga and M. Vidyasagar, "Bad data rejection properties of weighted least absolute value techniques applied to static state estimation," *IEEE Trans. Power App. and Sys.*, **PAS-101**, 844–853, April 1982.
63. M. Vidyasagar and A. Vannelli, "New relationships between input-output and Lyapunov stability," *IEEE Trans. Auto. Control*, **AC-27(2)**, 481–483, April 1982.
64. M. Vidyasagar, A. Boyarski and A. Vannelli, "On the stability properties of perturbed linear nonstationary systems," *J. Math. Anal. Appl.*, **38**, 245–256, July 1982.
65. M. Vidyasagar, H. Schneider and B. A. Francis, "Algebraic and topological aspects of feedback stabilization," *IEEE Trans. Auto.Control*, **AC-27(4)**, 880–894, August 1982.
66. M. Vidyasagar and N. Viswanadham, "Algebraic design techniques for reliable stabilization," *IEEE Trans. Auto. Control*, **AC-27(5)**, 1085–1095, October 1982.
67. B. A. Francis and M. Vidyasagar, "Algebraic and topological aspects of the regulator problem for lumped linear systems," *Automatica*, **19**, 87–90, January 1983.
68. K. R. Davidson and M. Vidyasagar, "Causal invertibility and stability of asymmetric half-plane digital filters," *IEEE Trans. Acoustics, Speech, Sig. Proc.*, **ASSP-31**, 195–201, February 1983.
69. M. Vidyasagar and N. Viswanadham, "Algebraic characterization of decentralized fixed modes," *Sys. Control Lett.*, **3**, 69–72, July 1983.
70. M. Vidyasagar, "A note on time-invariance and causality," *IEEE Trans. Auto. Control*, **AC-28(9)**, 929–931, September 1983.
71. M. Vidyasagar, "The graph metric for unstable plants and robustness estimates for feedback stability," *IEEE Trans. Auto. Control*, **AC-29(5)**, 403–418, May 1984.
72. H. Maeda and M. Vidyasagar, "Some results on simultaneous stabilization," *Sys. Control Lett.*, **5**, 205–208, September 1984.
73. A. Vannelli and M. Vidyasagar, "Maximal Lyapunov functions and domains of attraction for autonomous nonlinear systems," *Automatica*, **21**, 69–80, January 1985.
74. M. Vidyasagar, "Robust stabilization of singularly perturbed systems," *Sys. Control Lett.*, **5**, 413–418, May 1985.
75. H. Maeda and M. Vidyasagar, "Design of multivariable feedback systems with infinite gain margin and decoupling," *Sys. Control Lett.*, **6**, 127–130, July 1985.

76. M. Vidyasagar and N. Viswanadham, "Reliable stabilization using a multi-controller configuration," *Automatica*, **21**, 599–602, September 1985.
77. M. Vidyasagar and H. Kimura, "Robust controllers for uncertain linear multivariable systems," *Automatica*, **22**, 85–94, January 1986.
78. H. Maeda and M. Vidyasagar, "Infinite gain margin problem in multivariable feedback systems," *Automatica*, **22**, 131–133, January 1986.
79. C. C-H. Ma and M. Vidyasagar, "Nonpassivity of linear discrete-time systems," *Sys. Control Lett.*, **7**, 51–53, February 1986.
80. K. D. Minto and M. Vidyasagar, "A state-space approach to simultaneous stabilization," *Control: Theory and Advanced Technology*, **2**, 39–64, March 1986.
81. M. Vidyasagar, "On undershoot and nonminimum phase zeros," *IEEE Trans. Auto. Control*, **AC-31(5)**, 440, May 1986.
82. M. Vidyasagar "Optimal rejection of persistent bounded disturbances," *IEEE Trans. Auto. Control*, **AC-31(6)**, 527–534, June 1986.
83. M. Vidyasagar, "New directions of research in nonlinear system stability," (Invited Survey Paper), *Proc. IEEE*, **74(8)**, 1060–1091, August 1986.
84. M. Vidyasagar, B. C. Lévy and N. Viswanadham, "A note on the genericity of simultaneous stabilization and pole assignability," *Circ. Sys. Sig. Processing*, **5**, 371–387, 1986.
85. M. Vidyasagar N. Viswanadham, "Construction of inverses with prescribed nonzero minors and applications to decentralized stabilization," *Linear Alg. and Its Appl.*, **83**, 103–115, 1986.
86. C. C-H. Ma and M. Vidyasagar, "Direct globally convergent adaptive regulation of stable multivariable plants," *IEEE Trans. Auto. Control*, **AC-32(6)**, 543–547, June 1987.
87. M. W. Spong and M. Vidyasagar, "Robust linear compensator design for nonlinear robotic control," *IEEE J. Robotics and Automation*, **RA-3**, 345–351, August 1987.
88. M. Vidyasagar, "Some results on simultaneous stabilization with multiple domains of stability," *Automatica*, **23**, 535–540, July 1987.
89. C. C-H. Ma and M. Vidyasagar, "Parametric conditions for stability of reduced order linear time-varying control systems," *Automatica*, **23**, 625–634, September 1987.
90. M. Vidyasagar, "Normalized coprime factorizations for nonstrictly proper systems," *IEEE Trans. Auto. Control*, **AC-33(3)**, 300–301, March 1988.
91. M. Vidyasagar, R. Bertschmann and C. S. Sallaberger, "Some simplifications of the graphical Nyquist stability criterion," *IEEE Trans. Auto. Control*, **AC-33(3)**, 301–305, March 1988.
92. M. Vidyasagar, "A state space interpretation of simultaneous stabilization," *IEEE Trans. Auto. Control*, **AC-33(5)**, 506–508, May 1988.
93. K. H. Low and M. Vidyasagar, "A Lagrangian formulation of the dynamic model for flexible manipulators," *ASME Trans., J. Dynamic Sys., Meas. and Control*, **110(2)**, 175–181, June 1988.

94. A. Sankaranarayanan and M. Vidyasagar, "On the problem of contour following with applications to force control," *ASME Trans., J. Dynamic Sys., Meas. and Control*, **110(4)**, 443–448, December 1988.
95. M. Vidyasagar, "SFPACK—An interactive environment for control system analysis and synthesis," *Robotics and Computer Integrated Manufacturing*, **5 (4)**, 311–319, 1989.
96. M. Vidyasagar and B. D. O. Anderson, "Approximation and stabilization of distributed systems by lumped systems," *Sys. Control Lett.*, **12(2)**, 95–101, February 1989.
97. T. Sugie and M. Vidyasagar, "Further results on the robust tracking problem for two degree of freedom controllers," *Sys. Control Lett.*, **13**, 101–108, 1989.
98. D. McFarlane, K. Glover and M. Vidyasagar, "Order reduction using normalized coprime factorizations," *IEEE Trans. Auto. Control*, **AC-35(3)**, 369–373, March 1990.
99. K. A. Morris and M. Vidyasagar, "A comparison of different models for beam vibrations from the standpoint of controller design," *ASME Trans., J. Dyn. Sys., Meas. and Control*, **112(3)**, 349–356, September 1990.
100. G. S. Deodhare and M. Vidyasagar, "Some results on the ℓ_1-optimality of feedback control systems: The SISO discrete-time case," *IEEE Trans. Auto. Control*, **35(9)**, 1082–1085, September 1990.
101. Y. C. Chen and M. Vidyasagar, "Optimal control of robotic manipulators in the presence of obstacles," *J. Robotic Sys.*, **7(5)**, 721–740, October 1990.
102. M. Vidyasagar, "An analysis of the equilibria of neural networks with linear interconnections," *Sadhana*, bf 15(59–60), 283–300, December 1990.
103. M. Vidyasagar, "Further results on the optimal rejection of persistent bounded disturbances," *IEEE Trans. Auto. Control*, **AC-36(6)**, 642–652, June 1991.
104. G. S. Deodhare and M. Vidyasagar, "Every stabilising controller is H_∞ and ℓ_1-optimal," *IEEE Trans. Auto. Control*, **AC-36(9)**, 1070–1073, September 1991.
105. D. Wang and M. Vidyasagar, "Transfer functions for a single flexible link," *Int. J. Robotic Res.*, **10(5)**, 540–549, October 1991.
106. D. Wang and M. Vidyasagar, "Control of a class of manipulators with a single flexible link, Part I: Feedback linearization," *ASME Trans., J. Dyn. Sys., Meas. and Control*, **113(4)**, 655–661, December 1991.
107. D. Wang and M. Vidyasagar, "Control of a class of manipulators with a single flexible link, Part II: Observer-controller stabilization," *ASME Trans., J. Dyn. Sys., Meas. and Control*, **113(4)**, 662–668, December 1991.
108. D. Wang and M. Vidyasagar, "Modelling a class of multi-link manipulators with the last link flexible," *IEEE Trans. Robotics and Automation*, **8(1)**, 33–41, February 1992.
109. M. Vidyasagar, "Improved neural networks for analog to digital conversion," *Circ., Sys. and Sig. Proc.*, **11(3)**, 387–398, 1992.
110. G. S. Deodhare and M. Vidyasagar, "Control system design via infinite linear programming," *Int. J. Control*, **55(6)**, 1351–1380, June 1992.

111. D. Wang and M. Vidyasagar, "Passive control of a single flexible link," *Int. J. Robotic Res.*, **11(6)**, 572–579, December 1992.
112. H. Krishnan and M. Vidyasagar, "Bounded input H_2-optimal feedback control of linear systems with application to the control of a flexible beam," *Control: Theory and Advanced Technology*, **9(2)**, 381–403, June 1993.
113. M. Vidyasagar, "Location and stability of high-gain equilibria of nonlinear neural networks," *IEEE Trans. Neural Net.*, **NN-4(4)**, 660–672, July 1993.
114. M. Vidyasagar, "Convergence of higher-order two-state neural networks with modified updating," *Sadhana*, **19(2)**, 239–255, April 1994.
115. M. Vidyasagar, "Dynamical systems, gradient flows and optimization," *J. of Institute of Systems, Control and Information Engrg. (Japan)*, **39(1)**, 22–28, January 1995.
116. M. Vidyasagar, "Minimum-seeking properties of analog neural networks with multilinear objective functions," *IEEE Trans. Auto. Control*, **AC-40(8)**, 1359–1375, August 1995.
117. M. Vidyasagar, "A brief history of the graph topology," *J. Soc. Instrumentation and Control Engineers (Japan)*, **34(8)**, 621–628, August 1995.
118. M. Vidyasagar, "A brief history of the graph topology," *European J. Control*, **2**, 80–87, 1996.
119. S. R. Kulkarni and M. Vidyasagar, "Learning decision rules for pattern classification under a family of probability measures," *IEEE Trans. Info. Thy.*, **IT-43(1)**, 154–166, January 1997.
120. M. Vidyasagar, "Are analog neural networks better than binary neural networks?" *Circ., Sys. and Sig. Proc.*, **17(2)**, 243–269, 1998.
121. M. Vidyasagar, "An introduction to the statistical aspects of PAC learning theory," *Sys. Cont. Lett.*, **40**, 115–124, 1998.
122. D. Deodhare, M. Vidyasagar and S. S. Keerthi, "Design of fault-tolerant neural networks using minimax optimization," *IEEE Trans. Neural Net.*, **NN-9(5)**, 891–900, September 1998.
123. M. Vidyasagar and S. R. Kulkarni, "Some contributions to fixed distribution learning theory," *IEEE Trans. Auto. Control*, **45(2)**, 217–234, February 2000.
124. M. Vidyasagar, "Beyond H_∞ design: Robustness, disturbance rejection and Aizerman-Kalman type conjectures in general signal spaces," *Int. J. Robust and Nonlinear Control*, **10**, 961–982, 2000.
125. M. Vidyasagar, S. Balaji and B. Hammer, "Closure properties of uniform convergence of empirical means and PAC learnability under a family of probability measures," *Sys. Cont. Lett*, **42**, 151–157, 2001.
126. M. Vidyasagar and V. Blondel, "Probabilistic solutions to some NP-hard matrix problems," *Automatica*, **37**, 1397–1401, 2001.
127. M. Vidyasagar, "Randomized algorithms for robust controller synthesis using statistical learning theory," *Automatica*, **37**, 1515–1528, 2001.
128. M. Campi and M. Vidyasagar, "Learning with prior information," *IEEE Trans. Auto. Control*, **AC-41(11)**, 1682–1695, November 2001.

129. M. Vidyasagar, "A tutorial introduction to randomized algorithms for robust controller synthesis using statistical learning theory," *European J. Control*, **5–6**, 283–310, 2001.
130. R. L. Karandikar and M. Vidyasagar, "Rates of uniform convergence of empirical means with mixing processes," *Stat. and Prob. Let.*, **58**, 297–307, 1 June 2002.
131. M. Vidyasagar and R. L. Karandikar, "System identification—A learning theory approach," *Journal of Process Control*, **18**, 421–430, 2007.
132. M. Vidyasagar, S. S. Mande, C. V. S. K. Reddy and V. Raja Rao, "The 4M (mixed memory Markov model) algorithm for finding genes in prokaryotic genomes," *IEEE Transactions on Circuits and Systems*, **CAS-55(1)**, 26–37, January 2008 (Special Issue on Systems Biology).
133. M. Vidyasagar, "A tutorial introduction to financial engineering," *Current Trends in Science, (Platinum Jubilee Special, N. Mukunda Editor)*, Indian Academy of Sciences, 163–185, 2009.
134. M. Vidyasagar, "The complete realization problem for hidden Markov models: A survey and some new results," *Mathematics of Control, Signals and Systems*, 23(1), 1–65, 2011.
135. M. Vidyasagar, "Probabilistic methods in cancer biology," *European Journal of Control*, 17(5–6), 483–511, September–December 2011.
136. M. Vidyasagar, "A metric between probability distributions on finite sets of different cardinalities and applications to order reduction," *IEEE Transactions on Automatic Control*, 54(10), 2464–2477, October 2012.
137. M. Eren Ahsen and M. Vidyasagar, "Mixing coefficients between discrete and real random variables: Computation and properties," *IEEE Transactions on Automatic Control*, **59(1)**, 34–47, January 2014.
138. M. Vidyasagar, "An elementary derivation of the large deviation rate function for finite state Markov processes," *Asian Journal of Control*, **16(1)**, 1–19, January 2014.
139. M. Vidyasagar, "Machine learning methods in the cancer biology of cancer," *Proceeding of the Royal Society, Part A*, (Invited Paper), **470**, Item 20140081, 23 April 2014.
140. M. Vidyasagar, "Identifying predictive features in drug response using machine learning: Opportunities and challenges," *Annual Review of Pharmacology and Toxicology*, 55(1), 1–29, 2015.
141. Burook Misganaw and M. Vidyasagar, "Exploiting ordinal structure in multi-class classification: Application to ovarian cancer," *IEEE Life Sciences Letters*, 1(1), 15–18, 2015.
142. Burook Misganaw, Eren Ahsen, Nitin Singh, Keith A. Baggerly, Anna Unruh, Michael A. White and M. Vidyasagar, "Optimized Prediction of Extreme Treatment Outcomes in Ovarian Cancer," *Cancer Informatics*, 14 (Suppl5), 45–55, March 2016.
143. Nitin Singh and M. Vidyasagar, "bLARS: An algorithm to infer gene regulatory networks," *IEEE Transactions on Computational Biology and Bioinformatics*, 13(2), 301–314, March–April 2016.

144. M. Eren Ahsen and M. Vidyasagar, "Error bounds for compressed sensing algorithms with group sparsity: A unified approach," *Applied and Computational Harmonic Analysis*, accepted for publication.
145. Mehmet Eren Ahsen, Todd P Boren, Nitin K Singh, Burook Misganaw, David G Mutch, Kathleen N Moore, Floor J Backes, Carolyn K. McCourt, Jayanthi S Lea, David S Miller, Michael A White and Mathukumalli Vidyasagar, "Sparse Feature Selection for Classification and Prediction of Metastasis in Endometrial Cancer," *BMC Genomics*, accepted for publication.
146. Mehmet Eren Ahsen, Niharika Challapalli and Mathukumalli Vidyasagar, "Two New Approaches to Compressed Sensing Exhibiting Both Robust Sparse Recovery and the Grouping Effect," *Journal of Machine Learning Research*, accepted for publication.
147. Mathukumalli Vidyasagar, "Machine learning methods in the computational biology of cancer," *Annual Reviews in Control*, accepted for publication.

Pramod P. Khargonekar
Distinguished Professor of Electrical Engineering and
Computer Science and Vice Chancellor for Research
University of California
Irvine, CA, USA
e-mail: pramod.khargonekar@uci.edu

A Tribute to Roberto Tempo (1956–2017)

Photograph included with permission from M. Vidyasagar

The controls community lost an exemplary leader, researcher, and educator when Roberto Tempo passed away suddenly on 15 January, 2017. Roberto was not only a leading researcher but also a visionary when it came to developing and furthering all aspects of the controls community. As a researcher, he is perhaps best known for his contributions to formulating randomized algorithms for intractable design problems. Some of his former students have become leading researchers in their own right, a testimony to Roberto's mentorship. As a leader, he held many administrative positions, of which only a few are mentioned here. He was an Editor-in-Chief of Automatica and Senior Editor of IEEE Transactions on Automatic Control during 2011–2014. He served as the President of the IEEE Control Systems Society during 2010. In all of these capacities, he had a very clear idea of how to spot talent and nurture it. Many of the initiatives that he started as

Editor of Automatica and as President of IEEE CSS have now become permanent features. Aside from these, he also formulated alternative bibliographic metrics for the widely used (and widely misused) citation indices such as the h-index, that were less vulnerable to manipulation. These indices could be used to assess the reputations of individual researchers as well as journals.

Roberto was born in Torino in 1956, and in 1980 he graduated in Electrical Engineering from Politecnico di Torino. After a period spent at Politecnico di Torino, he joined the National Research Council of Italy (CNR) at the research institute IEIIT, Torino, where he was a Director of Research of Systems and Computer Engineering since 1991. He held many visiting appointments at leading universities around the world. He is survived by his wife Cristina, sister Raffaella, and daughter Giulia.

Beyond all of these dry statistics, those of us who were fortunate to have known him lost a true friend and a wonderful human being. He was easily approachable, selfless, and incredibly generous with his support and advice. He and I (and Cristina) became instant friends when we first met in 1995, and subsequently our families also became close friends. He will be dearly missed by all who knew him.

Roberto was originally supposed to have edited this volume, along with Steve Yurkovich. I am grateful to Pradeep Misra for stepping in and taking over the task of coediting the volume.

M. Vidyasagar
Systems Biology Science Department
The University of Texas at Dallas
Richardson, Dallas, TX, USA
e-mail: m.vidyasagar@utdallas.edu
April 2017

Contents

Chapter 1
Passivity-Based Ensemble Control for Cell Cycle Synchronization

Karsten Kuritz, Wolfgang Halter and Frank Allgöwer

Abstract We investigate the problem of synchronizing a population of cellular oscillators in their cell cycle. Restrictions on the observability and controllability of the population imposed by the nature of cell biology give rise to an ensemble control problem specified by finding a broadcast input based on the distribution of the population. We solve the problem by a passivity-based control law, which we derive from the reduced phase model representation of the population and the aim of sending the norm of the first circular moment to one. Furthermore, we present conditions on the phase response curve and circular moments of the population which are sufficient for synchronizing a population of cellular oscillators.

1.1 Introduction

The cell cycle is central to life. Every living organism relies on the cell division cycle for reproduction, tissue growth, and renewal. Malfunction in this highly controlled cell cycle machinery is linked to various diseases, including Alzheimer's disease and cancer [10, 26]. Cause and cure of these diseases are two sides of the same coin, and thus understanding of the cell cycle machinery and approaches to control it are subjects of ongoing research [20]. Mathematically, the cell cycle machinery can be described as dynamical system which obeys limit cycle behavior [3, 7] with dynamics of the general form

$$\dot{x} = f(x, u) \,. \tag{1.1}$$

Therein, the states x represent different molecular species in the cell which can be indirectly affected by external inputs u such as growth conditions, drugs, and other

This work is dedicated to Professor Muthukumalli Vidyasagar on his 70th birthday.

K. Kuritz · W. Halter · F. Allgöwer (✉)
Institute for Systems Theory and Automatic Control,
Pfaffenwaldring 9, 70569 Stuttgart, Germany
e-mail: allgower@ist.uni-stuttgart.de

R. Tempo et al. (eds.), *Emerging Applications of Control and Systems Theory*, Lecture Notes in Control and Information Sciences - Proceedings,
https://doi.org/10.1007/978-3-319-67068-3_1

environmental factors. Another control approach can be realized by directly regulating the expression levels of specific proteins, e.g., by optogenetics [14]. Besides the agent-based description, with each agent beeing a cellular oscillator with dynamics (1.1), proliferating cell populations are often represented by structured population models [2, 9]. The resulting dynamics are governed by partial differential equations, belonging to the *Liouville equations* [1] of the general form

$$\partial_t \rho(x,t) = -\langle \partial_x, f(x,u)\rho(x,t) \rangle \ . \tag{1.2}$$

The concept of reduced phase models connects the nonlinear dynamics in (1.1) with age-structured population models, thereby facilitating control approaches based on the phase distribution of nonlinear oscillators [12, 18]. Control of these oscillators is studied intensively, e.g., by the authors of [19, 22, 23].

In this article, we address the following control problem: Find a control input u for a population of identical cellular oscillators such that the agents are synchronized in their cell cycle. Several constraints imposed by the nature of cell biology complicate the task. (1) Experimental observation of the cell cycle state of individual agents over time is barely possible. A more realistic experimental observation is composed of representative samples drawn from the population from which the distribution of cells in the cell cycle must be reconstructed [13, 25]. (2) Two new agents arise by division at the end of the cell cycle, resulting in exponential growth of the number of controlled agents and non-smooth boundary conditions of the PDE. (3) Only broadcast input signals can be realized, giving rise to an ensemble control problem.

Our approach to solve the above stated control problem is organized as follows. Section 1.2 introduces the theoretic foundation of our control approach, compromising the classical input-output framework for passivity-based controller design and reduced phase models for the representation of weakly coupled oscillators. The control methodology is developed in Sect. 1.3. Section 1.4 examines the control methodology applied to a nonlinear ODE model of the mammalian cell cycle. Section 1.5 contains concluding comments.

1.2 Theoretical Foundation

As mentioned above, we are interested in controlling a population of many identical uncoupled dynamical systems (1.1). The dynamics of the population follows the aforementioned Liouville equation (1.2), so for a given input $u(t)$ and initial distribution $\rho(x,0) = \rho_0(x)$ we may find the solution of the PDE

$$\rho(\cdot,t) = \Upsilon(u,\rho_0,t) \ , \quad t > 0 \ . \tag{1.3}$$

An observable feature may for instance be the moments of (1.3), and the output of the system may be any function of these moments. More general, we consider any function which maps the solution of the PDE to a scalar value as a possible output function

$$y(t) = h(\rho(\cdot, t)) , \quad y(t) \in \mathbb{R} . \tag{1.4}$$

We will develop our control methodology on the fundamentals of classical input-output frameworks and the concept of reduced phase models, reviewed below.

1.2.1 Input/Output Mapping and Control Approach

With output (1.4) given, we note that the system can now be recast as an input-output mapping of an input signal u to an output signal y. Following the formal framework treated in [4], let $x : \mathbb{R}^+ \to \mathbb{R}$ be a scalar function of time and

$$x_T = \begin{cases} x(t), & t \leq T \\ 0, & t > T \end{cases} \tag{1.5}$$

the T-truncated signal. Given the L^2 inner product

$$\langle x, y \rangle = \int_0^{\infty} x(t)y(t)\mathrm{d}t , \tag{1.6}$$

we let

$$\mathscr{L}_e \triangleq \left\{x : \forall T \in \mathbb{R}^+ , \ \langle x_T, x_T \rangle < \infty \right\} \tag{1.7}$$

be the space of signals x with the property that all truncations have finite L^2-norm and

$$\mathscr{L} \triangleq \{x : \ \langle x, x \rangle < \infty\} \tag{1.8}$$

the space of signals for which this holds for the complete signal.

We now define the mapping

$$\begin{aligned} H_1 : \quad \mathscr{L}_e &\to \mathscr{L}_e , \\ u &\mapsto y \end{aligned} \tag{1.9}$$

which takes an arbitrary input signal $u \in \mathscr{L}_e$ and returns the output signal $y \in \mathscr{L}_e$, depending on the initial distribution $\rho_0(x)$ and its evolution dynamics (1.2).

Given this approach, the passivity of such a system can be studied using the classical input-output framework treated in [4], avoiding the difficulties of formulating a proper state space for defining a storage function. Such a state space may for instance be found by taking the circular moments as state variables, however, moment closure might not be given. With the mapping H_1 defined, we want to apply an output feedback approach as depicted in Fig. 1.1.

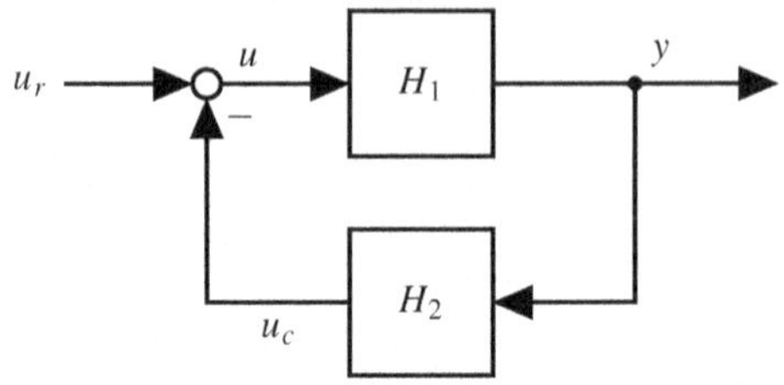

Fig. 1.1 The output is chosen such that it is connected to our goal of synchronizing (or balancing) the population of agents. If the mapping H_1 is passive and the controller H_2 is strictly passive one concludes that $y \in \mathscr{L}$

1.2.2 Reduced Phase Models

In the following, we review the basic concept of reduced phase models and phase response curves briefly and refer the interested reader to the excellent book [11] and references therein. The notion of reduced phase models greatly simplifies the system to be controlled. The main statement of the concept of reduced phase models is the following: Consider a family of dynamical systems of the form

$$\dot{\xi}(t) = f(\xi(t)) , \quad \xi(t) \in \mathbb{R}^n \tag{1.10}$$

having an exponentially stable limit cycle $\gamma \subset \mathbb{R}^n$ with period T_d. Then

$$\dot{\theta}(t) = \omega , \quad \theta(t) \in S^1 \tag{1.11}$$

is a local canonical model for such oscillators, where $\theta(t)$ is called the phase of the oscillator with frequency $\omega = \frac{2\pi}{T_d}$. This statement is based on the notion of *isochrons* introduced by Winfree [24] and its basic idea, illustrated in Fig. 1.2, is to find a neighborhood W of γ and a function $\psi: W \to S^1$, such that $\theta(t) = \psi(\xi(t))$ is a solution of (1.11). Winfree called the set of all initial conditions $z(0) \in \mathbb{R}^n$ of which the solution $z(t)$ approaches the solution $\xi(t)$, with $\xi(0) \in \gamma$ an *isochron* of $\xi(0)$

$$M_{\xi(0)} = \{z(0) \in W\colon \|\xi(t) - z(t)\| \to 0 \text{ as } t \to \infty\} . \tag{1.12}$$

Furthermore, Guckenheimer [8] showed, that there always exists a neighborhood W of a limit cycle that is invariantly foliated by the isochrons $M_\xi, \xi \in \gamma$ in the sense that the flow maps isochrons to isochrons. Consider the function $\psi_2\colon W \to \gamma$, sending a point in the neighborhood $z \in M_\xi \subset W$ to the generator of its isochron $\xi \in \gamma$. Additionally, the periodic orbit of an oscillator is homeomorphic to the unit circle. One can, therefore, define the function $\psi_1 : \gamma \to S^1$ which maps the solution $\xi(t)$ with $\xi(0) \in \gamma$ to the solution of (1.11). The function $\psi : W \to S^1$ is a composition of ψ_1 and ψ_2, $\psi = \psi_1 \circ \psi_2$, mapping $\xi(t) \in W$ uniquely to its corresponding phase $\theta(t)$ of the reduced phase model (Fig. 1.2).

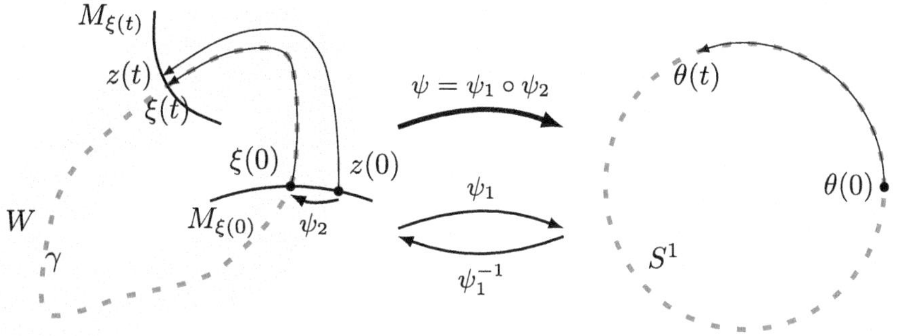

Fig. 1.2 A neighborhood W of the limit cycle γ of an oscillator is invariantly foliated by isochrons M_ξ. The flow maps isochrons to isochrons. The function $\psi = \psi_1 \circ \psi_2$ maps an oscillator $\xi(t) \in W$ uniquely to its phase on the unit circle $\theta(t) = \psi(\xi(t))$

Applying the theory of reduced phase models to a weakly forced oscillator

$$\dot{\xi}(t) = f(\xi(t)) + u(t) \ , \quad \xi(0) = \xi_0 \in W \tag{1.13}$$

where the term $u(t) = \varepsilon v(t)$ denotes an exogenous input, one obtains the reduced phase model of the form

$$\dot{\theta}(t) = \omega + Z(\theta(t))\, u(t) \ . \tag{1.14}$$

Here, weakly forced is in the sense that ε is sufficiently small such that $\xi(t)$ stays inside the neighborhood W for all $t > 0$. The function Z is called phase response curve (PRC) and describes the magnitude of phase changes after perturbing an oscillatory system. Based on *Malkins Theorem* [15, 16], the PRC is the solution of the adjoint problem $\mathrm{d}Z(t)/\mathrm{d}t = -\left(\mathrm{D}f(\xi(t))\right)^\top Z(t)$, with the normalization condition $Z(t)f(\xi(t)) = 1$ for any t, where $\mathrm{D}f$ is the Jacobian matrix which is evaluated along the periodic orbit, $\xi(t) \in \gamma$.

1.2.3 From Reduced Phase Model to Age-Structured Population Models

To simplify the notation, we replace the phase variable θ in the remainder by the variable $x \in S^1$. Given a family of weakly coupled identical oscillators in its reduced phase representation (1.14), the corresponding Liouville equation for the time evolution of the number density $n(x, t)$ of oscillators on the unit circle reads

$$\partial_t n(x, t) + \partial_x \left(\kappa(x, u) n(x, t)\right) = 0 \ . \tag{1.15}$$

The vector field equals the reduced phase model $\kappa(x, u) = \omega + Z(x)u$. In case of a cell population, a division of a mother cell into two daughter cells results in the boundary condition

$$n(0, t) = 2\,n(2\pi, t)\ . \tag{1.16}$$

The model (1.15) and (1.16), with $u(t) = 0$, belongs to the model class of age-structured population models, based on the well-known von Foerster–McKendrick models [6, 17], which are widely used to study cell cycle-related processes. The distribution of cells $q(x, t) = n(x, t)/N(t)$, obtained by normalizing the number density with the total cell number $N(t) = \int_0^{2\pi} n(x, t)\mathrm{dx}$ admits a time-invariant distribution

$$\bar{q}(x) = 2\gamma \mathrm{e}^{-\gamma x}\ , \tag{1.17}$$

where $\gamma = \frac{\log 2}{T_d}$ is the growth rate of the population [21].

We further define the k-th circular moment of some distribution ρ as

$$m_k(\rho(\cdot, t)) = \int_0^{2\pi} \mathrm{e}^{ikx}\rho(x, t)\mathrm{d}x\ . \tag{1.18}$$

By omitting the argument in (1.18), we refer to the complex number $m_k = r\mathrm{e}^{ik\phi}, r \in [0, 1], \phi \in S^1$, obtained by evaluating $m_k(\rho(\cdot, t))$ with some specified distribution. In a synchronized population corresponding to a Dirac delta distribution, the length of the first circular moment $|m_1| = r$ is equal to one. The control problem to synchronize (or balance) the agents in the population can now be stated as:

Problem 1.1 Given the system defined by (1.15) and (1.16), find a control input u, such that $|m_1(q(\cdot, t))| \to 1(\text{or } 0)$.

1.3 Results

We will first elaborate how to choose an output function h such that it is connected to our goal of synchronizing (or balancing) the population of agents. At the same time, the mapping H_1 is passive under this choice of output and by applying a strictly passive controller H_2 in the control approach of Fig. 1.1, we conclude that $y \in \mathcal{L}$. We will then study invariance properties and conditions of our system under the proposed control law. An interpretation of this result along with some further considerations indicate that the control law indeed synchronizes (or balances) the population of agents, thereby solving Problem 1.1.

1.3.1 Enabling Passivity-Based Controller Design

The controller design based on the theory of passive systems benefits from a system model for which the control objective remains constant whenever $u = 0$. This property is not met by the model (1.15). In the following section, we propose state transformations $n(x,t) \rightarrow p(x,t)$ such that $|m_1(p(\cdot,t))|$ remains constant whenever $u = 0$. The first transformation employing (1.17) eliminates the discontinuity at the boundary by defining $\tilde{n}(x,t) = n(x,t)/\bar{q}(x)$, resulting in

$$\partial_t \tilde{n}(x,t) + \partial_x \left(\kappa(x,u)\tilde{n}(x,t)\right) = \gamma\kappa(x,u)\tilde{n}(x,t)\,, \qquad \tilde{n}(0,t) = \tilde{n}(2\pi,t)\,. \tag{1.19}$$

Next, we define $p(x,t) = \tilde{n}(x,t)/\int_0^{2\pi}\tilde{n}(x,t)\mathrm{d}x$ which is a proper probability distribution with PDE

$$\begin{aligned} &\partial_t p(x,t) + \partial_x \left(\kappa(x,u)p(x,t)\right) = u\gamma p(x,t)\left(Z(x) - \int_0^{2\pi} Z(x)p(x,t)\mathrm{d}x\right), \\ &p(0,t) = p(2\pi,t)\,. \end{aligned} \tag{1.20}$$

The system (1.20) has now the favorable properties that facilitate the feedback approach for synchronization of the population: (1) $p(x,t)$ is a proper probability distribution, (2) $p(x,t)$ is smooth over the boundary, and (3) the length of the first circular moment $|m_1(p(\cdot,t))|$ remains constant whenever $u = 0$. Furthermore, if $|m_1| = 1$, then the agents are synchronized.

1.3.2 Synchronization of the Population

With the model (1.20) given, it remains to define an appropriate output and a suitable output feedback control law which synchronizes the population. This will be achieved by choosing the output $y = h(p(\cdot,t))$ such that: (1) $y = 0$ whenever the population is synchronized, and (2) the map $H_1 : u \mapsto y$ is passive. As synchrony is equivalent to $|m_1| = 1$, we first study the time derivative of $|m_1(p(\cdot,t))|$ evolving under (1.20), viz.

$$\begin{aligned} \frac{\mathrm{d}}{\mathrm{d}t}|m_1(p(\cdot,t))| = \Bigg(&(\gamma + \mathrm{i})m_{-1}\int_0^{2\pi} \mathrm{e}^{\mathrm{i}x} Z(x)p(x,t)\mathrm{d}x \\ &- 2\gamma m_1 m_{-1}\int_0^{2\pi} Z(x)p(x,t)\mathrm{d}x + (\gamma - \mathrm{i})m_1\int_0^{2\pi} \mathrm{e}^{-\mathrm{i}x}Z(x)p(x,t)\mathrm{d}x\Bigg)\, u\,. \end{aligned} \tag{1.21}$$

In the following, we define $d^p(x) = d(p(\cdot,t),x)$ with

$$
\begin{aligned}
d(p(\cdot,t),x) &= \sum_{l=-1}^{1} d_l \, \mathrm{e}^{\mathrm{i}lx} , \\
d_{-1} = (\gamma - \mathrm{i})m_1 , \quad d_0 &= -2\gamma m_1 m_{-1} , \quad d_1 = (\gamma + \mathrm{i})m_{-1} .
\end{aligned}
\tag{1.22}
$$

This leads to a more practical representation of (1.21) in terms of the inner product

$$
\frac{\mathrm{d}}{\mathrm{d}t}|m_1(p(\cdot,t))| = \langle Z, d^p p(\cdot,t)\rangle \, u , \tag{1.23}
$$

which is zero whenever $u = 0$. Thus, by choosing the output as

$$
h(p(\cdot,t)) = \langle Z, d^p p(\cdot,t)\rangle , \tag{1.24}
$$

we arrive at the following observations.

Lemma 1.1 *The system* H_1 *given by* (1.9) *with output* (1.24) *and internal dynamics* (1.20) *is passive.*

Proof Following the definition of [4], the system is passive if $\langle y, u\rangle_T \geq \beta, \forall u \in \mathscr{L}_e$, $\forall T \in \mathbb{R}^+$. We constructed y such that

$$
\langle y, u\rangle_T = \int_0^T y(t)u(t)\mathrm{d}t = \int_0^T \frac{\mathrm{d}}{\mathrm{d}t}|m_1(p(\cdot,t))| \, \mathrm{d}t = |m_1(p(\cdot,T))| - |m_1(p(\cdot,0))|
$$

and as the norm of the first circular moment of a probability distribution is upper bounded by 1, we can choose $\beta = -1$. □

Theorem 1.1 *If the output feedback* $u(t) = -y(t)$ *is chosen for system* H_1 *and the output* $y(t)$ *is given by* (1.24), $y(t)$ *converges to zero.*

Proof The result follows directly from the basic passivity theorem given in [4], namely that the output of a passive system H_1 lies in $\mathscr{L}$ if the output is fed back through a strictly passive system H_2. By Lemma 1.1, H_1 is a passive system. Further, we chose H_2 as the identity function $H_2 x = x$, which indeed is strictly passive, and $y \in \mathscr{L}$. With y being uniform continuous we know from Barbalat's Lemma, that $y(t) \to 0$, thereby concluding the proof. □

Next, we study the invariance properties of our system having zero output. Our study is based on the properties of Fourier series and the Fourier coefficients of Z, d^p, and $p(\cdot,t)$ in (1.24). The Fourier series of a function $F : x \mapsto F(x)$ in any 2π-interval is $F(x) = \sum_{k=-\infty}^{\infty} a_k \mathrm{e}^{\mathrm{i}kx}$ with Fourier coefficients $a_k = \frac{1}{2\pi}\int_0^{2\pi} \mathrm{e}^{-\mathrm{i}kx} F(x)\,\mathrm{d}x$. To keep the notation in accordance with the definition of the circular moments in (1.18), we introduce a modified series representation with $F(x) = \sum_{k=-\infty}^{\infty} \frac{b_k}{2\pi} \mathrm{e}^{-\mathrm{i}kx}$ and altered coefficients $b_k = \int_0^{2\pi} \mathrm{e}^{\mathrm{i}kx} F(x)\mathrm{d}x$. A distribution $p(\cdot,t)$ is contained in a forward invariant set in $E = \{p \colon h(p) = 0\}$ if and only if $h(p(\cdot,t+\tau)) = 0, \ \forall \tau \geq 0$. The modified series representation leads to the following lemma:

Lemma 1.2 *Let c_k, d_k, and m_k be the coefficients of Z, d^p and $p(\cdot, t)$, respectively. Then E is invariant if and only if*

$$kc_k\nu_k = 0\,, \quad \forall k \in \mathbb{Z}\,, \tag{1.25}$$

with $\nu_k = \sum_{l=-1}^{1} d_l m_{k-l}$.

Proof Due to the periodicity of the cell cycle we know that E is in an invariant set if and only if $h(p(\cdot, t+\tau)) = 0, \forall \tau \in [0, T]$, which is due to the constant propagation ($u = 0$) of $p(x,t)$ with $\frac{\mathrm{d}x}{\mathrm{d}t} = \omega$ equal to $h(p_{\omega\tau}(\cdot, t)) = h(p_\sigma(\cdot, t)) = 0,\ \forall \sigma \in [0, 2\pi]$, where we define $p_\sigma(x,t) = p(x-\sigma, t)$. We will use this notation to denote a shift in x also for Z later on. If $h(p_\sigma(\cdot, t)) = 0$, then this is also true for its derivative $\frac{\mathrm{d}}{\mathrm{d}t} h(p_\sigma(\cdot, t))$ resulting in the following condition for invariance

$$\frac{\mathrm{d}}{\mathrm{d}t}\langle Z, d^{p_\sigma}\, p_\sigma(\cdot, t)\rangle = 0\,, \quad \forall \sigma \in [0, 2\pi]\,. \tag{1.26}$$

The derivative is obtained by employing the identity from the PDE (1.20) with $u = 0$: $\partial_t p(x,t) = -\omega \partial_x p(x,t)$ and subsequently integrating by parts. Furthermore, the shift in x is transferred to the PRC by a change of variables $x = \xi + \sigma$, changing (1.26) to

$$\langle \frac{\mathrm{d}}{\mathrm{d}x} Z_{-\sigma}, d^p\, p(\cdot, t)\rangle = 0\,, \quad \forall \sigma \in [0, 2\pi]\,. \tag{1.27}$$

The last steps of the proof are: (1) substituting $p(\cdot, t)$ and $Z_{-\sigma}$ with its modified Fourier series and (2) employing Parseval's theorem. With $(c_k)_k$ being the coefficients of Z, the coefficients of the argument shifted derivative $\mathrm{d}Z_{-\sigma}/\mathrm{d}x$ in (1.27) are $\left(-ik\mathrm{e}^{-ik\sigma} c_k\right)_k$. The function d^p has Fourier coefficients d_{-1}, d_0, d_1, and all other coefficients equal zero. The product $d^p p(\cdot, t)$ has modified coefficients $(\nu_k)_k$. By Parseval's theorem, the inner product in (1.27) equals the sum of its coefficients

$$\langle \frac{\mathrm{d}}{\mathrm{d}x} Z_{-\sigma}, d^p\, p(\cdot, t)\rangle = \frac{-i}{(2\pi)^2} \sum_{k=-\infty}^{\infty} \mathrm{e}^{-ik\sigma} k c_k \nu_k \tag{1.28}$$

which can be written as inner product, and therefore the condition for invariance is

$$\langle (\mathrm{e}^{-ik\sigma})_k, (kc_k\nu_k)_k \rangle = 0\,, \quad \sigma \in [0, 2\pi]\,,\ k \in \mathbb{Z}\,. \tag{1.29}$$

The series $\left(\mathrm{e}^{-ik\sigma}\right)_k$ are basis functions of a complete orthogonal basis, hence the inner product (1.29) is zero if and only if $kc_k\nu_k = 0,\ \forall k \in \mathbb{Z}$. This equals (1.25), thereby concluding the proof of Lemma 1.2. □

With Lemma 1.2 at hand, we can identify conditions on the phase response curve Z, such that the synchronized and balanced population are the only invariant ones.

Theorem 1.2 *If the the output feedback $u(t) = -y(t)$ is chosen for system H_1 and the output $y(t)$ is given by* (1.24), *then*

$$\mathcal{M}_0 = \{p\colon |m_1(p)| = 0\} \ ,$$
$$\mathcal{M}_1 = \{p\colon |m_1(p)| = 1\}$$

are invariant sets in E. Furthermore, if the first moment of Z is not equal to zero, i.e., $c_1 \neq 0$, then no other invariant set exists.

Proof By Lemma 1.2, invariance of E requires Eq. (1.25) to be fulfilled. Invariance of $\mathcal{M}_0$ and $\mathcal{M}_1$ is then verified by showing that $\nu_k = 0$, $\forall k \in \mathbb{Z}$. As $|m_1| = 0$ implies $m_1 = m_{-1} = 0$, (1.25) is trivially met, and $\mathcal{M}_0$ is invariant. If $|m_1| = 1$, then all moments have length one and $m_k = \mathrm{e}^{\mathrm{i}k\phi}$. All terms in ν_k cancel out, hence $\mathcal{M}_1$ is invariant. To conclude the proof of Theorem 1.2 we verify that $c_1 = 0$ is a necessary condition for (1.25) by showing that $\nu_1 \neq 0$ whenever $|m_1| \notin \{0, 1\}$. m_1 and m_{-1} are again represented as complex numbers. Furthermore $p(\cdot, t)$ is a probability distribution with $m_0 = 1$ by definition and we get

$$\nu_1 = r\left(\mathrm{e}^{-\mathrm{i}\phi}\left(\gamma + \mathrm{i}\right) + \mathrm{e}^{\mathrm{i}\phi}\left((\gamma - \mathrm{i})m_2 - 2\gamma r\right)\right) \ . \tag{1.30}$$

From $|m_1| = r \neq 0$, $\mathrm{e}^{-\mathrm{i}\phi}$ and $\mathrm{e}^{\mathrm{i}\phi}$ are orthogonal and $\gamma > 0$ by definition, it follows that $\nu_1 \neq 0$. Hence, $\mathcal{M}_0$ and $\mathcal{M}_1$ are the only invariant sets in E if $c_1 \neq 0$. □

We will now discuss some aspects regarding the convergence to a synchronized (or balanced) population, given that the first moment of the phase response is not equal to zero. If the output is given by (1.24) and the output feedback $u(t) = -\varepsilon\, y(t)$, $\varepsilon > 0$, is chosen for system H_1, then

$$\frac{\mathrm{d}}{\mathrm{d}t}|m_1(p(\cdot, t))| = -\varepsilon\, h(p(\cdot, t))^2 \leq 0 \ , \qquad \forall t \geq 0 \ , \tag{1.31}$$

and $|m_1(p(\cdot, t))|$ decreases monotonically. Furthermore, the average of (1.31) over one period is strictly monotonically decreasing whenever $|m_1| \notin \{0, 1\}$. These observations suggest that $|m_1(p(\cdot, t))|$ approaches $\mathcal{M}_0$ from almost all initial conditions and the population is balanced. Synchronization of the population is achieved by sign reversal of the output function $y(t) = -h(p(\cdot, t))$ and the same output feedback. Sign reversal of the output preserves passivity of the system and by $\frac{\mathrm{d}}{\mathrm{d}t}|m_1(p(\cdot, t))| \geq 0$, p approaches $\mathcal{M}_1$ and a synchronized distribution is achieved.

Remark 1.1 Theorems 1.1 and 1.2 and the fact that an attractive set becomes a repelling set by sign reversal, strongly suggest that the population with output (1.24) and control input $u(t) = \varepsilon\, y(t)$ converges to a Dirac delta distribution. However, due to topological reasons, analysis of convergence of $p(\cdot, t)$ is difficult and beyond the scope of the present study.

1.4 Example

We conclude by demonstrating the developed control methodology on the reduced phase model (1.20). The underlying ODE model is a 5-state skeleton model of a mammalian cell cycle [7]. We extended the model by an additive input to the dynamics of Cyclin A

$$\dot{x}_{\text{CycA}} = f_{CycA}(x) + \frac{1.6\,(\alpha - x_{CycA})}{0.1 + \alpha - x_{CycA}}\, u(t) \,. \tag{1.32}$$

The input can be thought of, for e.g., an optogenetic signal causing a direct induction of Cyclin A expression with the total amount of Cyclin A being upper bounded by α. The phase response curve Z was obtained by solving the appropriate adjoint equation using the dynamic modeling program XPPAUT [5]. We take u according to Theorem 1.1 and simulated both the synchronizing and balancing scenario with h as defined in (1.24). The results are depicted in Fig. 1.3. In the synchronizing scenario, one observes how the first moment approaches the unit circle, indicating that the distribution of cells indeed converges to the Dirac distribution. This can also be observed in the simulation snapshots. By sign reversal of the output and starting with an imbalanced cell density, we further see that this process is reversed and the population approaches a uniform distribution.

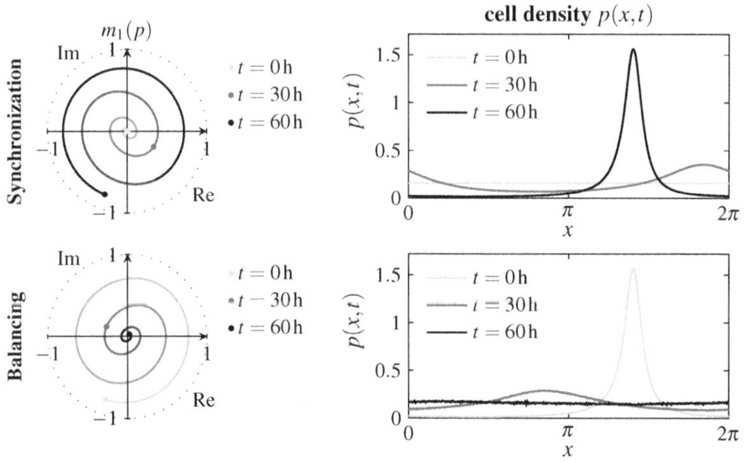

Fig. 1.3 Simulation of (1.20) derived from a 5-state cell cycle model with both a synchronizing (top) and a balancing (bottom) controller. On the left: temporal evolution of the first circular moment m_1 in the complex plane. On the right: snapshots of the cell density over the cell cycle position

1.5 Conclusion and Outlook

We studied the ensemble control problem of synchronizing a cell population in their cell cycle with restriction of the observation to representative samples of the population. Starting with a single cell as oscillator on a limit cycle, we developed a reduced phase model of the population with a broadcast input acting via the phase response curve. We then proposed state transformations for the age-structured population type model which enable controller design in the input-output framework for passive systems. Formulating the control problem in terms of the first circular moment of the population led to the desired output feedback which synchronizes the population. Finally, we derived sufficient conditions on the phase response curve for the synchronization of the population. We concluded by illustrating the controller action on a model of the mammalian cell cycle.

The present study solves the ensemble control problem of cell cycle synchronization by sending the first circular moment to one. However, we believe, that the here presented approach might be suitable to achieve any desired moment-determinate distribution by steering the circular moments of the population to the corresponding values of the target distribution.

Acknowledgements The authors thank the German Research Foundation (DFG) for financial support of the project under grant number AL316/14-1 and within the Cluster of Excellence in Simulation Technology (EXC 310/2) at the University of Stuttgart.

References

1. Brockett, R.: Notes on the control of the Liouville equation. In: Lecture Notes in Mathematics, vol. 2048, pp. 101–129. Springer, Berlin (2012)
2. Clairambault, J., Michel, P., Perthame, B.: A mathematical model of the cell cycle and its circadian control. Math. Model. Biol. Syst., 239–251 (2007). Birkhäuser Boston
3. Csikasz-Nagy, A.: Computational systems biology of the cell cycle. Brief. Bioinform. **10**(4), 424–434 (2009)
4. Desoer, C.A., Vidyasagar, M.: Feedback systems: input-output properties. Academic Press (1975)
5. Ermentrout, B.: Simulating, analyzing, and animating dynamical systems. Soc. Ind. Appl. Math. (2002)
6. von Foerster, H.: Some remarks on changing populations. In: Stohlman, J.F. (ed.) The Kinetics of Cellular Proliferation, pp. 382–407. Grune and Stratton, New York (1959)
7. Gérard, C., Gonze, D., Goldbeter, A.: Effect of positive feedback loops on the robustness of oscillations in the network of cyclin-dependent kinases driving the mammalian cell cycle. FEBS J. **279**(18), 3411–3431 (2012)
8. Guckenheimer, J.: Isochrons and phaseless sets. J. Math. Biol. **1**(3), 259–273 (1975)
9. Gyllenberg, M., Webb, G.F.: A nonlinear structured population model of tumor growth with quiescence. J. Math. Biol. **28**(6), 671–694 (1990)
10. Hanahan, D., Weinberg, R.A.: Hallmarks of cancer: the next generation. Cell **144**(5), 646–74 (2011)
11. Hoppensteadt, F.C., Izhikevich, E.M.: Weakly Connected Neural Networks, Applied Mathematical Sciences, vol. 126. Springer, New York (1997)

12. Kuramoto, Y.: Chemical Oscillations, Waves, and Turbulence, Springer Series in Synergetics, vol. 19. Springer, Berlin (1984)
13. Kuritz, K., Stöhr, D., Pollak, N., Allgöwer, F.: On the relationship between cell cycle analysis with Ergodic principles and age-structured cell population models. J. Theor. Biol. **414**(November 2016), 91–102 (2017)
14. Levskaya, A., Weiner, O.D., Lim, W.A., Voigt, C.A.: Spatiotemporal control of cell signalling using a light-switchable protein interaction. Nature **461**(7266), 1–5 (2010)
15. Malkin, I.G.: Methods of Poincare and Liapunov in Theory of Non-linear Oscillations. Gostexizdat, Moscow (1949)
16. Malkin, I.G.: Some Problems in Nonlinear Oscillation Theory. Gostexizdat, Moscow (1956)
17. McKendrick, A.G.: Applications of mathematics to medical problems. Proc. Edinburgh Math. Soc. **44**, 98–130 (1926)
18. Mirollo, R.E., Strogatz, S.H.: Synchronization of pulse-coupled biological oscillators. SIAM J. Appl. Math. **50**(6), 1645–1662 (1990)
19. Montenbruck, J.M., Bürger, M., Allgöwer, F.: Practical synchronization with diffusive couplings. Automatica **53**, 235–243 (2015)
20. Morgan, D.O.: The Cell Cycle: Principles of Control. New Science Press, London (2007)
21. Powell, E.O.: Growth rate and generation time of bacteria, with special reference to continuous culture. J. Gen. Microbiol. **15**(3), 492–511 (1956)
22. Scardovi, L., Arcak, M., Sontag, E.D.: Synchronization of interconnected systems with applications to biochemical networks: an input-output approach. IEEE Trans. Automat. Contr. **55**(6), 1367–1379 (2010)
23. Wilson, D., Moehlis, J.: Optimal chaotic desynchronization for neural populations. SIAM J. Appl. Dyn. Syst. **13**(1), 276–305 (2014)
24. Winfree, A.T.: Patterns of phase compromise in biological cycles. J. Math. Biol. **1**(1), 73–93 (1974)
25. Zeng, S., Waldherr, S., Ebenbauer, C., Allgower, F.: Ensemble observability of linear systems. IEEE Trans. Automat. Contr. **61**(6), 1452–1465 (2016)
26. Zhivotovsky, B., Orrenius, S.: Cell cycle and cell death in disease: past, present and future. J. Intern. Med. **268**(5), 395–409 (2010)

Chapter 2
Collective Formation Control of Multiple Constant-Speed UAVs with Limited Interactions

Brian D. O. Anderson, Zhiyong Sun, Georg S. Seyboth and Changbin Yu

Abstract In this chapter, we consider coordination control of a group of UAV agents with constant and in general nonidentical speeds. The control input is designed to steer their orientations and the control objective is to achieve a desired formation configuration for all the agents subject to constant-speed constraints. Through a formation feasibility analysis by a three-agent example, we show that it is generally impossible to control and maintain a formation by constant-speed agents if target formation shapes are defined by agents' actual positions. We then adopt a circular motion center as a *virtual position* for each agent to define the target formation shape. Two different formation design approaches, namely, a displacement-based approach and a distance-based approach, are discussed in detail to coordinate a group of constant-speed agents in achieving a target formation with stable circular motions via limited interactions.

G. S. Seyboth is now with Robert Bosch Automotive Steering GmbH.
Text and figures of this chapter were reproduced from Z. Sun and B.D.O. Anderson, Formation feasibility on coordination control of networked heterogeneous systems with drift terms, IEEE 55th Conference on Decision and Control (CDC), IEEE, 2016, 3462–3467. https://doi.org/10.1109/CDC.2016.7798788.

B. D. O. Anderson · Z. Sun (✉) · C. Yu
Research School of Engineering, The Australian National University,
Canberra, ACT 2601, Australia
e-mail: zhiyong.sun@anu.edu.au

B. D. O. Anderson
e-mail: brian.anderson@anu.edu.au

C. Yu
e-mail: brad.yu@anu.edu.au

B. D. O. Anderson · C. Yu
School of Automation, Hangzhou Dianzi University, Hangzhou 310018, China

G. S. Seyboth
Institute for Systems Theory and Automatic Control, University of Stuttgart,
Pfaffenwaldring 9, 70550 Stuttgart, Germany
e-mail: georg.seyboth@googlemail.com

R. Tempo et al. (eds.), *Emerging Applications of Control and Systems Theory*, Lecture Notes in Control and Information Sciences - Proceedings,
https://doi.org/10.1007/978-3-319-67068-3_2

2.1 Introduction

Collective coordination control of networked multi-agent systems has received considerable attention in recent years, partly motivated by its applications in many areas. In this chapter, we consider a particular class of cooperative tasks in multi-agent coordination, namely formation control, in which the control objective is to form or maintain a prescribed geometric shape for a group of agents. One of the most active and important challenges in this area is to control and coordinate a group of Unmanned Aerial Vehicles (UAVs) in a formation [1, 4, 14, 18]. A more realistic and complicated model other than single- or double-integrators that can describe the nonholonomic constraints of such vehicles is the unicycle model [7]. Early contributions on coordination and formation control of unicycle-type agents include the consensus-based formation control [10], the pursuit formation design [12], the rendezvous control [5], etc. Other recent papers include e.g. [3, 6, 11] with different control constraints, but all assume that not only the orientation but also the speed of individual agents are controllable.

A particular constraint in the cooperative control design for UAV agents is that on occasions the UAVs used in the control task (e.g. Aerosonde UAVs or other types of *fixed-wing* aircraft) usually fly most efficiently at fixed, nominal speeds [1, 25]. Further, agents within one formation may have different (but similar) speeds [19]. For collective control of unicycle-like agents with constant speed, the problem becomes more challenging. The two seminal papers [16, 17] provide comprehensive studies on how to coordinate different collective planar motions (e.g. parallel motions or circular motions) for such multi-unicycle systems with constant unit-speed constraints. More recently, the paper [19] has extended the results in [16] to control collective circular motions of heterogeneous unicycle-like agents with *nonidentical* constant speeds. It is shown in [19] that two kinds of circular motions are possible: a circular motion with a common angular frequency and different radii for each agent, or a circular motion with a common radius but different angular frequencies for each agent.

In this chapter, we are particularly interested in how to design controllers to achieve a target formation shape for a group of unicycle-type agents with constant and nonidentical speeds (the trajectory tracking control problem involving constant-speed agents has been discussed in a companion paper [22]). The results build on these previous papers including [16, 17, 19], but here we focus on formation shape control, instead of the circular motion stabilization problem as discussed in [16, 17, 19]. The main challenge for formation controller design comes from agent kinematic constraints, i.e., how to define the desired formations and how to design control laws which comply with the constraint of constant speeds. Of course, the constant-speed constraint indicates that all the agents are always moving, which significantly affects the formation maintenance task. To address this issue, the main idea on formation specification and control adopted in this chapter is to use circular motion center positions as *virtual positions* instead of agents' actual positions for defining a desired formation shape. To this end, the controller aims to drive each agent to reach a *stable circular motion* while achieving a target formation shape. We also note that circular motions are particularly useful in several real-life applications, including surveillance, circumnavigation, target circling and area monitoring [23, 26].

2.2 Agent Models

Before presenting model equations, we first introduce some special notation to be used in this chapter. The set $\mathbb{S}^1$ denotes the unit circle and an angle θ_i is a point in the unit circle space, i.e., $\theta_i \in \mathbb{S}^1$. The n-torus is the Cartesian product $\mathbb{T}^n = \mathbb{S}^1 \times \ldots \times \mathbb{S}^1$. For a complex variable $z \in \mathbb{C}$, we use $\bar{z}$ to denote its complex conjugate. For $z_1, z_2 \in \mathbb{C}^n$, the scalar product is defined by $\langle z_1, z_2 \rangle = \mathrm{Re}(\bar{z}_1^T z_2)$, i.e., the real part of the standard scalar product over $\mathbb{C}^n$.

In this chapter, we consider a group of n agents modeled by unicycle-like kinematics subject to nonholonomic dynamics and constant-speed constraint. The kinematic equations of agent k are described by

$$\begin{aligned} \dot{x}_k &= v_k \cos(\theta_k) \\ \dot{y}_k &= v_k \sin(\theta_k) \\ \dot{\theta}_k &= u_k \end{aligned} \tag{2.1}$$

where $x_k \in \mathbb{R}$, $y_k \in \mathbb{R}$ are the coordinates of agent k in the real plane and θ_k is the heading angle. The agents have fixed cruising speeds $v_k > 0$ which in general are distinct for different agents; u_k is the control input to be designed for steering the orientation of agent k.

With complex number notation, the complex variable $r_k(t) = x_k(t) + iy_k(t) := |r_k|e^{i\phi_k(t)} \in \mathbb{C}$ denotes the position of agent k in the complex plane. We also define the vectors $r = [r_1, r_2, \ldots, r_n]^T \in \mathbb{C}^n$, $\theta = [\theta_1, \theta_2, \ldots, \theta_n]^T \in \mathbb{T}^n$ and $e^{i\theta} = [e^{i\theta_1}, e^{i\theta_2}, \ldots, e^{i\theta_n}]^T \in \mathbb{C}^n$ to collect the positions and headings of all the agents. Then the above model (2.1) for agent k can be rewritten as

$$\begin{aligned} \dot{r}_k &= v_k e^{i\theta_k} \\ \dot{\theta}_k &= u_k(r, \theta). \end{aligned} \tag{2.2}$$

2.3 Formation Feasibility Analysis: An Example of Rigid Formation Maintenance by Constant-Speed Agents

In this section, we present a brief discussion on formation feasibility analysis for a group of constant-speed agents, based on an illustrative example. In this example, we suppose the formation is defined by a certain set of inter-agent distances between

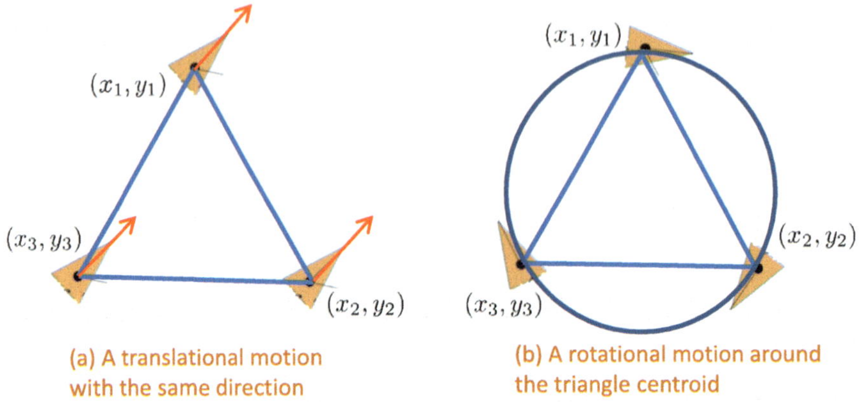

Fig. 2.1 Two feasible formations with a group of constant-speed agents in a triangular formation. (Reproduced with permission from © IEEE 2016, Z. Sun and B.D.O. Anderson [20])

agents' actual positions. This control task[1] is termed *rigid formation control*, which has received increasing attention in the research field of multi-agent coordination, in particular since the publication of [9].

Consider a group of three constant-speed agents in maintaining a triangular formation. The inter-agent distances are denoted by d_{ij} with $i, j = 1, 2, 3, i \neq j$, for which the three agents aim to achieve. If the distance error $e_{ij} = (x_i - x_j)^2 + (y_i - y_j)^2 - d_{ij}^2$ is zero for all the three edges, then the target formation is achieved and maintained. For the aim of demonstration, in the following analysis we assume that $d_{12} = d_{23} = d_{31} = d$, i.e., the target rigid formation is an equilateral triangle. In determining whether there exist feasible trajectories for all the agents which respect both the formation constraint and the kinematics constraint (i.e., constant-speed constraints), one needs to formulate a formation feasibility equation and to determine whether such an equation has solutions. We refer the readers to [20, 24] for the development of a formation feasibility theory under different constraints, while we omit the detailed calculations here.

In the case that all agents have identical cruising speeds, i.e., $v_1 = v_2 = v_3$, a simple calculation from the formation feasibility theory [20] shows that the motion solution is that either $\dot{\theta}_1 = \dot{\theta}_2 = \dot{\theta}_3 = 0, \theta_1 = \theta_2 = \theta_3$, or $\dot{\theta}_1 = \dot{\theta}_2 = \dot{\theta}_3, \theta_1 = \theta_2 + \frac{2\pi}{3} = \theta_3 + \frac{4\pi}{3}$, which correspond to a translational motion with the same direction or a rotational motion around the triangle centroid (see Fig. 2.1). In the case that $v_1 = 0$, $v_2 = v_3$, a feasible motion exists in which agents 2 and 3 rotate around agent 1 with the same angular velocity. In the case that all agents have nonidentical cruising speeds, there usually does not exist a solution to the feasibility condition except for some special cases, which agrees with our intuition that maintaining a

[1] Section 2.3 of this chapter includes material reproduced with permission from Sun, Z., Anderson, B.D.O.: Formation feasibility on coordination control of networked heterogeneous systems with drift terms. In: Proc. of the 55th Conference on Decision and Control, pp. 3462–3467. IEEE 2016.

rigid shape by a group of UAV agents with nonidentical constant speeds is generally impossible. This suggests that one needs to find an alternative way to define a target formation and to formulate different formation control approaches in coordinating multiple constant-speed agents.

2.4 Formulation of Target Formations for a Constant-Speed Agent Group

In formation shape control, the target formation shape is usually defined by some geometrical relationships between neighboring agents' positions among the group. However, as discussed in the preceding section, in the control problem with constant-speed unicycle-like agents the usual way of defining formation shapes in terms of agents' actual positions does not work in this context, since all the agents will always have motions due to the constant-speed constraints. Hence, we need to find alternative variables that are some kind of surrogate of the actual positions to define the desired formation shape.

Before presenting the formation control design, it is helpful to review the following observations ([16, 19]) on motion properties of constant-speed agents:

- If the control $u_k,\ k = 1, 2, \dots, n$ is identically zero, then each agent travels in a straight line (with the orientation determined by its initial heading $\theta_k(0)$);
- If the control $u_k = \omega_0,\ k = 1, 2, \dots, n$ where ω_0 is a nonzero constant, then all the agents travel in a circle of radius $v_k/|\omega_0|$, with the rotation direction determined by the sign of ω_0.
- For the case of circular motion generated by a constant control input $u_k = \omega_0 \neq 0$, the center of the circular motion for the k-th agent is described by

$$c_k = r_k + \frac{v_k}{\omega_0} i e^{i\theta_k} \tag{2.3}$$

  which could be regarded as the "state"(or "virtual position") of agent k in the formation shape control design.

In the following sections, we will describe the desired formation shapes by agents' virtual positions c_k. Thus, the control aim is to drive each agent to reach a *stable circular motion* and also a predefined formation shape specified by their circular motion centers. In the next two sections, we will present two different approaches to achieve a multi-agent formation for constant-speed agents with limited interactions, in contrast to the all-to-all interaction as assumed in e.g. [16, 19].

2.5 Formation Control Design with Limited Interaction: Displacement-Based Approach

2.5.1 Controller Design and Convergence Analysis

In this section, as well as the next section we assume the underlying interaction graph is undirected and connected but *not necessarily complete*. This implies each agent in the formation has *limited interaction* only to its neighboring agents, as opposed to the *all-to-all interaction* in which the underlying graph topology is complete.

In the displacement-based approach, the desired formation is described by a set of relative position vectors $\hat{c}_{kj}$ for each $(j, k) \in \mathscr{E}$ where $\mathscr{E}$ is the edge set of the underlying interaction graph. The control task now is to drive each relative virtual position $c_{kj} = c_k - c_j$ to converge to the desired formation shape described by $\hat{c}_{kj}$ where $(j, k) \in \mathscr{E}$. To achieve this formation control objective, we design the control law as

$$u_k = \omega_0 + \gamma\omega_0 \left\langle \sum_{j \in \mathscr{N}_k} \left(c_{kj} - \hat{c}_{kj}\right), v_k e^{i\theta_k} \right\rangle \tag{2.4}$$

where γ is a positive control gain and $\mathscr{N}_k$ denotes agent k's neighboring set.

The main result in this section is summarized in the following theorem.

Theorem 2.1 *For the designed controller* (2.4) *with an underlying undirected and connected graph, agent k's trajectory* $r_k(t)$ *of the closed-loop system* (2.2) *converges to a stable circular motion with angular velocity* ω_0 *and radius* $v_k/|\omega_0|$ *and all the agents form a desired formation shape defined by the desired relative center positions* $\hat{c}_{kj}$ *where* $(j, k) \in \mathscr{E}$.

Proof For the purposes of proof and analysis it is convenient to postulate that there are underlying variables $\hat{c}_j$, for all j with the property that $\hat{c}_{kj} = \hat{c}_k - \hat{c}_j$ with $(j, k) \in \mathscr{E}$ Further define $\tilde{c}_k = c_k - \hat{c}_k$ and a vector $\tilde{c} = [\tilde{c}_1, \tilde{c}_2, \ldots, \tilde{c}_m]^T$. Note that $\tilde{c}_k$ is not an available control term in (2.4) since $\hat{c}_k$ is not available for agent k. That is, the control input available in each agent's control term depends on *relative* vectors instead of *absolute* vectors, which are made clear in the expression of (2.4). The introduction of $\hat{c}_k$ and $\tilde{c}$ is for the convenience of proof and analysis. To this end, the control function (2.4) can be rewritten as

$$\begin{aligned} u_k &= \omega_0 + \gamma\omega_0 \left\langle \sum_{j \in \mathscr{N}_k} \left((c_k - c_j) - (\hat{c}_k - \hat{c}_j)\right), v_k e^{i\theta_k} \right\rangle \\ &= \omega_0 + \gamma\omega_0 \left\langle L_k(c - \hat{c}), v_k e^{i\theta_k} \right\rangle \end{aligned} \tag{2.5}$$

where L_k denotes the k-th row of the Laplacian matrix L for the underlying interaction graph which is assumed to be connected but not necessarily complete.

By the definition of $\tilde{c}_k$ and the control (2.4), one has

$$\dot{\tilde{c}}_k = \dot{c}_k - \dot{\hat{c}}_k = \frac{v_k}{\omega_0} e^{i\theta_k}(\omega_0 - u_k) = -\gamma v_k e^{i\theta_k} \left\langle L_k \tilde{c}, v_k e^{i\theta_k} \right\rangle. \tag{2.6}$$

Construct the following Lyapunov function candidate

$$V(\tilde{c}) = \frac{1}{2}\langle L\tilde{c}, \tilde{c}\rangle = \frac{1}{2}\langle H\tilde{c}, H\tilde{c}\rangle \tag{2.7}$$

where H is the incidence matrix for the underlying interaction graph. Note that for an undirected graph there holds $L = H^T H$ (see e.g. [13, Chap. 2]). The function $V(\tilde{c})$ is positive semi-definite in $\tilde{c}$ and positive definite in $H\tilde{c}$. Furthermore, there holds: (i) $V \geq 0$ for all $\tilde{c} \in \mathbb{C}^n$, (ii) $V = 0$ if and only if $H\tilde{c} = 0$, and (iii) $V \to \infty$ for $\|H\tilde{c}\| \to \infty$. Hence, the function V defined in (2.7) is a suitable Lyapunov function to assess the stability and convergence of the formation system consisting of constant-speed agents by the proposed control law (2.4). Note that $\tilde{c}$ is a vector function of (r, θ) and we may also rewrite $V(\tilde{c})$ as $V(r, \theta)$.

The time derivative of V along the solution of the formation system (2.2) with the control (2.4) can be computed as

$$\begin{aligned} \dot{V}(r,\theta) &= \langle L\tilde{c}, \dot{\tilde{c}}\rangle = \sum_{k=1}^{n} \left\langle L_k\tilde{c}, \dot{\tilde{c}}_k \right\rangle = \sum_{k=1}^{n} \left\langle L_k\tilde{c}, -\gamma v_k e^{i\theta_k} \left\langle L_k\tilde{c}, v_k e^{i\theta_k}\right\rangle\right\rangle \\ &= -\sum_{k=1}^{n} \left\langle L_k\tilde{c}, \gamma v_k e^{i\theta_k}\right\rangle \left\langle L_k\tilde{c}, v_k e^{i\theta_k}\right\rangle = -\gamma \sum_{k=1}^{n} \left\langle L_k\tilde{c}, v_k e^{i\theta_k}\right\rangle^2 \leq 0 \end{aligned} \tag{2.8}$$

The set on which $\dot{V} = 0$ is characterized by

$$\mathcal{O}(r,\theta) = \{(r,\theta) : u_k = \omega_0, \left\langle L_k\tilde{c}, v_k e^{i\theta_k}\right\rangle = 0, \forall k\}. \tag{2.9}$$

In the set $\mathcal{O}$ there holds either (i) $L_k\tilde{c} = 0$ or (ii) $L_k\tilde{c}$ and $v_k e^{i\theta_k}$ are perpendicular. Note that in $\mathcal{O}$, $u_k = \omega_0$ and hence $\dot{\theta}_k = \omega_0, \dot{c}_k = 0$. Consequently, because $v_k e^{i\theta_k}$ is not constant, $\langle L_k\tilde{c}, v_k e^{i\theta_k}\rangle = 0$ can be satisfied only if $L_k\tilde{c} = 0$ for all $k = 1, 2, \ldots, n$. Thus, by LaSalle's Invariance Principle, all trajectories converge to the largest invariant set contained in $\bar{\mathcal{O}}$ described as

$$\begin{aligned} \bar{\mathcal{O}}(r,\theta) &= \{(r,\theta) : u_k = \omega_0, L_k\tilde{c} = 0, \forall k\} \\ &= \{(r,\theta) : u_k = \omega_0, \tilde{c} \in \text{span}\{\mathbf{1}\}, \forall k\} \\ &= \{(r,\theta) : u_k = \omega_0, c_{kj} = \hat{c}_{kj}, (j,k) \in \mathcal{E}\}. \end{aligned} \tag{2.10}$$

In the set $\bar{\mathcal{O}}$ each agent reaches a stable circular motion with angular velocity ω_0 and radius $v_k/|\omega_0|$ (for agent k) which forms a stable formation described by $\hat{c}_{kj}$ as desired. The proof is completed. □

Remark 2.1 The above control law is a generalization of Theorem 2 (circular motion stabilization) and Corollary 2 (*SE*(2) symmetry breaking) in [16] which stabilize a group of *unit-speed* unicycles to a circular motion around a *single* and fixed beacon point. The main idea in the controller design is inspired by the consensus-based linear formation control [13, 15]. Note that in the above controller (2.4), the control input term involves the relative information of neighboring agents, i.e., the current displacement $c_k - c_j$ of virtual positions and the desired center displacement $\hat{c}_j - \hat{c}_k$ with respect to its neighbors, which can be calculated by using the formula (2.3).

2.5.2 Implementation Analysis

In this subsection, we discuss how the proposed controller (2.4) can be implemented. By denoting the relative virtual position vector as $c_{kj} = c_k - c_j := r_k - r_j + \frac{v_k}{\omega_0} i e^{i\theta_k} - \frac{v_j}{\omega_0} i e^{i\theta_j}$ with $r_{kj} := r_k - r_j = |r_{kj}| e^{i\phi_{r_{kj}}}$ and the desired relative position vector as $\hat{c}_{kj} = \hat{c}_k - \hat{c}_j = |\hat{c}_{kj}| e^{i\phi_{\hat{c}_{kj}}}$, we can obtain the following control term in real variables

$$
\begin{aligned}
&\left\langle \sum_{j\in\mathcal{N}_k} \left((c_k - c_j) - (\hat{c}_k - \hat{c}_j)\right), v_k e^{i\theta_k} \right\rangle \\
&= \left\langle \sum_{j\in\mathcal{N}_k} (r_k - r_j + \frac{v_k}{\omega_0} i e^{i\theta_k} - \frac{v_j}{\omega_0} i e^{i\theta_j} - \hat{c}_{kj}), v_k e^{i\theta_k} \right\rangle \\
&= \left\langle \sum_{j\in\mathcal{N}_k} (r_{kj} - \frac{v_j}{\omega_0} i e^{i\theta_j}), v_k e^{i\theta_k} \right\rangle - \sum_{j\in\mathcal{N}_k} \mathrm{Re}(\overline{\hat{c}_{kj}} v_k e^{i\theta_k}) \\
&= \sum_{j\in\mathcal{N}_k} \left(|r_{kj}| v_k \cos(\phi_{r_{kj}} - \theta_k) + \frac{v_j v_k}{\omega_0} \sin(\theta_k - \theta_j) \right) - \sum_{j\in\mathcal{N}_k} \left(|\hat{c}_{kj}| v_k \cos(\phi_{\hat{c}_{kj}} - \theta_k) \right).
\end{aligned}
\tag{2.11}
$$

Note that in the third line of the above derivation we have used the equality $\left\langle \frac{v_k}{\omega_0} i e^{i\theta_k}, v_k e^{i\theta_k} \right\rangle = 0$. As noted in Remark 2.1, only relative information is required for the control implementation.

Remark 2.2 (**Communication and measurement requirement**) It can be seen from the designed controller (2.4) and its real variable version (2.11) that each agent needs to measure the relative positions r_{kj} and relative headings $\theta_k - \theta_j$ with respect to its neighbors. We note that in practice information of relative headings $\theta_k - \theta_j$ can be obtained either by bearing sensors or by communication. Thus, the controller (2.4) is distributed since only neighboring information is involved.

We also note that in the control function (2.11) there is one term $\hat{c}_{kj}$ for defining the target formation shape, which implies each agent needs global knowledge of the orientation of a *common* coordinate frame such that these vectors can be correctly interpreted.

2.6 Formation Control Design with Limited Interaction: Distance-Based Approach

In this section, we aim to propose an alternative formation controller based on distance specifications. Suppose that the desired formation is specified by a set of m inter-agent distances d_{kj} for each $(j, k) \in \mathscr{E}$, which describes a desired *rigid* formation shape (for the definition of rigid formation, the reader is referred to [2]). We define the distance error as

$$\epsilon_{l_{kj}} = \|c_k - c_j\|^2 - d_{kj}^2$$

and the distance error vector for all the edges is constructed as $\epsilon = [\epsilon_1, \epsilon_2, \ldots, \epsilon_m]^T$. We also assume that any target formation that realizes the distances d_{kj} via the virtual position c is infinitesimally and minimally rigid [8]. By respecting the constant-speed constraint, the control in such a distance-based framework aims to drive all the agents to reach stable circular motions and to achieve the desired distances between their circular motion centers.

For distance-based rigid formation control, the rigidity matrix, denoted by R, plays a central role in shape specification and convergence analysis [2]. We show a nice structure of R with complex variable entries, regarding it as an extension of the standard rigidity matrix defined in real spaces. Denote the incidence matrix of the graph as H. Construct a block diagonal matrix $Z = \text{diag}\{z_1, z_2, \ldots, z_l, \ldots, z_m\}$, where z_l is the relative virtual position vector $z_l = c_k - c_j$ which relates to the relative centers of agent k and agent j with $(j, k) \in \mathscr{E}$. The construction of Z should be consistent with the direction specification of the incidence matrix H. Then the complex rigidity matrix R can be written as $R = ZH \in \mathbb{C}^{m \times n}$.

In order to achieve the target formation defined by prescribed distances d_{kj} between adjacent agents involving their virtual positions and to drive all agents to reach stable circular motions, we design the following distributed formation controller

$$\begin{aligned} u_k &= \omega_0 + \gamma\omega_0 \left\langle \sum_{j \in \mathscr{N}_k} \left(\epsilon_{l_{kj}}(c_k - c_j)\right), v_k e^{i\theta_k} \right\rangle \\ &= \omega_0 + \gamma\omega_0 \left\langle R_k^T \epsilon, v_k e^{i\theta_k} \right\rangle \end{aligned} \tag{2.12}$$

where R_k^T is the k-th row of the transposed rigidity matrix which involves complex variables. The main result in this section is summarized in the following theorem.

Theorem 2.2 *For the designed controller* (2.12), *agent k's trajectory $r_k(t)$ of the closed-loop system* (2.2) *converges to a stable circular motion with angular velocity ω_0 and radius $v_k/|\omega_0|$. Furthermore, all agents' circular motion centers converge locally to the desired formation shape defined by the distances d_{kj}.*

Proof Let us define the following Lyapunov function candidate

$$V = \frac{1}{4}\epsilon^T \epsilon = \frac{1}{4}\sum_{l=1}^{m} \epsilon_{l_{ij}}^2 = \frac{1}{4}\sum_{(k,j)\in\mathcal{E}} \left(\|c_k - c_j\|^2 - d_{kj}^2\right)^2. \tag{2.13}$$

The quadratic function V in (2.13) satisfies (i) $V \geq 0$ for all $c \in \mathbb{C}^n$, (ii) $V = 0$ if and only if $\epsilon = 0$, and (iii) $V \to \infty$ for $\epsilon \to \infty$. Hence, V defined in (2.13) is a suitable Lyapunov function to assess the stability and convergence of the formation control system with the proposed control law in (2.12). Note that ϵ is a vector function of (r, θ) and we may also rewrite $V(\epsilon)$ as $V(r, \theta)$. Its derivative can be calculated as

$$\dot{V}(r,\theta) = \frac{1}{2}\sum_{(k,j)\in\mathcal{E}} \left(\|c_k - c_j\|^2 - d_{kj}^2\right)\langle c_k - c_j, \dot{c}_k - \dot{c}_j\rangle. \tag{2.14}$$

By the definition of c_k and the control (2.12), one has

$$\dot{c}_k = -\gamma v_k e^{i\theta_k}\left\langle R_k^T \epsilon, v_k e^{i\theta_k}\right\rangle \tag{2.15}$$

Thus, the derivative of V in (2.13) along the solution of the formation system (2.1) with the controller (2.12) can be further written as

$$\dot{V}(r,\theta) = -\gamma \sum_{k=1}^{n} \left\langle R_k^T \epsilon, v_k e^{i\theta_k}\right\rangle^2 \leq 0 \tag{2.16}$$

The set on which $\dot{V} = 0$ is characterized by

$$\mathcal{O}(r,\theta) = \{(r,\theta) : u_k = \omega_0, \left\langle R_k^T \epsilon, v_k e^{i\theta_k}\right\rangle = 0\}. \tag{2.17}$$

By LaSalle's Invariance Principle and similar arguments as in Theorem 2.1, all trajectories converge to the largest invariant set contained in $\bar{\mathcal{O}}$ described as

$$\bar{\mathcal{O}}(\epsilon,\theta) = \{(\epsilon,\theta) : u_k = \omega_0, R^T \epsilon = 0\}. \tag{2.18}$$

Thus in the limit, the trajectory of each agent (agent k) converges to a stable circular motion with angular velocity ω_0 and radius $v_k/|\omega_0|$. Furthermore, the minimal and infinitesimal rigidity of the target formation implies that R is of full row rank for a formation close to the target formation, and therefore the null space of R^T is the zero vector. If the initial distances between the center positions (i.e., $\|c_k - c_j\|^2$) are close to the target distances d_{kj}, the limit set for which $R^T\epsilon = 0$ also implies $\epsilon = 0$, i.e., the distance error also converges to zero (see e.g. [21]). The remaining analysis is similar to that in Sect. 2.5. □

2.7 Simulation Examples

In this section, we show some simulation examples to illustrate the performance of the two proposed controllers. Consider a unicycle-like agent group consisting of four agents with constant speeds $v_1 = 1.0, v_2 = 1.1, v_3 = 1.2, v_4 = 1.3$. The parameters in the control laws are set as $\gamma = 0.1$ and $\omega_0 = 1$. First consider the displacement-based control law, which aims to stabilize a formation shape with desired displacement vectors $\hat{c}_{21} = 8 + 4i, \hat{c}_{32} = 5 - 3i, \hat{c}_{43} = -9 - 5i, \hat{c}_{14} = -4 + 4i, \hat{c}_{24} = 4 + 8i$, while the underlying graph is undirected and connected. The initial positions are chosen randomly in the simulation. The simulated trajectories by using the controller (2.4) are shown in Fig. 2.2, in which all four agents with constant speeds achieve their respective stable circular motions and also form the desired formation shape.

We then consider the distance-based formation control discussed in Sect. 2.6. Suppose four agents in a group are tasked to achieve a rigid rectangle shape with the desired distance set 3, 4, 5, 4, 3 under the control (2.12). The initial positions are chosen such that the initial relative center distances are close to the desired distances. The trajectories of each agent and the final shape are depicted in Fig. 2.3, which shows that stable circular motions for each agent and a rigid target formation are well achieved.

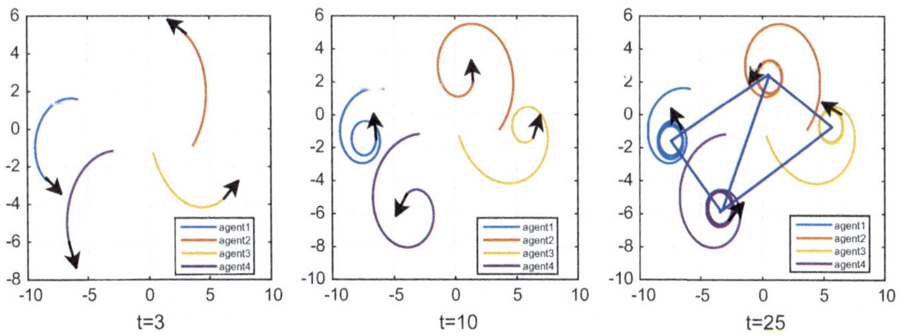

Fig. 2.2 Performance of the displacement-based formation controller (2.4) of four constant-speed unicycle agents with limited interaction

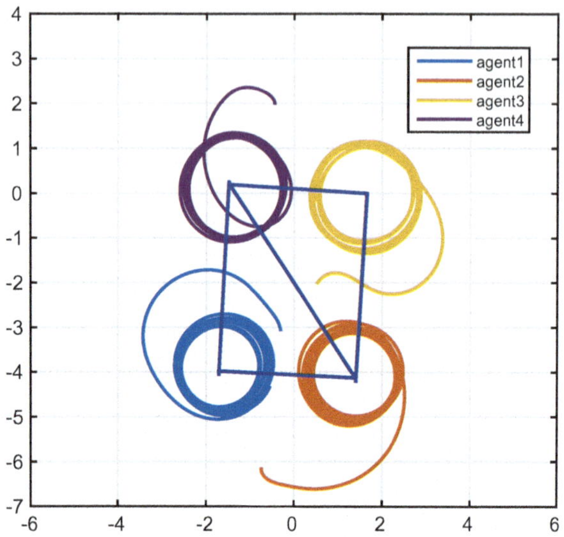

Fig. 2.3 Performance of the distance-based formation controller (2.12) of four constant-speed unicycle agents with limited communication

2.8 Conclusions

In this chapter, we have considered the formation stabilization problem for a group of unicycle-like agents with nonidentical and fixed cruising speeds. By respecting the dynamics constraints caused by constant speeds for each agent, the target formation shape is defined with respect to the rotation center arising from stable circular motions. Two different formation controllers based on different formation specifications and measurement requirements are proposed to coordinate multiple constant-speed agents in achieving a target formation shape and stable circular motions.

Acknowledgements This work was supported by the Australian Research Council under grant DP130103610 and DP160104500, and the DAAD-Go8 German–Australian Collaboration Project. Zhiyong Sun was supported by the Prime Minister's Australia Asia Incoming Endeavour Postgraduate Award.

References

1. Anderson, B.D.O., Fidan, B., Yu, C., Walle, D.: UAV formation control: theory and application. In: Recent Advances in Learning and Control, pp. 15–33. Springer (2008)
2. Anderson, B.D.O., Yu, C., Fidan, B., Hendrickx, J.M.: Rigid graph control architectures for autonomous formations. IEEE Control Syst. Mag. **28**(6), 48–63 (2008)
3. Briñón-Arranz, L., Seuret, A., Canudas-de Wit, C.: Cooperative control design for time-varying formations of multi-agent systems. IEEE Trans. Autom. Control **59**(8), 2283–2288 (2014)
4. Cao, Y., Yu, W., Ren, W., Chen, G.: An overview of recent progress in the study of distributed multi-agent coordination. IEEE Trans. Industr. Inf. **9**(1), 427–438 (2013)

5. Dimarogonas, D.V., Kyriakopoulos, K.J.: On the rendezvous problem for multiple nonholonomic agents. IEEE Trans. Autom. Control **52**(5), 916–922 (2007)
6. Fidan, B., Gazi, V., Zhai, S., Cen, N., Karatas, E.: Single-view distance-estimation-based formation control of robotic swarms. IEEE Trans. Industr. Electron. **60**(12), 5781–5791 (2013)
7. Francis, B.A., Maggiore, M.: Flocking and Rendezvous in Distributed Robotics. Springer (2016)
8. Hendrickson, B.: Conditions for unique graph realizations. SIAM J. Comput. **21**(1), 65–84 (1992)
9. Krick, L., Broucke, M.E., Francis, B.A.: Stabilisation of infinitesimally rigid formations of multi-robot networks. Int. J. Control **82**(3), 423–439 (2009)
10. Lin, Z., Francis, B.A., Maggiore, M.: Necessary and sufficient graphical conditions for formation control of unicycles. IEEE Trans. Autom. Control **50**(1), 121–127 (2005)
11. Liu, T., Jiang, Z.P.: Distributed formation control of nonholonomic mobile robots without global position measurements. Automatica **49**(2), 592–600 (2013)
12. Marshall, J.A., Broucke, M.E., Francis, B.A.: Pursuit formations of unicycles. Automatica **42**(1), 3–12 (2006)
13. Mesbahi, M., Egerstedt, M.: Graph theoretic methods in multiagent networks. Princeton University Press (2010)
14. Oh, K.K., Park, M.C., Ahn, H.S.: A survey of multi-agent formation control. Automatica **53**, 424–440 (2015)
15. Ren, W., Beard, R.W.: Distributed Consensus in Multi-Vehicle Cooperative Control. Springer, London (2008)
16. Sepulchre, R., Paley, D.A., Leonard, N.E.: Stabilization of planar collective motion: all-to-all communication. IEEE Trans. Autom. Control **52**(5), 811–824 (2007)
17. Sepulchre, R., Paley, D.A., Leonard, N.E.: Stabilization of planar collective motion with limited communication. IEEE Trans. Autom. Control **53**(3), 706–719 (2008)
18. Seyboth, G.S.: On distributed and cooperative control design for networks of dynamical systems. Ph.D. thesis, University of Stuttgart (2016)
19. Seyboth, G.S., Wu, J., Qin, J., Yu, C., Allgower, F.: Collective circular motion of unicycle type vehicles with nonidentical constant velocities. IEEE Trans. Control Netw. Syst. **1**(2), 167–176 (2014)
20. Sun, Z., Anderson, B.D.O.: Formation feasibility on coordination control of networked heterogeneous systems with drift terms. In: Proceedings of the 55th Conference on Decision and Control, pp. 3462–3467. IEEE (2016)
21. Sun, Z., Mou, S., Anderson, B.D.O., Cao, M.: Exponential stability for formation control systems with generalized controllers: a unified approach. Syst. Control Lett. **93**, 50–57 (2016)
22. Sun, Z., Seyboth, G.S., Anderson, B.D.O.: Collective control of multiple unicycle agents with non-identical constant speeds: tracking control and performance limitation. In: Proceedings of the 2015 IEEE Multi-Conference on Systems and Control (MSC), pp. 1361–1366. IEEE (2015)
23. Swartling, J.O., Shames, I., Johansson, K.H., Dimarogonas, D.V.: Collective circumnavigation. Unmanned Syst. **2**(03), 219–229 (2014)
24. Tabuada, P., Pappas, G.J., Lima, P.: Motion feasibility of multi-agent formations. IEEE Trans. Rob. **21**(3), 387–392 (2005)
25. Xargay, E., Dobrokhodov, V., Kaminer, I., Pascoal, A.M., Hovakimyan, N., Cao, C.: Time-critical cooperative control of multiple autonomous vehicles. IEEE Control Syst. Mag. **32**(5), 49 (2012)
26. Zheng, R., Liu, Y., Sun, D.: Enclosing a target by nonholonomic mobile robots with bearing-only measurements. Automatica **53**, 400–407 (2015)

Chapter 3
Control and Optimization Problems in Hyperpolarized Carbon-13 MRI

John Maidens and Murat Arcak

Abstract Hyperpolarized carbon-13 magnetic resonance imaging (MRI) is an emerging technology for probing metabolic activity in living subjects, which promises to provide clinicians new insights into diseases such as cancer and heart failure. These experiments involve an injection of a hyperpolarized substrate, often pyruvate labeled with carbon-13, which is imaged over time as it spreads throughout the subject's body and is transformed into various metabolic products. Designing these dynamic experiments and processing the resulting data requires the integration of noisy information across temporal, spatial, and chemical dimensions, and thus provides a wealth of interesting problems from an optimization and control perspective. In this work, we provide an introduction to the field of hyperpolarized carbon-13 MRI targeted toward researchers in control and optimization theory. We then describe three challenge problems that arise in metabolic imaging with hyperpolarized substrates: the design of optimal substrate injection profiles, the design of optimal flip angle sequences, and the constrained estimation of metabolism maps from experimental data. We describe the current state of research on each of these problems, and comment on aspects that remain open. We hope that these challenge problems will serve to direct future research in control.

3.1 Introduction

Carbon is arguably the most important element in biochemistry. It forms the basis of all organic molecules that make up the human body, yet only recently have we begun to be able to quickly image carbon in vivo using magnetic resonance imaging

J. Maidens (✉) · M. Arcak
Department of Electrical Engineering and Computer Sciences,
University of California, Cory Hall, Berkeley, CA 94720, USA
e-mail: maidens@eecs.berkeley.edu

M. Arcak
e-mail: arcak@eecs.berkeley.edu

R. Tempo et al. (eds.), *Emerging Applications of Control and Systems Theory*, Lecture Notes in Control and Information Sciences - Proceedings,
https://doi.org/10.1007/978-3-319-67068-3_3

(MRI). The emerging technology that makes this possible is known as hyperpolarized carbon-13 MRI, and it has enabled in vivo imaging with spatial, temporal and chemical specificity for the first time. This development is leading to new insights into the spatial distribution of metabolic activity through the analysis of dynamic image sequences.

The processes that are imaged in hyperpolarized carbon-13 MRI are inherently dynamic, resulting from blood flow, tissue perfusion, metabolic conversion, and polarization decay. Thus there is an opportunity for control researchers to improve the dynamic models, excitation inputs and estimation algorithms used in hyperpolarized carbon-13 MRI.

The remainder of this paper is organized as follows. In Sect. 3.2 we present the basics of hyperpolarized carbon-13 MRI. In Sect. 3.3 we present a dynamic model of metabolic flux and discuss methods of estimating model parameters from experimental MRI data. In Sect. 3.4 we discuss formulations of optimal design for dynamic experiments. Finally, in Sect. 3.5 we present three control and optimization problems that arise in metabolic MRI using hyperpolarized carbon-13 and discuss open questions.

3.2 Hyperpolarized Carbon-13 MRI for Imaging Metabolism

The measurable signal in MRI arises from radio-frequency electromagnetic waves generated by oscillating atomic nuclei. Nuclei containing an odd number of protons and/or neutrons possess a nuclear spin angular momentum, each giving rise to a small magnetic moment. Thus nuclei such as carbon (^{12}C) and oxygen (^{16}O) are invisible to MRI, while hydrogen (^{1}H) and the carbon-13 isotope (^{13}C) exhibit magnetic resonance (MR). Hydrogen MR, sometimes known as proton MR, is currently the most commonly-used in clinical settings due to the high abundance of hydrogen atoms in the human body (largely in the form of H_2O) and its high sensitivity [16]. Conventional hydrogen MRI is pervasive for noninvasive imaging of anatomic structure, but provides little functional information. In this work, we focus on carbon-13 MR, which can be used to provide information about metabolic function.

3.2.1 Chemical Shift

The unique aspect of hyperpolarized carbon-13 MRI, when compared to competing metabolic imaging technologies such as positron emission tomography (PET), is that it is the only technique that provides chemical specificity. It is possible to infer chemical information from MRI data due to a phenomenon known as chemical shift.

Chemical shift results in a small change in the resonant frequency of spins. This change is caused by shielding of the nuclei from the main magnetic field B_0 due to nearby electron orbitals [16]. The resulting frequency shift can be exploited to selectively excite specific metabolites [10], or distinguish between metabolites produced. This gives hyperpolarized carbon-13 MRI the unique ability to quantify metabolic flux in specific pathways.

3.2.2 Hyperpolarization Using DNP

Hyperpolarized carbon-13 MRI has been enabled by new technologies for hyperpolarizing carbon-13-containing substrates in liquid state, leading to a greater than 10000× increase in signal-to-noise ratio (SNR) when imaging carbon-13. This technology relies on dissolution dynamic nuclear polarization (D-DNP) to achieve significant polarization gains [1].

Dynamic nuclear polarization relies on transferring polarization to carbon-13 nuclei from electrons using microwave radiation. In this procedure, a sample is doped with a small quantity of stable electron radical. The sample is then cooled to cryogenic temperature and placed in a strong magnet. At this temperature and magnetic field strength, electrons become nearly 100% polarized. Then by irradiating the sample with microwaves, polarization is transferred from the electrons to the carbon-13 nuclei in a biochemical substrate of interest. To prepare the sample for injection and in vivo imaging, it is then rapidly dissolved in warm water, neutralized to a safe pH and the electron radical is removed before injection [15].

3.2.3 Polarization Decay in Hyperpolarized Substrates

Upon warming and removal from the magnet, the magnetization induced by hyperpolarization begins to decay over time toward the thermal equilibrium magnetization due to a phenomenon known as T_1 relaxation. The dynamics of the magnetization vector are governed by a system of state equations know as the rotating frame Bloch equations:

$$\frac{d}{dt}\begin{bmatrix} M_x \\ M_y \\ M_z \end{bmatrix} = \begin{bmatrix} -\frac{1}{T_2} & u_2 & 0 \\ -u_2 & -\frac{1}{T_2} & u_1 \\ 0 & -u_1 & \frac{1}{T_1} \end{bmatrix} \begin{bmatrix} M_x \\ M_y \\ M_z \end{bmatrix} + \begin{bmatrix} 0 \\ 0 \\ \frac{M_0}{T_1} \end{bmatrix} \tag{3.1}$$

with initial condition $M(0) = (0, 0, M_z(0))$. Here, the evolution of the state M is dependent on a sequence of control inputs u_1 and u_2 corresponding to the amplitude and frequency of the applied radio-frequency (RF) electromagnetic excitation pulse (known as the B_1 field) that rotates the vector M about the origin, and T_1 and T_2 parameters that govern the relaxation time in the longitudinal (z) and transverse (x, y) directions respectively.

When the sample is hyperpolarized we have $M_z(0) \gg M_0$, therefore the contribution of the affine term in (3.1) is negligible. Thus in the absence of RF excitation, the longitudinal magnetization exhibits exponential the decay

$$M_z(t) = M_z(0)e^{-t/T_1}.$$

In addition to T_1 relaxation, magnetization also decays due to repeated RF excitation. Throughout this paper we will assume that the RF pulse occurs on a time scale much faster than T_1 and T_2, therefore it can be modeled as an instantaneous state reset that rotates M to some angle α away from the z axis, known as the flip angle. We also assume that a spoiled gradient echo pulse sequence [2] is used, thus between RF pulses a strong magnetic field gradient is applied to dephase the transverse magnetization ensuring that $M_x = M_y = 0$. Thus at a time t^+ immediately after an RF pulse, the magnetization is given in terms of the magnetization at time t^- immediately before the RF pulse as

$$M_z(t^+) = \cos(\alpha)M_z(t^-)$$

$$M_{xy}(t^+) := \sqrt{M_x(t^+)^2 + M_y(t^+)^2} = \sin(\alpha)M_z(t^-).$$

It now follows that at a time t following a sequence of RF pulses with flip angles $\alpha_0, \ldots, \alpha_{N-1}$ the longitudinal magnetization remaining has decayed to

$$M_z(t) = M_z(0)e^{-t/T_1} \prod_{k=0}^{N-1} \cos(\alpha_k).$$

3.3 Quantifying Metabolic Flux

Hyperpolarized carbon-13 MRI enables dynamic experiments that show metabolic activity with spatial, temporal and chemical specificity. This enables quantifying the spatial distribution of the activity of specific metabolic pathways. In this section, we discuss model-based methods of fusing this information into spatial maps of metabolic activity. This is done by estimating kinetic parameters in a model describing the evolution of the MR signal observed in each spatial volume element (voxel).

3.3.1 *Kinetic Models of Hyperpolarized MRI Signal in a Single Voxel*

Hyperpolarized carbon-13 MRI researchers commonly rely on linear compartmental models for describing the evolution of signal in a voxel [4, 8, 9]. These models describe the magnetization exchange from the pool of injected hyperpolarized substrate to pools corresponding to various metabolic products. In its simplest form, this amounts to the irreversible metabolic conversion of the substrate S to a single product P performed at a characteristic kinetic rate k_{SP}:

$$S \xrightarrow{k_{SP}} P.$$

Throughout this article, we will focus on extremely simple pathways of this form, though extension to multiple products or bidirectional conversion is straightforward.

In the absence of external RF excitation, magnetization in a particular voxel i evolves via T_1 decay and label exchange according to the differential equations

$$\frac{d}{dt}\begin{bmatrix} M_{z,i,S}(t) \\ M_{z,i,P}(t) \end{bmatrix} = \begin{bmatrix} -R_{1,i,S} - k_{SP,i} & 0 \\ k_{SP,i} & -R_{1,i,P} \end{bmatrix} \begin{bmatrix} M_{z,i,S}(t) \\ M_{z,i,P}(t) \end{bmatrix} + \begin{bmatrix} k_{TRANS,i} \\ 0 \end{bmatrix} u(t) \tag{3.2}$$

where the states $M_{z,i,S}$ and $M_{z,i,P}$ represent the longitudinal magnetization in voxel i in the substrate and product compartments respectively, the input u models an arterial input function (AIF) describing the arrival of substrate from the circulatory system, and the parameters $k_{SP,i}$, $R_{1,i,S}$, $R_{1,i,P}$, and k_{TRANS} describe the metabolic rate, T_1 decay rate in the substrate pool, and T_1 decay rate in the product pool, and perfusion rate respectively.

When a constant flip angle excitation sequence and repetition time is used for imaging, decay due to RF excitation can be modeled by replacing $R_{1,i,X}$ by an effective decay rate

$$R_{1,i,X,\text{effective}} = R_{1,i,X} - \frac{\log(\cos\alpha)}{T_R}$$

where α is the flip angle and T_R is the repetition time, and X denotes an arbitrary compound (either S or P) [18]. However, when a variable flip angle sequence is used, signal decay due to RF excitation must be accounted for as in Sect. 3.2.3. This leads to a discrete time model for the transverse and longitudinal magnetization immediately preceding excitation k given by

$$\begin{bmatrix} M_{z,i,S}[k+1] \\ M_{z,i,P}[k+1] \end{bmatrix} = A_d \begin{bmatrix} \cos\alpha_S[k] & 0 \\ 0 & \cos\alpha_P[k] \end{bmatrix} \begin{bmatrix} M_{z,i,S}[k] \\ M_{z,i,P}[k] \end{bmatrix} + B_d u[k] \tag{3.3}$$

where A_d and B_d are computed by discretizing (3.2) assuming a zero order hold with sampling time T_R. A model for the transverse magnetization immediately following excitation k given by

$$M_{xy,i,X}[k] = \sin\alpha_X[k] M_{z,i,X}[k]. \tag{3.4}$$

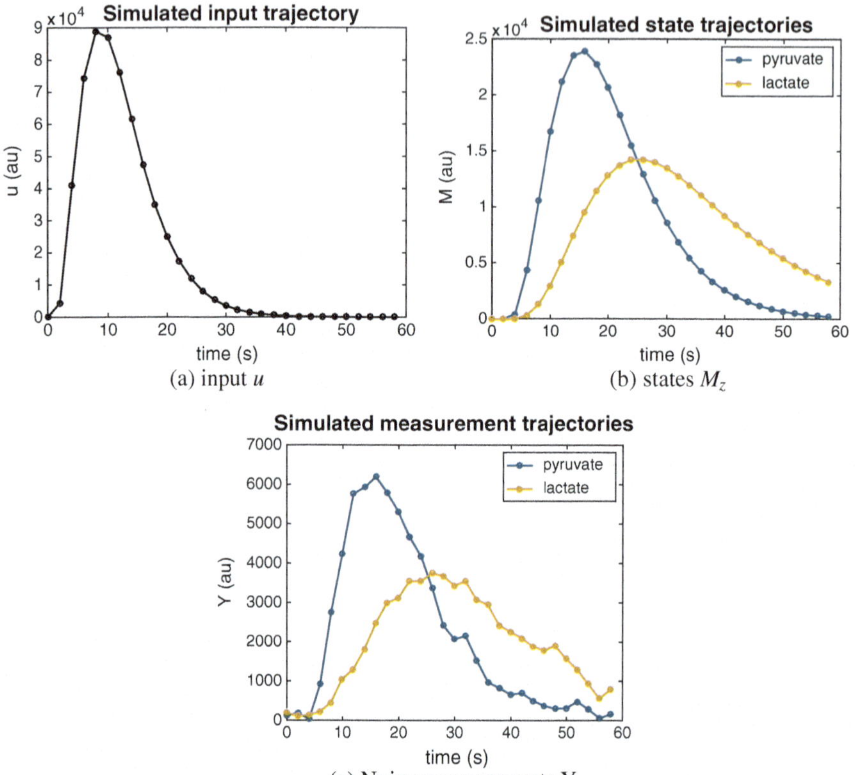

(a) input u

(b) states M_z

(c) Noisy measurements Y

Fig. 3.1 Simulated trajectories for a pyruvate to lactate conversion model using a constant flip angle sequence with $\alpha_S[k] = \alpha_P[k] = 15°$. (Adapted from J. Maidens, J.W. Gordon, M. Arcak, P.E.Z. Larson, IEEE Trans Med Imaging. 2016 Nov; 35(11): 2403–2412.) [13]

This transverse magnetization leads to the observable signal which we measure as an output from voxel i at time k. In the case of normally-distributed measurements, we model the generated data as

$$Y_{i,X}[k] \sim M_{xy,i,X}[k] + \varepsilon_{i,X}[k]$$

where ε is independent identically distributed gaussian noise with a known variance σ^2. Simulated trajectories of this model are shown in Fig. 3.1.

3.3.2 *Estimation of Unknown Model Parameters*

Estimating metabolic rate parameters θ_i from experimental data collected from voxel i involves minimizing a statistical loss function $L(\theta_i|Y_i)$ that describes how well a signal model fits the observed data Y_i. Using the model Eqs. (3.3)–(3.4) as the basis of a signal model describing the predicted measurement

$$y_i(\theta_i) = \big[[M_{xy,i,S}[1]\ M_{xy,i,P}[1]\ \ldots\ M_{xy,i,S}[N]\ M_{xy,i,P}[N]]\big]$$

in terms of the vector model parameters θ_i. Loss functions include:

- the least squares loss
$$L(\theta_i|Y_i) = \|Y_i - y_i(\theta)\|^2$$
which corresponds to a nonlinear least squares estimation problem and
- the negative log likelihood loss
$$L(\theta_i|Y_i) = -\log p_{\theta_i}(Y_i)$$
which corresponds to a maximum likelihood estimation problem. Unlike the least squares loss function, this loss requires a that a probability density function describing the joint distribution of Y_i be specified. Common choices are $Y_i \sim y_i + \varepsilon$ where ε is independent, identically distributed (iid) Gaussian noise or independent Rician noise with location parameters given by y_i [7].

3.4 Optimal Design of Dynamic Experiments

Two of the three problems we will discuss in this paper address the design of optimized experiments for estimating the value of unknown parameters in a mathematical model of a dynamical system from noisy output data. Thus, in this section we provide background on optimal experiment design.

In dynamical systems with noisy outputs, the reliability of the parameter estimates depends on the choice of input used to excite the system, as some inputs provide much greater information about the parameters than others. Much work has been done on the optimal experiment design problem in the last 50 years [5, 6, 11, 17, 19]. Historically, a great deal of work on this problem has taken a frequency domain approach, where the input to the system is designed based on its power spectrum. Here, we will approach this problem in the time domain, to be able to perform experiment design for systems with nonlinear dynamics.

3.4.1 *Problem Description*

We consider a discrete-time dynamical system with noisy observations

$$\begin{aligned} x_{t+1} &= f(t, x_t, u_t, \theta) \\ Y_t &\sim P_{x_t} \end{aligned} \tag{3.5}$$

where $x_t \in \mathbb{R}^n$ denotes the system's state, $u_t \in \mathbb{R}^m$ is a sequence of inputs to be designed and $\theta \in \mathbb{R}^p$ is a vector of unknown parameters that we wish to estimate. Observations are drawn independently from a known distribution that is parametrized by the system state x_t. We assume that for all $x_t \in \mathbb{R}^n$ the probability distribution P_{x_t} is absolutely continuous with respect to some measure μ and we denote its density with respect to μ by $p_{x_t}(y_t)$. We consider this system over a finite horizon $0 \leq t \leq N$. Our goal is to design a sequence u that provides a maximal amount of information about the unknown parameter vector θ. This problem can be addressed by maximizing the Fisher information about θ.

3.4.2 *Fisher Information*

An important notion in frequentist statistics is the Fisher information matrix for the vector of model parameters θ. The Fisher information is fundamental in the analysis of numerous statistical estimators from unbiased estimation to maximum likelihood estimation. We begin with a definition.

Definition 1 Let $\mathscr{P} = \{P_\theta : \theta \in \Omega\}$ be a family of probability distributions parametrized by θ in an open set $\Omega \subseteq \mathbb{R}^p$ and dominated by some measure μ. Denote the probability densities with respect to μ by p_θ and assume that the densities are differentiable with respect to θ. We define the Fisher information matrix as the $p \times p$ matrix $\mathscr{I}(\theta)$ with (i, j)-th entry defined as

$$\mathscr{I}(\theta)_{i,j} = \mathbb{E}\left[\frac{\partial \log p_\theta(Y)}{\partial \theta_i} \frac{\partial \log p_\theta(Y)}{\partial \theta_j}\right]$$

where $Y \sim P_\theta$.

3.5 Control and Optimization Problems in Hyperpolarized Carbon-13 MRI

We now present three optimization problems that arise in the design of hyperpolarized carbon-13 MRI experiments and the subsequent data analysis. The first involves the design of substrate injection inputs to generate maximally informative data, a problem in which the control input enters linearly. The second involves the design of optimized flip angle sequences, again for generating maximally informative data. In contrast with Problem 1, this problem involves a nonlinear control system model, which is significantly more difficult to analyze globally. The third problem involves estimating the spatial distribution of metabolic flux parameters from the acquired data. Problem 3 completes the experimental sequence from experimental design to data acquisition to data analysis.

Problem 1: Substrate Injection Design

Data collected in MRI experiments is typically noisy due to thermal movement of electrons in the receiver coil and the object being imaged. This makes it challenging to estimate model parameters from dynamic data sets when the signal-to-noise ratio is small. This challenge can be addressed by designing experimental parameters with the goal of maximizing the information about unknown parameters contained in the data collected.

The first problem we consider is the optimal design of the injection input subject to constraints on the maximum injection rate and volume. This results in a dynamic optimal experiment design problem of the form discussed in Sect. 3.4. More formally, we consider the dynamic model defined in Eq. (3.3) with an output defined in Eq. (3.4) which is corrupted by iid additive Gaussian noise. Problem 1 is to design an injection input $u[k]$ to maximize the Fisher information about the parameter of interest k_{SP} contained in the data generated from a finite number of samples under this model. The input is constrained such that both the maximum injection rate $\|u\|_\infty$ and the maximum injection volume $\|u\|_1$ are upper bounded by some positive constant.

We first formulated this problem in [12], where we showed that this problem can be reformulated as a nonconvex quadratic program (QP). We then developed a procedure for approximating the global solution of the QP using a semidefinite programming relaxation. This method allowed us to compute approximate solutions to particular instances of this problem as well as bounds on the global solution. In particular, for an instance with realistic values for model parameters, we found that the optimal input consists of a bolus applied at the beginning of the experiment injected at the maximum rate until the volume budget is reached (Fig. 3.2). Based on the semidefinite relaxation, we then show that this input achieves an objective function value at least 98.7% of the global optimum, for these particular values of the model parameters.

We conjecture that all optimal solutions are of the form shown in Fig. 3.2: an injection at the maximum rate until the volume budget is reached. We expect this

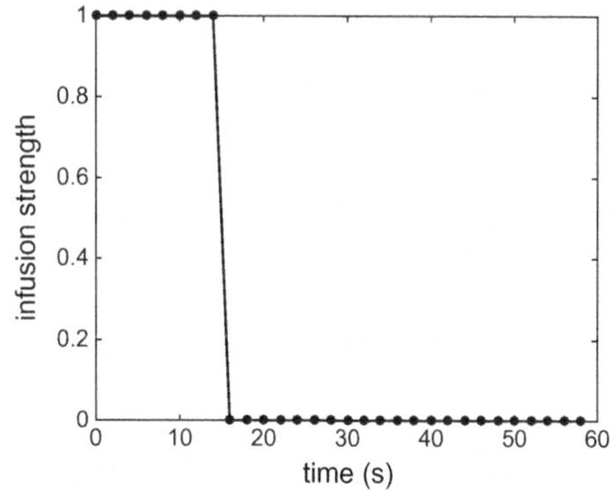

Fig. 3.2 Conjectured solution to a particular instance of Problem 1. The optimal input sequence $u[k]$ applies a bolus injection at the maximum allowable rate until the total input budget is reached

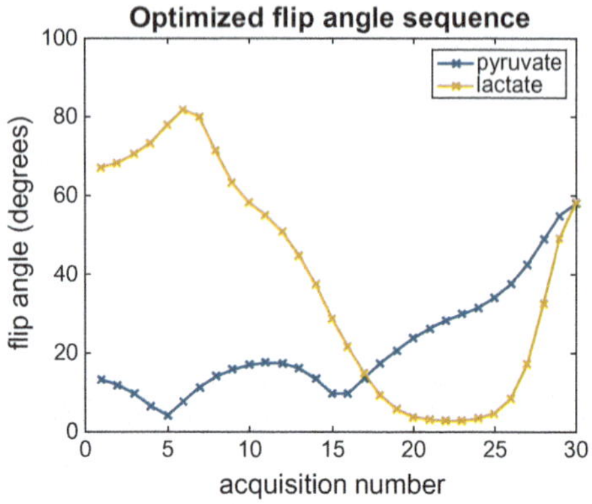

Fig. 3.3 Optimized input sequence for the flip angle sequence design problem. Reproduced with permission from John Maidens, et al. IEEE Trans Med Imaging;35(11):2403–2412 [13]

to hold independent of the choice of model parameters, as well as in more complex metabolic networks. However, this conjecture remains unproven.

Problem 2: Flip Angle Sequence Design

Similarly to the first problem, the second problem we consider involves designing experimental parameters to maximize the Fisher information about unknown rates in the model. Here we consider the problem of designing optimal RF flip angle excitation sequences.

Again we use the model defined in Eq. (3.3) with an output defined in Eq. (3.4) corrupted by iid noise. In Problem 2, we wish to select a sequence of flip angles $\alpha_S[k]$ and $\alpha_P[k]$ used to excite each of the chemical species. Here the choice of $\alpha_S[k]$ and $\alpha_P[k]$ at each time is unconstrained. Since the flip angles enter the model in a nonlinear fashion, the resulting optimization problem is no longer a QP, so other optimization techniques must be used.

This problem is solved to local optimality under additional smoothness constraints in [13] using a nonlinear programming approach. The resulting optimized flip angle sequence is shown in Fig. 3.3. This flip angle sequence results in a 20% decrease in the uncertainty of metabolic rate estimates, when compared against the best existing sequences.

These results demonstrate that optimal experiment design can help to improve the quality of parameter estimates in dynamic MRI experiments. But they could be further improved by the development of techniques for computing global solutions to this optimization problem.

Problem 3: Constrained Parameter Mapping

The third problem involves computing maps of metabolic activity from the experimental data collected. Here we assume that we are given a statistical model for the data as well as a loss function, as described in Sect. 3.3.2. The challenge is to summarize the spatial, temporal and chemical information contained in the dynamic experimental data into a single spatial map of metabolic activity. We do so by estimating a value for the metabolic rate parameter $\theta_i = k_{SP,i}$ for each voxel i in space.

Since the objects imaged often contain spatial structure, this structure can be exploited to improve the quality of the estimated parameter maps. This can be achieved by adding regularization to the objective function that is optimized. Formally, we solve an optimization problem of the form

$$\textbf{minimize} \sum_i L(\theta_i | Y_i) + \lambda r(\theta)$$

where L is a loss function that depends on the data Y_i collected in each voxel i, and r is a regularization term that couples nearby voxels thereby enforcing spatial structure in the estimated maps. Possible choices of regularization used to enforce smoothness, sparsity and edge preservation include ℓ_2, ℓ_1 and total variation penalties. By including such penalties to exploit spatial correlations in the data, we have shown that better image quality can be achieved compared with independently fitting each voxel [14].

Both choices of loss function described in Sect. 3.3.2 are nonconvex. However, we have observed that despite the nonconvexity of the problem satisfactory solutions can be found using convex optimization algorithms such as ADMM [3]. Problem 3 is to better understand the convergence of this algorithm for estimating parameters in spatially-distributed dynamical system models. Why does this algorithm successfully converge to the same optimum for various initial conditions? And can we provide any formal guarantees of global convergence?

References

1. Ardenkjær-Larsen, J.H., Fridlund, B., Gram, A., Hansson, G., Hansson, L., Lerche, M.H., Servin, R., Thaning, M., Golman, K.: Increase in signal-to-noise ratio of >10,000 times in liquid-state NMR. Proc. Nat. Acad. Sci. **100**(18), 10,158–10,163 (2003)
2. Bernstein, M.A., King, K.F., Zhou, X.J.: Basic pulse sequences. In: Handbook of MRI Pulse Sequences, pp. 579–647. Academic Press (2004)
3. Boyd, S., Parikh, N., Chu, E., Peleato, B., Eckstein, J.: Distributed optimization and statistical learning via the alternating direction method of multipliers. Found. Trends Mach. Learn. **3**(1), 1–122 (2011)
4. Brindle, K.M.: NMR methods for measuring enzyme kinetics in vivo. Prog. Nucl. Magn. Reson. Spectrosc. **20**(3), 257–293 (1988)
5. Gevers, M., Bombois, X., Hildebrand, R., Solari, G.: Optimal experiment design for open and closed-loop system identification. Commun. Inf. Syst. **11**(3), 197–224 (2011)
6. Goodwin, G., Payne, R.: Dynamic System Identification: Experiment Design and Data Analysis. Academic Press (1977)
7. Gudbjartsson, H., Patz, S.: The rician distribution of noisy MRI data. Magn. Reson. Med. **34**(6), 910–914 (1995)
8. Harrison, C., Yang, C., Jindal, A., Deberardinis, R., Hooshyar, M., Merritt, M., Dean Sherry, A., Malloy, C.: Comparison of kinetic models for analysis of pyruvate-to-lactate exchange by hyperpolarized ^{13}C NMR. NMR Biomed. **25**(11), 1286–1294 (2012)
9. Kazan, S.M., Reynolds, S., Kennerley, A., Wholey, E., Bluff, J.E., Berwick, J., Cunningham, V.J., Paley, M.N., Tozer, G.M.: Kinetic modeling of hyperpolarized ^{13}C pyruvate metabolism in tumors using a measured arterial input function. Magn. Reson. Med. **70**(4), 943–953 (2013)

10. Larson, P.E., Kerr, A.B., Chen, A.P., Lustig, M.S., Zierhut, M.L., Hu, S., Cunningham, C.H., Pauly, J.M., Kurhanewicz, J., Vigneron, D.B.: Multiband excitation pulses for hyperpolarized ^{13}C dynamic chemical-shift imaging. J. Magn. Reson. **194**(1), 121–127 (2008)
11. Ljung, L.: System Identification: Theory for the User. Pearson Education (1999)
12. Maidens, J., Arcak, M.: Semidefinite relaxations in optimal experiment design with application to substrate injection for hyperpolarized MRI. In: Proceedings of the American Control Conference (ACC), pp. 2023–2028 (2016)
13. Maidens, J., Gordon, J.W., Arcak, M., Larson, P.E.Z.: Optimizing flip angles for metabolic rate estimation in hyperpolarized carbon-13 MRI. IEEE Trans. Med. Imaging **35**(11), 2403–2412 (2016)
14. Maidens, J., Gordon, J.W., Arcak, M., Chen, H.Y., Park, I., Criekinge, M.V., Milshteyn, E., Bok, R., Aggarwal, R., Ferrone, M., Slater, J.B., Kurhanewicz, J., Vigneron, D.B., Larson, P.E.Z.: Spatio-temporally constrained reconstruction for hyperpolarized carbon-13 MRI using kinetic models. In: Proceedings of the ISMRM Annual Meeting. http://submissions.mirasmart.com/ISMRM2017/ViewSubmissionPublic.aspx?sei=GH0eaFQTF (2017)
15. Nelson, S.J., Kurhanewicz, J., Vigneron, D.B., Larson, P.E.Z., Harzstark, A.L., Ferrone, M., van Criekinge, M., Chang, J.W., Bok, R., Park, I., Reed, G., Carvajal, L., Small, E.J., Munster, P., Weinberg, V.K., Ardenkjaer-Larsen, J.H., Chen, A.P., Hurd, R.E., Odegardstuen, L.I., Robb, F.J., Tropp, J., Murray, J.A.: Metabolic imaging of patients with prostate cancer using hyperpolarized [1-^{13}C]pyruvate. Sci. Transl. Med. **5**(198), 198ra108 (2013)
16. Nishimura, D.G.: Principles of Magnetic Resonance Imaging. Lulu (2010)
17. Pukelsheim, F.: Optimal Design of Experiments. Probability and mathematical statistics. Wiley, New York (1993)
18. Søgaard, L.V., Schilling, F., Janich, M.A., Menzel, M.I., Ardenkjaer-Larsen, J.H.: In vivo measurement of apparent diffusion coefficients of hyperpolarized ^{13}C-labeled metabolites. NMR Biomed. **27**(5), 561–569 (2014)
19. Walter, É., Pronzato, L.: Identification of Parametric Models from Experimental Data. Communications and Control Engineering. Springer (1997)

Chapter 4
Parameter Selection and Preconditioning for a Graph Form Solver

Christopher Fougner and Stephen Boyd

Abstract In the chapter "Block splitting for distributed optimization", Parikh and Boyd describe a method for solving a convex optimization problem, where each iteration involves evaluating a proximal operator and projection onto a subspace. In this chapter, we address the critical practical issues of how to select the proximal parameter in each iteration, and how to scale the original problem variables, so as to achieve reliable practical performance. The resulting method has been implemented as an open-source software package called POGS (Proximal Graph Solver), that targets multi-core and GPU-based systems, and has been tested on a wide variety of practical problems. Numerical results show that POGS can solve very large problems (with, say, a billion coefficients in the data), to modest accuracy in a few tens of seconds, where similar problems take many hours using interior-point methods.

4.1 Introduction

We consider the convex optimization problem

$$\begin{array}{ll} \text{minimize} & f(y) + g(x) \\ \text{subject to} & y = Ax, \end{array} \tag{4.1}$$

where $x \in \mathbf{R}^n$ and $y \in \mathbf{R}^m$ are the variables, and the (extended real-valued) functions $f : \mathbf{R}^m \to \mathbf{R} \cup \{\infty\}$ and $g : \mathbf{R}^n \to \mathbf{R} \cup \{\infty\}$ are convex, closed and proper. The matrix $A \in \mathbf{R}^{m \times n}$, and the functions f and g are the problem data. Infinite values

C. Fougner
DeepMind, Stanford, USA

S. Boyd (✉)
Department of Electrical Engineering, Stanford University,
Packard 254, Stanford, CA 94305, USA
e-mail: boyd@stanford.edu

R. Tempo et al. (eds.), *Emerging Applications of Control and Systems Theory*, Lecture Notes in Control and Information Sciences - Proceedings,
https://doi.org/10.1007/978-3-319-67068-3_4

of f and g allow us to encode convex constraints on x and y, since any feasible point (x, y) must satisfy

$$x \in \{x \mid g(x) < \infty\}, \qquad y \in \{y \mid f(y) < \infty\}.$$

We will be interested in the case when f and g have simple proximal operators, but for now we do not make this assumption. The problem form (4.1) is known as *graph form* [39], since the variable (x, y) is constrained to lie in the graph $\mathcal{G} = \{(x, y) \in \mathbf{R}^{n+m} \mid y = Ax\}$ of A. We denote $p^\star$ as the optimal value of (4.1), which we assume is finite.

The graph form includes a large range of convex problems, including linear and quadratic programming, general conic programming [8, Sect. 11.6], and many more specific applications such as logistic regression with various regularizers, support vector machine fitting [29], portfolio optimization [8, Sect. 4.4.1] [25] [4], signal recovery [16], and radiation treatment planning [38], to name just a few.

In [39], Parikh and Boyd described an operator splitting method for solving (a generalization of) the graph form problem (4.1), based on the alternating direction method of multipliers (ADMM) [5]. Each iteration of this method requires a projection (either exactly or approximately via an iterative method) onto the graph $\mathcal{G}$, and evaluation of the proximal operators of f and g. Theoretical convergence was established in those papers, and basic implementations were demonstrated. However, it has been observed that practical convergence of the algorithm depends very much on the choice of algorithm parameters (such as the proximal parameter ρ), and scaling of the variables (i.e., preconditioning).

The purpose of this chapter is to explore these issues, and to add some critical variations on the algorithm that make it a relatively robust general purpose solver, at least for modest accuracy levels. The algorithm we propose, which is the same as the basic method described in [39], with modified parameter selection, diagonal preconditioning, and modified stopping criterion, has been implemented in an open-source software project called POGS (for **Pro**ximal **G**raph **S**olver), and tested on a wide variety of problems. Our CUDA implementation reliably solves (to modest accuracy) problems $10^3\times$ larger than those that can be handled by interior-point methods; and for those that can be handled by interior-point methods, $100\times$ faster.

4.1.1 Outline

In Sect. 4.1.2 we describe related work. In Sect. 4.2 we derive the graph form dual problem, and the primal-dual optimality conditions, which we use to motivate the stopping criterion and to interpret the iterates of the algorithm. In Sect. 4.3 we describe the ADMM-based graph form algorithm, and analyze the properties of its iterates, giving some results that did not appear in [39]. In Sect. 4.4 we address the topic of preconditioning, and suggest novel preconditioning and parameter selection techniques. In Sect. 4.5 we describe our implementation POGS, and in Sect. 4.6 we report performance results on various problem families.

4.1.2 Related Work

Many generic methods can be used to solve the graph form problem (4.1), including projected gradient descent [12], projected subgradient methods [42, Chap. 5] [47], operator splitting methods [32] [20], interior-point methods [35, Chap. 19] [7, Chap. 6], and many more. (Of course many of these methods can only be used when additional assumptions are made on f and g, e.g., differentiability or strong convexity.) For example, if f and g are separable and smooth, the problem (4.1) can be solved by Newton's method, with each iteration requiring the solution of a set of linear equations that requires $O(\max\{m, n\} \min\{m, n\}^2)$ floating point operations (flops) when A is dense. If f and g are separable and have smooth barrier functions for their epigraphs, the problem (4.1) can be solved by an interior-point method, which in practice always takes no more than a few tens of iterations, with each iteration involving the solution of a system of linear equations that requires $O(\max\{m, n\} \min\{m, n\}^2)$ flops when A is dense [8, Chap. 11] [35, Chap. 19].

We now turn to first-order methods for the graph form problem (4.1). In [1] Briceño-Arias and Combettes describe methods for solving a generalized version of (4.1), including a forward–backward–forward algorithm and one based on Douglas–Rachford splitting [17]. Their methods are especially interesting in the case when A represents an abstract operator, and one only has access to A through Ax and $A^T y$. In [37] O'Connor and Vandenberghe propose a primal-dual method for the graph form problem where A is the sum of two structured matrices. They contrast it with methods such as Spingarn's method of partial inverses [49], Douglas–Rachford splitting, and the Chambolle–Pock method [14].

Davis and Yin [18] analyze convergence rates for different operator splitting methods, and in [24] Giselsson proves the tightness of linear convergence for the operator splitting problems considered [22]. Goldstein et al. [26] derive Nesterov type acceleration, and show $O(1/k^2)$ convergence for problems where f and g are both strongly convex.

Nishihara et al. [34] introduce a parameter selection framework for ADMM with over relaxation [19]. The framework is based on solving a fixed-size semidefinite program (SDP). They also make the assumption that f is strongly convex. Ghadimi et al. [27] derive optimal parameter choices for the case when f and g are both quadratic. In [22], Giselsson and Boyd show how to choose metrics to optimize the convergence bound, and in [21] Giselsson and Boyd suggest a diagonal preconditioning scheme for graph form problems based on semidefinite programming. This scheme is primarily relevant in small to medium scale problems, or situations where many different graph form problems, with the same matrix A, are to be solved. It is clear from these papers (and indeed, a general rule) that the practical convergence of first-order methods depends heavily on algorithm parameter choices.

GPUs are used extensively for stochastic gradient descent-based optimization when training neural networks [11, 31, 33], and they are slowly gaining popularity in convex optimization as well [13, 41, 52].

4.2 Optimality Conditions and Duality

4.2.1 Dual Graph Form Problem

The Lagrange dual function of (4.1) is given by

$$\inf_{x,y} f(y) + g(x) + \nu^T(Ax - y) = -f^*(\nu) - g^*(-A^T\nu),$$

where $\nu \in \mathbf{R}^n$ is the dual variable associated with the equality constraint, and f^* and g^* are the conjugate functions of f and g respectively [8, Chap. 4]. Introducing the variable $\mu = -A^T\nu$, we can write the dual problem as

$$\begin{array}{ll} \text{maximize} & -f^*(\nu) - g^*(\mu) \\ \text{subject to} & \mu = -A^T\nu. \end{array} \tag{4.2}$$

The dual problem can be written as a graph form problem, if we negate the objective and minimize rather than maximize. The dual graph form problem (4.2) is related to the primal graph form problem (4.1) by switching the roles of the variables, replacing the objective function terms with their conjugates, and replacing A with $-A^T$.

The primal and dual objectives are $p(x, y) = f(y) + g(x)$ and $d(\mu, \nu) = -f^*(\nu) - g^*(\mu)$, respectively, giving us the duality gap

$$\eta = p(x, y) - d(\mu, \nu) = f(y) + f^*(\nu) + g(x) + g^*(\mu). \tag{4.3}$$

We have $\eta \geq 0$, for any primal and dual feasible tuple (x, y, μ, ν). The duality gap η gives a bound on the suboptimality of (x, y) (for the primal problem) and also (μ, ν) for the dual problem:

$$f(y) + g(x) \leq p^\star + \eta, \qquad -f^*(\nu) - g^*(\mu) \geq p^\star - \eta.$$

4.2.2 Optimality Conditions

The optimality conditions for (4.1) are readily derived from the dual problem. The tuple (x, y, μ, ν) satisfies the following three conditions if and only it is optimal:

Primal feasibility:

$$y = Ax. \tag{4.4}$$

Dual feasibility:

$$\mu = -A^T \nu. \tag{4.5}$$

Zero gap:

$$f(y) + f^*(\nu) + g(x) + g^*(\mu) = 0. \tag{4.6}$$

If both (4.4) and (4.5) hold, then the zero gap condition (4.6) can be replaced by the Fenchel equalities

$$f(y) + f^*(\nu) = \nu^T y, \quad g(x) + g^*(\mu) = \mu^T x. \tag{4.7}$$

We refer to a tuple (x, y, μ, ν) that satisfies (4.7) as *Fenchel feasible*. To verify the statement, we add the two equations in (4.7), which yields

$$f(y) + f^*(\nu) + g(x) + g^*(\mu) = y^T \nu + x^T \mu = (Ax)^T \nu - x^T A^T \nu = 0.$$

The Fenchel equalities (4.7) are also equivalent to

$$\nu \in \partial f(y), \quad \mu \in \partial g(x), \tag{4.8}$$

where ∂ denotes the subdifferential, which follows because

$$\nu \in \partial f(y) \Leftrightarrow \sup_z \left(z^T \nu - f(z)\right) = \nu^T y - f(y) \Leftrightarrow f(y) + f^*(\nu) = \nu^T y.$$

In the sequel we will assume that strong duality holds, meaning that there exists a tuple $(x^\star, y^\star, \mu^\star, \nu^\star)$ which satisfies all three optimality conditions.

4.3 Algorithm

4.3.1 Graph Projection Splitting

In [39] Parikh et al. apply ADMM [5, Sect. 5] to the problem of minimizing $f(y) + g(x)$, subject to the constraint $(x, y) \in \mathcal{G}$. This yields the *graph projection splitting* Algorithm 1.

Algorithm 1 Graph projection splitting

Input: A, f, g
1: Initialize $(x^0, y^0, \tilde{x}^0, \tilde{y}^0) = 0,\ k = 0$
2: **repeat**
3: $\quad (x^{k+1/2},\ y^{k+1/2}) := \left(\mathbf{prox}_g(x^k - \tilde{x}^k),\ \mathbf{prox}_f(y^k - \tilde{y}^k)\right)$
4: $\quad (x^{k+1}, y^{k+1}) := \Pi(x^{k+1/2} + \tilde{x}^k,\ y^{k+1/2} + \tilde{y}^k)$
5: $\quad (\tilde{x}^{k+1}, \tilde{y}^{k+1}) := (\tilde{x}^k + x^{k+1/2} - x^{k+1},\ \tilde{y}^k + y^{k+1/2} - y^{k+1})$
6: $\quad k := k + 1$
7: **until** converged

The variable k is the iteration counter, $x^{k+1}, x^{k+1/2} \in \mathbf{R}^n$ and $y^{k+1}, y^{k+1/2}, \in \mathbf{R}^m$ are primal variables, $\tilde{x}^{k+1} \in \mathbf{R}^n$ and $\tilde{y}^{k+1} \in \mathbf{R}^m$ are scaled dual variables, Π denotes the (Euclidean) projection onto the graph $\mathcal{G}$,

$$\mathbf{prox}_f(v) = \underset{y}{\operatorname{argmin}} \left(f(y) + (\rho/2)\,\|y - v\|_2^2\right)$$

is the proximal operator of f (and similarly for g), and $\rho > 0$ is the proximal parameter. The projection Π can be explicitly expressed as the linear operator

$$\Pi(c, d) = K^{-1} \begin{bmatrix} c + A^T d \\ 0 \end{bmatrix}, \qquad K = \begin{bmatrix} I & A^T \\ A & -I \end{bmatrix}. \tag{4.9}$$

Roughly speaking, in steps 3 and 5, the x (and $\tilde{x}$) and y (and $\tilde{y}$) variables do not mix; the computations can be carried out in parallel. The projection step 4 mixes the $x, \tilde{x}$ and $y, \tilde{y}$ variables.

General convergence theory for ADMM [5, Sect. 3.2] guarantees that (with our assumption on the existence of a solution)

$$(x^{k+1}, y^{k+1}) - (x^{k+1/2}, y^{k+1/2}) \to 0, \quad f(y^k) + g(x^k) \to p^\star, \quad (\tilde{x}^k, \tilde{y}^k) \to (\tilde{x}^\star, \tilde{y}^\star), \tag{4.10}$$

as $k \to \infty$.

4.3.2 Extensions

We discuss three common extensions that can be used to speed up convergence in practice: over-relaxation, approximate projection, and varying penalty.

Over-relaxation. Replacing $x^{k+1/2}$ by $\alpha x^{k+1/2} + (1 - \alpha)x^k$ in the projection and dual update steps is known as over-relaxation if $\alpha > 1$ or under-relaxation if $\alpha < 1$. The algorithm is guaranteed to converge [19] for any $\alpha \in (0, 2)$; it is observed in practice [36] that using an over-relaxation parameter in the range [1.5, 1.8] can improve practical convergence.

Approximate projection. Instead of computing the projection Π exactly one can use an approximation $\tilde{\Pi}$, with the only restriction that

$$\sum_{k=0}^{\infty} \|\Pi(x^{k+1/2}, y^{k+1/2}) - \tilde{\Pi}(x^{k+1/2}, y^{k+1/2})\|_2 < \infty$$

must hold. This is known as approximate projection [36], and is guaranteed to converge [1]. This extension is particularly useful if the approximate projection is computed using an indirect or iterative method.

Varying penalty. Large values of ρ tend to encourage primal feasibility, while small values tend to encourage dual feasibility [5, Sect. 3.4.1]. A common approach is to adjust or vary ρ in each iteration, so that the primal and dual residuals are (roughly) balanced in magnitude. When doing so, it is important to re-scale $(\tilde{x}^{k+1}, \tilde{y}^{k+1})$ by a factor ρ^k / ρ^{k+1}.

4.3.3 *Feasible Iterates*

In each iteration, Algorithm 1 produces sets of points that are either primal, dual, or Fenchel feasible. Define

$$\mu^k = -\rho\tilde{x}^k, \quad \nu^k = -\rho\tilde{y}^k, \quad \mu^{k+1/2} = -\rho(x^{k+1/2} - x^k + \tilde{x}^k), \quad \nu^{k+1/2} = -\rho(y^{k+1/2} - y^k + \tilde{y}^k).$$

The following statements hold.

1. The pair (x^{k+1}, y^{k+1}) is primal feasible, since it is the projection onto the graph $\mathcal{G}$.
2. The pair (μ^{k+1}, ν^{k+1}) is dual feasible, as long as (μ^0, ν^0) is dual feasible and (x^0, y^0) is primal feasible. Dual feasibility implies $\mu^{k+1} + A^T\nu^{k+1} = 0$, which we show using the update equations in Algorithm 1:

$$\begin{aligned}\mu^{k+1} + A^T\nu^{k+1} &= -\rho(\tilde{x}^k + x^{k+1/2} - x^{k+1} + A^T(\tilde{y}^k + y^{k+1/2} - y^{k+1})) \\ &= -\rho(\tilde{x}^k + A^T\tilde{y}^k + x^{k+1/2} + A^T y^{k+1/2} - (I + A^T A)x^{k+1}),\end{aligned}$$

where we substituted $y^{k+1} = Ax^{k+1}$. From the projection operator in (4.9) it follows that $(I + A^T A)x^{k+1} = x^{k+1/2} + A^T y^{k+1/2}$, therefore

$$\mu^{k+1} + A^T\nu^{k+1} = -\rho(\tilde{x}^k + A^T\tilde{y}^k) = \mu^k + A^T\nu^k = \mu^0 + A^T\nu^0,$$

where the last equality follows from an inductive argument. Since we made the assumption that (μ^0, ν^0) is dual feasible, we can conclude that (μ^{k+1}, ν^{k+1}) is also dual feasible.

3. The tuple $(x^{k+1/2}, y^{k+1/2}, \mu^{k+1/2}, \nu^{k+1/2})$ is Fenchel feasible. From the definition of the proximal operator,

$$x^{k+1/2} = \operatorname*{argmin}_x \left(g(x) + (\rho/2)\left\|x - x^k + \tilde{x}^k\right\|_2^2\right) \Leftrightarrow 0 \in \partial g(x^{k+1/2}) + \rho(x^{k+1/2} - x^k + \tilde{x}^k)$$
$$\Leftrightarrow \mu^{k+1/2} \in \partial g(x^{k+1/2}).$$

By the same argument $\nu^{k+1/2} \in \partial f(y^{k+1/2})$.

Applying the results in (4.10) to the dual variables, we find $\nu^{k+1/2} \to \nu^\star$ and $\mu^{k+1/2} \to \mu^\star$, from which we conclude that $(x^{k+1/2}, y^{k+1/2}, \mu^{k+1/2}, \nu^{k+1/2})$ is primal and dual feasible in the limit.

4.3.4 Stopping Criteria

In Sect. 4.3.3 we noted that either (4.4, 4.5, 4.6) or (4.4, 4.5, 4.7) are sufficient for optimality. We present two different stopping criteria based on these conditions.

Residual-based stopping. The tuple $(x^{k+1/2}, y^{k+1/2}, \mu^{k+1/2}, \nu^{k+1/2})$ is Fenchel feasible in each iteration, but only primal and dual feasible in the limit. Accordingly, we propose the residual-based stopping criterion

$$\|Ax^{k+1/2} - y^{k+1/2}\|_2 \le \varepsilon^{\text{pri}}, \quad \|A^T\nu^{k+1/2} + \mu^{k+1/2}\|_2 \le \varepsilon^{\text{dual}}, \tag{4.11}$$

where the ε^{pri} and ε^{dua} are positive tolerances. These should be chosen as a mixture of absolute and relative tolerances, such as

$$\varepsilon^{\text{pri}} = \varepsilon^{\text{abs}} + \varepsilon^{\text{rel}}\|y^{k+1/2}\|_2, \quad \varepsilon^{\text{dual}} = \varepsilon^{\text{abs}} + \varepsilon^{\text{rel}}\|\mu^{k+1/2}\|_2.$$

Reasonable values for ε^{abs} and ε^{rel} are in the range $[10^{-4}, 10^{-2}]$.

Gap-based stopping. The tuple (x^k, y^k, μ^k, ν^k) is primal and dual feasible, but only Fenchel feasible in the limit. We propose the gap-based stopping criteria

$$\eta^k = f(y^k) + g(x^k) + f^*(\nu^k) + g^*(\mu^k) \le \varepsilon^{\text{gap}},$$

where ε^{gap} should be chosen relative to the current objective value, i.e.,

$$\varepsilon^{\text{gap}} = \varepsilon^{\text{abs}} + \varepsilon^{\text{rel}}|f(y^k) + g(x^k)|.$$

Here too, reasonable values for ε^{abs} and ε^{rel} are in the range $[10^{-4}, 10^{-2}]$.

Although the gap-based stopping criteria is very informative, since it directly bounds the suboptimality of the current iterate, it suffers from the drawback that

f, g, f^*, and g^* must all have full domain, since otherwise the gap η^k can be infinite. Indeed, the gap η^k is almost always infinite when f or g represent constraints.

4.3.5 Implementation

Projection. There are different ways to evaluate the projection operator Π, depending on the structure and size of A.

One simple method that can be used if A is sparse and not too large is a direct sparse factorization. The matrix K is quasi-definite, and therefore the LDL^T decomposition is well defined [51]. Since K does not change from iteration to iteration, the factors L and D (and the permutation or elimination ordering) can be computed in the first iteration (e.g., using CHOLMOD [9]) and reused in subsequent iterations. This is known as *factorization caching* [5, Sect. 4.2.3] [39, Sect. A.1]. With factorization caching, we get a (potentially) large speedup in iterations, after the first one.

If A is dense, and $\min(m, n)$ is not too large, then block elimination [8, Appendix C] can be applied to K [39, Appendix A], yielding the reduced update

$$\begin{aligned} x^{k+1} &:= (A^T A + I)^{-1}(c + A^T d) \\ y^{k+1} &:= A x^{k+1} \end{aligned}$$

if $m \geq n$, or

$$\begin{aligned} y^{k+1} &:= d + (AA^T + I)^{-1}(Ac - d) \\ x^{k+1} &:= c - A^T(d - y^{k+1}) \end{aligned}$$

if $m < n$. Both formulations involve forming and solving a system of $\min(m, n)$ equations with $\min(m, n)$ unknowns. Since the coefficient matrix is symmetric positive definite, we can use the Cholesky decomposition. Forming the coefficient matrix $A^T A + I$ or $AA^T + I$ dominates the computation. Here too, we can take advantage of factorization caching.

The regular structure of dense matrices allows us to analyze the computational complexity of each step. We define $q = \min(m, n)$ and $p = \max(m, n)$. The first iteration involves the factorization and the solve step; subsequent iterations only require the solve step. The computational cost of the factorization is the combined cost of computing $A^T A$ (or AA^T, whichever is smaller), at a cost of pq^2 flops, in addition to the Cholesky decomposition, at a cost of $(1/3)q^3$ flops. The solve step consists of two matrix-vector multiplications at a cost of $4pq$ flops and solving a triangular system of equations at a cost of q^2 flops. The total cost of the first iteration is $O(pq^2)$ flops, while each subsequent iteration only costs $O(pq)$ flops, showing that we obtain savings by a factor of q flops, after the first iteration, by using factorization caching.

For very large problems direct methods are no longer practical, at which point indirect (iterative) methods can be used. Fortunately, as the primal and dual variables converge, we are guaranteed that $(x^{k+1/2}, y^{k+1/2}) \to (x^{k+1}, y^{k+1})$, meaning that we will have a good initial guess we can use to initialize the iterative method to (approximately) evaluate the projection. One can either apply CGLS (conjugate gradient least-squares) [28] or LSQR [45] to the reduced update or apply MINRES (minimum residual) [44] to K directly. It can be shown the latter requires twice the number of iterations as compared to the former, and is therefore not recommended.

Proximal operators. Since the $x, \tilde{x}$ and $y, \tilde{y}$ components are decoupled in the proximal step and dual variable update step, both of these can be done separately, and in parallel for x and y. If either f or g is separable, then the proximal step can be parallelized further. Combettes and Pesquet [15, Sect. 10.2] contains a table of proximal operators for a wide range of functions, and the monograph [40] details how proximal operators can be computed efficiently, in particular for the case where there is no analytical solution. Typically, the cost of computing the proximal operator will be negligible compared to the cost of the projection. In particular, if f and g are separable, then the cost will be $O(m+n)$, and completely parallelizable.

4.4 Preconditioning and Parameter Selection

The practical convergence of the algorithm (i.e., the number of iterations required before it terminates) can depend greatly on the choice of the proximal parameter ρ, and the scaling of the variables. In this section we analyze these, and suggest a method for choosing ρ and for scaling the variables that (empirically) speeds up practical convergence.

4.4.1 Preconditioning

Consider scaling the variables x and y in (4.1), by E^{-1} and D respectively, where $D \in \mathbf{R}^{m \times m}$ and $E \in \mathbf{R}^{n \times n}$ are non-singular matrices. We define the scaled variables

$$\hat{y} = Dy, \quad \hat{x} = E^{-1}x,$$

which transforms (4.1) into

$$\begin{array}{ll} \text{minimize} & f(D^{-1}\hat{y}) + g(E\hat{x}) \\ \text{subject to} & \hat{y} = DAE\hat{x}. \end{array} \tag{4.12}$$

This is also a graph form problem, and for notational convenience, we define

$$\hat{A} = DAE, \quad \hat{f}(\hat{y}) = f(D^{-1}\hat{y}), \quad \hat{g}(\hat{x}) = g(E\hat{x}),$$

so that the problem can be written as

$$\begin{array}{ll} \text{minimize} & \hat{f}(\hat{y}) + \hat{g}(\hat{x}) \\ \text{subject to} & \hat{y} = \hat{A}\hat{x}. \end{array}$$

We refer to this problem as the preconditioned version of (4.1). Our goal is to choose D and E so that (a) the algorithm applied to the preconditioned problem converges in fewer steps in practice, and (b) the additional computational cost due to the preconditioning is minimal.

Graph projection splitting applied to the preconditioned problem (4.12) can be interpreted in terms of the original iterates. The proximal step iterates are redefined as

$$x^{k+1/2} = \underset{x}{\text{argmin}} \left(g(x) + (\rho/2)\|x - x^k + \tilde{x}^k\|^2_{(EE^T)^{-1}} \right),$$

$$y^{k+1/2} = \underset{y}{\text{argmin}} \left(f(y) + (\rho/2)\|y - y^k + \tilde{y}^k\|^2_{(D^TD)} \right),$$

and the projected iterates are the result of the weighted projection

$$\begin{array}{ll} \text{minimize} & (1/2)\|x - x^{k+1/2}\|^2_{(EE^T)^{-1}} + (1/2)\|y - y^{k+1/2}\|^2_{(D^TD)} \\ \text{subject to} & y = Ax, \end{array}$$

where $\|x\|_P = \sqrt{x^TPx}$ for a symmetric positive-definite matrix P. This projection can be expressed as

$$\Pi(c, d) = \hat{K}^{-1} \begin{bmatrix} (EE^T)^{-1}c + A^TD^TDd \\ 0 \end{bmatrix}, \quad \hat{K} = \begin{bmatrix} (EE^T)^{-1} & A^TD^TD \\ D^TDA & -D^TD \end{bmatrix}.$$

Notice that graph projection splitting is invariant to orthogonal transformations of the variables x and y, since the preconditioners only appear in terms of D^TD and EE^T. In particular, if we let $D = U^T$ and $E = V$, where $A = U\Sigma V^T$, then the preconditioned constraint matrix $\hat{A} = DAE = \Sigma$ is diagonal. We conclude that any graph form problem can be preconditioned to one with a diagonal nonnegative constraint matrix Σ. For analysis purposes, we are therefore free to assume that A is diagonal. We also note that for orthogonal preconditioners, there exists an analytical relationship between the original proximal operator and the preconditioned proximal operator. With $\phi(x) = \varphi(Qx)$, where Q is any orthogonal matrix ($Q^TQ = QQ^T = I$), we have

$$\mathbf{prox}_\phi(v) = Q^T\mathbf{prox}_\varphi(Qv).$$

While the proximal operator of ϕ is readily computed, orthogonal preconditioners destroy separability of the objective. As a result, we cannot easily combine them with other preconditioners.

Multiplying D by a scalar α and dividing E by the same scalar has the effect of scaling ρ by a factor of α^2. It however has no effect on the projection step, showing that ρ can be thought of as the relative scaling of D and E.

In the case where f and g are separable and both D and E are diagonal, the proximal step takes the simplified form

$$\begin{aligned} x_j^{k+1/2} &= \operatorname*{argmin}_{x_j} \left(g_j(x_j) + (\rho_j^E/2)(x_j - x_j^k + \tilde{x}_j^k)^2\right) \qquad j = 1, \ldots, n \\ y_i^{k+1/2} &= \operatorname*{argmin}_{y_i} \left(f_i(y_i) + (\rho_i^D/2)(y_i - y_i^k + \tilde{y}_i^k)^2\right) \qquad i = 1, \ldots, m, \end{aligned}$$

where $\rho_j^E = \rho/E_{jj}^2$ and $\rho_i^D = \rho D_{ii}^2$. Since only ρ is modified, any routine capable of computing $\mathbf{prox}_f$ and $\mathbf{prox}_g$ can also be used to compute the preconditioned proximal update.

4.4.1.1 Effect of Preconditioning on Projection

For the purpose of analysis, we will assume that $A = \Sigma$, where Σ is a nonnegative diagonal matrix. The projection operator simplifies to

$$\Pi(c,d) = \begin{bmatrix} (I + \Sigma^T\Sigma)^{-1} & (I + \Sigma^T\Sigma)^{-1}\Sigma^T \\ (I + \Sigma\Sigma^T)^{-1}\Sigma & (I + \Sigma\Sigma^T)^{-1}\Sigma\Sigma^T \end{bmatrix} \begin{bmatrix} c \\ d \end{bmatrix},$$

which means the projection step can be written explicitly as

$$\begin{aligned} x_i^{k+1} &= \frac{1}{1+\sigma_i^2}(x_i^{k+1/2} + \tilde{x}_i^k + \sigma_i(y_i^{k+1/2} + \tilde{y}_i^k)) \quad & i &= 1, \ldots, \min(m,n) \\ x_i^{k+1} &= x_i^{k+1/2} + \tilde{x}_i^k & i &= \min(m,n)+1, \ldots, n \\ y_i^{k+1} &= \frac{\sigma_i}{1+\sigma_i^2}(x_i^{k+1/2} + \tilde{x}_i^k + \sigma_i(y_i^{k+1/2} + \tilde{y}_i^k)) \quad & i &= 1, \ldots, \min(m,n) \\ y_i^{k+1} &= 0 & i &= \min(m,n)+1, \ldots, m, \end{aligned}$$

where σ_i is the ith diagonal entry of Σ and subscripted indices of x and y denote the ith entry of the respective vector. Notice that the projected variables x_i^{k+1} and y_i^{k+1} are equally dependent on $(x_i^{k+1/2} + \tilde{x}_i^k)$ and $\sigma_i(y_i^{k+1/2} + \tilde{y}_i^k)$. If σ_i is either significantly smaller or larger than 1, then the terms x_i^{k+1} and y_i^{k+1} will be dominated by either $(x_i^{k+1/2} + \tilde{x}_i^k)$ or $(y_i^{k+1/2} + \tilde{y}_i^k)$. However if $\sigma_i = 1$, then the projection step exactly averages the two quantities

$$x_i^{k+1} = y_i^{k+1} = \frac{1}{2}(x_i^{k+1/2} + \tilde{x}_i^k + y_i^{k+1/2} + \tilde{y}_i^k) \qquad i = 1, \dots, \min(m, n).$$

As pointed out in Sect. 4.3, the projection step mixes the variables x and y. For this to approximately reduce to averaging, we need $\sigma_i \approx 1$.

4.4.1.2 Choosing D and E

The analysis suggests that the algorithm will converge quickly when the singular values of DAE are all near one, i.e.,

$$\mathbf{cond}(DAE) \approx 1, \quad \|DAE\|_2 \approx 1. \tag{4.13}$$

(This claim is also supported in [23], and is consistent with our computational experience.) Preconditioners that exactly satisfy these conditions can be found using the singular value decomposition of A. They will, however, be of little use, since such preconditioners generally destroy our ability to evaluate the proximal operators of $\hat{f}$ and $\hat{g}$ efficiently.

So we seek choices of D and E for which (4.13) holds (very) approximately, and for which the proximal operators of $\hat{f}$ and $\hat{g}$ can still be efficiently computed. We now specialize to the special case when f and g are separable. In this case, diagonal D and E are candidates for which the proximal operators are still easily computed. (The same ideas apply to block separable f and g, where we impose the further constraint that the diagonal entries within a block are the same.) So we now limit ourselves to the case of diagonal preconditioners.

Diagonal matrices that minimize the condition number of DAE, and therefore approximately satisfy the first condition in (4.13), can be found exactly, using semidefinite programming [3, Sect. 3.1]. But this computation is quite involved, and may not be worth the computational effort since the conditions (4.13) are just a heuristic for faster convergence. (For control problems, where the problem is solved many times with the same matrix A, this approach makes sense; see [21].)

A heuristic that tends to minimize the condition number is to equilibrate the matrix, i.e., choose D and E so that the rows all have the same p-norm, and the columns all have the same p-norm. (Such a matrix is said to be equilibrated.) This corresponds to finding D and E so that

$$|DAE|^p \mathbf{1} = \alpha \mathbf{1}, \qquad \mathbf{1}^T |DAE|^p = \beta \mathbf{1}^T,$$

where $\alpha, \beta > 0$. Here the notation $|\cdot|^p$ should be understood in the elementwise sense. Various authors [6, 13, 36] suggest that equilibration can decrease the number of iterations needed for operator splitting and other first-order methods. One issue that we need to address is that not every matrix can be equilibrated. Given that equilibration is only a heuristic for achieving $\sigma_i(DAE) \approx 1$, which is in turn a heuristic for fast convergence of the algorithm, partial equilibration should serve the same purpose just as well.

Sinkhorn and Knopp [48] suggest a method for matrix equilibration for $p < \infty$, and A is square and has full support. In the case $p = \infty$, the Ruiz algorithm [46] can be used. Both of these methods fail (as they must) when the matrix A cannot be equilibrated. We give below a simple modification of the Sinkhorn–Knopp algorithm, modified to handle the case when A is non-square, or cannot be equilibrated.

Choosing preconditioners that satisfy $\|DAE\|_2 = 1$ can be achieved by scaling D and E by $\sigma_{\max}(DAE)^{-q}$ and $\sigma_{\max}(DAE)^{q-1}$ respectively for $q \in \mathbf{R}$. The quantity $\sigma_{\max}(DAE)$ can be approximated using power iteration, but we have found it is unnecessary to exactly enforce $\|DAE\|_2 = 1$. A more computationally efficient alternative is to replace $\sigma_{\max}(DAE)$ by $\|DAE\|_F/\sqrt{\min(m,n)}$. This quantity coincides with $\sigma_{\max}(DAE)$ when $\mathbf{cond}(DAE) = 1$. If DAE is equilibrated and $p = 2$, this scaling corresponds to $(DAE)^T(DAE)$ (or $(DAE)(DAE)^T$ when $m < n$) having unit diagonal.

4.4.2 Regularized Equilibration

In this section, we present a self-contained derivation of our matrix-equilibration method. It is similar to the Sinkhorn–Knopp algorithm, but also works when the matrix is non-square or cannot be exactly equilibrated.

Consider the convex optimization problem with variables u and v,

$$\text{minimize} \quad \sum_{i=1}^{m}\sum_{j=1}^{n} |A_{ij}|^p e^{u_i+v_j} - n\mathbf{1}^T u - m\mathbf{1}^T v + \gamma\left[n\sum_{i=1}^{m} e^{u_i} + m\sum_{j=1}^{n} e^{v_j}\right], \tag{4.14}$$

where $\gamma \geq 0$ is a regularization parameter. The objective is bounded below for any $\gamma > 0$. The optimality conditions are

$$\sum_{j=1}^{n} |A_{ij}|^p e^{u_i+v_j} - n + n\gamma e^{u_i} = 0, \quad i = 1, \dots, m,$$

$$\sum_{i=1}^{m} |A_{ij}|^p e^{u_i+v_j} - m + m\gamma e^{v_j} = 0, \quad j = 1, \dots, n.$$

By defining $D_{ii} = e^{u_i/p}$ and $E_{jj}^p = e^{v_j/p}$, these conditions are equivalent to

$$|DAE|^p\mathbf{1} + n\gamma D\mathbf{1} = n\mathbf{1}, \quad \mathbf{1}^T|DAE|^p + m\gamma\mathbf{1}^T E = m\mathbf{1}^T,$$

where $\mathbf{1}$ is the vector with all entries one. When $\gamma = 0$, these are the conditions for a matrix to be equilibrated. The objective may not be bounded when $\gamma = 0$, which exactly corresponds to the case when the matrix cannot be equilibrated. As $\gamma \to \infty$, both D and E converge to the scaled identity matrix $(1/\gamma)I$, showing that γ can be thought of as a regularizer on the elements of D and E. If D and E are optimal, then the two equalities

$$\mathbf{1}^T |DAE|^p \mathbf{1} + n\gamma \mathbf{1}^T D\mathbf{1} = mn, \qquad \mathbf{1}^T |DAE|^p \mathbf{1} + m\gamma \mathbf{1}^T E\mathbf{1} = mn$$

must hold. Subtracting the one from the other, and dividing by γ, we find the relationship

$$n\mathbf{1}^T D\mathbf{1} = m\mathbf{1}^T E\mathbf{1},$$

implying that the average entry in D and E is the same.

There are various ways to solve the optimization problem (4.14), one of which is to apply coordinate descent. Minimizing the objective in (4.14) with respect to u_i yields

$$\sum_{j=1}^{n} e^{u_i^k + v_j^{k-1}} |A_{ij}|^p + n\gamma e^{u_i^k} = n \Leftrightarrow e^{u_i^k} = \frac{n}{\sum_{j=1}^n e^{v_j^{k-1}} |A_{ij}|^p + n\gamma}$$

and similarly for v_j

$$e^{v_i^k} = \frac{m}{\sum_{i=1}^n e^{u_i^{k-1}} |A_{ij}|^p + m\gamma}.$$

Since the minimization with respect to u_i^k is independent of u_{i-1}^k, the update can be done in parallel for each element of u, and similarly for v. Repeated minimization over u and v will eventually yield values that satisfy the optimality conditions.

Algorithm 2 summarizes the equilibration routine. The inverse operation in steps 4 and 5 should be understood in the element-wise sense.

Algorithm 2 Regularized Sinkhorn-Knopp

Input: $A, \varepsilon > 0, \gamma > 0$
1: Initialize $e^0 := \mathbf{1},\ k := 0$
2: **repeat**
3: $\quad k := k + 1$
4: $\quad d^k := n\,(|A|^p e^{k-1} + n\gamma \mathbf{1})^{-1}$
5: $\quad e^k := m\,(|A^T|^p d^k + m\gamma \mathbf{1})^{-1}$
6: **until** $\|e^k - e^{k-1}\|_2 \le \varepsilon$ and $\|d^k - d^{k-1}\|_2 \le \varepsilon$
7: **return** $D := \mathbf{diag}(d^k)^{1/p}, E := \mathbf{diag}(e^k)^{1/p}$

4.4.3 Adaptive Penalty Update

The projection operator Π does not depend on the choice of ρ, so we are free to update ρ in each iteration, at no extra cost. While the convergence theory only holds for fixed ρ, it still applies if one assumes that ρ becomes fixed after a finite number of iterations [5].

As a rule, increasing ρ will decrease the primal residual, while decreasing ρ will decrease the dual residual. The authors in [5, 30] suggest adapting ρ to balance the primal and dual residuals. We have found that substantially better practical convergence can be obtained using a variation on this idea. Rather than balancing the primal and dual residuals, we allow either the primal or dual residual to approximately converge and only then start adjusting ρ. Based on this observation, we propose the following adaptive update scheme.

Algorithm 3 Adaptive ρ update

Input: $\delta > 1$, $\tau \in (0, 1]$,

Initialize $l := 0$, $u := 0$

repeat

  Apply graph projection splitting

  if $\|A^T \nu^{k+1/2} + \mu^{k+1/2}\|_2 < \varepsilon^{\text{dual}}$ and $\tau k > l$ **then**

    $\rho^{k+1} := \delta \rho^k$

    $u := k$

  else if $\|Ax^{k+1/2} - y^{k+1/2}\|_2 < \varepsilon^{\text{pri}}$ and $\tau k > u$ **then**

    $\rho^{k+1} := (1/\delta)\rho^k$

    $l := k$

until $\|A^T \nu^{k+1/2} + \mu^{k+1/2}\|_2 < \varepsilon^{\text{dual}}$ and $\|Ax^{k+1/2} - y^{k+1/2}\|_2 < \varepsilon^{\text{pri}}$

Once either the primal or dual residual converges, the algorithm begins to steer ρ in a direction so that the other residual also converges. By making small adjustments to ρ, we will tend to remain approximately primal (or dual) feasible once primal (dual) feasibility has been attained. Additionally by requiring a certain number of iterations between an increase in ρ and a decrease (and vice versa), we enforce that changes to ρ do not flip-flop between one direction and the other. The parameter τ determines the relative number of iterations between changes in direction.

4.5 Implementation

Proximal Graph Solver (POGS) is an open-source (BSD-3 license) implementation of graph projection splitting, written in C++. It supports both GPU and CPU platforms and includes wrappers for C, MATLAB, and R. POGS handles all combinations of sparse/dense matrices, single/double precision arithmetic, and direct/indirect solvers, with the exception (for now) of sparse indirect solvers. The only dependency is a tuned BLAS library on the respective platform (e.g., cuBLAS or the Apple Accelerate Framework). The source code is available at

`https://github.com/foges/pogs`

In lieu of having the user specify the proximal operators of f and g, POGS contains a library of proximal operators for a variety of different functions. It is currently assumed that the objective is separable, in the form

$$f(y) + g(x) = \sum_{i=1}^{m} f_i(y_i) + \sum_{j=1}^{n} g_j(x_j),$$

where $f_i, g_j : \mathbf{R} \to \mathbf{R} \cup \{\infty\}$. The library contains a set of base functions, and by applying various transformations, the range of functions can been greatly extended. In particular we use the parametric representation

$$f_i(y_i) = c_i h_i(a_i y_i - b_i) + d_i y_i + (1/2) e_i y_i^2,$$

where $a_i, b_i, d_i \in \mathbf{R}$, $c_i, e_i \in \mathbf{R}_+$, and $h_i : \mathbf{R} \to \mathbf{R} \cup \{\infty\}$. The same representation is also used for g_j. It is straightforward to express the proximal operators of f_i in terms of the proximal operator of h_i using the formula

$$\mathbf{prox}_f(v) = \frac{1}{a}\left(\mathbf{prox}_{h,(e+\rho)/(ca^2)}\left(a\,(v\rho - d)\,/(e+\rho) - b\right) + b\right),$$

where for notational simplicity we have dropped the index i in the constants and functions. It is possible for a user to add their own proximal operator function, if it is not in the current library. We note that the separability assumption on f and g is a simplification, rather than a limitation of the algorithm. It allows us to apply the proximal operator in parallel using either CUDA or OpenMP (depending on the platform).

The constraint matrix is equilibrated using Algorithm 2, with a choice of $p = 2$ and $\gamma = \frac{m+n}{mn}\sqrt{\varepsilon^{\text{cmp}}}$, where ε^{cmp} is machine epsilon. Both D and E are rescaled evenly, so that they satisfy $\|DAE\|_F/\sqrt{\min(m,n)} = 1$. The projection Π is computed as outlined in Sect. 4.3.5. We work with the reduced update equations in all versions of POGS. In the indirect case, we chose to use CGLS. The parameter ρ is updated according to Algorithm 3. Empirically, we found that $(\delta, \tau) = (1.05, 0.8)$ works well. We also use over-relaxation with $\alpha = 1.7$. POGS supports warm starting, whereby an initial guess for x^0 and/or ν^0 may be supplied by the user. If only x^0 is provided, then ν^0 will be estimated, and vice versa. The warm-start feature allows any cached matrices to be used to solve additional problems with the same matrix A. POGS returns the tuple $(x^{k+1/2}, y^{k+1/2}, \mu^{k+1/2}, \nu^{k+1/2})$, since it has finite primal and dual objectives. The primal and dual residuals will be nonzero and are determined by the specified tolerances. Future plans for POGS include extension to block-separable

f and g (including general cone solvers), additional wrappers for Julia and Python, support for a sparse direct solver, and a multi-GPU extension.

4.6 Numerical Results

To highlight the robustness and general purpose nature of POGS, we tested it on 9 different problem classes using random but realistic data. We considered the following 9 problem classes: basis pursuit, entropy maximization, Huber fitting, lasso, logistic regression, linear programming, nonnegative least-squares, portfolio optimization, and support vector machine fitting. For each problem class, the number of nonzeros in A was varied on a logarithmic scale from 100 to 1 Billion. The aspect ratio of A also varied from 1:1.25 to 1:10, with the orientation (wide or tall) chosen depending on what was reasonable for each problem. We report running time averaged over all aspect ratios. These problems and the data generation methods are described in detail in a longer version of this chapter [10]. All experiments were performed in single precision arithmetic on a machine equipped with an Intel Core i7-870, 16GB of RAM, and a TitanX GPU. Timing results include the data copy from CPU to GPU.

We compare POGS to SDPT3 [50], an open-source solver that handles linear, second-order, and positive semidefinite cone programs. Since SDPT3 uses an interior-point algorithm, the solution returned will be of high precision, allowing us

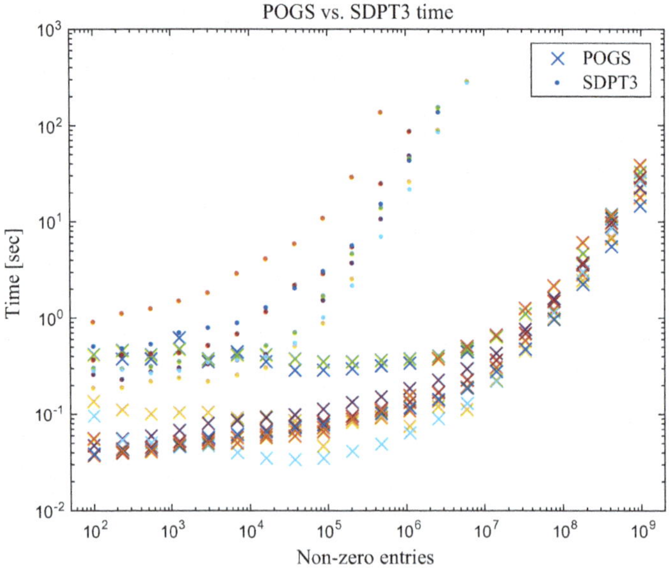

Fig. 4.1 POGS (GPU version) versus SDPT3 for dense matrices (color represents problem class)

to verify the accuracy of the solution computed by POGS. Problems that took SDPT3 more than 200 seconds (of which there were many) were aborted.

The maximum number of iterations was set to 10^4, but all problems converged in fewer iterations, with most problems taking a couple of hundred iterations. The relative tolerance was set to 10^{-3}, and where solutions from SDPT3 were available, we verified that the solutions produced by both solvers matched to 3 decimal places. We omit SDPT3 running times for problems involving exponential cones, since SDPT3 does not support them.

Figure 4.1 compares the running time of POGS versus SDPT3, for problems where the constraint matrix A is dense. We can make several general observations.

- POGS solves problems that are 3 orders of magnitude larger than SDPT3 in the same amount of time.
- Problems that take 200 s in SDPT3 take 0.5 s in POGS.
- POGS can solve problems with 1 Billion nonzeros in 10–40 s.
- The variation in solve time across different problem classes was similar for POGS and SDPT3, around one order of magnitude.

In summary, POGS is able to solve much larger problems, much faster (to moderate precision).

References

1. Briceno-Arias, L.M., Combettes, P.L.: A monotone + skew splitting model for composite monotone inclusions in duality. SIAM J. Optim. **21**(4), 1230–1250 (2011)
2. Briceno-Arias, L.M., Combettes, P.L., Pesquet, J.C., Pustelnik, N.: Proximal algorithms for multicomponent image recovery problems. J. Math. Imaging Vis. **41**(1–2), 3–22 (2011)
3. Boyd, S., El Ghaoui, L., Feron, E., Balakrishnan, V.: Linear Matrix Inequalities in System and Control Theory, vol. 15. SIAM (1994)
4. Boyd, S., Mueller, M., O'Donoghue, B., Wang, Y.: Performance bounds and suboptimal policies for multi-period investment. Found. Trends Optim. **1**(1), 1–69 (2013)
5. Boyd, S., Parikh, N., Chu, E., Peleato, B., Eckstein, J.: Distributed optimization and statistical learning via the alternating direction method of multipliers. Found. Trends Mach. Learn. **3**(1), 1–122 (2011)
6. Bradley, A.M.: Algorithms for the equilibration of matrices and their application to limited-memory quasi-Newton methods. Ph.D. thesis. Stanford University (2010)
7. Ben-Tal, A., Nemirovski, A.: Lectures on Modern Convex Optimization: Analysis, Algorithms, and Engineering Applications, vol. 2. SIAM (2001)
8. Boyd, S., Vandenberghe, L.: Convex Optimization. Cambridge University Press (2004)
9. Chen, Y., Davis, T.A., Hager, W.W., Rajamanickam, S.: Algorithm 887: CHOLMOD, supernodal sparse Cholesky factorization and update/downdate. ACM Trans. Math. Softw. **35**(3), 22 (2008)
10. Boyd, S., Fougner, C.: Parameter Selection and Pre-conditioning for a Graph form Solver (2015). www.stanford.edu/~boyd/papers/pogs.html
11. Coates, A., Huval, B., Wang, T., Wu, D., Catanzaro, B., Ng, A.Y.: Deep learning with COTS HPC systems. In: Proceedings of the 30th International Conference on Machine Learning, pp. 1337–1345 (2013)

12. Calamai, P.H., Moré, J.J.: Projected gradient methods for linearly constrained problems. Math. Program. **39**(1), 93–116 (1987)
13. Chu, E., O'Donoghue, B., Parikh, N., Boyd, S.: A primal-dual operator splitting method for conic optimization (2013)
14. Chambolle, A., Pock, T.: A first-order primal-dual algorithm for convex problems with applications to imaging. J. Math. Imaging Vis. **40**(1), 120–145 (2011)
15. Combettes, P.L., Pesquet, J.C.: Proximal splitting methods in signal processing. In: Fixed-Point Algorithms for Inverse Problems in Science and Engineering, pp. 185–212. Springer (2011)
16. Combettes, P.L., Wajs, V.R.: Signal recovery by proximal forward-backward splitting. Multiscale Model. Simul. **4**(4), 1168–1200 (2005)
17. Douglas, J., Rachford, H.H.: On the numerical solution of heat conduction problems in two and three space variables. Trans. Am. Math. Soc. 421–439 (1956)
18. Davis, D., Yin, W.: Convergence rate analysis of several splitting schemes (2014). arXiv:1406.4834
19. Eckstein, J., Bertsekas, D.P.: On the Douglas-Rachford splitting method and the proximal point algorithm for maximal monotone operators. Math. Program. **55**(1–3), 293–318 (1992)
20. Eckstein, J., Svaiter, B.F.: A family of projective splitting methods for the sum of two maximal monotone operators. Math. Program. **111**(1–2), 173–199 (2008)
21. Giselsson, P., Boyd, S.: Diagonal scaling in Douglas-Rachford splitting and ADMM. In: 53rd IEEE Conference on Decision and Control (2014)
22. Giselsson, P., Boyd, S.: Metric selection in Douglas-Rachford splitting and ADMM (2014). arXiv:1410.8479
23. Giselsson, P., Boyd, S.: Preconditioning in fast dual gradient methods. In: 53rd IEEE Conference on Decision and Control (2014)
24. P. Giselsson. Tight linear convergence rate bounds for Douglas-Rachford splitting and ADMM (2015). arXiv:1503.00887
25. Glowinski, R., Marroco, A.: Sur l'approximation, par élḿents finis d'ordre un, et la résolution, par pénalisation-dualité d'une classe de problémes de Dirichlet non linéaires. Math. Model. Numer. Anal. **9**(R2), 41–76 (1975)
26. Goldstein, T., O'Donoghue, B., Setzer, S., Baraniuk, R.: Fast alternating direction optimization methods. SIAM J. Imaging Sci. **7**(3), 1588–1623 (2014)
27. Ghadimi, E., Teixeira, A., Shames, I., Johansson, M.: Optimal parameter selection for the alternating direction method of multipliers (ADMM): quadratic problems. IEEE Trans. Autom. Control **60**, 644–658 (2013)
28. Hestenes, M.R., Stiefel, E.: Methods of conjugate gradients for solving linear systems. J. Res. Natl. Bur. Stand. **49**(6), 409–436 (1952)
29. Hastie, T., Tibshirani, R., Friedman, T.: The Elements of Statistical Learning. Springer (2009)
30. He, B.S., Yang, H., Wang, S.L.: Alternating direction method with self-adaptive penalty parameters for monotone variational inequalities. J. Optim. Theory Appl. **106**(2), 337–356 (2000)
31. Krizhevsky, A., Sutskever, I., Hinton, G.E.: Imagenet classification with deep convolutional neural networks. Adv. Neural Inf. Process. Syst. 1097–1105 (2012)
32. Lions, P.L., Mercier, B.: Splitting algorithms for the sum of two nonlinear operators. SIAM J. Numer. Anal. **16**(6), 964–979 (1979)
33. Ngiam, J., Coates, A., Lahiri, A., Prochnow, B., Le, Q.V., Ng, A.Y.: On optimization methods for deep learning. In: Proceedings of the 28th International Conference on Machine Learning, pp. 265–272 (2011)
34. Nishihara, R., Lessard, L., Recht, B., Packard, A., Jordan, M.I.: A general analysis of the convergence of ADMM (2015). arXiv:1502.02009
35. Nocedal, J., Wright, S.: Numerical Optimization, vol. 2. Springer (1999)
36. O'Donoghue, B., Stathopoulos, G., Boyd, S.: A splitting method for optimal control. IEEE Trans. Control Syst. Technol. **21**(6), 2432–2442 (2013)
37. O'Connor, D., Vandenberghe, L.: Primal-dual decomposition by operator splitting and applications to image deblurring. SIAM J. Imaging Sci. **7**(3), 1724–1754 (2014)

38. Olafsson, A., Wright, S.: Efficient schemes for robust IMRT treatment planning. Phys. Med. Biol. **51**(21), 5621–5642 (2006)
39. Parikh, N., Boyd, S.: Block splitting for distributed optimization. Math. Program. Comput. 1–26 (2013)
40. Parikh, N., Boyd, S.: Proximal algorithms. Found. Trends Optim. **1**(3), 123–231 (2013)
41. Pock, T., Chambolle, A.: Diagonal preconditioning for first order primal-dual algorithms in convex optimization. IEEE Int. Conf. Comput. Vis. 1762–1769 (2011)
42. Polyak, B.: Introduction to Optimization. Optimization Software Inc., Publications Division, New York (1987)
43. Pesquet, J.C., Pustelnik, N.: A parallel inertial proximal optimization method. Pac. J. Optim. **8**(2), 273–305 (2012)
44. Paige, C.C., Saunders, M.A.: Solution of sparse indefinite systems of linear equations. SIAM J, Numer. Anal. **12**(4), 617–629 (1975)
45. Paige, C.C., Saunders, M.A.: LSQR: An algorithm for sparse linear equations and sparse least squares. ACM Trans. Math. Softw. **8**(1), 43–71 (1982)
46. Ruiz, D.: A scaling algorithm to equilibrate both rows and columns norms in matrices. Technical report, Rutherford Appleton Laboratory (2001) (Technical Report RAL-TR-2001-034)
47. Shor, N.Z.: Nondifferentiable Optimization and Polynomial Problems. Kluwer Academic Publishers (1998)
48. Sinkhorn, R., Knopp, P.: Concerning nonnegative matrices and doubly stochastic matrices. Pac. J. Math. **21**(2), 343–348 (1967)
49. Spingarn, J.E.: Applications of the method of partial inverses to convex programming: decomposition. Math. Program. **32**(2), 199–223 (1985)
50. Toh, K., Todd, M.J., Tütüncü, R.H.: SDPT3-a MATLAB software package for semidefinite programming, version 1.3. Optim. Methods Softw. **11**(1–4), 545–581 (1999)
51. Vanderbei, R.J.: Symmetric quasidefinite matrices. SIAM J. Optim. **5**(1), 100–113 (1995)
52. Wang, H., Banerjee, A.: Bregman alternating direction method of multipliers. Adv. Neural Inf. Process. Syst. 2816–2824 (2014)

Chapter 5
Control and Systems Theory for Advanced Manufacturing

Joel A. Paulson, Eranda Harinath, Lucas C. Foguth and Richard D. Braatz

Abstract This chapter describes systems and control theory for advanced manufacturing. These processes have (1) high to infinite state dimension; (2) probabilistic parameter uncertainties; (3) time delays; (4) unstable zero dynamics; (5) actuator, state, and output constraints; (6) noise and disturbances; and (7) phenomena described by combinations of algebraic, ordinary differential, partial differential, and integral equations (that is, generalizations of descriptor/singular systems). Model predictive control formulations are described that have the flexibility to handle dynamical systems with these characteristics by employing polynomial chaos theory and projections. Implementations of these controllers on multiple advanced manufacturing processes demonstrate an order-of-magnitude improved robustness and decreased computational cost. Some promising directions are proposed for future research.

5.1 Introduction

Many countries and companies in recent years have become interested in advanced manufacturing, which is the highly efficient, effective, and reliable manufacturing of complex products to satisfy tight specifications. Such products include microelectronic devices, biomedical devices, and pharmaceutical products. This chapter

J. A. Paulson (✉)
Department of Chemical and Biomolecular Engineering, University of California, Berkeley, CA 94720, USA
e-mail: joelpaulson@berkeley.edu

E. Harinath · L. C. Foguth · R. D. Braatz
Department of Chemical Engineering, Massachusetts Institute of Technology, Cambridge, MA 02139, USA
e-mail: eranda@mit.edu

L. C. Foguth
e-mail: lcfoguth@mit.edu

R. D. Braatz
e-mail: braatz@mit.edu

R. Tempo et al. (eds.), *Emerging Applications of Control and Systems Theory*, Lecture Notes in Control and Information Sciences - Proceedings,
https://doi.org/10.1007/978-3-319-67068-3_5

considers the systems and control theory associated with the generation of tools for the design of control systems that satisfy all of the needs of advanced manufacturing. For broader descriptions of advanced manufacturing and closely related activities such as Industry 4.0, Smart Manufacturing, industrial internet of things, cyberphysical systems, and cloud computing, the readers are encouraged to read some of the many recent publications on the topics.

Advanced manufacturing systems have many characteristics that break existing control theories, which include (1) high to infinite state dimension; (2) probabilistic parameter uncertainties; (3) time delays; (4) unstable zero dynamics; (5) actuator, state, and output constraints; (6) noise and disturbances; and (7) phenomena described by combinations of algebraic, ordinary differential, partial differential, and integral equations (that is, generalizations of descriptor/singular systems). While many theories have been developed to design control systems that address processes with some of these characteristics, attempting to address all of these characteristics results in computational costs that are too high to be implemented, a lack of theoretical guarantees, and/or conservatism. While industry has been able to hobble along with the existing control technology, more powerful systems and control theory have the potential to lead to step improvements in manufacturing efficiency and reliability and product quality and consistency. This chapter expands upon [1].

Notation. Hereafter, $\mathbb{R}$ and $\mathbb{N} = \{1, 2, \ldots\}$ are the sets of real and natural numbers, respectively; $\mathbb{R}_{\geq 0}$ and $\mathbb{N}_0$ are the sets of non-negative real and integer numbers; $\mathbb{N}_a^b$ is the set of integers from a to b; $\mathbb{E}\{\cdot\}$ is the expected value; $\text{Var}\{\cdot\}$ is variance; $\Pr(A)$ denotes probability of event A; $p(x)$ is the probability density function (pdf) of a random variable x; the symbol $\sim$ means "distributed as"; $\mathcal{N}(\mu, \Sigma)$ is the normal distribution with mean μ and covariance Σ; and $\mathbb{1}_A(\cdot)$ is the indicator function defined on set A.

5.2 General Problem Formulation

Consider a stochastic discrete-time nonlinear dynamical system described by

$$x_{k+1} = f(x_k, u_k, \theta), \quad y_k = h(x_k, v_k), \tag{5.1}$$

with states $x \in \mathbb{R}^{n_x}$, control inputs $u \in \mathbb{R}^{n_u}$, parameters $\theta \in \mathbb{R}^{n_\theta}$, measured outputs $y \in \mathbb{R}^{n_y}$, measurement noise $v \in \mathbb{R}^{n_v}$, and time index $k \in \mathbb{N}_0$. The parameters θ are time invariant with known probability density function (pdf) $p(\theta)$, and the initial states x_0 are have known pdf $p(x_0)$. The noise sequence $\{v_k\}_{k\in\mathbb{N}_0}$ are realizations of an independent and identically distributed (i.i.d.) zero-mean random variable V with known pdf $p(V)$.

This formulation contrasts with most model-based control algorithms for stochastic systems, which consider time-varying uncertainty $\{\theta_k\}_{k\in\mathbb{N}_0}$ assumed to be realizations of i.i.d. random variables so that the system is a Markov process. These θ_k, which replace θ in (5.1), can be interpreted as time-varying parameters or exogenous disturbances. Also, although this chapter considers discrete-time systems (5.1), the results straightforwardly extend to continuous-time systems, $\dot{x}(t) = f(x(t), u(t), \theta)$. Some algorithms reviewed in Sect. 5.5 are formulated in continuous time, and readers are referred to those papers for details.

From applying (5.1) recursively, the states x_k at time k are functions of the initial condition x_0, control input sequence $u_0, \ldots, u_{k-1}$, and random parameters θ. These explicit functional dependencies are dropped for notational convenience. At any given time k, past inputs and outputs are available to the controller which are grouped into an *information vector* I^k defined by

$$I^k = \begin{cases} (y_0) & \text{if } k = 0 \\ (y_0, y_1, u_0) & \text{if } k = 1 \\ \vdots & \vdots \end{cases}$$

Then, $I^k \in \mathbb{R}^{(k+1)n_y + kn_u}$ with size that is a function of the current time index, and $I^{k+1} = (I^k, y_{k+1}, u_k)$ can be defined recursively where $I^k \subset I^{k+1}$.

Of importance in manufacturing are hard input constraints, $u_k \in \mathbb{U},\ k \in \mathbb{N}_0$, in which the set $\mathbb{U} \subseteq \mathbb{R}^{n_u}$ can be any set in theory but is closed and bounded in practice. Important in advanced manufacturing are probabilistic state constraints of the form

$$\Pr(x_{k+1} \in \mathbb{X}) \geq 1 - \alpha, \quad k \in \mathbb{N}_0, \tag{5.2}$$

where $\alpha \in [0, 1)$ is the maximum allowed probability of violation. A control problem may have a number of constraints of the form (5.2) with possibly different probability levels α. Since (5.2) includes the special case of $\alpha = 0$, this formulation includes a mixture of probabilistic ($\alpha > 0$) and robust ($\alpha = 0$) constraints.

Of interest are methods that (approximately) solve an optimal control problem over a horizon N. Define the policy Π_N to be a finite sequence of functions $\pi_k : \mathbb{R}^{(k+1)n_y + kn_u} \to \mathbb{R}^{n_u}$ for $k = 0, \ldots, N-1$ such that $\Pi_N = \{\pi_0(\cdot), \pi_1(\cdot), \ldots, \pi_{N-1}(\cdot)\}$. The class of *admissible* control policies $\bar{\Pi}_N$ is defined by

$$\bar{\Pi}_N = \left\{ \Pi_N \;\middle|\; \begin{array}{ll} x_{k+1} = f(x_k, \pi_k(I^k), \theta), & k \in \mathbb{N}_0^{N-1} \\ y_k = h(x_k, v_k), & k \in \mathbb{N}_0^N \\ \Pr(\pi_k(I^k) \in \mathbb{U}) = 1, & k \in \mathbb{N}_0^{N-1} \\ \Pr(x_k \in \mathbb{X}) \geq 1 - \alpha, & k \in \mathbb{N}_1^N \end{array} \right\}, \tag{5.3}$$

where the hard input constraints $\Pr(\pi_k(I^k) \in \mathbb{U}) = 1$ can be equivalently written as $\pi_k(I^k) \in \mathbb{U}$ for all reachable I^k.

Problem 5.1 *(Closed-loop stochastic optimal control)* Given the pdfs $p(x_0)$, $p(\theta)$, and $p(V)$ of the initial states, parameters, and measurement noise, respectively, determine the optimal policy $\Pi_N^\star$, belonging to admissible set $\bar{\Pi}_N$, that minimizes the multistage cost function

$$J_N(\Pi_N) = \mathbb{E}\left\{F(x_N) + \sum_{k=0}^{N-1} L(x_k, \pi_k(I^k))\right\}, \tag{5.4}$$

where $L : \mathbb{R}^{n_x} \times \mathbb{R}^{n_u} \to \mathbb{R}_{\geq 0}$ is the stage cost function, and $F : \mathbb{R}^{n_x} \to \mathbb{R}_{\geq 0}$ is the terminal cost function. Then, the optimal control policy is defined by

$$\Pi_N^\star = \underset{\Pi_N \in \bar{\Pi}_N}{\text{argmin}} \; J_N(\Pi_N), \tag{5.5}$$

with optimal cost $J_N^\star = J_N(\Pi_N^\star) = \min_{\Pi_N \in \bar{\Pi}_N} J_N(\Pi_N)$. ◀

Under the assumptions, the quantity $F(x_N) + \sum_{k=0}^{N-1} L(x_k, u_k)$ is a random variable. This formulation is flexible enough to penalize moments or probability directly by defining L in terms of powers of x_k and u_k or in terms of the indicator function, taking advantage of the fact that $\mathbb{E}\{\mathbb{1}_A(X)\} = \Pr(X \in A)$. The Π_N in Problem 5.1 are a sequence of *functions*, and the optimal policy (5.5) cannot be determined using standard optimization solvers. Some considerations to address in the derivation of a more tractable approximate solution of Problem 5.1 are as follows: (1) choice of feedback parametrization, (2) accurate uncertainty propagation, (3) chance constraint enforcement, (4) online implementation cost, and (5) stability and recursive feasibility. The following sections discuss these considerations and provide an overview of some algorithms for stochastic control under time-invariant uncertainty.

5.3 Challenges and Requirements

5.3.1 *Incorporation of Feedback*

In theory, $\Pi_N^\star$ could be solved by dynamic programming (DP), which breaks Problem 5.1 into smaller subproblems based on Bellman's principle of optimality. Application of this technique leads to the Bellman equations, whose solution methods suffer from the well-known curse of dimensionality which limits the approach to simple problems. Approximate DP (ADP) approximately solves the Bellman equations through intelligent sampling/discretization schemes [2]. ADP is too computationally expensive for advanced manufacturing applications, and a more tractable approach is to incorporate feedback through parametrization of the control policy.

The popular parametrization is linear state feedback $\pi_k(I^k) = K_k\hat{x}_k + c_k$, where $K_k \in \mathbb{R}^{n_u \times n_x}$ and $c_k \in \mathbb{R}^{n_u}$ for $k = 0, \ldots, N-1$ become the real-valued decision variables in the optimization (5.5), and $\hat{x}_k$ denotes the state estimate at time k. When the states are perfectly measured $y_k = x_k$, then $\hat{x}_k = x_k$ and the current state is exactly known to the controller (i.e., *full state feedback*). Rarely are all x_k measurable in advanced manufacturing applications and instead $\hat{x}_k$ must be estimated from I^k

(i.e., *output feedback*). Some common choices for determining $\hat{x}_k = \mathbb{E}\{x_k \mid I^k\}$ include particle filtering [3] and moving horizon estimation [4].

The explicit parametrization in the state feedback law makes the optimization more tractable but suboptimal. For a nonlinear dynamical system, a simple functional form for the state feedback law that provides nearly optimal performance can be difficult to select *a priori* or may not even exist.

A popular approximation to Problem 5.1 is *receding-horizon control*, aka model predictive control (MPC). MPC directly handles most of the characteristics of advanced manufacturing systems including constraints, time delays, unstable zero dynamics, and complex dynamics. Stochastic MPC (SMPC) is an extension of classical MPC to handle probabilistic uncertainty. In this case, the control policy is defined as the solution to a constrained optimization with real-valued decision variables.

Problem 5.2 *(SMPC)* Let $\mathbf{u}_{N|k} = (u_{0|k}, u_{1|k}, \ldots, u_{N-1|k}) \in \mathbb{R}^{n_u N}$ be the set of "open-loop" control inputs over the horizon N and $\hat{x}_k \in \mathbb{R}^{n_x}$ be a point estimate of the states at time k (i.e., some function of I^k). Given the pdf $p(\theta)$ of the parameters, define the optimization to be solved at each time k as

$$\min_{\mathbf{u}_{N|k}} \ \mathbb{E}_k\left\{F(x_{N|k}) + \sum_{i=0}^{N-1} L(x_{i|k}, u_{i|k})\right\} \tag{5.6a}$$

$$\text{subject to:} \quad x_{i+1|k} = f(x_{i|k}, u_{i|k}, \theta), \qquad i \in \mathbb{N}_0^{N-1} \tag{5.6b}$$

$$u_{i|k} \in \mathbb{U}, \qquad i \in \mathbb{N}_0^{N-1} \tag{5.6c}$$

$$\Pr_k(x_{i|k} \in \mathbb{X}) \geq 1 - \alpha, \qquad i \in \mathbb{N}_1^{N} \tag{5.6d}$$

$$x_{0|k} \overset{a.s.}{=} \hat{x}_k, \quad \theta \sim p(\theta), \tag{5.6e}$$

where $x_{i|k}$ are the (uncertain) predicted values of x_{k+i} at time k. Denote the solution to the optimization (5.6) as $\mathbf{u}^\star_{N|k}(\hat{x}_k)$, which is implicitly a function of I^k through the state estimate $\hat{x}_k$. Then, the SMPC control law $\kappa_N : \mathbb{R}^{n_x} \to \mathbb{R}^{n_u}$ is given by $\kappa_N(\hat{x}_k) = u^\star_{0|k}(\hat{x}_k)$, based on the first part of the optimal trajectory. ◀

For the SMPC to be computationally tractable, N must be finite. The system behavior from N to infinity is usually lumped into the terminal cost $F(x_{N|k})$. In other words, the terminal cost should provide information about the desirability of any state reachable by finite-horizon SMPC. The terminal cost ideally would be selected as the value function for the infinite-horizon problem, as defined by the Bellman equation. Since its computation is not possible in most practical applications, approximations of the value function are typically employed. Readers are referred to [5] and citations therein for more details.

The x_k from (5.1) and $x_{i|k}$ from (5.6) have significant differences. System (5.1) describes the evolution of a non-Markovian stochastic process wherein the states are elements of a function space with some probability measure. Controller design for (5.1) is complicated because the full ensemble of trajectories must be considered simultaneously. In contrast, SMPC considers only specific state realizations that can be measured (or estimated) online, which leads to predictions modeled as random variables that are used in the control design to steer to a particular sample path.

SMPC defines an implicit control law in terms of an optimization, with no explicit functional form for κ_N as a function of the state, which makes the analysis of properties of the closed-loop system more challenging even in the full state feedback case. In particular, as shown in Sect. 5.6, the probabilistic constraints (5.2) are not guaranteed to be satisfied by the closed-loop system controlled by SMPC even when the optimization (5.6) is feasible at every time.

5.3.2 *Uncertainty Propagation*

Another consideration in solving Problems 5.1 or 5.2 is the propagation of probabilistic uncertainty through the system dynamics. To illustrate, consider $x_1 = A(\theta)x_0 + B(\theta)u_0$. For a particular value of u_0, the cumulative distribution function (cdf) of x_1 is defined as $\Pr(x_1 \leq x) = \int_{\{\theta, x_0 | A(\theta)x_0 + B(\theta)u_0 \leq x\}} p(\theta)p(x_0)d\theta dx_0$. The calculation of state cdfs or pdfs can be rarely carried out in closed form, and calculations quickly grow to be intractable for higher dimensions. Additionally, the distributions are a function of the control input so any closed-form expression would need to include this dependency.

The intractability of propagating entire distributions exactly through uncertain systems for all but the simplest systems has motivated the development of two main classes of approximate methods. This first class of methods involves sampling techniques such as Monte Carlo (MC) or scenario approaches [6]. The second class of methods represents the distributions in terms of basis function expansions such as power series [7] or polynomial chaos [8]. The latter approach optimally selects the basis functions based on the parameter distributions, which allows for high accuracy with a relatively low number of terms, and hence lower computational complexity. Polynomial chaos is reviewed in the next section as a technique for providing tractable solutions for Problem 5.2.

5.3.3 *Chance Constraints*

In Problem 5.1, evaluation of the probabilistic constraints requires the pdf of the states at every time k to be known, $\Pr(x_k \in \mathbb{X}) = \int_{\mathbb{X}} p(x_k)dx_k$, which generally cannot be found in closed form since this pdf is a complicated nonconvex function of the control policies $\pi_0(\cdot), \ldots, \pi_{k-1}(\cdot)$, system functions $f(\cdot)$ and $h(\cdot)$, parameter pdf $p(\theta)$, and noise pdf $p(V)$. Similarly, the enforced chance constraints (5.6d) in SMPC are nonconvex and difficult to implement.

As an aside, note that constraints on the marginal probability of the states (5.2) are different from constraints on the one-step conditional probability of the states (5.6d). The former are typically hard to address due to involving all possible state trajectories. The latter, on the other hand, are based on particular sample paths of the system (5.1) as they have been conditioned on all available measurements. In fact, (5.6d) is only a sufficient (not necessary) condition for (5.2), making this one-step constraint potentially conservative as seen from the law of total probability:

$$\Pr(x_{k+1} \in \mathbb{X}) = \int \underbrace{\Pr(x_{k+1} \in \mathbb{X} \mid y_k, \ldots, y_0)}_{\geq 1-\alpha \text{ for all } y_k, \ldots, y_0} p(y_k, \ldots, y_0) dy_k, \ldots, dy_0 \geq 1 - \alpha.$$

The general intractability and nonconvexity of these probabilistic constraints have inspired the use of well-known probabilistic bounds for approximating these constraints. Two of the most common examples include the Markov inequality $\Pr(X \geq a) \leq \mathbb{E}\{X\}/a$, which holds for any random variable $X \in \mathbb{R}_{\geq 0}$, and the Chebyshev inequality $\Pr(|X - \mu| \geq a) \leq \sigma^2/a^2$, where X is any random variable with mean μ and variance σ^2. Boole's inequality for a countable number of events has also been used to approximate joint chance constraint as a series of individual chance constraints, but can be conservative in certain applications [9].

5.3.4 *Fast Implementation*

In theory, Problem 5.1 can be solved offline for the optimal control policy $\Pi_N^\star$. The cost of implementing the controller online is then related to how fast the map can be evaluated at the current information state I^k, which can be fast depending on how $\Pi_N^\star$ is stored (for example, interpolation over a finitely discretized grid). However, as discussed above, Problem 5.1 is too expensive to solve for advanced manufacturing systems so SMPC (Problem 5.2) has been proposed as a less expensive alternative. SMPC control κ_N requires that the optimization (5.6) can be solved fast enough online for real-time implementation. Sampling algorithms require a large number of simulations in order to obtain information about distributions, so such implementations of SMPC have high online computational cost.

Ensuring convexity of the optimization is useful when implementing receding-horizon control, as efficient algorithms have been developed for convex optimizations that are guaranteed to find the global optimum.

5.3.5 *Stability and Recursive Feasibility*

Problems 5.1 and 5.2 do not explicitly ensure that the closed-loop system is stable. In robust model predictive control (RMPC), robust stability is guaranteed by ensuring stability for all possible realizations of the parameters. This guarantee can be provided by using, for example, robustly positively invariant sets, tubes, or in some cases by guaranteeing stability for each vertex of the uncertainty set (e.g.,

[10]). These methods are conservative for the SMPC analog of the RMPC problem. These considerations are further complicated by hard constraints on the input that must be always be satisfied. Based on this discussion, it may be beneficial to require stability in a probabilistic sense (for example, that the closed-loop states are mean-square bounded). A nonconservative analysis, however, is expected to be at least as challenging as implementing a chance constraint nonconservatively.

The optimal control problem in SMPC is solved recursively at every sampling time so that the constraints may become infeasible at a certain point. Potential infeasibility can be avoided if the optimization (5.6) can be shown to be *recursively feasible*, which occurs if, whenever there is a solution to (5.6) at time k, a solution to (5.6) is guaranteed to exist at time $k + 1$. Ways of ensuring recursive feasibility in SMPC have been explored for linear systems subject to additive white noise using probabilistic tubes [11] and a first step constraint (e.g., [12]); however, little work has been published for systems (5.1) subject to time-invariant uncertainty.

5.4 Uncertainty Propagation Using Polynomial Chaos

Polynomial chaos theory (PCT) is a collection of computationally efficient methods for propagating time-invariant uncertainty through dynamic equations. This method has gained interest in the control community as an efficient alternative to sampling approaches, such as MC or Latin hypercube sampling, which require repeated simulation of the system model for a large number of uncertainty realizations.

The term *polynomial chaos* was introduced by Norbert Wiener [13], in which a generalized harmonic analysis was applied to Brownian motion. The basic idea was to expand finite-variance random variables by an infinite series of Hermite polynomials that are functions of a normally distributed input random variable. These expansions then allow for easier computation and analysis. Later, the convergence of the polynomial chaos expansions was established [14].

Although Wiener–Hermite chaos expansions converge as long as the random variable η is measurable with respect to the Gaussian Hilbert space, their convergence rate can be slow for non-Gaussian input random variables [15]. This behavior is a result of the fact that, when expressed as functions of Gaussian input random variables, η can be highly nonlinear so that high-order expansions are required. *Generalized polynomial chaos* (gPC) [8] replaces the Hermite polynomials with polynomials that are orthogonal with respect to possibly non-Gaussian input random variables that are more closely related to η.

5.4.1 Generalized Polynomial Chaos

Let $(\Omega, \mathscr{F}, \Pr)$ be a probability space where Ω is the abstract set of elementary events (sample space), $\mathscr{F}$ is a σ-algebra composed of subsets of Ω, and $\Pr : \mathscr{F} \to \mathbb{R}$ is the probability measure. Let $\xi = (\xi_1, \ldots, \xi_n)$ be a vector composed of *independent* real-valued random variables $\xi_i : \Omega \to \mathbb{R}$, each of which induces a probability measure on the real line with probability space $(\mathbb{R}, \mathfrak{B}(\mathbb{R}), F_{\xi_i}(dx))$, where $\mathfrak{B}(\mathbb{R})$ denotes the Borel σ-algebra on $\mathbb{R}$, and $F_{\xi_i}(x) = \Pr(\xi_i \leq x)$ is the cdf of ξ_i.

For any random variable $\eta \in L_2(\Omega, \sigma(\xi), \Pr)$ that is measurable with respect to ξ, some measurable function $\psi : \mathbb{R}^n \to \mathbb{R}$ with $\eta = \psi(\xi)$ must exist from the Doob–Dynkin lemma (e.g., [16]). Whenever F_ξ is continuous and has finite moments of all orders, η can be written as an abstract Fourier series expansion, $\eta \overset{L_2}{=} \sum_{i=0}^{\infty} a_i \Phi_i(\xi)$, where $\overset{L_2}{=}$ denotes mean-square convergence, $a_i = \langle \eta \Phi_i \rangle / \langle \Phi_i^2 \rangle$ are the expansion coefficients with inner product $\langle h, g \rangle = \mathbb{E}[h(\xi)g(\xi)] = \int h(\xi)g(\xi)p(\xi)d\xi$, and $\{\Phi_i\}_{i \in \mathbb{N}_0}$ are the set of polynomials orthogonal to the distribution of ξ, satisfying $\langle \Phi_i \Phi_j \rangle = \int \Phi_i(\xi)\Phi_j(\xi)p(\xi)d\xi = \langle \Phi_i^2 \rangle \delta_{ij}$, with δ_{ij} being the Kronecker delta ($\delta_{ij} = 1$ when $i = j$ and $\delta_{ij} = 0$ otherwise).

Due to the independence of the elements of ξ, the multivariate basis can be constructed from products of each univariate basis, $\Phi_i(\xi) = \prod_{j=1}^{n} \phi_{\alpha_j^{(i)}}^{(j)}(\xi_j)$, where $\phi_k^{(j)}$ is the univariate polynomial basis function for the jth random variable of order k, and $\alpha_j^{(i)} \in \mathbb{N}_0$ is a *multi-index* that represents all possible combinations of the univariate basis functions that can be thought of as the order of ξ_j in the ith expansion term. Basis sets that are orthogonal with respect to commonly used distributions are available in many references (e.g., [8]).

To implement the method numerically, the infinite PC expansion is truncated at some finite order N_{PC}. Let d denote the highest order of polynomial kept in the truncated expansion, then the total number of terms is given by $N_{\text{PC}} = \frac{(n+d)!}{n!d!} - 1$, which imposes a constraint on the multi-index, $\sum_{j=1}^{n} \alpha_j^{(i)} \leq d$ for all $i \in \mathbb{N}_0^{N_{\text{PC}}}$. By taking advantage of the orthogonality property, the moments of η can be calculated directly from the gPC coefficients. For example, expressions for the first two moments are $\mathbb{E}\{\eta\} = a_0$ and $\mathbb{E}\{\eta^2\} = \sum_{i=0}^{\infty} a_i^2 \langle \Phi_i^2 \rangle$.

5.4.2 Galerkin Projection for Dynamic Systems

PCT can be used for the propagation of parameter and/or initial condition uncertainty through nonlinear dynamical systems (5.1) by approximating the states by a truncated gPC expansion

$$x_k \approx \sum_{i=0}^{N_{\text{PC}}} a_i(k)\Phi_i(\xi), \tag{5.7}$$

where $a_i(k)$ is the ith time-dependent coefficient of the expansion at time $k \in \mathbb{N}_0$, and ξ is a collection of independent standard random variables related to the uncertain parameters through diffeomorphism T such that $\theta = T(\xi)$.

In general, a collection of correlated random variables, such as θ, can be related to a set of independent uniform random variables through the method described in [17] which provides a way to construct the diffeomorphism T. A more detailed discussion on this topic is given in [18].

Several methods are available for determining $a_i(k)$ over time. A popular method is collocation wherein a set of grid points is generated over the support of ξ, the difference equation (5.1) is solved at each grid point, and then these solutions are fit to (5.7) using least-squares [19]. Quadrature methods numerically approximate the integrals defining the inner products $\langle\cdot\rangle$ using the Gauss quadrature rule [20]. Since the coefficients $a_i(k)$ are functions of the inputs $u_0, \ldots, u_{k-1}$, these methods often lead to a large numerical cost in optimization-based control as many simulations must be performed for each candidate input sequence tried by the optimizer.

An alternative to grid-based approaches is Galerkin projection, which substitutes (5.7) into (5.1) and derives deterministic equations for the coefficients by projecting their error onto each of the basis functions [21]. This procedure leads to

$$a_j(k+1) = \frac{1}{\langle \Phi_j^2 \rangle} \left\langle f\left(\sum_{i=0}^{N_{\text{PC}}} a_i(k)\Phi_i(\xi), u_k, T(\xi) \right), \Phi_j(\xi) \right\rangle, \quad j \in \mathbb{N}_0^{N_{\text{PC}}}. \tag{5.8}$$

By evaluating these inner products offline and stacking into vectors, closed-form expressions for the evolution of the gPC coefficients,

$$\mathbf{a}_{k+1} = \mathbf{f}(\mathbf{a}_k, u_k), \tag{5.9}$$

can be determined where $\mathbf{a}_k = (a_0(k), \ldots, a_{N_{\text{PC}}}(k))$ is the concatenated vector of state gPC coefficients at time k, and $\mathbf{f}$ is the projected function with elements (5.8).

Once $\mathbf{a}_k$ have been calculated, (5.7) can be used as a surrogate model for x_k to generate approximate pdfs (by substituting randomly drawn samples of ξ) or to efficiently calculate moments. Although the dimension has been increased to $(N_{\text{PC}} + 1)n_x$ in (5.9) in order to capture the state pdf evolution, this increase can be lower than sampling approaches or other series expansions due to the fact that the basis functions were chosen optimally and the projection operator is typically sparse.

5.5 Stochastic Model Predictive Control Approaches

This section reviews some promising approaches for control of systems subject to time-invariant uncertainty (5.1). Since Problem 5.1 is intractable, we focus on the SMPC approaches formulated similarly to Problem 5.2. Due to the computational advantages of PCT described in Sect. 5.4, there is a particular emphasis on methods that tackle solving (5.6) with PCT.

One of the first applications of PCT to control problems analyzed the distribution of closed-loop finite-time processes with uncertain parameters, which was demonstrated for a batch crystallization process [7]. A second-order PCE provided more accurate pdfs than a power series expansion of the same order and provided nearly identical pdfs as direct MC at three orders of magnitude lower computational cost.

Problem 5.2 has been tackled using PCT [19]. Individual chance constraints were considered with $\mathbb{X} = \{x \in \mathbb{R}^{n_x} : c^\top x \leq d\}$. Since chance constraints are nonconvex and intractable (Sect. 5.3), (5.6d) was replaced by a conservative approximation of the form [22, Thm. 3.1] $c^\top \mathbb{E}_k\{x_{i|k}\} + \sqrt{\frac{1-\alpha}{\alpha}}\sqrt{c^\top \mathrm{Var}_k\{x_{i|k}\}c} \leq d$, which only requires knowledge of the first two moments of the distribution. This result is *distributionally robust* in that, whenever this inequality is satisfied, the probabilistic constraint (5.6d) is met for all possible distributions with that mean and variance. PCT can then be used to approximate the moments of x_k as discussed in Sect. 5.4. Since Galerkin projection can be difficult to execute for general nonlinearities, [19] suggested using collocation to compute the gPC coefficients. Full state feedback was assumed.

The algorithm was applied to control the manufacturing of crystals modeled by partial differential equation (PDEs) that were converted to nonlinear ordinary differential equations (ODEs) using the method of moments. The pdfs of five kinetic parameters were determined from parameter estimation. The control objective was to minimize the nucleation of crystals of a desired form while preventing any nucleation or growth of crystals of an undesired form and avoiding the dissolution of the desired crystal form. The latter requirement was enforced through a 95% chance constraint. SMPC provided tighter distributions of the control objectives and much lower constraint violation than nominal NMPC (see [19, Figs. 2–3]).

Advanced manufacturing systems are typically composed of many different complex interacting units that collectively are described by a system of mixed differential and algebraic equations (DAEs) of high state dimension. Quadratic dynamic matrix control (QDMC) is an input–output formulation of MPC that is widely applied in manufacturing in part because its online computational cost is *independent* of the number of states in the original model. QDMC does not explicitly account for uncertain parameters, however, which can lead to closed-loop performance degradation.

A fast SMPC algorithm that combines the concepts of QDMC with PCT has been derived that handles uncertain nonlinear DAE models of high (including infinite) state dimension [21]. The high-order DAE is replaced by a low-dimensional surrogate for prediction of the output distribution as a function of the control inputs by propagating uncertainty through a local linearization of the DAE using PCT and Galerkin projection. The Galerkin-projected model is a deterministic linear DAE that is sparse. To reduce dimensionality, a finite step response (FSR) is computed that maps the control inputs directly to the output PCE coefficients, $Y_{k+1}^{\mathrm{PC}} = MY_k^{\mathrm{PC}} + S\Delta u_k$, where Y^{PC} are the collection of PCE coefficients for the output over the model length horizon (i.e., number of FSR coefficients retained) and M and S are known FSR matrices [21, 23]. Let $\mathbf{y}_{N|k}^{\mathrm{PC}}$ denote the predictions of the output PCE coefficients over the horizon N. Using the equation for Y_{k+1}^{PC}, the $\mathbf{y}_{N|k}^{\mathrm{PC}}$ can be computed as a function of the future input moves, $\mathbf{y}_{N|k}^{\mathrm{PC}} = M_p Y_k^{\mathrm{PC}} + G\Delta\mathbf{u}_{N|k} + \mathscr{I}(y_k - NY_k^{\mathrm{PC}})$, where expressions for the matrices M_p, G, $\mathscr{I}$, and N are given in [23]. The last term $\mathscr{I}(y_k - NY_k^{\mathrm{PC}})$ is a feedback correction term based on the classical *disturbance update rule* in QDMC.

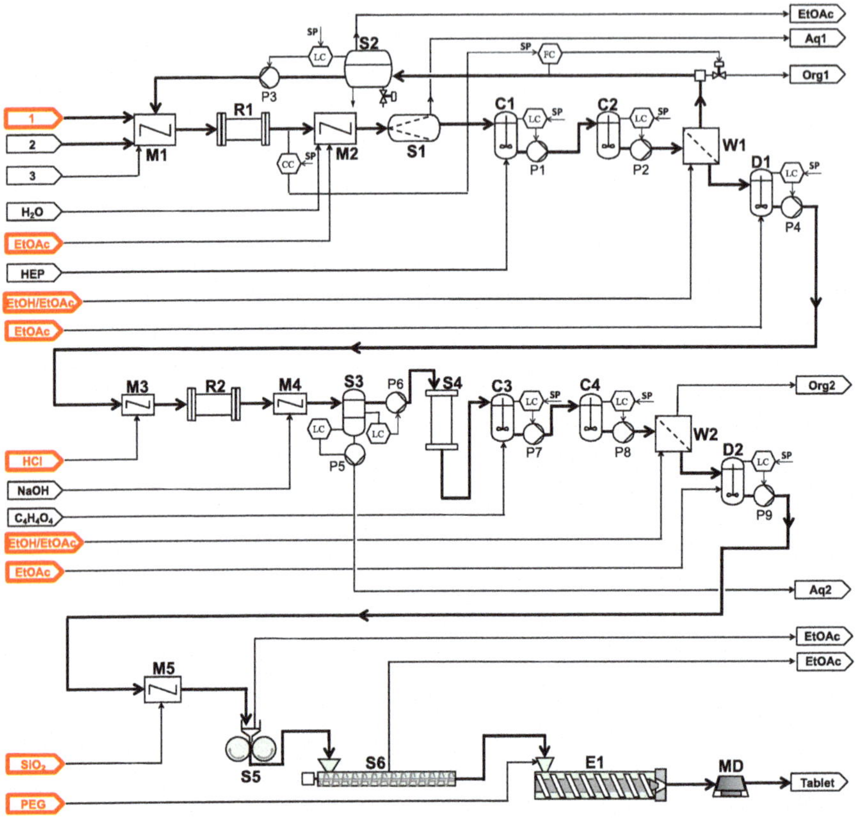

Fig. 5.1 Integrated continuous pharmaceutical manufacturing plant process flow diagram. Red boxes indicate inputs to which the outputs are most sensitive based on dynamic sensitivity analysis

The expression for the PCE moments can be used to derive $\mathbb{E}_k\{\mathbf{y}_{N|k}(\theta)\} = E\mathbf{y}^{\text{PC}}_{N|k}$ and $\mathbb{E}_k\{\mathbf{y}^{\top}_{N|k}(\theta)Q\mathbf{y}_{N|k}(\theta)\} = (\mathbf{y}^{\text{PC}}_{N|k})^{\top}Q^{\text{PC}}\mathbf{y}^{\text{PC}}_{N|k}$, with E and Q^{PC} defined in [23]. The predicted expected output value and its variance are linear and quadratic in the output coefficients, respectively. For a quadratic stage L and terminal cost F, SMPC using this model can posed as a quadratic program (QP) that can efficiently be solved to global optimality.

This efficient SMPC method was applied for control of an end-to-end continuous pharmaceutical manufacturing plant with nearly 8000 states (see Fig. 5.1). The fast SMPC algorithm yielded much lower variability in the active pharmaceutical ingredient (API) and production rate than nominal QDMC. The fast SMPC optimization was solved in less than one second at every iteration and can easily be implemented in real time in advanced manufacturing systems (see [21] for details).

5.6 Future Outlook

Any MPC algorithm can be categorized in terms of practical value and theoretical rigor. For example, QDMC is one of the most widely applied advanced control methodologies for industrial processes [24] due to the simplicity of implementation/maintenance combined with improved performance (compared to classical control methods) on a wide variety of real-life applications. Over the past two decades, MPC has been rigorously characterized (from a theoretical perspective), resulting in a number of elegant proofs regarding stability and/or recursive feasibility in a plethora of different scenarios (see, for example, [5] for a theoretical overview of MPC). Simultaneous research on both practical and theoretically rigorous algorithms is necessary to advance the field of control as these two areas can inform and complement one another.

In SMPC with time-invariant uncertainty, the focus has been on developing practical algorithms with low online cost. Methods have been proposed that, in simulation studies, outperform their nominal counterparts under similar tuning metrics [19, 25], but the theoretical properties of these algorithms have not been well characterized. For example, little is known of the effect of truncation error of the gPC expansion on closed-loop performance. Even if $p(x_{i|k})$ could be exactly determined at every $i \in \mathbb{N}_1^N$, it is not clear that chance constraints (5.2) will be satisfied by the closed-loop system. These points, as well as some possible interesting future directions for SMPC, are discussed below.

5.6.1 Chance Constraints in SMPC

Here, we prove by counterexample that satisfaction of chance constraints in standard receding-horizon control formulations, such as Problem 5.2, does not imply satisfaction of these constraints by the closed-loop system of interest.

Theorem 5.1 *Consider the system* (5.1) *and the SMPC control law* κ_N *(specified by Problem 5.2) in the case of full state feedback. Then, the stochastic closed-loop system*

$$x_{k+1} = f(x_k, \kappa_N(x_k), \theta) \tag{5.10}$$

may not satisfy the probabilistic constraints $Pr(x_k \in \mathbb{X}) \geq 1 - \alpha$ *even when constraints* (5.6d) *are feasible at each time k.*

Proof This result is shown by selecting a system function f, parameter pdf $p(\theta)$, constraint sets $\mathbb{X}$ and $\mathbb{U}$, probability violation α, horizon N, and costs L and F such that $\Pr(x_k \in \mathbb{X}) \geq 1 - \alpha$ is not satisfied by (5.10). Consider the choices

$$f(x_k, u_k, \theta) = \theta x_k + u_k, \quad x_0 = 1, \quad p(\theta) = 0.5\delta(\theta - 1) + 0.5\delta(\theta - 2),$$
$$\mathbb{X} = \{x \mid x \le 2\}, \quad \mathbb{U} = \mathbb{R}, \quad \alpha = 0.5, \quad N = 2, \quad L(x_k, u_k) = 0, \quad F(x_N) = 0.$$

Clearly, the control law $\kappa_N(x_k)$ is time invariant and simply maps the current states to the current inputs. Let's evaluate this map for different x_k values. For the specific case defined above, it can be shown that

$$\kappa_N(1) = -1.1, \quad \kappa_N(-0.1) = 2.2, \quad \kappa_N(0.9) = 1.1$$

is a feasible SMPC control law. We only show these three values of the map as they are the only values required in this proof (note that κ_N could be replaced by any smooth function that goes through these three points). To show that the constraints are satisfied in (5.6) for this κ_N, note that the predicted state pdfs from any $x_k = x$ can be derived to be

$$p(x_{1|k}) = 0.5\delta(x_{1|k} - x - u_{0|k}) + 0.5\delta(x_{1|k} - 2x - u_{0|k}),$$
$$p(x_{2|k}) = 0.5\delta(x_{2|k} - x - u_{0|k} - u_{1|k}) + 0.5\delta(x_{2|k} - 4x - 2u_{0|k} - u_{1|k}).$$

For $x = 1$, $x = -0.1$, and $x = 0.9$, $\mathbf{u}_{N|k} = [-1.1, 0.3]^\top$, $\mathbf{u}_{N|k} = [2.2, -1]^\top$, and $\mathbf{u}_{N|k} = [1.1, 0]^\top$ are, respectively, feasible input sequences that can be selected under some objective (the trivial case of $J = 0$ is chosen here).

Now let's determine how (5.10) evolves under this control law for the chosen system dynamics. Two cases for the parameter must be considered:

$$\theta = 1 : x_1 = 1(1) + \kappa_N(1) = -0.1 \quad \longrightarrow \quad x_2 = 1(-0.1) + \kappa_N(-0.1) = 2.1,$$
$$\theta = 2 : x_1 = 2(1) + \kappa_N(1) = 0.9 \quad \longrightarrow \quad x_2 = 2(0.9) + \kappa_N(0.9) = 2.9.$$

The closed-loop state distribution $p(x_2) = 0.5\delta(x_2 - 2.1) + 0.5\delta(x_2 - 2.9)$ indicates that the true system reaches $x_2 = 2.1$ with 50% probability and $x_2 = 2.9$ with 50% probability. Since $\Pr(x_2 \le 2) = 0$, the probabilistic constraint is violated by the closed-loop system and the assertion directly follows. □

Theorem 5.1 shows that chance constraints may not be satisfied when a system is controlled using SMPC. Clearly, this theorem directly extends to the case of output feedback, which is a more challenging problem. This result can be understood intuitively from the fact that distributions predicted along sample individual paths do not equal the marginal distributions that consider all uncertainty simultaneously. This result does not show that all systems will behave in this manner. Notice, for example, that the closed-loop simulations presented in [19] do meet chance constraints even though this issue was not considered. Chance constraint satisfaction in [19] was likely aided by its conservative approximation, and this result may not have been observed with a tighter approximation of (5.6d).

5.6.2 *Frequentist Versus Bayesian Viewpoint*

The result of Theorem 5.1 is especially important for advanced manufacturing due to the stringent requirements imposed on these types of systems. For these cases, guaranteed performance and constraint satisfaction are often required. Two approaches are suggested to be pursued for a more rigorous treatment of chance constraints based on two different interpretations of the pdf $p(\theta)$.

The first potential direction involves interpreting $p(\theta)$ using the frequentist viewpoint, such as a collection of batches or a number of parallel runs of continuous-flow processes. An example application is batch crystallization of pharmaceutical products, in which the parameters (e.g., growth and nucleation) are different from batch to batch, but essentially constant within a given batch. Marginal chance constraints can be difficult to enforce in a nonconservative manner using SMPC as realizations are treated independently as shown explicitly in Theorem 5.1.

An alternative approach is to interpret $p(\theta)$ as a quantification of subject belief according to the Bayesian viewpoint. Most continuous-flow processes would behave in this way, as the true system behavior is time invariant (or slowly time varying) but unknown. Assuming the model structure is accurate, the real system would be deterministic of the form (5.1) with unknown parameters $\theta = \theta^\star$. The pdf $p(\theta)$ is then a prior that can be determined using parameter estimation techniques. Chance constraints in this case would not have physical meaning since the true deterministic system either meets constraints or does not. Problem 5.2 does not take this approach either as the prior should be recursively updated with new measurements using Bayes' rule. This pdf $p(\theta \mid y_0, \ldots, y_k)$ should then replace the $p(\theta)$ in (5.6). An efficient algorithm for solving this estimation problem using PCT in conjunction with SMPC has been developed [26]. Significant improvements were observed on a benchmark chemical reactor example by updating the parameter distribution.

5.6.3 *New Directions and Open Questions*

An interesting future direction involves analysis of the closed-loop properties of (5.10) using PCT. Similar to the ideas discussed in Sect. 5.5, gPC expansions could be used as cheap surrogates of the true system. Then, the performance and constraint violation of the marginal states could be efficiently calculated. In addition, this analysis could be performed iteratively to aid in the selection of important tuning parameters such as weight matrices or violation levels. An open question involves the approximation error of the PCE as a function of the order retained in the truncated expansion. Theory related to the convergence of these expansions must be developed to systematically provide guaranteed performance bounds. Ways for ensuring stability and recursive feasibility (by design) also need to be developed. These tasks are complicated by the gPC truncation error, which must be properly taken into account in SMPC.

Recently, a number of papers have been published on SMPC for time-varying parameters and disturbances (see [27] for a recent review of these methods). While considering i.i.d. noise may not be conservative, treating the parameters as i.i.d. will often be conservative for describing complex advanced manufacturing systems. A possible method for reducing the conservatism of this assumption is to place low-pass filters in series with the time-varying parameters so that their variation is bounded by the choice of the filter time constant λ_f. This approach may allow for simpler theoretical analysis and algorithm design, and requires that λ_f is mapped to a bound on the time variation of the parameters. An open question is to whether these methods would hold for $\lambda_f \to \infty$, which corresponds to the parameters being time invariant.

Lastly, as recently explored in the literature, these issues become more complicated for distributed parameter and mixed continuous-discrete systems (aka *hybrid systems*), which are commonplace in many industries. Handling uncertainty within these complex systems is not well established in the literature.

References

1. Paulson, J.A., Harinath, E., Foguth, L.C., Braatz, R.D.: Nonlinear model predictive control of systems with probabilistic time-invariant uncertainties. In: Proceedings of the IFAC Conference on Nonlinear Model Predictive Control, pp. 937–943, Seville, September 2015
2. Bertsekas, D.P.: Dynamic Programming and Optimal Control. Athena Scientific, Belmont, MA (1995)
3. Chen, Z.: Bayesian filtering: from Kalman filters to particle filters, and beyond. Statistics **182**, 1–69 (2003)
4. Rao, C.V., Rawlings, J.B., Lee, J.H.: Constrained linear state estimation—A moving horizon approach. Automatica **37**, 1619–1628 (2001)
5. Mayne, D.Q., Rawlings, J.B., Rao, C.V., Scokaert, P.O.M.: Constrained model predictive control: stability and optimality. Automatica **36**(6), 789–814 (2000)
6. Schildbach, G., Fagiano, L., Frei, C., Morari, M.: The scenario approach for stochastic model predictive control with bounds on closed-loop constraint violations. Automatica **50**, 3009–3018 (2014)
7. Nagy, Z.K., Braatz, R.D.: Distributional uncertainty analysis using power series and polynomial chaos expansions. J. Process Control **17**, 229–240 (2007)
8. Xiu, D., Karniadakis, G.E.: The Wiener-Askey polynomial chaos for stochastic differential equations. SIAM J. Sci. Comput. **24**, 619–644 (2002)
9. Paulson, J.A., Buehler, E.A., Braatz, R.D., Mesbah, A.: Stochastic predictive control with joint chance constraints. Int. J. Control (2017)
10. Langson, W., Chryssochoos, I., Raković, S.V., Mayne, D.Q.: Robust model predictive control using tubes. Automatica **40**, 125–133 (2004)
11. Kouvaritakis, B., Cannon, M., Raković, S.V., Cheng, Q.: Explicit use of probabilistic distributions in linear predictive control. Automatica **46**, 1719–1724 (2010)
12. Lorenzen, M., Dabbene, F., Tempo, R., Allgower, F.: Constraint-tightening and stability in stochastic model predictive control. IEEE Trans. Autom. Control (2016)
13. Wiener, N.: The homogeneous chaos. Am. J. Math. **60**, 897–936 (1938)
14. Cameron, R.H., Martin, W.T.: The orthogonal development of non-linear functionals in series of Fourier-Hermite functionals. Ann. Math. **48**, 385–392 (1947)
15. Ernst, O.G., Mugler, A., Starkloff, H., Ullmann, E.: On the convergence of generalized polynomial chaos expansions. ESAIM. Math. Modell. Numer. Anal. **46**, 317–339 (2012)

16. Rao, M.M., Swift, R.J.: Probability Theory with Applications. Springer Science & Business Media (2006)
17. Rosenblatt, M.: Remarks on a multivariate transformation. Ann. Math. Stat. **23**, 470–472 (1952)
18. Kim, K.K., Braatz, R.D.: Generalised polynomial chaos expansion approaches to approximate stochastic model predictive control. Int. J. Control **86**, 1324–1337 (2013)
19. Mesbah, A., Streif, S., Findeisen, R., Braatz, R.D.: Stochastic nonlinear model predictive control with probabilistic constraints. In: Proceedings of the American Control Conference, pp. 2413–2419, Portland (2014)
20. Eldred, M.S., Burkardt, J.: Comparison of non-intrusive polynomial chaos and stochastic collocation methods for uncertainty quantification. In: Proceedings of the 47th AIAA Aerospace Sciences Meeting, pp. 1–20, Orlando (2009)
21. Paulson, J.A., Mesbah, A., Streif, S., Findeisen, R., Braatz, R.D.: Fast stochastic model predictive control of high-dimensional systems. In: Proceedings of the IEEE Conference on Decision and Control, pp. 2802–2809, Los Angeles, CA, USA (2014)
22. Calafiore, G.C., El Ghaoui, L.: On distributionally robust chance-constrained linear programs. J. Optim. Theory Appl. **130**, 1–22 (2006)
23. Paulson, J.A.: Modern control methods for chemical process systems. PhD thesis, Massachusetts Institute of Technology, Cambridge, Massachusetts (2016)
24. Qin, S.J., Badgwell, T.A.: A survey of industrial model predictive control technology. Control Eng. Pract. **11**, 733–764 (2003)
25. Fisher, J.R.: Stability analysis and control of stochastic dynamic systems using polynomial chaos. PhD thesis, Texas A&M University, College Station, Texas (2008)
26. Muhlpfordt, T., Paulson, J.A., Braatz, R.D., Findeisen, R.: Output feedback model predictive control with probabilistic uncertainties for linear systems. In: Proceedings of the American Control Conference, pp. 2035–2040, Boston (2016)
27. Mesbah, A.: Stochastic model predictive control: an overview and perspectives for future research. IEEE Control Syst. Mag. **36**, 30–44 (2016)

Chapter 6
Robustness Sensitivities in Large Networks

T. Sarkar, M. Roozbehani and M. A. Dahleh

Abstract This article focuses on developing a framework to assess robustness in large interconnected networks that arise frequently in many socioeconomic networks such as transportation, economics, and opinion dynamics. We first introduce the idea of "asymptotic" resilience, i.e., a measure of robustness to disturbances, as the network size increases, keeping the underlying structure invariant. We argue that such a notion of robustness is different from existing ideas in robust control theory that do not account for network topology and dimension. Under this new framework, we formulate a hierarchy of resilience for different network topologies. We present examples of commonly encountered network topologies and comment on their resilience. We then provide a formal characterization of how edge link perturbation affects resilience in large networks. Further, we show how each node of the network contributes to its resilience and identify critical nodes that become "fragile" as the network dimension grows. A major contribution of our work is that the analysis is no longer limited to undirected networks, as in previous literature.

6.1 Introduction

In recent years, substantial interdisciplinary research has been devoted to understanding socioeconomic systems like transportation and consensus in decision-making. Representing such systems as networks has been a widely adopted modeling technique (see [6, 9, 20] and references therein), however, despite the preponderance of such network formulations, a unified framework to analyze the properties of such an abstraction is still lacking.

T. Sarkar (✉) · M. Roozbehani · M. A. Dahleh
Massachusetts Institute of Technology, Cambridge, MA, USA
e-mail: tsarkar@mit.edu

M. Roozbehani
e-mail: mardavij@mit.edu

M. A. Dahleh
e-mail: dahleh@mit.edu

R. Tempo et al. (eds.), *Emerging Applications of Control and Systems Theory*, Lecture Notes in Control and Information Sciences - Proceedings,
https://doi.org/10.1007/978-3-319-67068-3_6

A real-life network is characterized by its topology and the dynamical behavior on the topology. Another important characteristic of such socioeconomic systems is the network dimension, e.g., a transportation network may be composed of thousands of nodes and edges, or a social opinion dynamics system is comprised of millions of agents. As a result, questions of how robustness in such systems varies with topology and dimension become important. This requires a union of tools from standard control theory, where the robustness of systems with complex dynamics in response to external disturbances is well understood, and algebraic graph theory, where notions of criticality, centrality, and robustness in graphs are well studied.

There is a great body of interdisciplinary research on the fragility, or the lack of robustness, in large interconnected networks (see [3, 7, 9, 10, 12, 16, 22], for example). Examples of networks that grow fragile as the network dimension increases can be found in [1–3, 9]. Different measures to quantify fragility in general networks are introduced in [10, 12]. $\mathscr{H}_2$ norm-based robustness measures to quantify network volatility are studied in [10]; however, the analyses there are limited to symmetric, or undirected, networks. Robustness in consensus problems, where the spectral radius of the system is unity, has also been investigated previously in [10, 12].

Here, we build on the work introduced in [17]. In Sect. 6.3, we briefly review the framework studied in [17], based on that we build a hierarchy of resilience. Section 6.4 describes how modifying edge links affect the robustness of networks and prove that a resilient network can have no "fragile" links. In Sect. 6.5, we identify the critical nodes and formalize the notion of nodal volatility before concluding.

6.2 Mathematical Notation

Matrix Theory: A vector $v \in \mathbb{R}^{n \times 1}$ is of the form $[v_1, \ldots, v_n]^T$, where v_i denotes the ith element, unless specified otherwise. The vector $\mathbf{1}$ is the all 1s vector of appropriate dimension; to specify the dimension, we sometimes refer to it as $\mathbf{1}_n$, where it is a $n \times 1$ vector. Similarly, for a $m \times n$ matrix, A, we refer to it as $A_{m \times n}$ when we want to specify dimension. We refer to $A_{n \times n}$ as A_n for shorthand. For a matrix, A, we denote by $\rho(A)$ its spectral radius, and by $\sigma_i(A)$ the ith largest singular value of A. I is the identity matrix of appropriate dimension. The $\mathscr{L}_p$ norm of a matrix, A, is given by $||A||_p = \sup_v ||Av||_p/||v||_p$. Finally, $\Pi_n = I_n - \mathbf{1}_n\mathbf{1}_n^T/n$ is the projection matrix.

Order Notation: For functions, $f(\cdot), g(\cdot)$, we have $f(n) = O(g(n))$, when there exist constants C, n_0 such that $f(n) \leq Cg(n)$ for all $n \in \mathbb{N} > n_0$. Further, if $f(n) = O(g(n))$, then $g(n) = \Omega(f(n))$. For functions $g(\cdot), h(\cdot)$, we have $g(n) = \Theta(h(n))$ when there exist constants C_1, C_2, n_1 such that $C_1 h(n) \leq g(n) \leq C_2 h(n)$ for all $n \in \mathbb{N} > n_1$. Finally, for functions $h_1(\cdot), h_2(\cdot)$, we have $h_1(n) = o(h_2(n))$ when $\lim_{n\to\infty} |h_1(n)/h_2(n)| = 0$.

Graph Theory: A graph is the tuple $\mathscr{G} = (\mathscr{V}_\mathscr{G}, \mathscr{E}_\mathscr{G}, w_\mathscr{G})$, where $\mathscr{V}_\mathscr{G} = \{v_1, v_2, \ldots, v_n\}$ represents the set of nodes, and $\mathscr{E}_\mathscr{G} \subseteq \mathscr{V}_\mathscr{G} \times \mathscr{V}_\mathscr{G}$ represents the set of edges or communication links. An edge or link from node i to node j is denoted by $e[i, j] =$

$(v_i, v_j) \in \mathcal{E}_{\mathcal{G}}$, and $w_{\mathcal{G}} : \mathcal{E}_{\mathcal{G}} \to \mathbb{R}$. Denote by $\mathbb{A}_{\mathcal{G}}$ the incidence matrix of $\mathcal{G}$. A graph, $\mathcal{G}$, is symmetric or undirected if $w_{\mathcal{G}}(v_i, v_j) = w_{\mathcal{G}}(v_i, v_j)$ for all $1 \leq i, j \leq |\mathcal{V}_{\mathcal{G}}|$. $\mathcal{G}$ is induced by a matrix, $A_{n\times n}$ if $\mathcal{V}_{\mathcal{G}} = \{1, \ldots, n\}$, and $(i, j) \in \mathcal{E}_{\mathcal{G}}$ if $[A]_{ij} \neq 0$, and $w_{\mathcal{G}}(i, j) = [A]_{ij}$.

Miscellaneous: Denote by $\mathcal{P}_d$ is the family of polynomials with maximum degree $d \in \mathbb{N}$.

6.3 Resilience Measures

Consider the following discrete time LTI dynamics:

$$x(k+1) = A\, x(k) + \omega\, \delta(0, k), \quad k \in \{0, 1, 2, \ldots\} \tag{6.1}$$

Here, $x(k) = [x_1(k), \ldots, x_n(k)]^T$ is the vector of state variables. A is the $n \times n$ state transition matrix. $\delta(0, k)$ is the Kronecker delta function, with $\delta(0, 0) = 1$, and $\delta(0, k) = 0\ \ \forall k \neq 0$ and $\omega = [\omega_1, \ldots, \omega_n]^T$ is an input disturbance exogenous to the system.

We impose the following distributional assumption on the noise, ω.

Assumption 6.1 ω is an $n \times 1$ random vector with $\mathbb{E}[\omega\omega^T] = I_{n\times n}$ and $\mathbb{E}[\omega] = \mathbf{0}$.

Definition 6.1 A network, $\mathcal{N}(A; \mathcal{G})$, is a graph $\mathcal{G} = (\mathcal{V}_{\mathcal{G}}, \mathcal{E}_{\mathcal{G}}, w_{\mathcal{G}})$, where $\mathcal{V}_{\mathcal{G}} = \{1, 2, \ldots, n\}$, and for each node, $i \in V_{\mathcal{G}}$, there is an associated dynamical behavior, $i \to x_i(\cdot)$, given by Eq. (6.1). Further, $w_{\mathcal{G}}(i, j) = [A]_{ij}$ for all $1 \leq i, j \leq n$.

Resilience, or robustness, analysis in large networks differs from standard control theory in one major way—any performance/robustness/resilience measure for a network should explicitly include its dimension and topology. Therefore, we will need a clear idea of what it means to be a "large" network. To visualize such a network, one needs to keep the topology fixed and progressively increase the dimension. For example, to evaluate the robustness of a "large" line network, we would compute the robustness measure on a sequence of line networks, of dimensions $\{1, 2, \ldots\}$. This idea is described in greater detail in [3], [18].

The following discussion will revisit measures of (asymptotic) robustness for large networks introduced in [17].

6.3.1 Resilience in Stable Networks

Consider the network given in Fig. 6.1, with some state matrix, A. At time $k = 0$, each node i is hit with a shock, ω_i as shown, then we are interested in the following questions:

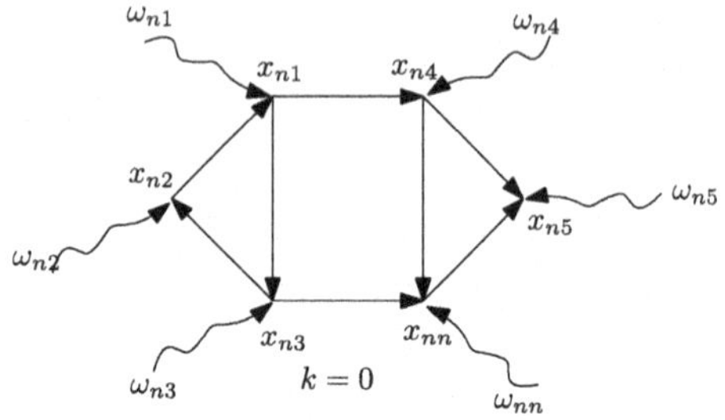

Fig. 6.1 Network with noise on every node, $k = 0$

- If $\omega = [\omega_1, \ldots, \omega_n]^T$ is a deterministic shock, with $||\omega||_2 \leq K$, then what is the maximum effect of the shock on the network ?
- If ω is a random shock, then what is the effect of the shock on the network, on average ?

We measure the energy of the network by the quantity, $E_\infty = \sum_{k=0}^{\infty} x^T(k)x(k)$, where we assume without loss of generality that $x(0) = \mathbf{0}$. Then, depending on the stochasticity of the input, we have:

Definition 6.2 Given a network, $\mathcal{N}(A; \mathcal{G})$, with a stable A, as in Eq. (6.1) and a deterministic shock, ω, at time $k = 0$, to the system, the max norm, $\mathcal{M}(A)$, or just $\mathcal{M}$, is given by

$$\mathcal{M}(A) = \sup_{||\omega||_2=1} E_\infty \tag{6.2}$$

Definition 6.3 Given a network, $\mathcal{N}(A; \mathcal{G})$, with a stable A, as in Eq. (6.1) and a random shock, ω, at time $k = 0$, that satisfies Property 6.1, the average norm, $\mathcal{E}(A)$, or just $\mathcal{E}$, is defined as the following:

$$\mathcal{E}(A) = \mathbb{E}_\omega[E_\infty] \tag{6.3}$$

Proposition 6.1 *The max norm, $\mathcal{M}$, is $\sigma_{max}(P)$, and the average norm, $\mathcal{E}$, is $tr(P)$ where*

$$A^T PA + I = P \tag{6.4}$$

Here, $\sigma_{max}(P)$ is the largest singular value of P and $tr(P)$ is the trace of P. Alternatively, $P(A) = P = \sum_{k=0}^{\infty}(A^T)^k A^k$.

Assumption 6.2 For a sequence of networks, with network matrices $\{A_n\}_{n=1}^{\infty}$, we have that $\limsup_{n\to\infty} \max_{i,j} |[A_n]_{ij}| < \infty$. Further, $\limsup_{n\to\infty} \rho(A_n) < 1$.

Definition 6.4 A sequence of networks, $\{\mathcal{N}(A_n, \mathcal{G}_n)\}_{n=1}^{n}$, with network matrices $\{A_n\}_{n=1}^{\infty}$, is asymptotically robust, or resilient, if we have

- Network matrix, A_n, is stable for each n.
- $||P(A_n)||_2 = O(p(n))$.

Here, $p(\cdot) \in \mathscr{P}_d$ for some $d \in \mathbb{N}$. Fragility is the lack of resilience in the sense of super-polynomial or exponential scaling of $||P(A_n)||_2$, for the network matrix sequence $\{A_n\}_{n=1}^{\infty}$.

6.3.2 Resilience in Consensus Graphs

For a consensus network, we are interested in the amplification of the state vector, $x(\cdot)$, in a subspace perpendicular to $\mathbf{11}^T/n$ (see [17] for details).

Define the consensus system as

$$x(k+1) = Ax(k) + \omega\delta(0,k) \tag{6.5}$$

Here, $\delta(\cdot,\cdot)$ is the Kronecker delta function, and ω is a $n \times 1$ input vector. Next, we define the projection matrix, Π, that is perpendicular to the vector $[1,\ldots,1]^T$:

$$\Pi = I - \mathbf{11}^T/n \tag{6.6}$$

Further, we will impose the following assumption on A in a consensus system for the rest of this paper.

Assumption 6.3 The state transition matrix for a consensus system is an aperiodic and irreducible stochastic matrix.

We study the Π projection of the state $x(k)$ at each time point k, where $x(k)$ is generated as in Eq. (6.5). Roughly, for resilient large stochastic networks, we expect that the "total amplification" of the projection should not grow uncontrolled in the dimension of the network. We have $x_\Pi(k) = \Pi\, x(k)$, and we redefine our average and max norms as

$$\mathscr{M}^\Pi(A) = \sup_{||\omega||_2=1} \left(\sum_{k=0}^{\infty} x_\Pi^T(k) x_\Pi(k) \right)$$

$$\mathscr{E}^\Pi(A) = \mathbb{E}_\omega \left[\left(\sum_{k=0}^{\infty} x_\Pi^T(k) x_\Pi(k) \right) \right]$$

where ω is deterministic in the definition of $\mathscr{M}^\Pi$, and stochastic in the definition of $\mathscr{E}^\Pi$, with distributional assumptions as in Assumption 6.1.

Proposition 6.2 *Under Assumption 3 on the network matrix, A, we have the following equations:*

$$\begin{aligned} \mathscr{E}^\Pi(A) &= tr(P_\Pi) \\ \mathscr{M}^\Pi(A) &= \sigma_{max}(P_\Pi) \end{aligned} \tag{6.7}$$

Here, $P_{\Pi} = A^T P_{\Pi} A + \Pi$. *Further,* $P_{\Pi} = P_{\Pi}(A)$ *can be represented by*

$$P_{\Pi}(A) = \sum_{k=1}^{\infty} (A^T)^k \; \Pi \; A^k + \Pi \tag{6.8}$$

Definition 6.5 A sequence of stochastic network matrices, $\{A_n\}_{n=1}^{\infty}$, is asymptotically robust, or resilient if

- $\mathscr{E}^{\Pi}(A_n) = O(p(n))$

Here, $p(\cdot) \in \mathscr{P}_d$ for some $d \in \mathbb{N}$.

Definition 6.4 is general in the sense that resilience is a property of the sequence $\{P(A_n)\}_{n=1}^{\infty}$, and hence $\{A_n\}_{n=1}^{\infty}$, and not specific to the $\mathscr{L}_p$ norm that we use. This follows from the following fact about vector norms, for $p > r > 0$:

$$||x||_p \leq ||x||_r \leq n^{1/r-1/p} ||x||_p \tag{6.9}$$

Specifically, Eq. (6.9) implies that if an induced norm of P_n is polynomial in dimension, then so are all other induced norms of P_n. Further, fixing an $\mathscr{L}_p$ norm gives us a hierarchy of resilience, in that norm, which we formally define below.

Definition 6.6 For a given polynomial, $p(n)$, we call a network sequence, $\{A_n\}_{n=1}^{\infty}$, $\mathscr{L}_q - p(n)$ order if $||P(A_n)||_q = \Theta(p(n))$.

It follows that $\mathscr{M}(A_n) = ||P(A_n)||_2$, i.e., is a measure of resilience in the $\mathscr{L}_2$ norm. Figure 6.2 shows some common network topologies. The undirected line in the figure means that the network is undirected. Further, for the regular network, we

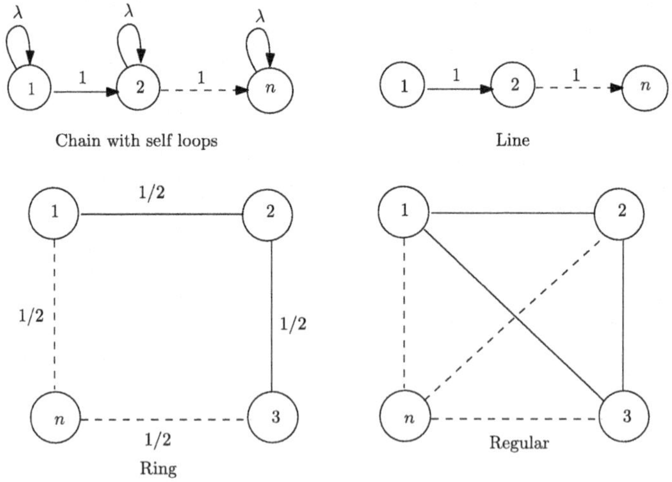

Fig. 6.2 Different network topologies

Table 6.1 Hierarchy of resilience under $\mathscr{L}_2$ norm

Topology	Resilient	Order
Line	Yes	$\mathscr{L}_2$–linear order
Ring	Yes	$\mathscr{L}_2$–constant order
Regular	Yes	$\mathscr{L}_2$–constant order
Chain with self-loops	No	$\mathscr{M}_n = \Omega(\exp(an))$

assume that the edge weight is $1/(n-1)$. In Table 6.1, we present the examples of resilience order that we will commonly visit in this work.

6.4 Edge Link Effects on Resilience Measures

In previous sections, we presented why quantifying resilience as a function of network dimension is important. We showed how different network topologies behave under an input shock. Specifically, we used the tools we developed in Sect. 6.3 to analyze the resilience in these topologies. In this section, we will demonstrate how individual edge links affect the resilience properties of a network topology.

The primary focus of this section is to determine if in a fragile network sequence, does their exist a set of bottleneck links, i.e., those links that give substantial improvement when modified. Trivially, this set includes removing *all* the links of the network; however, the challenge is to find a nontrivial set and formalize this notion of "substantial improvement."

When is an edge link fragile?

Definition 6.7 A link between nodes i, j in a network, A_n, in the network sequence, $\{A_n\}_{n-1}^{\infty}$ is fragile if

$$\frac{\partial \mathrm{tr}(P(A_n))}{\partial a_{ij}^n} = \Omega(\mathrm{SP}(n))$$

Here, $a_{ij}^n = [A_n]_{ij}$ and $\mathrm{SP}(\cdot)$ is some super-polynomial function in n.

If a network sequence is fragile (or a large network is fragile), one might expect that there exists a critical link, or a set of critical links, in every network of the network sequence. Definition 6.7 attempts to capture this notion of criticality. Then, this leads to the question whether this definition of edge link fragility is consistent with the definition of network resilience, i.e.,

Do we find fragile links only when a large network is fragile?

Lemma 6.1 *For each network, A_n, in the network sequence, $\{A_k\}_{k=1}^{\infty}$, with $a_{ij}^n \geq 0$ we have:*

$$\frac{\partial tr(P(A_n))}{\partial a_{ij}^n} \geq 2[P_n A_n]_{ij} \tag{6.10}$$

Here $a_{ij}^n = [A_n]_{ij}$.

Theorem 6.1 *There exists a fragile link in a large network, A_n, if and only if it is fragile.*

As an example for Theorem 6.1, we show in Fig. 6.3 the fragile links in the directed chain with self-loops. In the extreme case when the link weight of all edges from $i \to i+1$ is 0, this reduces to the disconnected network. However, such a reduction is in no way unique, since if we removed all the self-loops, we would obtain a simple chain network that we already know is resilient. We find that from Chain 1 to Chains 2, 3 in Fig. 6.3, the log edge link sensitivities,

$$\frac{\partial \text{tr}(P_n)}{\partial a_{ij}^n} = \Omega(\exp{(cn)}),\ c > 0$$

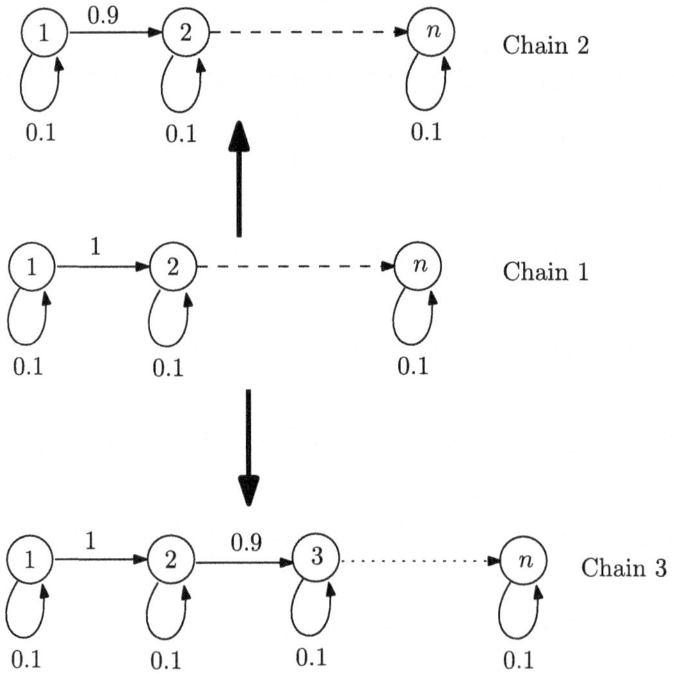

Fig. 6.3 Different chain networks

6.5 Nodal Dependence of Performance Measure

In many practical situations, there maybe a subset of nodes that are observable and of interest (see [20]). It then becomes important to study the effect of a subset of nodes on the performance measure, for example, the nodal volatilities, i.e., effect of one node on the robustness. Here, we consider the canonical form of a discrete time LTI system:

$$\begin{aligned} x(k+1) &= Ax(k) + w(k) \\ y(k) &= Cx(k) \end{aligned} \tag{6.11}$$

Here, $A \in \mathbb{R}^{n \times n}$, $x(k)$, $w(k) \in \mathbb{R}^{n \times 1}$, $C \in \mathbb{R}^{m \times n}$, where $m \leq n$. Following [10, 20], we have the discrete time Lyapunov equation:

$$A^T P_C A + C^T C = P_C \tag{6.12}$$

For brevity, call $P_C = \text{LYAP}(A, C)$, then P in Eq. 6.1 is given by $P = \text{LYAP}(A, I)$, where I is the identity matrix of appropriate dimension. Informally, one should think of C as a matrix that "mixes" the outputs of each node, or "hides" away a few nodes.

Theorem 6.2 *For the canonical discrete time linear system in Eq.* (6.11) *and the corresponding Lyapunov equation in Eq.* (6.12)*, with* $||C||_F = K$*, we have that*

$$trace\Big(LYAP(A, C)\Big) = \sum_{i=1}^{n}\sum_{j=1}^{n} b_{ij}^2 \sigma_j$$

where A *is an* $n \times n$ *matrix,* $\sum_{j=1}^{n}\sum_{i=1}^{n} b_{ij}^2 = K^2$ *for all* i*, and* $\sigma_1 \geq \ldots \geq \sigma_n$ *are the eigenvalues of* $LYAP(A, I)$*. Further, if* $(\pi^1, \ldots, \pi^n)$ *be the right eigenvectors (such that* $||\pi^i||_2 = 1$ *for all* $i \geq 1$*) of* $LYAP(A^T, I)$ *corresponding to the eigenvalues* $\sigma_1 \geq \ldots \geq \sigma_n$*, then* $c_i = \sum_{j=1}^{n} b_{ij}\pi^j$ *where* c_i *is the* i*th column of* C*.*

Theorem 6.2 gives a closed-form dependence of different nodes on the robustness of a network matrix. For example, consider the sequence of networks, $\{A_n\}_{n=1}^{\infty}$, where A_n has the following form:

$$A_n = \mu \times \begin{pmatrix} 0 & \frac{1}{n-1} & \cdots & \frac{1}{n-1} \\ 1 & 0 & \cdots & 0 \\ \vdots & \vdots & \ddots & \vdots \\ 1 & 0 & \cdots & 0 \end{pmatrix} \tag{6.13}$$

Here, $0 < \mu < 1$, and this network looks as in Fig. 6.4. Now, for each n, and dynamics as in Eq. (6.14), one can show that $\sigma_1(\text{LYAP}(A, I)) = \mathcal{M}(A_n) = \Theta(n)$. Further, for $\text{LYAP}(A_n^T, I)$, the leading eigenvector corresponding to $\mathcal{M}(A_n)$ is

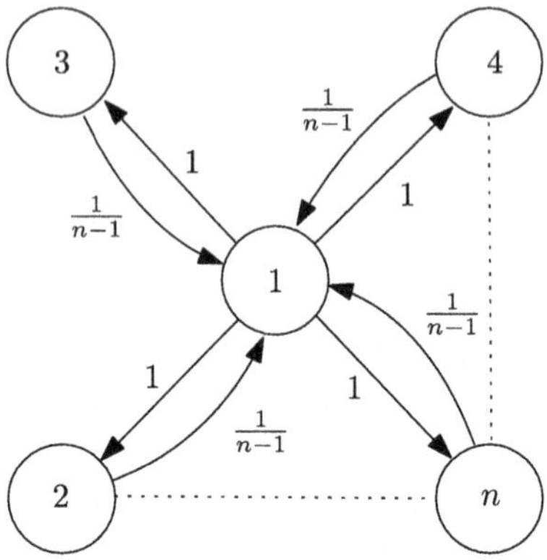

Fig. 6.4 Weighted star network

$\pi^1 = [1, 0, \ldots, 0]^T$. Therefore, to reduce $\mathscr{M}(A_n)$, by obfuscating the effects of $\sigma_1(\text{LYAP}(A, I))$, one needs to pick the columns of C from an orthogonal subspace to π^1. This coincides with the idea of "hiding" the most vulnerable node, which in this case is node 1 in Fig. 6.4. Now, if the columns C lie in an orthogonal subspace to π^1, then we have that $\sigma_1(\text{LYAP}(A, C)) = \Theta(1)$. Similarly, in Definition 6.4 of asymptotic robustness, we take $C = I$, i.e., all nodes are equally important (Table 6.2).

$$x(k+1) = A_n x(k) + \omega\delta(0, k) \tag{6.14}$$

Nodal volatilities in a network can be intuitively thought as the effect on the robustness of the network, when the observability of the node is changed. More formally,

Definition 6.8 For a node, i, the nodal volatility is defined as

$$V_i = \lim_{\varepsilon \to 0^+} \frac{\Big(\text{trace}(\text{LYAP}(A, I)) - \text{trace}(\text{LYAP}(A, I_\varepsilon^i))\Big)}{\varepsilon}$$

Here, $I_\varepsilon^i = \text{diag}(1, 1, \ldots, 1 - \varepsilon, 1, \ldots, 1)$, with $1 - \varepsilon$ at the ith position.

Table 6.2 Nodal volatilities

Topology	Maximum nodal volatility
Line	$\Theta(n)$
Ring	$\Theta(1)$
Regular	$\Theta(1)$
Weighted star	$\Theta(n)$
Chain with self-loops	$\Omega(\exp(an))$

Then, as a consequence of Theorem 6.2, we have the following result on nodal volatility, studied in [10].

Corollary 6.1 *The nodal volatility at node, i, is given by $V_i = P_{ii}$, where P_{ii} is the ith diagonal element of $P = A^T PA + I$.*

6.6 Conclusion

In this work, we studied the dependence of transient dynamics of networks on their topology. Motivated by applications in socioeconomic networks such as economics, transportation systems, and social networks, we provided a framework to assess resilience in general network topologies. Specifically, we showed how resilience varies with edge weights and nodal degrees. We found that a network was fragile if and only if it had fragile links, which showed that the notion of resilience developed here consistently captures the effects of interconnections on the transient behavior of a networked system. We also derived a formal characterization of nodal volatilities in any general network.

Although we show that our measures of resilience capture the effect of interconnections in a network, finding the critical links in a general network is a future direction of research. An important unexplored area is when the dynamics of the network are linear but time varying.

References

1. Acemoglu, D., Ozdaglar, A., Tahbaz-Salehi, A.: Cascades in networks and aggregate volatility. National Bureau of Economic Research (2010)
2. Acemoglu, D., Ozdaglar, A., Tahbaz-Salehi, A.: Microeconomic origins of macroeconomic tail risks. National Bureau of Economic Research (2015)
3. Acemoglu, D., Carvalho, V.M., Ozdaglar, A., Tahbaz-Salehi, A.: The network origins of aggregate fluctuations. Econometrica **80**(5), 1977–2016 (2012)
4. Borovykh, N., Spijker, M.N.: Resolvent conditions and bounds on the powers of matrices, with relevance to numerical stability of initial value problems. J. Comput. Appl. Math. **125**, 41–56 (2000)
5. Cao, M., Spielman, D.A., Morse, S.: A lower bound on convergence of a distributed network consensus algorithm. In: IEEE Decision and Control, and European Control Conference, CDC-ECC, pp. 2356–2361 (2005)
6. Como, G., Fagnani, F.: From local averaging to emergent global behaviors: the fundamental role of network interconnections. http://arxiv.org/abs/1509.08572 (2015)
7. Demetrius, L., Manke, T.: Robustness and network evolution-an entropic principle. Phys. A Stat. Mech. Appl. **346**(3), 682–696 (2005)
8. Dullerud, G.E., Paganini, F.: A Course in Robust Control Theory, vol. 6. Springer, New York (2000)
9. Herman, I., Martinec, D., Hurák, Z., Šebek, M.: Nonzero bound on Fiedler eigenvalue causes exponential growth of h-infinity norm of vehicular platoon. IEEE Trans. Autom. Control **60**(8), 2248–2253 (2015)

10. Huang, Q., Yuan, Y., Gonçalves, J., Dahleh, M.A.: $\mathcal{H}_2$ norm based network volatility measures. In: American Control Conference (ACC), pp. 3310–3315. IEEE (2014)
11. Kwon, W.H., Moon, Y.S., Ahn, S.C.: Bounds in algebraic Riccati and Lyapunov equations: a survey and some new results. Int. J. Control **64**(3), 377–389 (1996)
12. Leonard, N.E., Scardovi, L., Young, G.F.: Robustness of noisy consensus dynamics with directed communication. In: American Control Conference (ACC), pp. 6312–6317. IEEE (2010)
13. Levin, D.A., Peres, Y., Wilmer, E.L.: Markov Chains and Mixing Times. American Mathematical Society (2009)
14. Olfati-Saber, R., Fax, J.A., Murray, R.M.: Consensus and cooperation in networked multi-agent systems. Proc. IEEE **95**(1), 215–233 (2007)
15. Olshevsky, A., Tsitsiklis, J.N.: Degree fluctuations and the convergence time of consensus algorithms. IEEE Trans. Autom. Control **58**, 2626–2631 (2013)
16. Sandhu, R., Georgiou, T., Tannenbaum, A.: Market Fragility, Systemic Risk, and Ricci Curvature. http://arxiv.org/abs/1505.05182 (2015)
17. Sarkar, T., Roozbehani, M., Dahleh, M.A.: Robustness scaling in large networks. In: American Control Conference (ACC), pp. 197–202. IEEE (2016)
18. Sarkar, T.: Understanding resilience in large networks. SM thesis. http://dspace.mit.edu/handle/1721.1/107374 (2016)
19. Scardovi, L., Sepulchre, R.: Synchronization in networks of identical linear systems. Automatica **45**, 2557–2562 (2009)
20. Siami, M., Motee, N.: Fundamental limits and tradeoffs on disturbance propagation in linear dynamical networks. IEEE Trans. Autom. Control **61**, 4055–4062 (2016)
21. Xia, T., Scardovi, L.: Output-feedback synchronizability of linear time-invariant systems. Syst. Control Lett. **94**, 152–158 (2016)
22. Zhao, Y., Minero, P., Gupta, V.: On disturbance propagation in vehicle platoon control systems. In: American Control Conference (ACC), pp. 6041–6046 (2012)

Chapter 7
Feedback Control for Distributed Massive MIMO Communication

S. Dasgupta, R. Mudumbai and A. Kumar

Abstract We present a distributed nullforming algorithm where a set of transmitters transmit at full power to minimize the received power at a designated receiver. Each transmitter adjusts the phase and frequency of its transmitted RF signal in a purely distributed fashion as it uses only an estimate of its own channel gain to the receiver, and a feedback signal from the receiver, that is common across all the transmitters. This assures its scalability; in contrast any noniterative approach to the nullforming problem requires that each transmitter know every other transmitter's channel gain. We prove that the algorithm practically, globally converges to a null at the designated receiver. By practical convergence we mean that the algorithm always converges to a stationary trajectory, and though some of these trajectories may not correspond to a minimum, those that do not are locally unstable, while those that do are locally stable. Unlike its predecessors the paper does not assume prior frequency synchronization among the transmitters, but asymptotically secures frequency consensus.

7.1 Introduction

Multiple Input Multiple Output (MIMO) techniques [1–4] have played an important part in the remarkable explosion of wireless communication systems in the past two decades, and MIMO is now an integral component of all recent WiFi and cellular

---

This work was partly supported by US NSF grants CNS-1239509, CAREER award ECCS-1150801, and CCF-1302456, and ONR grant N00014-13-1-0202.

S. Dasgupta (✉) · R. Mudumbai · A. Kumar
Department of Electrical and Computer Engineering, University of Iowa,
Iowa City, IA 52242, USA
e-mail: dasgupta@engineering.uiowa.edu

R. Mudumbai
e-mail: rmudumbai@engineering.uiowa.edu

A. Kumar
e-mail: amy-kumar@uiowa.edu

R. Tempo et al. (eds.), *Emerging Applications of Control and Systems Theory*, Lecture Notes in Control and Information Sciences - Proceedings,
https://doi.org/10.1007/978-3-319-67068-3_7

standards e.g. 802.11ac [2] and 5G [3]. Specifically, MIMO communication involves transceivers equipped with multiple antennae. This allows directional transmission, permitting control of the interference produced by wireless transmitters, and orders of magnitude increases in energy and spectral efficiency.

Yet the applicability of MIMO is in practice severely limited by constraints such as form factors, and the size and the number of antennae that can be realistically supported. An attractive alternative, springing from the pioneering work of [5], is *distributed MIMO (DMIMO)*, [6] where instead of deploying centralized antenna systems, groups of single antennae transceivers collaborate to form a *virtual antennae* system that mimics the functionality of a centralized multi-antennae system. Studied extensively by theoreticians, until recently, this concept has been dismissed by practioners as being beyond the realm of practicality for several reasons. Among these are issues engendered by uncertain geometries and the fact that unlike centralized systems, by its very definition, each constituent of a DMIMO system carries its own clock and oscillator. These suffer significant and rapid drift modeled as a second-order stochastic process, [7] that induces DMIMO units to migrate from synchrony to complete incoherence within mere hundreds of milliseconds.

Over the last decade several authors, [8–22], have sought to realize this decades-old concept, by addressing two salient components of DMIMO: distributed beamforming, [8–16] and distributed nullforming, [17–22]. In the former, groups of transmitters, collaborate to form a beam of maximum possible power using constructive interference at a receiver. In the latter their transmissions cancel each other at the designated receiver. On the other hand [21] simultaneously forms nulls at some receivers and beams at others. Physical implementations of both beam and nullforming algorithms on software defined radio (SDR) platforms, are described in [12, 13, 20]. Apart from being important constituents of the overall DMIMO architecture, beamforming is key to power efficient communication, just as nullforming has applications in interference avoidance for increased spatial spectrum reuse [23], cognitive radio [24] and cyber security [25].

This brings us to the role of *feedback control* in these schemes. Ultimately in all these applications all transmissions must arrive at their target receiver synchronized in frequency and with precise phase, and for nullforming, amplitude relationships. Uncertain geometries and mobility, make it impossible for the transmitters to determine the phases, and amplitudes of their transmission at the receiver. Thus all the references cited above rely on some receiver cooperation. This takes the form of feedback from the receiver to the transmitters, and possibly between the transmitters, using which the transmitters must adjust the phase, amplitude and frequency of their transmissions to achieve synchronization at the receiver. A key issue, of course, is what type of feedback is needed, the minimum information exchange required to achieve one's objective.

In this light, among the DMIMO papers we have cited here, barring [11, 19], all assume, prior frequency synchronization, presumably through information exchange among the transmitters. The earliest among the DMIMO papers, [8], assumes prior frequency synchronization, and requires that the receiver feed back to *each transmitter a separate feedback signal throughout its operation.* Such an algorithm is thus not scalable.

A breakthrough idea introduced in [9], and used in several subsequent papers, is the notion of *common aggregate feedback*. This involves the receiver broadcasting to all the transmitters either the complex baseband version of its total received signal or some attribute there of. In either case the burden of repeatedly feeding back a separate signal to each transmitter is alleviated. To wit [9], assumes prior frequency synchronization among the nodes and executes a *1-bit feedback* algorithm to achieve beamforming. The algorithm itself is in the mold of randomized ascent. Each transmitter updates its phase according to a preselected distribution. The 1-bit feedback is whether or not the net received power declines as a result of these updates. If the power declines, the updates are discarded. Else they are retained. Under mild assumptions on the distribution from which the phase updates are chosen, this algorithm is provably convergent.

In this paper we consider distributed nullforming without prior frequency synchronization with only phase and frequency updates. The algorithm we formulate is akin to *Lyapunov redesign* in the controls and adaptive systems literature, [26, 27]. We observe that distributed nullforming algorithms can be found in [17–22]. Each of these, however assumes prior frequency synchronization at the outset of operation, presumably through information exchange among transmitters. While this is reasonable, oscillator frequencies also undergo drift modeled as Brownian motion, albeit at orders of magnitude slower rates than oscillator phases. On the other hand drift in oscillator frequencies has a more dramatic impact on performance than has phase drift. Furthermore Doppler shift occurs at receivers, thus receiver feedback should be used to guide the adjustment of frequencies at transmitters.

Nullforming is fundamentally more difficult than is beamforming, [22], for two reasons. First, it is much more sensitive to phase drift. Because of this, a 1-bit algorithm like in [9] is unable to adjust quickly enough to achieve a decent null. Second, while beamforming only requires frequency and phase alignment at the receiver, nullforming requires the precise control of the amplitude and phase of the radio-frequency signal transmitted by each cooperating transmitter to ensure that they cancel each other. Accordingly, [17, 18], requires that in addition to the common aggregate feedback of the total baseband received signal, each transmitter must also learn the *channel state information* of *every* transmitter to the receiver. This latter requirement is substantially relaxed in [22], where each channel, at the point of setup requires only the knowledge of its channel to the receiver. Simulations in [22] show that the gradient descent algorithm it employs is robust to substantial channel estimation errors, while capable of tracking significant oscillator drift.

This paper extends [22] to the case where the transmitters even if initially frequency synchronized undergo small frequency drifts. These frequency drifts even if small, can destroy a good null very quickly. As in [22], we assume that each user knows its complex channel gain to the receiver. Unlike [17, 18] it does not have the CSI seen by the other transmitters. Like [22], the receiver feeds back the net baseband signal. In [22] this feedback is used by each transmitter to perform a distributed gradient descent minimization of the total received power. The minimization is distributed, as each transmitter can perform it using only the aggregate feedback and its own complex channel gain to the receiver. Similarly, in this paper each transmitter adjusts its phase and *frequency* knowing only its CSI to the receiver and the aggregate feedback signal, to asymptotically drive the received signal to zero. However, we show that the lack of frequency synchronization precludes the use of gradient descent. Instead a Lyapunov redesign is needed.

We observe that [11, 19] also eschew the assumption of prior frequency synchronization. Among these, [11] uses a 1-bit type algorithm to beamform. We have however, discovered a conceptual error in the algorithm. On the other hand the preliminary paper, without proofs, [19], has the important difference that it critically assumes that each transmitter equalizes its complex channel gain. In this paper we equalize only the phase and *not the magnitude* of the channel. This is in the vein of [22] and permits a key application of distributed nullforming: *Namely, permitting transmission at full power, thus maximizing incoherent power pooling gains, while protecting the null target.* As explained in [22] this opens up the prospect of both *Space Division Multiple Access (SDMA)* and cyber security.

Section 7.2 provides the algorithm. Section 7.3 has some preliminary analysis. Section 7.4 presents a stability analysis. Received power which must be minimized is a nonconvex function of the transmitter phases and frequencies. Unsurprisingly our Lyapunov redesign yields a distributed algorithm that has multiple stationery points/trajectories that may not correspond to a minimum. Yet we show that only those stationary points are locally stable that do correspond to global minima. Section 7.5 is the conclusion.

7.2 The Algorithm

We now describe a scalable gradient descent algorithm for distributed nullforming in a node. As in [22] and unlike [17, 18], we assume that at the beginning of a nullforming epoch, each transmitter has access only to its own complex channel gain to the receiver, using which it equalizes the *phase rather than also the magnitude* of the channel to the receiver. Assume there are N transmitter nodes.

Denote $\theta_i(t)$ to be the equalized phase of the i-th transmitter and $r_i > 0$ is the magnitude of the received signal. Assume that $\omega_i(t)$, is a frequency offset of the i-th transmitter, from a nominal frequency to which each transmitter should ideally have been synchronized, but oscillator drift prevents the maintenance of such synchronization.

Then the complex baseband signal received at the cooperating receiver is:

$$s(t) = R(t) + jI(t) \tag{7.1}$$

where

$$R(t) = \sum_{i=1}^{N} r_i \cos(\omega_i(t)t + \theta_i(t)) \tag{7.2}$$

and

$$I(t) = \sum_{i=1}^{N} r_i \sin(\omega_i(t)t + \theta_i(t)). \tag{7.3}$$

As in [22], the transmission process is thus, equivalent to each transmitter transmitting, the phase equalized complex baseband signal, $e^{j((\omega_i(t)t+\theta_i(t))}$. The baseband signal the receiver sees is

$$s(t) = \sum_{i=1}^{N} r_i e^{j((\omega_i(t)t+\theta_i(t))} \tag{7.4}$$

see (7.1). Indeed it is $s(t)$ that the receiver broadcasts to all transmitters, which they must use to adjust their frequency and phase. Define $\theta(t) = [\theta_1(t), \cdots, \theta_N(t)]^\top$ and $\omega(t) = [\omega_1(t), \cdots, \omega_N(t)]^\top$. We note the key difference with [19], which apart from not providing any proofs, assumes that all $r_i = 1$. The received power is:

$$J(\theta, \omega, t) = I^2(t) + R^2(t). \tag{7.5}$$

The receiver feeds back $s(t)$. The i-th transmitter uses $s(t)$ to adjust its $\theta_i(t)$ and $\omega_i(t)$ to asymptotically force $J(\theta, \omega, t)$ to zero. In reality both the adjustment and feedback will be discrete time. However, should the continuous time algorithm be uniformly asymptotically stable (u.a.s), then averaging theory, [28] ensures that for high enough feedback rates, and small enough gains, an obvious discretized version will also be u.a.s.

As noted above, [22] where $\omega = 0$, uses a gradient descent algorithm. The resulting J is autonomous in [22]. The frequency offsets here make the cost function nonautonomous, as it may change its value even if the phases and frequencies are not adjusted. To amplify this point observe that a pure gradient descent algorithm would take the form:

$$\dot{\theta}(t) = -\frac{\partial J(\theta, \omega, t)}{\partial \theta}; \ \dot{\omega}(t) = -\frac{\partial J(\theta, \omega, t)}{\partial \omega}. \tag{7.6}$$

Now observe that under (7.6)

$$\begin{aligned} \dot{J} &= \frac{\partial J}{\partial t} + \frac{\partial J}{\partial \theta}\dot{\theta} + \frac{\partial J}{\partial \omega}\dot{\omega} \\ &= \frac{\partial J}{\partial t} - \left\| \frac{\partial J}{\partial \theta} \right\|^2 - \left\| \frac{\partial J}{\partial \omega} \right\|^2. \end{aligned}$$

Should $\frac{\partial J}{\partial t}$ be zero, as is the case is in [22], this guarantees a nonincreasing $J(\theta, \omega, t)$. However, with ω_i potentially nonzero under (7.2), $\frac{\partial J}{\partial t}$ need not be zero, preventing guaranteed decrescence of $J(\theta, \omega, t)$. Thus a Lyapunov redesign of the nullforming algorithm is needed.

In Sect. 7.3 we present a Lyapunov function, and show that its decrescence is guaranteed by the following algorithm.

$$\dot{\theta} = -\frac{\partial J}{\partial \theta} - \frac{\omega}{2} \tag{7.7}$$

$$\dot{\omega} = -\frac{1}{2}\frac{\partial J}{\partial \theta}. \tag{7.8}$$

Thus the i-th node can implement these as long as it has access to its frequency, phase, CSI to the receiver and the common feedback signals $I(t)$ and $R(t)$ permitting a totally distributed implementation, as

$$\frac{\partial J}{\partial \theta_i} = -2R(t)r_i \sin(\omega_i(t)t + \theta_i(t)) + 2I(t)r_i \cos(\omega_i(t)t + \theta_i(t)), \tag{7.9}$$

and

$$\frac{\partial J}{\partial \omega} = t\frac{\partial J}{\partial \theta}. \tag{7.10}$$

Also observe that unlike (7.6) there is an additional corrective term $\frac{\omega}{2}$ in (7.7), that accounts for the frequency offsets. In keeping with our mandate for phase and frequency only updates the gains r_i are not updated.

7.3 Preliminaries of the Stability Analysis

In this section, we present certain preliminary results that among other things show the uniform convergence of the gradient of J with respect to θ, and explore the properties of the stationary points of (7.7, 7.8).

But first, a result used in [29].

Lemma 7.1 *Suppose on a closed interval $\mathscr{I} \subset \mathbb{R}$ of length T, a signal $w : \mathscr{I} \to \mathbb{R}$ is twice differentiable and for some ε and M'*

$$|w(t)| \leq \varepsilon_1 \text{ and } |\ddot{w}(t)| \leq M' \quad \forall \quad t \in \mathscr{I}.$$

Then for some M independent of ε_1, $\mathscr{I}$ and M', and $M'' = \max(M', 2\varepsilon_1 T^{-2})$ one has:

$$|\dot{w}(t)| \leq M(M''\varepsilon_1)^{1/2} \quad \forall \quad t \in \mathscr{I}.$$

We begin by deriving a preliminary relation.

$$\begin{aligned}\frac{\partial J}{\partial t} &= -2R(t)\sum_{i=1}^{N}\omega_i(t)r_i\sin(\omega_i(t)t+\theta_i(t)) + 2I(t)\sum_{i=1}^{N}r_i\omega_i(t)\cos(\omega_i(t)t+\theta_i(t)),\\ &= \omega^\top(t)\frac{\partial J}{\partial \theta}.\end{aligned} \tag{7.11}$$

We now show that under (7.7, 7.8) there is a Lyapunov function that is nonincreasing and the gradient of J with respect to θ converges uniformly to zero.

Lemma 7.2 *Consider (7.7, 7.8) with (7.2–7.5) initial time $t_0 \geq 0$. Then the following hold:*

(a) For all $t \geq t_0$,

$$V(t) = J(t) + \frac{\|\omega(t)\|^2}{2} \tag{7.12}$$

is nonincreasing.

(b) The following occurs uniformly in t_0.

$$\lim_{t\to\infty}\frac{\partial J}{\partial \theta}(t) = 0. \tag{7.13}$$

Proof Because of (7.5), (7.9–7.10) and (7.7, 7.8), there holds:

$$\begin{aligned}\dot{J} + \frac{d}{dt}\left\{\frac{\omega^\top\omega}{2}\right\} &= \frac{\partial J}{\partial t} + \dot{\theta}^\top\frac{\partial J}{\partial \theta} + \dot{\omega}^\top\frac{\partial J}{\partial \omega} + \omega^\top\dot{\omega}\\ &= \omega^\top\frac{\partial J}{\partial \theta} - \left\|\frac{\partial J}{\partial \theta}\right\|^2 - \frac{\omega^\top}{2}\frac{\partial J}{\partial \theta}\\ &\quad -\frac{\omega^\top}{2}\frac{\partial J}{\partial \theta} - \frac{t}{2}\left\|\frac{\partial J}{\partial \theta}\right\|^2\\ &= -\left(1+\frac{t}{2}\right)\left\|\frac{\partial J}{\partial \theta}\right\|^2\end{aligned} \tag{7.14}$$

Consequently (a) holds. Further, ω is uniformly bounded. Consequently from (7.9) there is an M_1, independent of t_0, such that for all $t \geq t_0$

$$\left\|\frac{d}{dt}\left\{\frac{\partial J}{\partial \theta}(t)\right\}\right\| \leq M_1.$$

Equally, there exists an M_2, also independent of t_0, such that for all $t \geq t_0$,

$$\left\| \frac{\partial J}{\partial \theta}(t) \right\| \leq M_2.$$

Further, since the initial time $t_0 \geq 0$, from (7.14) and $V(t)$ is nonnegative, one obtains that for all $t \geq t_0$:

$$\begin{aligned}\int_{t_0}^{t} \left\| \frac{\partial J}{\partial \theta}(s) \right\|^2 \mathrm{d}s &\leq \int_{t_0}^{t} \left(1 + \frac{s}{2}\right) \left\| \frac{\partial J}{\partial \theta}(s) \right\|^2 \mathrm{d}s \\ &\leq V(t_0).\end{aligned}$$

Thus, for every $\varepsilon > 0$, there exists a T independent of t_0 such that for all $t \geq T + t_0$,

$$\int_{T+t_0}^{t} \left\| \frac{\partial J}{\partial \theta}(s) \right\|^2 \mathrm{d}s \leq \varepsilon.$$

Then from Lemma 7.1, there is a K independent of t_0 such that for all $\varepsilon > 0$, there exists a T independent of t_0 such that for all $t \geq T + t_0$,

$$\left\| \frac{\partial J}{\partial \theta}(s) \right\|^2 \leq \varepsilon, \quad \forall s \geq T + t_0.$$

Thus indeed (b) holds uniformly in t_0.

Thus, (7.7, 7.8) converges uniformly to a trajectory where:

$$\frac{\partial J}{\partial \theta} = 0. \tag{7.15}$$

Because of (7.2, 7.3), (7.9) there holds:

$$\begin{aligned}\frac{1}{2}\frac{\partial J}{\partial \theta_i} &= -R(t)\sum_{i=1}^{N} r_i \sin(\omega_i(t)t + \theta_i(t)) + I(t)\sum_{i=1}^{N} r_i \cos(\omega_i(t)t + \theta_i(t)) \\ &= \sum_{i=1}^{N}\sum_{l=1}^{N} r_i r_l \{\cos(\omega_i(t)t + \theta_i(t)) \sin(\omega_l(t)t + \theta_l(t)) \\ &\quad - \sin(\omega_i(t)t + \theta_i(t)) \cos(\omega_l(t)t + \theta_l(t))\} \\ &= \sum_{i=1}^{N}\sum_{l=1}^{N} r_i r_l \sin((\omega_l(t) - \omega_i(t))t + \theta_l(t) - \theta_i(t)).\end{aligned} \tag{7.16}$$

Thus (7.15) also implies that for some constant ω^* on this trajectory the frequency offsets

$$\omega_i = \omega^*, \ \forall i \in \{1, \cdots, N\}. \tag{7.17}$$

Observe from (7.5), (7.2) and (7.3),

$$\begin{aligned} J(\theta, \omega, t) &= \left(\sum_{i=1}^{N} r_i \cos(\omega_i(t)t + \theta_i(t))\right)^2 + \left(\sum_{i=1}^{N} r_i \sin(\omega_i(t)t + \theta_i(t))\right)^2 \\ &= \sum_{i=1}^{N} r_i^2 + 2\sum_{i=1}^{N}\sum_{\substack{l=1 \\ l\neq i}}^{N} r_i r_l \cos((\omega_l(t) - \omega_i(t))t + \theta_l(t) - \theta_i(t)) \end{aligned} \tag{7.18}$$

This shows that these are in fact *manifolds* rather than points.

We now turn to a major nontrivial consequence of having potentially nonunity r_i, a problem absent in [19] where all r_i are 1. There are sets of positive r_i for which a null may not be possible. It is thus useful to first characterize conditions on the r_i that ensure that a null is possible. The theorem below provides this characterization. We note it is similar to a result in [22] where the ω_i are all fixed to zero. The theorem also characterizes the global minimum value

$$J^* = \max_{t\geq 0} \min_{\theta,\omega} J(\theta, \omega, t). \tag{7.19}$$

Theorem 7.1 *Consider $J(\theta, \omega, t)$ defined in (7.2, 7.3, 7.5) and J^* as in (7.19), with all $r_i > 0$. Without loss of generality label r_i, so that $r_i \geq r_{i+1}$. Then the following hold:*

(i) $J^* = 0$ *iff*

$$r_1 \leq \sum_{i=2}^{N} r_i. \tag{7.20}$$

(ii) If (7.20) is violated

$$J^* = \left(r_1 - \sum_{i=2}^{N} r_i\right)^2. \tag{7.21}$$

Proof Observe $J(\theta, \omega, t) = |s(t)|^2$, with $s(t)$ defined in (7.4). Suppose (7.20) is violated. Clearly, under (7.4), as $r_i > 0$,

$$J(\theta, \omega, t) \geq \left(r_1 - \sum_{i=2}^{N} r_i\right)^2.$$

Thus $J^* > 0$. Further this minimum is attained by choosing $\omega_i = \omega_l$, for all i, l, $\theta_1 = 0$ and $\theta_i = \pi$, $\forall\, i > 1$. This proves (ii) and the "only if" part of (i).

To prove the "if" part of (i), set all ω_i to zero and assume that (7.20) holds. We use induction. Consider $N = 2$. Then $r_1 = r_2$. Thus with $\theta_1 = 0, \theta_2 = \pi$, $J(\theta, 0, t) = |r_1 - r_2|^2 = 0$, $\forall t$. Now suppose the result holds for some $N = n \geq 2$. Consider $N = n + 1$.

Observe with

$$J(0, 0, t) = \sum_{i=1}^{n+1} r_i > 0. \tag{7.22}$$

Consider two cases.

Case I: $r_2 > \sum_{i=3}^{n+1} r_i$: In this case, by hypothesis $0 < r_2 - \sum_{i=3}^{n+1} r_i < r_1$. Thus,

$$J(\pi[1, 0, [1, \cdots, 1]]^T, 0, t) = -r_1 + r_2 - \sum_{i=3}^{n+1} r_i < 0.$$

As for every t, $J(\theta, 0, t)$ is continuous in θ, and (7.22) holds, there exist a θ for which $J(\theta, 0, t) = 0$, $\forall t$. Thus $J^* = 0$.

Case II: $r_2 \leq \sum_{i=3}^{n+1} r_i$: From the induction hypothesis, there exist $\hat{\theta}_2, \cdots, \hat{\theta}_{n+1}$ such that $\sum_{i=2}^{n+1} r_i e^{j\hat{\theta}_i} = 0 < r_1$. Moreover, by assumption $\sum_{i=2}^{n+1} r_i \geq r_1$. Since $\left|\sum_{i=2}^{n+1} r_i e^{j\theta_i}\right|$ is continuous in θ, moving continuously between $[\theta_2, \cdots, \theta_{n+1}]$ between 0 and $[\hat{\theta}_2, \cdots, \hat{\theta}_{n+1}]$ one can find a set of phases $[\theta_2^*, \cdots, \theta_{n+1}^*]$ such that $\left|\sum_{i=2}^{n+1} r_i e^{j\theta_i^*}\right| = r_1$. Thus, for some δ, $\sum_{i=2}^{n+1} r_i e^{j\theta_i^*} = r_1 e^{j\delta}$. Then $J([\pi + \delta, \theta_2^*, \cdots, \theta_{n+1}^*]^T, 0, t) = 0$, completing the proof.

Returning to the stationary trajectories, given by (7.15) and (7.17), some of these correspond to the desired null, or in their absence (7.19). Others do not, and will be called *false stationary* points. We show below that the latter are locally unstable and are thus, rarely attained, and even if attained cannot be practically maintained as noise would drive the trajectories away from them. Thus, by showing the local stability of the global minimum, we will demonstrate the practical uniform convergence of the algorithm to the global minimum.

In the stationary trajectory, (7.15) and (7.17) are not nulls, i.e., $s(t) \neq 0$, then from (7.9) for all i, l, and $t > 0$, $\tan(\omega^* t + \theta_i) = \tan(\omega^* t + \theta_l)$. Consequently on stationary trajectories that are not nulls,

$$\theta_i - \theta_l = k_{il}\pi, \ \forall \{i, l\} \subset \{1, \cdots, N\} \tag{7.23}$$

where k_{il} are integers.

The local analysis of these stationary trajectories, will require the examination of the Hessian with respect to θ on these trajectories. Consider again (7.16). Due to

(7.17), along (7.15) and (7.17) the il-th element of the Hessian along the stationary trajectory is given by:

$$[H(\theta)]_{il}|_{\omega_i=\omega_l} = \frac{\partial^2 J(\theta)}{\partial\theta_i \partial\theta_l} = \begin{cases} -2\sum_{i\neq l} r_i r_l \cos(\theta_i - \theta_l) & i = l \\ 2r_{il}\cos(\theta_i - \theta_l) & i \neq l \end{cases} \tag{7.24}$$

7.4 Stability Analysis

Armed with the preliminary results in Sect. 7.3 we now complete our stability analysis. Lemma 7.2 shows that uniform convergence to a stationary trajectory is guaranteed. Some of these trajectories correspond to a null. Other do not. In this section, we show that *only those that correspond to a null are locally stable. The others are not.* Consequently, one is assured of practical uniform convergence in the sense that stationary trajectories that do not correspond to the desired nulls if at all attained, cannot be practically maintained. Thus for all practical purposes, the algorithm defined in (7.7, 7.8) achieves a desired null.

First we demonstrate the local instability of spurious stationary trajectories. To this end we present need the following lemma.

Lemma 7.3 *The linear system below with scalar $a > 0$ is unstable:*

$$\dot{\eta} = \begin{bmatrix} a & ta + \frac{1}{2} \\ \frac{a}{2} & \frac{at}{2} \end{bmatrix} \eta \tag{7.25}$$

Proof Consider the initial condition $\eta(0) = [0, 1]^\top$. Then it is evident that both elements of the state are nonegative for all $t > 0$. Then the first element of the state vector is

$$\eta_1(t) \geq e^{at}.$$

Thus the system is unstable.

Consider next the Hessian with respect to θ at a critical trajectory. As given in (7.24) this is identical to the Hessian studied in [22]. From [22] we have the following lemma.

Lemma 7.4 *Assume all $r_i > 0$. Consider a false stationary manifold i.e., a stationary manifold on which $J \neq 0$ and $J \neq J^*$. Then $H(\theta)|_{\forall i, \omega_i = \omega^*}$ has at least one negative eigenvalue.*

We now prove that a false stationary trajectory is unstable.

Theorem 7.2 *Assume all $r_i > 0$. Consider (7.7, 7.8), and a stationary manifold defined by (7.15), (7.17) such that along this trajectory $J \neq 0$ and $J \neq J^*$. Then this trajectory is unstable.*

Proof Observe, that for all i, l

$$\frac{\partial^2 J(\theta)}{\partial \theta_i \partial \omega_l} = t \frac{\partial^2 J(\theta)}{\partial \theta_i \partial \theta_l}$$

(7.7, 7.8), linearized around (7.15), (7.17) is given by:

$$\dot{x} = \begin{bmatrix} -H(\theta) & -tH(\theta) + \frac{I}{2} \\ -\frac{H(\theta)}{2} & -\frac{t}{2}H(\theta) \end{bmatrix} x. \tag{7.26}$$

In view of Lemma 7.4 and the symmetry of $H(\theta)$, there is an orthogonal matrix Ω and real λ_i, with $\lambda_1 > 0$, such that with

$$\Lambda = \text{diag}\{-\lambda_1, \cdots, \lambda_N\},$$

$$H(\theta) = \Omega \Lambda \Omega^T$$

Define $\beta = \text{diag}\{\Omega, \Omega\}x$. Then the linearized systems is equivalent to:

$$\dot{\beta} = \begin{bmatrix} -\Lambda & -t\Lambda + \frac{I}{2} \\ -\frac{\Lambda}{2} & -\frac{t}{2}\Lambda \end{bmatrix} \beta.$$

Then instability follows from Lemma 7.3.

Thus indeed false stationary manifolds are unstable. To complete the result, it suffices to show that the global minima, nulls or otherwise, are locally stable. As the Hessian is indefinite this is a nonhyperbolic system and a linearization approach will be inconclusive. Thus we use a more direct approach to proving local stability.

Because of (7.18) and (7.23) at false stationary points the cost function takes the values

$$\sum_{i=1}^{N} r_i^2 + 2\sum_{i=1}^{N}\sum_{\substack{l=1 \\ l\neq i}}^{N} r_i r_l \mu_{il}, \; \mu_{il} \in \{-1, 1\}. \tag{7.27}$$

Assume again $r_i \geq r_{i+1} > 0$. With $r = [r_1 \cdots, r_N]^T$, define:

$$f(r) = \begin{cases} \min\left\{\left\{\left[\sum_{i=1}^{N} r_i^2 + 2\sum_{i=1}^{N}\sum_{\substack{l=1 \\ l\neq i}}^{N} r_i r_l \mu_{il} | \mu_{il} \in \{-1,1\}\right\} \setminus \{0\}\right\} & \text{if } r_1 \leq \sum_{i=2}^{N} r_i \\ \min\left\{\left\{\left[\sum_{i=1}^{N} r_i^2 + 2\sum_{i=1}^{N}\sum_{\substack{l=1 \\ l\neq i}}^{N} r_i r_l \mu_{il} | \mu_{il} \in \{-1,1\}\right\} \setminus \{r_1 - \sum_{i=2}^{N} r_i\}\right\} & \text{else} \end{cases} \tag{7.28}$$

In other words $f(r)$ is the smallest value that J can take at a false stationary point. We can now prove local stability of the null manifold.

Theorem 7.3 *Suppose $r_i \geq r_{i+1} > 0$ and with $f(r)$ defined in (7.28). Then with positive initial time t_0, (7.7, 7.8) uniformly converges to a global minimum of $J(\theta, \omega, t)$*

if $V(t_0) < f(r)$. *Further for a constant* ω^* *and all* $i \in \{1, \cdots, N\}$,

$$\lim_{t\to\infty} \omega_i(t) = \omega^* \tag{7.29}$$

Proof Item (b) of Lemma 7.2 holds uniformly in t_0. Further for all $t \geq t_0$

$$J(\theta(t), \omega(t), t) \leq V(t) \leq V(t_0) < f(r).$$

As Lemma 7.2 guarantees convergence to a stationary manifold, and only stationary manifold at which $J(\theta(t), \omega(t), t) < f(r)$, convergence occurs to a global minimum. Finally, (7.29) follows from (7.15) and (7.8).

As convergence to a stationary manifold is guaranteed, and all false stationary points are locally unstable, this thus proves practical uniform convergence to a global minimum. Observe in addition the transmitters attain *frequency consensus.*

In principle the phases need not converge to a fixed point. However, there is a subtlety. On a stationary trajectory, (7.7) and (7.8) reduces to, (7.17) and for each i,

$$\dot{\theta}_i(t) = -\frac{\omega^*}{2}.$$

Thus for suitable α_i the i-th transmitted signal becomes

$$e^{j(\frac{\omega^*}{2}t+\alpha_i)}.$$

thus effectively, the attained phases *are constants*, and *de facto* frequency synchronization obtains.

7.5 Conclusion

We have provided a new distributed nullforming algorithm that is guaranteed to achieve a null, through phase and frequency adapatation. Unlike [22] this paper does not assume prior frequency synchronization. Its effect though is the achievement of frequency synchronization. That all this can be achieved with *no cooperation* between the transmitters, a feedback of the aggregate received signal by the receiver, and the knowledge to each transmitter of only its CSI is in our opinion an intriguing fact. Also intriguing is the fact that an otherwise nonconvex cost function has the attractive property that all its local minima are global minima. Further studies that delineate the minimum information needed to achieve distributed nullforming are warranted.

References

1. Foschini, G.J: Layered space-time architecture for wireless communication in a fading environment when using multi-element antennas. Bell Labs Tech. J. **1**, 41–59 (1996)
2. Bejarano, O., Knightly, E.W., Park, M.: IEEE 802.11ac: from channelization to multi-user mimo. IEEE Commun. Mag. **51**, 84–90 (2013)
3. Wang, C.X., Haider, F., Gao, X., You, X.H., Yang, Y., Yuan, D., Aggoune, H.M., Haas, H., Fletcher, S., Hepsaydir, E.: Cellular architecture and key technologies for 5G wireless communication networks. IEEE Commun. Mag. **52**, 122–130 (2014)
4. Larsson, E.G., Edfors, O., Tufvesson, F., Marzetta, T.L.: Massive MIMO for next generation wireless systems. IEEE Commun. Mag. **52**, 186–195 (2014)
5. Cover, T., Gamal, A.E.L.: Capacity theorems for the relay channel. IEEE Trans. Inf. Theory **25**(5), 572–584 (1979)
6. Madhow, U., Brown, D.R., Dasgupta, S., Mudumbai, R.: Distributed massive MIMO: algorithms, architectures and concept systems. In: Proceedings of Information Theory and Applications Workshop (ITA), pp. 1–7 (2014)
7. Brown, D.R., Mudumbai, R., Dasgupta, S.: Fundamental limits on phase and frequency estimation and tracking in drifting oscillators. In: Proceedings of ICASSP, Invited Paper. Kyoto, Japan, March 2012
8. Tu, Y.-S., Pottie, G.J.: Coherent cooperative transmission from multiple adjacent antennas to a distant stationary antenna through AWGN channels. In: Proceedings of IEEE VTC Spring 02. Birmingham, Alabama, May 2002
9. Mudumbai, R., Barriac, G., Madhow, U.: On the feasibility of distributed beamforming in wireless networks. IEEE Trans. Wirel. Commun. **6**(5), 1754–1763 (2007)
10. Mudumbai, R., Brown III, D.R., Madhow, U., Poor, H.V.: Distributed transmit beamforming: challenges and recent progress. IEEE Commun. Mag. **47**(2), 102–110 (2009)
11. Seo, M., Rodwell, M., Madhow, U.: A feedback-based distributed phased array technique and its application to 60-gGHz wireless sensor network. In: Microwave Symposium Digest, 2008 IEEE MTT-S International, pp. 683–686, June 2008
12. Rahman, M.M., Baidoo-Williams, H.E., Mudumbai, R., Dasgupta, S.: Fully wireless implementation of distributed beamforming on a software-defined radio platform. In: Proceedings of the The 11th ACM/IEEE Conference on Information Processing in Sensor Networks, IPSN'12, pp. 305–316. Beijing, China (2012)
13. Quitin, F., Rahman, M.M.U., Mudumbai, R., Madhow, U.: Distributed beamforming with software-defined radios: frequency synchronization and digital feedback. In: IEEE Globecom 2012, Dec 2012 (to appear)
14. Mudumbai, R., Hespanha, J., Madhow, U., Barriac, G.: Scalable feedback control for distributed beamforming in sensor networks. In: IEEE International Symposium on Information Theory (ISIT), pp. 137–141. Adelaide, Australia, Sept 2005
15. Mudumbai, R., Bidigare, P., Pruessing, S., Dasgupta, S., Oyarzun, M., Raeman, D.: Scalable feedback algorithms for distributed transmit beamforming in wireless networks. In: Acoustics, Speech and Signal Processing (ICASSP), 2012 IEEE International Conference on, pp. 5213–5216, Mar 2012
16. Brown, D.R., Wang, R., Dasgupta, S.: Channel state tracking for large-scale distributed MIMO communication systems. IEEE Trans. Signal Process. **63**(10), 2559–2571 (2015). https://doi.org/10.1109/TSP.2015.2407316
17. Brown, D.R., Madhow, U., Bidigare, P., Dasgupta, S.: Receiver-coordinated distributed transmit nullforming with channel state uncertainty. In: Information Sciences and Systems (CISS), 2012 46th Annual Conference on, pp. 1–6, Mar 2012
18. Brown, D.R., Bidigare, P., Dasgupta, S., Madhow, U.: Receiver-coordinated zero-forcing distributed transmit nullforming. In: Statistical Signal Processing Workshop (SSP), 2012 IEEE, pp. 269–272, Aug 2012
19. Kumar, A., Dasgupta, S., Mudumbai, R.: Distributed nullforming without prior frequency synchronization. In: Proceedings of Australian Control Conference, Nov 2013

20. Peiffer, B., Mudumba, R., Goguri, S., Dasgupta, S., Kruger, A.: Experimental demonstration of nullforming from a fully wireless distributed array. In: Proceedings of ICASSP. New Orlans, LA, Mar 2017
21. Kumar, A., Mudumbai, R., Dasgupta, S., Madhow, U., Brown, D.R.: Distributed MIMO multicast with protected receivers: a scalable algorithm for joint beamforming and nullforming. IEEE Trans. Wirel. Commun. **16**(1), 512–525 (2017). https://doi.org/10.1109/TWC.2016.2625784
22. Kumar, A., Mudumbai, R., Dasgupta, S., Rahman, M.M., Brown, D.R., Madhow, U., Bidigare, P.: A scalable feedback mechanism for distributed nullforming with phase-only adaptation. IEEE Trans. Signal Inf. Process. over Netw. **1**, 58–70 (2015). https://doi.org/10.1109/TSIPN.2015.2442921
23. Ozgur, A., Lévêque, O., Tse, D.N.C.: Hierarchical cooperation achieves optimal capacity scaling in ad hoc networks. IEEE Trans. Inf. Theory **53**(10), 3549–3572 (2007)
24. Yucek, T., Arslan, H.: A survey of spectrum sensing algorithms for cognitive radio applications. IEEE Commun. Surv. Tutor. **11**(1), 116–130 (2009)
25. Dong, L., Han, Z., Petropulu, A.P., Poor, H.V.: Cooperative jamming for wireless physical layer security. In: Statistical Signal Processing, 2009: SSP'09. IEEE/SP 15th Workshop on, pp. 417–420. IEEE (2009)
26. Dasgupta, S., Anderson, B.D.O., Kaye, R.J.: Output error identification methods for partially known systems. Int. J. Control **43**, 177–191 (1986)
27. Fu, M., Dasgupta, S., Soh, Y. C.: Integral quadratic constraint approach vs. multiplier approach. Automatica, 281–287, Feb 2005
28. Anderson, B.D.O., Bitmead, R.R., Jr. Johnson, Kokotovic, P.V., Kosut, R.L., Mareels, I.M.Y., Praly, L., Riedle, B.D.: Stability of Adaptive Systems: Passivity and Averaging Analysis. MIT Press, Cambridge (1986)
29. Dasgupta, S., Anderson, B.D.O., Tsoi, A.C.: Input conditions for continuous time adaptive systems problems. IEEE Trans. Autom. Control, 78–82 (1990)
30. Rahman, M.M., Dasgupta, S., Mudumbai, R.: A scalable feedback-based approach to distributed nullforming. In: Proceedings of WICON. Shanghai, China, Apr 2013
31. Cao, M., Yu, C., Morse, A.S., Anderson, B.D.O., Dasgupta, S.: Generalized controller for directed triangle formations. In: Preprinits of IFAC World Congress. Seoul, Korea, July 2008
32. Dasgupta, S., Anderson, B.D.O.: Physically based parameterizations for designing adaptive algorithms. Automatica **23**, 469–477 (1987)
33. Hahn, W.: Stability of Motion. Springer, Heidelberg (1967)

Chapter 8
The "Power Network" of Genetic Circuits

Yili Qian and Domitilla Del Vecchio

Abstract Synthetic biology is an emergent research field whose aim is to engineer gene regulatory networks (GRNs) in living cells for useful functionalities. However, due to the unwanted interactions among nodes and with the cellular "chassis", behavior of a GRN is often context dependent. One source of context dependence that has received much attention recently is the competition among nodes in a GRN for a limited amount of resources provided by the host cell. In this paper, we review our recent research outcomes on the modeling and mitigation of resource competition in GRNs from a control theoretic perspective. In particular, we demonstrate that resource competition gives rise to hidden interactions among nodes that change the intended network topology. By regarding hidden interactions as disturbances, we formulate a network disturbance decoupling problem that can be solved by implementing decentralized feedback controllers in the nodes. Our results provide examples of how introducing concepts in systems and control theory can lead to solutions to pressing practical problems in synthetic biology.

8.1 Introduction

Synthetic biology is an emergent research field whose aim is to engineer novel genetic circuits in living cells. It has potential applications ranging from increasing biofuel production [14], to sensing environmental hazards [2], as well as detecting and/or killing cancer cells [6]. However, the efforts to engineer large scale and complex genetic circuits are often impeded by context dependence. Context dependence is the phenomenon in which functional genetic modules behave differently when they are connected in a circuit as opposed to when they are in isolation, often

Y. Qian · D. Del Vecchio (✉)
Department of Mechanical Engineering, Massachusetts Institutie of Technology, Cambridge, MA, USA
e-mail: ddv@mit.edu

Y. Qian
e-mail: yiliqian@mit.edu

R. Tempo et al. (eds.), *Emerging Applications of Control and Systems Theory*, Lecture Notes in Control and Information Sciences - Proceedings,
https://doi.org/10.1007/978-3-319-67068-3_8

due to unintended interactions among genes and with the cellular "chassis" [5]. These unintended interactions lead to *ad-hoc* design processes and unpredictable outcomes, largely hindering our capability to scale up genetic circuits for real-world applications. Therefore, much of the current research in the field is devoted to the mitigation of context dependence in genetic circuits [7]. These research questions can be formulated as classical systems and control theory problems such as disturbance attenuation and output regulation [8]. However, their solutions are often challenged by physical constraints in living cells [8]. In particular, feedback control needs to be realized by a limited set of available biomolecular processes. In this paper, we review our latest research outcomes on the modeling, analysis and mitigation of a specific form of context dependence that has received much attention recently: the competition among synthetic genes for a limited amount of cellular resources provided by the host cell [4, 11, 16].

Gene expression in vivo relies on the key processes of transcription and translation. Transcription is initiated by RNA polymerases (RNAPs) binding with the promoter sites of DNAs to produce messenger RNAs (mRNAs), which are then translated by ribosomes to produce target proteins. Genetic circuits are gene regulatory networks (GRNs) that are constructed by allowing transcription of one gene to be regulated by proteins expressed by other genes. These proteins are called transcription factors (TFs), which can either activate or repress transcription by binding with promoter sites on DNAs (Fig. 8.1a). Therefore, naturally, the functionality of any GRN requires the availability of RNAPs and ribosomes [9].

Recent experimental results have suggested that the limited amount of ribosomes is the main bottleneck for gene expression in *E. coli* bacteria in exponential growth phase [11, 12]. In particular, it has been demonstrated experimentally that, due to resource competition, expression of an unregulated gene can be reduced by more than 60% when expression of another gene is activated [4, 11, 19]. However, resource limitation is often neglected in standard gene expression models [1, 9], and rarely considered in circuit design, leading to unexpected design outcomes. Therefore, it is desirable to develop a systematic framework to model and mitigate the effects of resource competition in genetic circuits.

In this paper, we review our recent progress towards these goals [15–17]. In [15, 16], we developed and experimentally validated a general gene expression model accounting for resource competition. These results are briefly reviewed in Sects. 8.2 and 8.3. In Sect. 8.4, we review our recent theoretical work on mitigating the effects of resource competition by introducing a system concept and formulating the problem as a disturbance decoupling problem for networks [17]. The problem can be solved by implementing decentralized feedback controllers which are described in Sect. 8.5. Section 8.6 discusses the stability of the decentralized control scheme. Simulation examples are provided in Sect. 8.7.

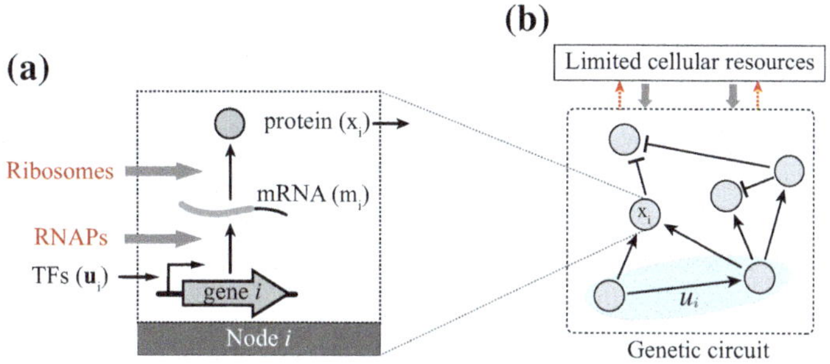

Fig. 8.1 **a**: Schematic of gene expression process in a node i. **b**: In a GRN, all nodes are competing for a limited amount of resources in the host cell

8.2 Modeling Resource Competition in Gene Networks

A GRN is an interconnection of gene expression cassettes, which we call *nodes*. Each node i is an input/output system that takes l_i TFs as inputs to regulate the production of a TF $\mathrm{x_i}$ as output (Fig. 8.1a). We denote the set of TF inputs to node i by $\mathcal{U}_i$ (Fig. 8.1b), and use $\mathbf{u}_i = [u_1, \cdots, u_{l_i}]^T$ to represent their concentrations. A series of chemical reactions take place in node i. The input TFs bind with its DNA promoter site to regulate mRNA ($\mathrm{m_i}$) transcription; free ribosomes then bind with mRNAs to initiate translation and produce output TF $\mathrm{x_i}$. Assuming that binding reactions are much faster than transcription and translation [1, 9], the state of node i can be described by the concentrations of its mRNA transcript m_i and TF output x_i (*italics*). In a standard gene expression model [1, 9], the free amount of ribosomes available for translation is assumed to be constant, yielding the following node dynamics:

$$\dot{m}_i = T_i F_i(\mathbf{u}_i) - \delta_i m_i, \qquad \dot{x}_i = R_i m_i - \gamma_i x_i, \tag{8.1}$$

where T_i is the maximum transcription rate constant, R_i is the translation rate constant proportional to the free concentration of ribosomes, and δ_i and γ_i are the mRNA and protein decay rate constants, respectively. The function $F_i(\mathbf{u}_i)$ describes regulatory effects of the input TFs on node i, and can be written as a standard Hill function [1, 9]. Specifically, based on reaction rate equations, when node i takes a single activator with concentration u_i as input, $F_i(u_i) = \frac{\beta_i + (u_i/k_i)^{n_i}}{1+(u_i/k_i)^{n_i}}$; if the TF input is a repressor, then $F_i(u_i) = \frac{1}{(u_i/k_i)^{n_i}}$. Parameter k_i is the dissociation constant of TF $\mathrm{u_i}$ binding with the promoter site of node i. The stronger the binding affinity, the smaller the dissociation constant. Parameter n_i is the binding Hill coefficient, and $\beta_i \in [0, 1)$ characterizes basal expression (i.e. expression when $u_i = 0$).

In a GRN, nodes are connected through regulatory interactions, where the output of node i ($\mathrm{x_i}$) can regulate gene expression in node j. In a GRN composed of n nodes, let $\mathbf{x} = [x_1, \cdots, x_n]^T$ represent the concentrations of all TFs in the network, we have $\mathbf{u}_i = Q_i \mathbf{x}$, where Q_i is a binary selection matrix, whose (j, k)-th element is 1 if x_k is the j-th input to node i, and 0 otherwise. When ribosomes are limited, the constant free ribosome assumption used to derive (8.1) fails. Since the host cell produces a limited amount of ribosomes [3], resource availability depends on the extent to which different nodes in the network demand them (Fig. 8.1b). We use z_T, z and z_i to denote the concentrations of total ribosomes, free ribosomes, and that of ribosomes bound to $\mathrm{m_i}$, respectively. In particular, we obtain $z_i = zm_i/\kappa_i$ from reaction rate equations, where smaller κ_i indicates stronger capability of $\mathrm{m_i}$ to sequester free ribosomes (z). Therefore, the concentration of ribosomes in a GRN follows the conservation law [3]

$$z_T = z + \sum_{i=1}^{n} z_i = z \cdot (1 + \sum_{i=1}^{n} m_i/\kappa_i), \tag{8.2}$$

indicating that the total ribosome concentration is the summation of its free concentration (z) and its concentration bound in the nodes (z_i). From Eq. (8.2), the free concentration of ribosomes z can be found as $z = \frac{z_T}{1+\sum_{i=1}^{n} m_i/\kappa_i}$, and by replacing the constant free ribosome concentration in (8.1) with this state-dependent free amount, the node dynamics in (8.1) can be modified as:

$$\dot{m}_i = T_i F_i(\mathbf{u}_i) - \delta_i m_i, \quad \dot{x}_i = R_i \frac{m_i/\kappa_i}{1 + m_i/\kappa_i + \sum_{j \neq i} m_j/\kappa_j} - \gamma_i x_i. \tag{8.3}$$

In the next section, we demonstrate that (8.3) implies that hidden interactions arise due to ribosome competition in GRNs.

8.3 Hidden Interactions and Effective Interaction Graphs

Synthetic biologists often analyze and design genetic circuits based on interaction graphs, which use directed edges to represent regulatory interactions among nodes. A standard interaction graph is drawn based on regulatory interactions among nodes. In particular, the *regulatory interaction* from $\mathrm{x_j}$ to $\mathrm{x_i}$ is determined by $\mathrm{sign}(\partial F_i/\partial x_j)$. If $\mathrm{sign}(\partial F_i/\partial x_j) > 0 (< 0)$, then the regulatory interaction is an activation (repression), represented by a $\mathrm{x_j} \rightarrow \mathrm{x_i}$ ($\mathrm{x_j} \dashv \mathrm{x_i}$). Here, we expand the concept of interaction graph to incorporate hidden interactions due to resource competition. We call the resultant graph the *effective interaction graph*. We then illustrate its applications to predict the behavior of two simple GRNs.

Since an interaction graph represents the interactions among TFs in a GRN and since mRNA dynamics are often much faster than those of the TFs [1, 9], we set the mRNA dynamics in (8.3) to quasi-steady state [1, 9] to obtain:

$$\dot{x}_i = \underbrace{\frac{\bar{T}_i F_i(\mathbf{u}_i)}{1 + J_i F_i(\mathbf{u}_i) + \sum_{k \neq i} J_k F_k(\mathbf{u}_k)}}_{G_i(\mathbf{x})\text{:node } i \text{ effective production rate}} - \gamma_i x_i, \tag{8.4}$$

where parameter $\bar{T}_i := R_i T_i / \delta_i \kappa_i$ represents the maximum protein production rate and parameter $J_i := T_i / \delta_i \kappa_i$ represents the resource sequestration capability in node i, which we call its *resource demand coefficient*.

The effective production rate of node i, $G_i(\mathbf{x})$, encapsulates the joint effects of regulatory interactions and hidden interactions due to resource competition. The *effective interaction* from node j to node i is determined based on sign$(\partial G_i / \partial x_j)$, representing how $\mathrm{x_j}$ affects the production rate of $\mathrm{x_i}$. In particular, we draw $\mathrm{x_j} \rightarrow \mathrm{x_i}$ ($\mathrm{x_j} \dashv \mathrm{x_i}$) if sign$(\partial G_i / \partial x_j) > 0$ (< 0). We draw $\mathrm{x_j} \multimap \mathrm{x_i}$ if sign$(\partial G_i / \partial x_j)$ is undetermined, that is, it depends on parameters and/or the steady state the network is operating at. We say there is a *hidden interaction* from $\mathrm{x_j}$ to $\mathrm{x_i}$ if $\partial F_i / \partial x_j = 0$ but $\partial G_i / \partial x_j \neq 0$.

Based on (8.4), we identify the effective interaction graphs of two simple GRNs shown in Fig. 8.2, where we use black solid edges to represent regulatory interactions and red dashed edges to represent hidden interactions. Figure 8.2a shows a single-input motif [1], where a TF input u represses two targets $\mathrm{x_1}$ and $\mathrm{x_2}$. According to a standard model of this network [1], the steady states of both x_1 and x_2 decrease with u. However, due to resource competition the effective interactions from u to its targets become undetermined and steady state x_1 increases with u for some u levels (Fig. 8.2a). In Fig. 8.2b, we demonstrate that the steady state i/o response of an activation cascade can become biphasic due to hidden interactions. This prediction has been verified experimentally [16]. We refer the readers to [15, 16] for detailed discussions on graphical rules to draw effective interactions and on how to mitigate hidden interactions by tuning resource demand coefficients.

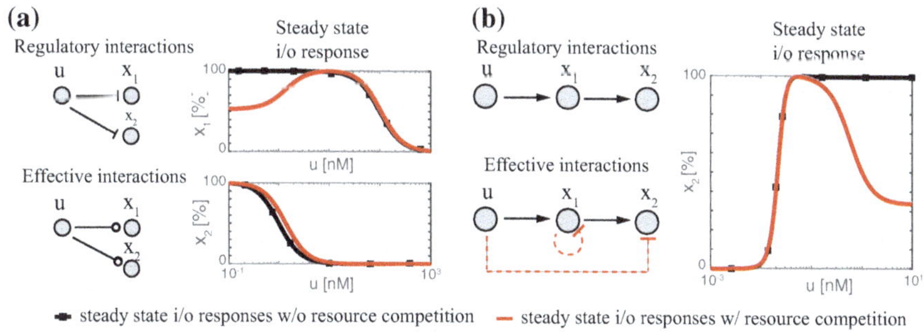

Fig. 8.2 Effects of resource competition on a single-input motif (**a**) and a two-stage activation cascade (**b**). Steady state i/o responses with resource competition are simulated according to model (8.3), and steady state i/o responses without resource competition are obtained by assuming $\sum_{j \neq i} m_j / \kappa_j = 0$ for all i in (8.3)

Due to the global feature of resource competition, its effects must be re-examined when additional nodes are added to the network, complicating the design process. Therefore, it is desirable to design a feedback controller that can automatically mitigate the effects of hidden interactions so that networks can be designed solely based on regulatory interactions, and additional nodes can be included in a "plug-and-play" fashion. In the next section, we formulate this problem as a network disturbance decoupling problem.

8.4 Mitigation of Hidden Interactions Through Network Disturbance Decoupling

In this section, we introduce a system concept that allows us to formulate the mitigation of hidden interactions as a network disturbance decoupling problem. The main difference between model (8.3) and the standard model in (8.1) is that in (8.3), x_i dynamics depend on the mRNA concentrations of other nodes ($m_j,\ j \neq i$). With reference to Fig. 8.3a, according to model (8.3), each node in the network can be regarded as a two-input-two-output system that takes a reference input $v_i := F_i(\mathbf{u}_i)$, a disturbance input w_i, a reference output $y_i := x_i$ and a disturbance output d_i. The disturbance input w_i represents how resource demand by the rest of the network affects node i, and the disturbance output d_i quantifies resource demand by node i (Fig. 8.3b). In particular, they follow

$$w_i := \sum_{j \neq i} m_j/\kappa_j, \qquad d_i := m_i/\kappa_i, \qquad w_i = \sum_{j \neq i} d_j. \tag{8.5}$$

We would like that the steady state reference output of each node (y_i) be only dependent on its own reference input v_i and essentially independent of disturbances among them (w_i). This concept, which we call *static network disturbance decoupling*, is visualized in Fig. 8.3c and described as follows.

We consider a GRN where the steady state i/o responses of each node are parametrized by a small parameter ε, and can be written as:

$$y_i = h_i(v_i, w_i, \varepsilon), \qquad d_i = g_i(v_i, w_i, \varepsilon). \tag{8.6}$$

Functions $h_i(\cdot)$ and $g_i(\cdot)$ are C^2 in ε for $(v_i, w_i, \varepsilon) \in \mathcal{V}_i \times \mathcal{W}_i \times (-\varepsilon^*, \varepsilon^*)$ with $\mathcal{V}_i \times \mathcal{W}_i \subseteq \mathbb{R}^2_+$, and $0 < \varepsilon^* \ll 1$. We further assume that each node is a positive i/o system such that for all $v_i, d_i > 0$, we have $y_i, w_i > 0$. We use the following notations: $\mathbf{V} := \mathcal{V}_1 \times \cdots \times \mathcal{V}_n$, $\mathbf{W} := \mathcal{W}_1 \times \cdots \times \mathcal{W}_n$, $\mathbf{v} := [v_1, \cdots, v_n]^T$, $\mathbf{w} := [w_1, \cdots, w_n]^T$, $\mathbf{y} := [y_1, \cdots, y_n]^T$ and $\mathbf{d} := [d_1, \cdots, d_n]^T$.

Definition 1 (*Network disturbance decoupling*) A network is said to have *local ε-static network disturbance decoupling property* in $\mathcal{V} \times \tilde{\mathcal{W}}$ if there exists an ε^* sufficiently small and an open set $\tilde{\mathcal{W}} \subseteq \mathbf{W}$ such that for all $i = 1, \cdots, n$,

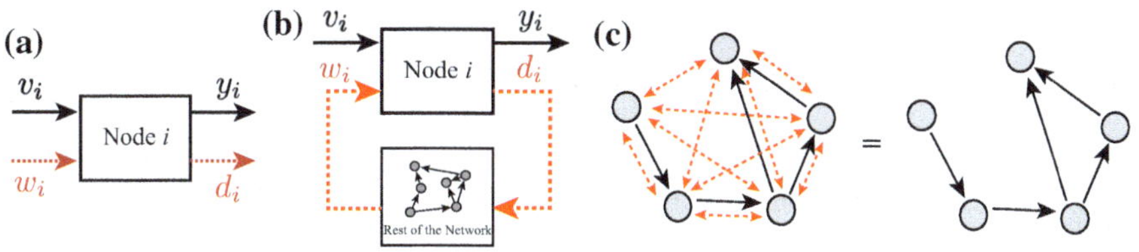

Fig. 8.3 **a**: Each node i in a GRN with resource competition can be treated as a two-input-two-output system. Black edges represent reference input/outputs due to regulatory interactions, and red dashed edges represent hidden interactions modeled as disturbances. **b**: In a network setting, disturbance input to node i (w_i) is produced by resource demand by the rest of the network; resource demand of node i (d_i) becomes disturbance inputs to the other nodes. **c** In a network with disturbance decoupling property, the steady state behavior of the network is essentially the same as that of a network without disturbances

$$y_i = h_i(v_i, w_i(\mathbf{v}, \varepsilon), \varepsilon) = h_i(v_i, 0, 0) + \mathcal{O}(\varepsilon). \tag{8.7}$$

In principle, network disturbance decoupling requires each node to possess some disturbance attenuation property. In addition, since $w_i = w_i(\mathbf{v}, \varepsilon)$ depends on resource demand by other nodes (d_j), we need to ensure that w_i is not amplified while we increase disturbance attenuation capability in each node. In what follows, we give algebraic conditions on the node and the network such that static network disturbance decoupling can be achieved. We first state the definition of *ε-static disturbance attenuation*, which is the property required for each node.

Definition 2 (*Node disturbance attenuation*) Node i has ε-static disturbance attenuation property in $\mathcal{V}_i \times \mathcal{W}_i$ if $h_i(v_i, w_i, 0) = h_i(v_i, 0, 0)$ for all $(v_i, w_i) \in \mathcal{V}_i \times \mathcal{W}_i$.

For a node with ε-static disturbance attenuation property, the effect of disturbance input on reference output is attenuated by a factor of ε, and can be written as $y_i = h_i(v_i, 0, 0) + \mathcal{O}(\varepsilon)$ for ε sufficiently small. This property does not require any information on the network, and is a characterization of node i in isolation. It can be achieved, for example, by implementing decentralized controllers in the nodes. When node i is part of the network, since w_i also depends on ε, it may grow unbounded as $\varepsilon \to 0$, in principle. The next property on the network excludes this possibility and guarantees that w_i is smooth in ε as $\varepsilon \to 0$.

Definition 3 (*Network ε -well-posedness*) Consider a network where the nodes are connected according to $w_i = \sum_{j \neq i} d_j$, and the steady state i/o response of each node follows (8.6). It is *locally ε-well-posed* in $\mathcal{V} \times \mathcal{W} \subseteq \mathbf{V} \times \mathbf{W}$ if there is an open set $\mathcal{W} \subseteq \mathbf{W}$ and $\varepsilon^* > 0$ such that there exists $\mathbf{w}(\mathbf{v}, \varepsilon) \in \mathcal{W}$ that satisfies

$$w_i = \sum_{j \neq i} g_j(v_j, w_j, \varepsilon), \text{ for all } i \in \{1, \cdots, n\}. \tag{8.8}$$

Furthermore, $\mathbf{w}(\mathbf{v}, \varepsilon)$ is $\mathcal{C}^1$ in ε for all $(\mathbf{v}, \mathbf{w}, \varepsilon) \in \mathcal{V} \times \mathcal{W} \times (-\varepsilon^*, \varepsilon^*)$.

The following result provides sufficient conditions for local ε-static network disturbance decoupling (see [17] for details).

Theorem 1 *A network has local ε-static network disturbance decoupling property in $\mathcal{V} \times \mathcal{W}$ if (i) each node i has ε-disturbance attenuation property in $\mathcal{V}_i \times \mathcal{W}_i$, and (ii) the network is locally ε-well-posed in $\mathcal{V} \times \mathcal{W}$.*

While condition (i) in Theorem 1 can be obtained by implementing decentralized controllers in the nodes, condition (ii) needs to be certified by exploring more properties of the network. We further assume that when $\varepsilon = 0$, steady state disturbance output of each node (d_i) is affine in its disturbance input (w_i) for all $(\mathbf{v}, \mathbf{w}) \in \mathcal{V} \times \mathcal{W}$

$$g_i(v_i, w_i, 0) = \bar{g}_i(v_i) + \hat{g}_i(v_i)w_i. \tag{8.9}$$

In this case, the local ε-well-posedness property of a network can be certified by the diagonal dominance of an interconnection matrix A, whose (j, k)-th element is defined to be 1 if $j = k$ and $-\hat{g}_k(v_k)$ if $j \neq k$. This result is stated in the following theorem, and we refer the readers to [17] for detailed proofs and discussions.

Theorem 2 *Assume that the steady state disturbance i/o response of each node i follows (8.9) and that the nodes are connected by $w_i = \sum_{j \neq i} d_j$, if the interconnection matrix A is diagonally dominant for all $\mathbf{v} \in \mathcal{V}$, then there exists an open set $\mathcal{W}$ such that the network is locally ε-well-posed in $\mathcal{V} \times \mathcal{W}$.*

The diagonal dominance condition in Theorem 2 resembles a network small-gain theorem for stability test [18]. However, in the context of [18], the elements in the interconnection matrix are dynamic i/o gains of the individual nodes, while here, the elements in interconnection matrix A are static i/o gains. Therefore, instead of guaranteeing stability, Theorem 2 guarantees, roughly speaking, that a steady state of the network remains $\mathcal{O}(1)$ as $\varepsilon \to 0$. In the next section, we propose a biomolecular feedback controller design that guarantees ε-static network disturbance decoupling in resource-competing GRNs. We then discuss its stability in Sect. 8.6.

8.5 Biomolecular Realization of Disturbance Decoupling Through Decentralized sRNA-Based Feedback

Small RNAs (sRNAs) are short non-coding RNAs that can bind to complementary RNAs to induce rapid degradation [13]. In this section, we demonstrate that decentralized sRNA-based feedback controllers (Fig. 8.4) can achieve ε-static network disturbance decoupling. We first introduce the biomolecular mechanism of this controller and then show that each node has ε-static disturbance attenuation property and the resulting network is ε-well-posed.

A diagram of node i equipped with the proposed sRNA-based feedback is shown in Fig. 8.4b. In node i, the output protein $\mathrm{x_i}$ activates production of sRNA $\mathrm{s_i}$. sRNA

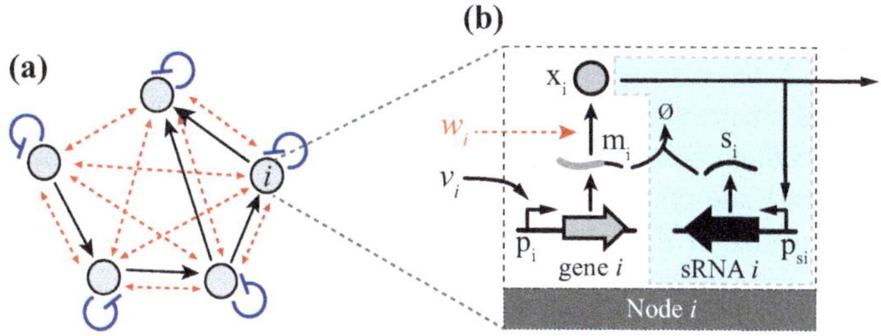

Fig. 8.4 **a**: Decentralized implementation of sRNA-based feedback controllers in a GRN. **b**: Biomolecular mechanism of the controller, where elements of the controller are shaded in blue

s_i binds with mRNA m_i to form a complex which degrades rapidly [13]. When ribosome availability decreases, less x_i is produced, resulting in reduced production of s_i, which in turn increases mRNA concentration to compensate for reduction in ribosome concentration. Using m_i, s_i and x_i to represent the concentrations of mRNA, sRNA and TF produced in node i, respectively, an ODE model of node i can be written as follows:

$$\dot{m}_i = GT_i v_i - \delta m_i - G\frac{m_i s_i}{k_i}, \quad \dot{s}_i = GT_{si}F_{si}(x_i) - \delta s_i - G\frac{m_i s_i}{k_i}, \quad \dot{x}_i = R_i H_i(m_i, w_i) - \gamma x_i, \tag{8.10}$$

where $F_{si}(x_i) = \frac{x_i/k_{si}}{1+x_i/k_{si}}$, $H_i(m_i, w_i) = \frac{m_i/\kappa_i}{1+m_i/\kappa_i+w_i}$ and $w_i = \sum_{j\neq i} m_j/\kappa_j$ from (8.5). In (8.10), δ and γ are the decay rate constants of mRNAs and proteins, respectively, which we assume to be the same for all nodes without loss of generality; k_i and k_{si} characterizes mRNA-sRNA binding, and the binding of x_i with the sRNA promoters, respectively; and $G \gg \delta$ represents the rapid degradation of the mRNA-sRNA complex. To compensate for the decrease in gene expression due to rapid degradation (G) of mRNAs and sRNAs, we set their transcription rates (GT_i and GT_{si}) to scale with G. By letting $\varepsilon := \delta/G$, the steady state of m_i and x_i can be written as:

$$m_i = \frac{T_i\kappa_i k_{si}\gamma v_i(1+w_i)}{T_{si}R_i - (\gamma k_{si} + R_i)T_i v_i} + \mathcal{O}(\varepsilon), \quad x_i = \frac{T_i k_{si} v_i}{T_{si} - T_i v_i} + \mathcal{O}(\varepsilon), \tag{8.11}$$

when v_i belongs to the *node admissible input set* $\mathcal{V}_i = \left\{v_i : 0 < v_i < \frac{T_{si}R_i}{T_i(\gamma k_{si}+R_i)}\right\}$. Since the zero-th order approximation of steady state output x_i in (8.11) is independent of disturbance input w_i, node i has ε-disturbance attenuation property for all $v_i \in \mathcal{V}_i$ and w_i positive. To verify if the GRN with decentralized sRNA-based controllers is locally ε-well-posed, we note that the steady state disturbance output of node i is: $d_i = \frac{m_i}{\kappa_i} = \frac{T_i k_{si}\gamma v_i(1+w_i)}{T_{si}R_i-(\gamma k_{si}+R_i)T_i v_i} + \mathcal{O}(\varepsilon)$, which is affine in w_i when $\varepsilon = 0$. This satisfies the assumption in (8.9) with $\hat{g}_i(v_i) = \frac{T_i k_{si}\gamma v_i}{T_{si}R_i-(\gamma k_{si}+R_i)T_i v_i}$.

Thus, according to Theorem 2, the network is ε-well-posed in a *network admissible input set* $\mathcal{V}$, if the interconnection matrix A is diagonally dominant for all $\mathbf{v} \in \mathcal{V}$.

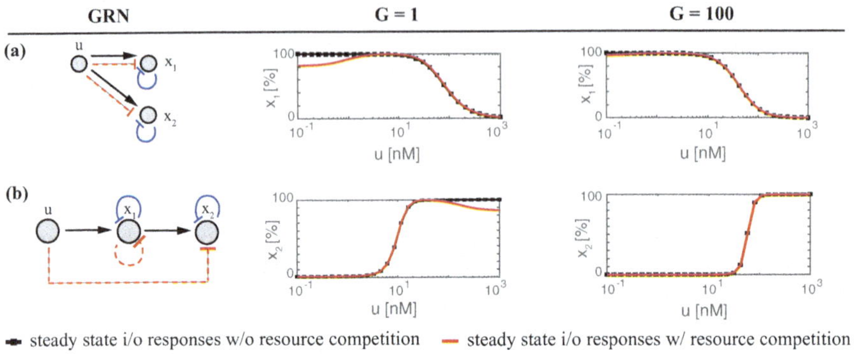

Fig. 8.5 Steady state i/o responses of the single-input motif (**a**) and the activation cascade (**b**) with different G values. We use model (8.5) and (8.10) for simulation, and assume $w_i = 0$ for all i to obtain the responses without resource competition

closer to the hypothetical case where disturbance inputs to all nodes are assumed to be zero (i.e. $w_i = 0$ for all i). These simulations support our claim that decentralized sRNA-based feedback can increase network's robustness to resource competition.

8.8 Discussion and Conclusions

In this paper, we demonstrate that resource competition creates hidden interactions in GRNs that change the network's intended topology. To mitigate the effects of these hidden interactions, we take a control theoretic perspective and formulate a static network disturbance decoupling problem. We demonstrate that this problem can be solved by implementing decentralized sRNA-based feedback controllers in vivo. While these results are promising, we still need to tackle the case in which the inputs to the nodes are time-varying and the results we obtain are global instead of being local. These problems are particularly difficult due to the nonlinear and singular structure of the dynamics when feedback gains are increased, and a solution will likely require the development of novel control theoretic methods.

Our results provide examples of how introducing systems and control concepts can help address concrete problems in synthetic biology. In general, synthetic biology is an exciting platform to leverage existing control theoretic tools and to introduce new ones. Due to various sources of uncertainties and disturbances present in living cells, the advancement of synthetic biology relies heavily on our capabilities to engineer robust genetic circuits. While such capabilities are still largely missing at this stage, they can be significantly improved by synthesizing feedback controllers in cells.

Acknowledgements We would like to thank Ms. Rushina Shah who assisted in the proof-reading of the manuscript. This work was supported by AFOSR grant FA9550-14-1-0060.

References

1. Alon, U.: An Introduction to Systems Biology: Design Principles of Biological Circuits. Chapman & Hall/CRC Press (2006)
2. Bereza-Malcolm, L.T., Mann, G., Franks, A.E.: Environmental sensing of heavy metals through whole cell microbial biosensors: a synthetic biology approach. ACS Synth. Biol. **4**(5), 535–546 (2015)
3. Bremer, H., Dennis, P.P.: Modulation of chemical composition and other parameters of the cell by growth rate. In: Neidhardt, F.C. (ed.) Escherichia coli and Salmonella: Cellular and Molecular Biology. ASM Press (1996)
4. Carbonell-Ballestero, M., Garcia-Ramallo, E., Montañez, R., Rodriguez-Caso, C., Macía, J.: Dealing with the genetic load in bacterial synthetic biology circuits: convergences with the ohm's law. Nucleic Acids Res. **44**(1), 496–507 (2015)
5. Cardinale, S., Arkin, A.P.: Contextualizing context for synthetic biology- identifying causes of failure of synthetic biological systems. Biotechnol. J. **7**, 856–866 (2012)
6. Danino, T., Prindle, A., Kwong, G.A., Skalak, M., Li, H., Allen, K., Hasty, J., Bhatia, S.N.: Programmable probiotics for detection of cancer in urine. Sci. Transl. Med. **7**(289), 289ra84–289ra84 (2015)
7. Del Vecchio, D.: Modularity, context-dependence, and insulation in engineered biological circuits. Trends Biotechnol. **33**(2), 111–119 (2015)
8. Del Vecchio, D., Dy, A.J., Qian, Y.: Control theory meets synthetic biology. J. R. Soc. Interface **13**, 20160380 (2016)
9. Del Vecchio, D., Murray, R.M.: Biomolecular Feedback Systems. Princeton University Press, Princeton (2014)
10. Fax, J.A., Murray, R.M.: Graph Laplacians and stabilization of vehicle formations. In: Proceedings of the 15th IFAC World Congress, vol. 35, pp. 55–60 (2002)
11. Gyorgy, A., Jiménez, J., Yazbek, J., Huang, H.-H., Chung, H., Weiss, R., Del Vecchio, D.: Isocost lines describe the cellular economy of gene circuits. Biophys. J. **109**, 639–646 (2015)
12. Klumpp, S., Dong, J., Hwa, T.: On ribosome load, codon bias and protein abundance. PLoS One **7**(11), e48542 (2012)
13. Levine, E., Zhang, Z., Kuhlman, T., Hwa, T.: Quantitative characteristics of gene regulation by small RNA. PLoS Biol. **5**(9), e229 (2007)
14. Peralta-Yahya, P.P., Zhang, F., del Cardayre, S.B., Keasling, J.D.: Microbial engineering for the production of advanced biofuels. Nature **488**, 320–328 (2012)
15. Qian, Y., Del Vecchio, D.: Effective interaction graphs arising from resource limitations in gene networks. In: Proceedings of the American Control Conference (ACC), pp. 4417–4423 (2015)
16. Qian, Y., Huang, H.-H., Jiménez, J.I., Del Vecchio, D.: Resource competition shapes the response of genetic circuits. ACS Synth. Biol. **6**(7), 1263–1272 (2017)
17. Qian, Y., Del Vecchio, D.: Mitigation of ribosome competition through distributed sRNA feedback. In: IEEE 55th Conference on Decision and Control (CDC), pp. 758–763 (2016)
18. Vidyasagar, M.: Input-output analysis of large-scale interconnected systems. Lecture Notes in Control and Information Sciences, vol. 29. Springer-Verlag, New York (1981)
19. Vind, J., Sørensen, M.A., Rasmussen, M.D., Pedersen, S.: Synthesis of proteins in Escherichia coli is limited by the concentration of free ribosomes: expression from reporter gene does not always reflect functional mRNA levels. J. Mol. Biol. **231**, 678–688 (1993)

Chapter 9
Controlling Biological Time: Nonlinear Model Predictive Control for Populations of Circadian Oscillators

John H. Abel, Ankush Chakrabarty and Francis J. Doyle III

Abstract In mammals, circadian regulation of gene expression is accomplished within each cell through a transcriptional oscillator commonly modeled by a limit cycle. There has been recent interest in regulating this oscillator by delivering doses of pharmaceuticals or light in a systematic manner. Generally, controller design for circadian manipulation has been formulated by considering the dynamics of a single oscillator representing the average dynamics of the population. We illustrate in this paper that such an approximation can result in desynchronization of circadian oscillators even if the mean dynamics attain desired behavior, due to the range of dynamic responses elicited among oscillators in a population with nonidentical phases. To address this issue, we present herein nonlinear MPC for control of phase and synchrony within a population of uncoupled circadian oscillators, by explicitly predicting the evolution of the phase probability density function. We then demonstrate in silico phase shifting of an example oscillator population while maintaining a high degree of synchrony. The MPC strategy formulated herein is a step toward a detailed, systems approach integrating population effects, pharmacokinetics and pharmacodynamics, and interactions between different oscillator populations.

J. H. Abel
Department of Systems Biology, Harvard Medical School,
Boston, MA 02115, USA
e-mail: johnhabel@g.harvard.edu

A. Chakrabarty
Harvard John A. Paulson School of Engineering and Applied Sciences,
Harvard University, Cambridge, MA 02138, USA
e-mail: achakrabarty@g.harvard.edu

F. J. Doyle III (✉)
Harvard John A. Paulson School of Engineering and Applied Sciences,
Harvard University, Cambridge, MA 02138, USA
e-mail: frank_doyle@seas.harvard.edu

F. J. Doyle III
Division of Sleep Medicine, Harvard Medical School, Boston, MA 02115, USA

R. Tempo et al. (eds.), *Emerging Applications of Control and Systems Theory*, Lecture Notes in Control and Information Sciences - Proceedings,
https://doi.org/10.1007/978-3-319-67068-3_9

9.1 Introduction

Disruption to the circadian clock, such as misaligned light or feeding, or extended night shift work, has been associated with a wide range of disorders such as decreased cognitive function [6, 13, 15]. Circadian rhythms are endogenous oscillatory processes involved in biological timekeeping and temporal regulation of biological function in nearly all forms of life [7]. Although the mechanism driving these oscillations varies between organisms, circadian rhythms share the characteristics of being endogenously generated, entrainable to environmental rhythms, and maintain a near-24h periodicity. In mammals, a transcriptional–translational oscillator present in nearly every cell type regulates transcriptional architecture, and along with further posttranscriptional regulation, intricately coordinates cellular and tissue function [23]. These cell-autonomous oscillators regulate gene expression within local tissue and, in turn, are coordinated by signals from the hypothalamic suprachiasmatic nucleus (SCN) which responds to light cues to entrain to the environment [26]. The sleep-wake cycle is the most immediately recognizable feature of circadian regulation, however, it is estimated that 5–20% of all transcripts within any mammalian cell type oscillate with a circadian frequency [28], and circadian regulation is broadly integrated with metabolic function [6, 16].

Low-amplitude circadian oscillation within tissue may result from perturbations to the oscillator driving lower amplitudes within individual cells, or driving desynchronization of the oscillators comprising the tissue population [22, 25]. Meanwhile, high-amplitude circadian oscillations are associated with good metabolic health [17]. Thus, there is a need to develop therapies to mitigate the effects of disruption to the circadian clock, to rapidly realign circadian phase with the environment following jet lag or shift work to avoid cognitive difficulty, and to promote high-amplitude circadian oscillation. Light, and more recently, small-molecule pharmaceuticals [10] provide possible two paths toward control of the clock. Since circadian rhythms are highly complex phenomena, phase and amplitude control of the clock may be achieved by a control theoretic approach to timing the delivery of drugs or light [5, 18, 27].

Control strategies devised for the manipulation of mammalian circadian rhythms may analogously be applied to other oscillatory biological systems, provided they are well described by limit cycle oscillators. Circadian oscillators in other species such as the KaiABC system in the cyanobacterium *Synechococcus* [12], or the *Period-Timeless* oscillator in *Drosophila* [8] have been modeled as limit cycle oscillators for more than a decade. More generally, genetic or phosphorylation-driven oscillators are a ubiquitous biological motif involved in the metabolic processes of numerous organisms from prokaryotes to mammals [9, 11], and the development of strategies for manipulating these systems is broadly desirable.

Several previous studies have examined the phase control of circadian rhythms. Several recent studies have focused on the application of light to drive phase shifts in the *Drosophila* [19, 20] or mammalian [18, 27] clock using an optimal control approach. However, solving the resulting optimal control problem is computationally prohibitive due to the inherent nonlinearity of the model, and the computed optimal control trajectory is susceptible to modeling errors and disturbances. Iteratively solving finite-horizon optimal control problems (as in model predictive control (MPC)) has provided demonstrably superior control performance. For example, MPC was used for shifting of the *Drosophila* clock [4, 5] or the mammalian clock [2]. This work uses nonlinear MPC as in [2] as a starting point for developing control of an oscillator population. These studies of the control of biological oscillators have focused on control of a single oscillator despite notable mismatch between single-cell and population dynamics [14]. Additionally, these studies have primarily focused on the application of light to the clock. Recently, pharmaceuticals have presented advantages for control of the clock, as they are much less invasive than strict control of an individual's light environment, and may be delivered at any time of day. A more detailed approach to control of circadian dynamics would integrate tissue or population-scale effects, pharmacokinetics and pharmacodynamics, and interactions between different oscillator populations.

Herein, we begin to approach this detailed formulation by presenting a MPC framework for manipulating phase and synchrony within populations of uncoupled biological oscillators using a pharmaceutical agent, and demonstrate the efficacy of such an approach by in silico simulations of phase shifting in the mammalian clock. We do so by describing the population of oscillators as a phase probability density function (PDF), as in [22], and using the parametric infinitesimal phase response curve (PRC) to calculate how the phase PDF evolves in response to control input. By using a predictor that accounts explicitly for the variability in phase within a synchronized population, we are able to maintain synchrony of oscillators while performing a phase shift, unlike the single-oscillator case.

Chapter Overview

In this chapter, we present a framework for the control of a population of biological oscillators, motivated by the example of the mammalian circadian clock. In Sect. 9.2.1, we present an established model of the mammalian circadian oscillator and its response to the small molecule KL001, an early [10, 21]. In Sect. 9.2.2, we motivate and formulate the phase shifting control problem. We use a previously derived simplification of oscillator dynamics using the parametric phase response to a control input to reduce the dynamical model to a phase-only representation [2]: this is advantageous in reducing the dimensionality of the MPC problem, thereby curbing computational effort. In Sect. 9.2.3, we describe a controller for the case for a single oscillator, and demonstrate its function in silico in Sect. 9.2.4. In Sect. 9.3.1, we demonstrate that although this formulation may successfully control a single oscillator, applying this controller to manipulate mean phase of a population of oscillators may effect a desynchronization detrimental to biological function. We then modify the MPC problem in Sect. 9.3.2 using a probability density function of population phase in conjunction with the simplified dynamics to exert simultaneous control over phase and synchrony within an oscillator population. We demonstrate in Sect. 9.3.3

the in silico efficacy of this approach in maintaining synchrony throughout a phase realignment. Finally in Sect. 9.4, we discuss limitations of this approach, and challenges to its implementation in vivo and in vitro.

9.2 Control of a Single Circadian Oscillator

A standard approximation in the control of the circadian oscillator is describing the targeted circadian system (i.e., the population of cells comprising clocks in brain or peripheral tissues such as the liver) as a single limit cycle oscillator [2, 4, 5, 18–20, 27]. We begin our treatment of applying control to phase shift the circadian oscillator by exploiting this approximation, as in our previous work [2], to illustrate where it fails to capture population-scale phenomena.

9.2.1 *Modeling the Circadian Oscillator*

The mammalian circadian oscillator within an individual cell is comprised of interlocked transcriptional–translational feedback loops. The core negative feedback loop involves isoforms of the genes *Period* (*Per*) and *Cryptochrome* (*Cry*), which form PER-CRY heterodimers and enter the nucleus to bind to BMAL1-CLOCK E-box activators to repress their own transcription. As these repressors are degraded, BMAL1-CLOCK dimers activate transcription of *Per*s and *Cry*s, resulting in a self-sustained oscillation. Downstream of the circadian feedback loops, clock components regulate transcriptional architecture through D-box, E-box, and ROR-binding elements. For an excellent review of the genetic components of the mammalian oscillator, see [23].

Numerous dynamical models of the circadian oscillator have been proposed. In this work, we selected the model from [10, 21], as it was created to identify the effect of the small-molecule KL001 on the mammalian oscillator. This model consists of 8 nonlinear coupled ODEs and 21 kinetic parameters, fully described in the supplement to [21] and not reproduced here due to space considerations. Because KL001 was found to stabilize nuclear PER-CRY transcription factors, control is implemented in the model by modifying the ODEs describing the degradation of PER-CRY dimers as follows:

$$\frac{d\mathrm{C1P}}{dt} = \nu_{\mathrm{a,CP}}\mathrm{P}\cdot\mathrm{C1} - \nu_{\mathrm{d,CP}}\mathrm{C1P} - \frac{(vdCn - u(t))\mathrm{C1P}}{k_{\mathrm{deg,Cn}} + \mathrm{C1P} + \mathrm{C2P}} \tag{9.1a}$$

$$\frac{d\mathrm{C2P}}{dt} = \nu_{\mathrm{a,CP}}\mathrm{P}\cdot\mathrm{C2} - \nu_{\mathrm{d,CP}}\mathrm{C2P} - \frac{(vdCn - u(t))m_{\mathrm{C2N}}\mathrm{C2P}}{k_{\mathrm{deg,Cn}} + \mathrm{C2P} + \mathrm{C1P}}, \tag{9.1b}$$

where $u(t) \in [0, \bar{u}]$ is control input at time t, which reduces degradation rate $vdCn$. Generally $\bar{u}$ should not be greater than $vdCn$, as a degradation reaction cannot be reversed so far as to synthesize new PER-CRY. Throughout, we set this value to 0.08. The states present in this model, and the effect of KL001, are shown in the schematic in Fig. 9.1a.

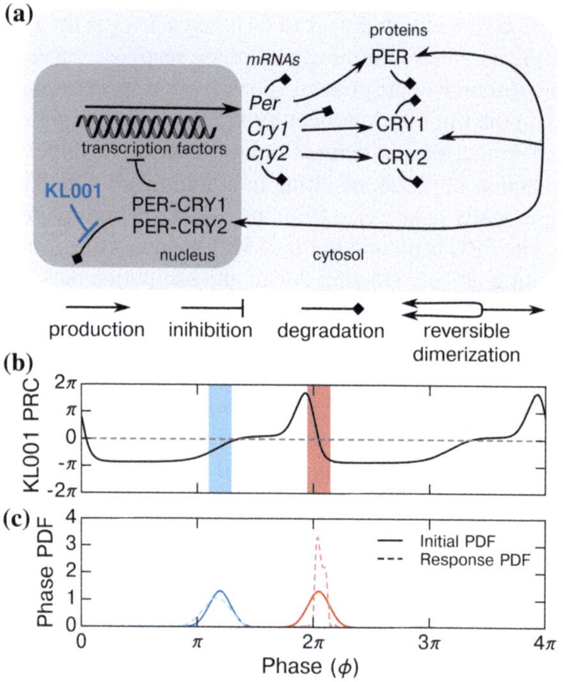

Fig. 9.1 Schematic of mammalian circadian oscillator and effect of KL001. **a** Diagram of the core negative feedback loop driving mammalian circadian rhythms. All states and reactions shown are included explicitly in the model [21]. **b** Parametric infinitesimal PRC describing response to KL001 (reduction in parameter $vdCn$). Note that this is double plotted to allow visualization. **c** Synchrony of a population is affected by timing of KL001 application. Two example probability density functions (PDFs) of phase are plotted before (solid) and after (dashed) application of KL001 in silico at the phases shown. In regions of positive slope (blue), the phase PDF is dispersed, whereas in regions of negative slope (red), the PDF is condensed, even though the mean phase is unchanged

9.2.2 *Model Reduction and the Phase Control Problem*

Control of the circadian clock is primarily focused on shifting the phase of the circadian oscillator. A unique phase $\phi \in [0, 2\pi)$ may be assigned to each unique point on the limit cycle. An unperturbed oscillator on the limit cycle will advance in phase at a constant rate. Thus, points in state space that are not on the limit cycle may be assigned the phase of the point on the limit cycle to which they converge asymptotically in time [2].

The 8-ODE oscillator may be reduced to a single-ODE describing phase [24]:

$$\frac{d\phi}{dt} = \frac{2\pi}{\tau} - u(t)\,\frac{d}{dt}\frac{d\phi}{d(vdCn)}\bigg|_{\mathbf{x}^{\gamma}}, \tag{9.2}$$

where τ is the period of oscillation (set to 24 h here), $u(t)$ is the control input, and $-\frac{d}{dt}\frac{d\phi}{d(vdCn)}\Big|_{\mathbf{x}^{\gamma}}$ is the parametric infinitesimal phase response curve (PRC), the first-order Taylor approximation of the phase response with respect to changing parameter $vdCn$, evaluated on the limit cycle trajectory $\mathbf{x}^{\gamma}$. The negative sign here is included in the PRC to reflect that $u(t)$ is defined to have a positive value. Importantly, the PRC is itself a function of phase, resulting in a nonlinear ODE. This function may be calculated numerically in advance from the 8-ODE model by previously defined methods [2, 24]. The PRC is plotted in Fig. 9.1b. One may consider the control input to be either "speeding up" or "slowing down" the oscillation depending on the sign of the PRC. We opted to set the reference $\phi = 0$ where the concentration of *Per* mRNA is at a maximum.

As in [2], we are aiming to align the phase of the circadian oscillator (ϕ) with an external tracking phase (denoted ϕ_e), for example, the phase of the environment before and after a plane flight through multiple time zones or before and after beginning a shift work cycle. This externally imposed phase may be captured by a single ODE as well:

$$\frac{d\phi_e}{dt} = \frac{2\pi}{\tau_e} + \Delta\phi\,\delta(t_{shift}), \tag{9.3}$$

where τ_e is the period of the environment (set to 24 h), and $\Delta\phi \in [-\pi, \pi)$ is the phase shift of the environment that occurs at time t_{shift}, for example, when disembarking from the flight or starting a period of shift work. We are searching for a control trajectory that will drive the oscillator ϕ to external phase ϕ_e and maintain it at ϕ_e for all subsequent time. That is:

$$\lim_{t\to\infty} \|\phi(t, u) - \phi_e(t)\| = 0.$$

9.2.3 Control of Single-Oscillator Phase

While light may be applied or removed instantaneously, pharmaceutical perturbation of the clock is constrained by pharmacokinetics. To alleviate numerical complications arising due to unidentified (and likely nonlinear) pharmacokinetics, we selected a piecewise constant parameterization of the control:

$$u(t) = u(t_k)\ \forall\ t_k \le t < t_{k+1}, \tag{9.4}$$

and denote the sampling time $t_{k+1} - t_k$ as τ_u. For a predictive horizon of N_p steps of duration τ_u, we define

$$U \triangleq \left[u(t_k)\ u(t_k + \tau_u)\ \cdots\ u\left(t_k + (N_p - 1)\tau_u\right)\right]^{\top} \tag{9.5}$$

as the knots of the control trajectory defined at each of the N_p steps. To estimate the phase at the end of each step, the phase dynamics in (9.2) may be integrated over each of the $\ell \in [1 \cdots N_p]$ steps to yield:

$$\hat{\phi}(t_i + \ell\tau_u) = \phi(t_i) + \frac{2\pi\,\ell\tau_u}{\tau} - \sum_{k=0}^{\ell-1} \int_{t_{k+i}}^{t_{k+1+i}} u(t_k) \frac{d}{dt} \frac{d\phi}{d(vdCn)}\bigg|_{\mathbf{x}^\gamma} dt, \tag{9.6}$$

where $\hat{\phi}$ is the predicted phase, t_i is the current time, $\phi(t_i)$ is measured, and the PRC is a function of $\hat{\phi}$. We define the phase error to be the magnitude of the phase difference between the predicted phase and the environmental phase:

$$e_\phi(\cdot) \triangleq \left| \angle \left(\exp(\mathrm{i}\hat{\phi}(\cdot) - \mathrm{i}\phi_e(\cdot)) \right) \right|, \tag{9.7a}$$

where $\mathrm{i} = \sqrt{-1}$. Computationally, driving phase error to 0 is numerically unstable, so we relax the terminal constraint by ignoring phase error below a constant δ_ϕ:

$$g_\phi(\cdot) \triangleq \begin{cases} 0 & \text{if } e_\phi(\cdot) < \delta_\phi \\ e_\phi(\cdot) & \text{otherwise} \end{cases} \tag{9.7b}$$

so that numerical imprecisions do not result in controller action. Thus, the finite-horizon optimal control problem at each time t_j may be solved for the optimal trajectory $u^\star$:

$$u^\star = \arg\min_U \sum_{\ell=1}^{N_p} w_\ell^\phi g_\phi^2(t_i + \ell\tau_u) + w_{\ell-1}^u u^2(t_i + (\ell - 1)\tau_u)$$

subject to:

$$\hat{\phi}(t_i + \ell\tau_u) = \phi(t_i) + \frac{2\pi\,\ell\tau_u}{\tau} - \sum_{k=0}^{\ell-1} \int_{t_{k+i}}^{t_{k+1+i}} u(t_k) \frac{d}{dt} \frac{d\phi}{d(vdCn)}\bigg|_{\mathbf{x}^\gamma} dt, \tag{9.8}$$

$$0 \leq u_{\ell-1} \leq \bar{u},$$

for all $\ell = 1, \cdots, N_p$, where w_ℓ^ϕ and $w_{\ell-1}^u$ are positive weighting scalars evaluated at the end of the time step and start of the time step, respectively, as phase error is calculated after the control is applied for that step. After identifying the optimal piecewise control trajectory $u^\star$, we applied $u^\star(t_i)$ to the full 8-state ODE model for $t \in (t_i, t_i + \tau_u]$ as is standard in model predictive control.

For this MPC formulation, w^ϕ, w^u, N_p, and τ_u are design parameters which may vary. We have selected:

$$w_\ell^\phi = \ell,$$
$$w_{\ell-1}^u = 1,$$

for $\ell = 1, \cdots, N_p$. We have elsewhere investigated optimal selection of design variables N_p and τ_u for this exact formulation of MPC [1]. Based on these findings, we set $N_p = 3$ (and $N_u = N_p$ where N_u is the number of length of the control horizon) and $\tau_u = 2\,\text{h}$, and instead studied the how phase control of a single oscillator differs from an oscillator population.

9.2.4 Case Study #1: Nonlinear MPC for a Single Oscillator

To demonstrate the behavior of this controller, we applied it for the phase shifting control problem where a phase delay of 6 h occurs at 12h (a phase delay of $\pi/4$ at a phase of π). The environmental phase for this example was given by (9.3), where $\Delta\phi = -\pi/2$, and $t_{shift} = 12\,\text{h}$. We used the Python language to solve the MPC problem, specifically, we used CasADi [3] and SciPy for formulation, and PySwarm to solve the nonlinear programming optimization problem. Here, the numerical error in calculating phase was $\mathscr{O}(10^{-2})$ and so design variable $\delta_\phi = 0.1$.

The results of this simulation are shown in Fig. 9.2. Briefly, the controller began its action at $t = 12\,\text{h}$ to achieve a phase delay. The full 6 h phase delay could not be achieved within the initial negative region, and so the remainder of the shift was accomplished when the PRC returned to a negative value near 30h. A decrease in amplitude occurred near $t = 30\,\text{h}$ due to transient deviation from the limit cycle, and the full amplitude returned for the following peak.

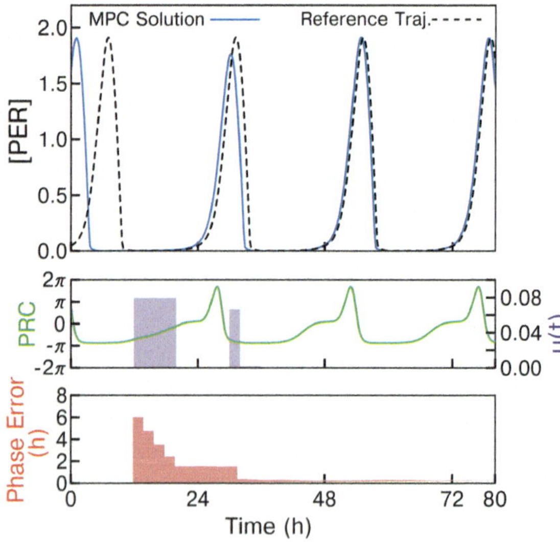

Fig. 9.2 Application of the controller detailed in Sect. 9.2.3 for a phase delay of 6 h ($\Delta\phi = -\pi/2$) with $t_{shift} = 12\,\text{h}$. The controller acts immediately beginning at $t = 12\,\text{h}$, and completes the phase delay to align with the desired reference trajectory after the PRC returns to a negative value. Note that the initial negative region of the PRC is elongated due to control slowing the advance of phase and maintaining the oscillator in a phase with a negative PRC for longer

9.3 Control of Population Phase and Synchrony

The circadian oscillator is cell-autonomous, and each tissue is comprised of many thousands of individual oscillators. While the SCN master pacemaker maintains its synchrony through intercellular communication, other tissues such as the liver that lack paracrine signaling are kept synchronized through a variety of identified and as-yet unidentified means, as guided by the SCN [26]. The application of a pharmaceutical to any of these populations will affect this synchrony, and thus the amplitude of oscillation [22, 25].

While transient deviations from the limit cycle eventually return to the limit cycle amplitude, reduction in amplitude due to desynchrony will persist in the absence of signaling, making populations with weak coupling susceptible to long-term desynchrony from mistimed control. The change in synchrony in response to perturbation may be calculated from the PRC [22]. In Fig. 9.1c, we show how this may be intuitively understood based on the slope of the PRC: oscillator populations that lie on regions of positive slope result in the cells which are ahead in phase advancing further (or being delayed less) than oscillators which lag to begin with, broadening the probability density function (PDF) of phase. Inversely, populations lying on a region of negative slope are condensed in phase by a similar argument.

Here, we first apply the previously described controller to a population of uncoupled oscillators to demonstrate the deleterious effect of a population-agnostic nonlinear MPC on synchrony. We then modify the MPC controller to explicitly penalize population desynchronization and demonstrate the ability to simultaneously control phase and synchrony of a population.

9.3.1 Case Study #2: Limitation of Single-Oscillator Assumption

We first applied the controller from Sect. 9.2.3 to a population of oscillators, with slight modification for tracking the mean phase of the population. We modified the predictor in (9.6) to use the mean phase of the population $\bar{\phi}$ rather than the phase of an individual oscillator:

$$\hat{\phi}(t_i + \ell\tau_u) = \bar{\phi}(t_i) + \frac{2\pi\ell\tau_u}{\tau} - \sum_{k=0}^{\ell-1} \int_{t_{k+i}}^{t_{k+1+i}} u(t_k) \frac{d}{dt} \frac{d\phi}{d(vdCn)}\bigg|_{\mathbf{x}^\gamma} dt. \tag{9.9}$$

Here, we calculated $\bar{\phi}(t_i)$ from the phases of each oscillator ϕ_n in the population by the parameter $z \in \mathbb{C}$ describing the population:

$$z = \rho \exp(\mathrm{i}\bar{\phi}) = \frac{1}{N} \sum_{n=1}^{N} \exp(\mathrm{i}\phi_n), \tag{9.10}$$

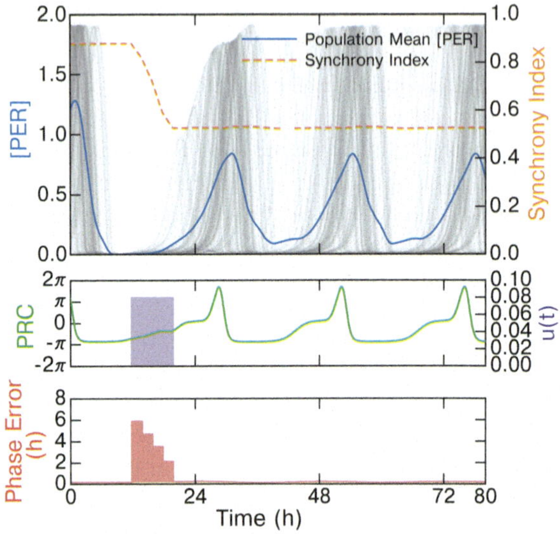

Fig. 9.3 Application of the controller from Sect. 9.2.3 for a phase delay of 6 h ($\Delta\phi = -\pi/2$) with $t_{shift} = 12$, for shifting mean phase of a population of 200 circadian oscillators (individual trajectories plotted in gray). While control is applied in nearly the same regions as Fig. 9.1 and a −6 h shift was attained, this resulted in a dispersion of phase, a decrease in synchrony index, and a reduction in mean oscillatory amplitude

where ρ is the Kuramoto order parameter, or colloquially, the synchrony index. We emphasize that the controller did not observe any information about the population aside from its mean phase, and as such, the predictor (9.6) used in the finite horizon optimal control problem was imprecise due to slight differences between single-cell and population mean phase response [22]. The controller was otherwise parameterized identically to that in Sect. 9.2.3.

Figure 9.3 shows the result of applying this controller to a population of 200 identical uncoupled oscillators with initial phases sampled from the PDF:

$$p(\phi) = f_{WN}(\phi; \phi_0, \sigma), \tag{9.11}$$

where f_{WN} indicates a wrapped normal distribution with mean ϕ_0 (set to 0, to match the mean with the single oscillator case), and standard deviation σ (set to $\pi/12$, approximately capturing the distribution of phases observed) [22]. As in the single-cell case, control was applied immediately starting at $t = 12$ h to begin correcting for the 6 h phase delay. In the population case, this corresponds to a region of positive slope of the PRC, and intuitively resulted in a desynchronization of phase as evidenced by the decline in synchrony index for the duration of the KL001 pulse. Despite the amplitudes of the individual oscillators (gray) returning after a transient to their pre-pulse levels, the population mean [PER] amplitude was reduced by approximately one-third due to desynchronization of the population. Because there was no intercellular or external communication driving synchrony, the amplitude of the mean remained diminished for the duration of the simulation.

9.3.2 Population MPC Algorithm for Phase Coherence

A more sophisticated approach to control of an oscillator population involves predicting the evolution of the PDF itself rather than only mean phase. Methods have been developed previously to compute the change in PDF directly in response to stimulation [22]. A change in variables allows the numerical computation of the new phase PDF:

$$\hat{p}(\phi, t)dh(\phi) = p(\phi, t)d\phi, \tag{9.12}$$

where $\hat{p}(\phi, t)$ is the predicted PDF at time t for phases ϕ, and $h(\phi) = \phi + \Delta\phi$, called the phase transition curve (PTC), is the total response to perturbation. We may therefore revise the prediction model to explicitly involve calculation of the phase PDF at each step of the predictive horizon. The PTC for each step within the predictive horizon may be calculated in a similar fashion to the predictor for the single-oscillator case:

$$h_k(\phi) = \phi + \frac{2\pi\tau_u}{\tau} - u(t_k)\int_{t_k}^{t_{k+1}} \frac{d}{dt}\frac{d\phi}{d(vdCn)}\Big|_{\mathbf{x}^\gamma} dt, \tag{9.13}$$

where u is piecewise constant and the integrand is a function of ϕ which may be calculated numerically in advance as previously defined. This function may be calculated numerically for each step, and used to calculate the evolution of the PDF:

$$\hat{p}(\phi, t_{k+1})dh_k(\phi) = \hat{p}(\phi, t_k)d\phi. \tag{9.14}$$

To calculate the first step, $\hat{p}(\phi, t) = p(\phi, t)$ is measured from the population under control. The population predicted mean phase $\hat{\phi}$ and synchrony index $\hat{\rho}$ may then be calculated from the predicted PDF:

$$\hat{\rho}(t_k)\exp(i\hat{\phi}(t_k)) = \int_0^{2\pi} \hat{p}(\theta, t_k)\exp(i\theta)d\theta, \tag{9.15}$$

where θ is a dummy variable of phase. The phase error term g_ϕ may remain as defined in (9.7), and we define a similar error term penalizing desynchrony

$$g_\rho(\cdot) \triangleq \begin{cases} 0 & \text{if } \rho > \delta_\rho \\ (1 - \rho(\cdot)) & \text{otherwise} \end{cases} \tag{9.16}$$

which reduces to 0 for a satisfactorily synchronized population.

Thus, the population control problem is:

$$u^\star = \arg\min_{U} \mathcal{J}$$

subject to:

$$\begin{aligned}
\mathcal{J} &= \sum_{\ell=1}^{N_p} w_\ell^\phi g_\phi^2(t_i + \ell\tau_u) + w_\ell^\rho g_\rho^2(t_i + \ell\tau_u) + w_{\ell-1}^u u^2(t_i + (\ell-1)\tau_u), \\
h_k(\phi) &= \frac{2\pi\,\tau_u}{\tau} - u(t_k)\int_{t_k}^{t_{k+1}} \frac{d}{dt}\frac{d\phi}{d(vdCn)}\bigg|_{\mathbf{x}^\gamma} dt, \\
\hat{p}(\phi, t_{k+1})dh_k(\phi) &= \hat{p}(\phi, t_k)d\phi \\
\hat{\rho}(t_k)\exp(\mathrm{i}\hat{\phi}(t_k)) &= \int_0^{2\pi} \hat{p}(\theta, t_k)\exp(\mathrm{i}\theta)d\theta \\
0 &\le u_{\ell-1} \le \bar{u},
\end{aligned} \tag{9.17}$$

where phase error and synchrony are calculated at the end of each step within the predictive horizon.

9.3.3 *Case Study #3: Implementation of Population Nonlinear MPC Controller*

As before, w^ϕ, w^u, w^ρ, N_p and τ_u are design parameters. These parameters were set as in Sect. 9.2.3, with the exception of the new weighting of synchrony:

$$w^\rho = 10(\ell + 1),$$

which was set such that the synchrony term is of the same order of magnitude as the phase term, and increases to allow flexibility of synchrony early in the horizon. Tuning this parameter will adjust the sensitivity of the controller to temporary desynchrony. As $w^\rho \to \infty$, the controller will take no action unless it results in no loss of synchrony, i.e., the population lies completely on a region of negative slope of the PRC. For lower but nonzero w^ρ, as in the case here, some flexibility is permitted, in that the controller may apply action that desynchronizes the population if it results in a large reduction in phase error. Correspondingly, the controller will later resynchronize the population to account for this early desynchrony. For $w^\rho = 0$, the controller will behave as the population-agnostic case. After calculating the finite-horizon optimal control $u^\star$, we apply the first step $u^\star(t_j)$ to all oscillators within the population for $t \in [t_j, t_j + \tau_u)$, and repeat this process.

We applied the controller described in Sect. 9.3.2 to the same phase shifting problem as the previous controllers: $t_{shift} = 12$ h, $\Delta\phi = -6$ h. The initial phase of the 200 uncoupled oscillators comprising the population was sampled from the PDF:

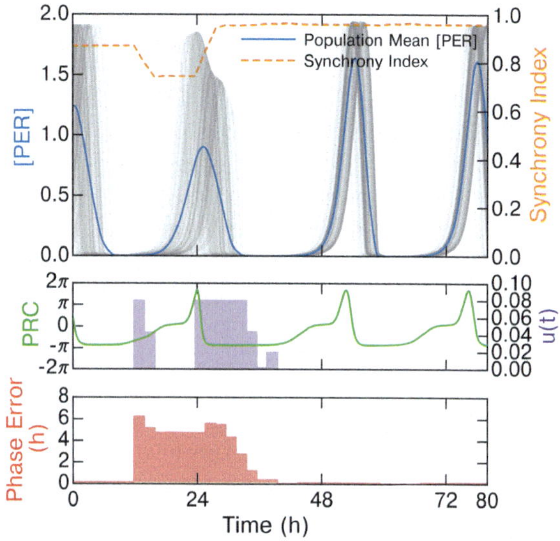

Fig. 9.4 Application of the controller from Sect. 9.3.2 for a phase delay of 6 h ($\Delta\phi = -\pi/2$) at $t_{shift} = 12$ for a population of 200 oscillators (individual trajectories plotted in gray). This controller explicitly accounted for synchrony of the population. After a brief input to begin the shift, the controller delayed the majority of its input to find a region where population synchrony would be maintained. Indeed, synchrony is slightly improved by the control action, and a phase delay of $6h$ was achieved

$$p(\phi) = f_{WN}(\phi; \phi_0, \sigma), \tag{9.18}$$

where f_{WN} indicates a wrapped normal distribution with mean $\phi_0 = 0$ and standard deviation $\sigma = \pi/12$.

Results from this simulation are shown in Fig. 9.4. The controller first applied input briefly in a slight desynchronizing region of the PRC, then paused for the remainder of the first negative PRC region due to the likelihood of further desynchronizing the population. Once the population PDF returned to a region of negative PRC and a negative first derivative of the PRC, the controller resumed input driving the population to its 6 h phase delay and restoring full synchrony. Strikingly, changing the time of input resulted in a slight increase in the synchrony of the population in comparison to its original state, evidenced by an increase in the synchrony index. Visually, this is evident from the clear alignment of individual oscillators within the population (plotted in gray) in comparison to the dispersion in phase evident in Fig. 9.3 where synchrony was ignored.

9.4 Conclusion

We have presented a modification of nonlinear MPC for phase manipulation of circadian oscillator populations, in which a PDF of phase is used in solving the finite-horizon optimal control problem, allowing mean phase and population synchrony to be regulated simultaneously. For many PRCs that have been calculated, there exists a region where the PRC or first derivative of the PRC is zero [7], it is therefore possible to manipulate phase and synchrony independently for a population of circadian oscillators through a single control input. In reality, the ability to target these regions is limited by precision of the measurements of the PRC and population phase PDF.

One significant challenge in implementing this control algorithm in vitro or in vivo is the construction of an observer with sufficient accuracy to accurately reconstruct the phase PDF. Current methods of assessing the phase of circadian oscillators rely on either noisy single-cell bioluminescent markers in vitro or system level metrics in vivo such as melatonin, activity, or body temperature. However, even a simplistic assumption such as a wrapped normal PDF with an arbitrary estimate of standard deviation of phase could help avoid delivering control inputs where the slope of the PRC is expected to be positive, and thus help avoid desynchrony. In this case, a long predictive horizon would quickly become inaccurate, however, necessitating careful selection of design variables τ_u and N_p.

Another challenge toward implementing such an algorithm is incorporating the as-yet uncharacterized pharmacokinetics and pharmacodynamics (PK/PD) for a small molecule such as KL001. This could potentially be achieved by including these terms directly in the prediction step of the MPC. Uncertainty and individual variability in PK/PD measurements may reduce the accuracy of such an approach, however, further study is necessary to determine the extent of this variability, and how these inaccuracies affect controller performance.

It is our hope that these and similar control theoretic methods will inform the discovery of circadian therapies, and enable novel experimental design toward better understanding the dynamics of cellular populations and communication.

References

1. Abel, J.H., Chakrabarty, A., Doyle III, F.J.: Nonlinear model predictive control for circadian entrainment using small-molecule pharmaceuticals. Proceedings of 20th IFAC World Congress. 9864–9870 (2017)
2. Abel, J.H., Doyle III, F.J.: A systems theoretic approach to analysis and control of mammalian circadian dynamics. Chem. Eng. Res. Des. **116**, 48–60 (2016)
3. Andersson, J., Akesson, J., Diehl, M.: Recent Advances in Algorithmic Differentiation, vol. 87. Springer, Berlin Heidelberg (2012)
4. Bagheri, N., Stelling, J., Doyle III, F.J.: Circadian phase entrainment via nonlinear model predictive control. Int. J. Robust Nonlinear Control **17**(May), 1555–1571 (2007)
5. Bagheri, N., Taylor, S.R., Meeker, K., Petzold, L.R., Doyle III, F.J.: Synchrony and entrainment properties of robust circadian oscillators. J. R. Soc. Interface **5**, S17–S28 (2008)

6. Bass, J., Takahashi, J.S.: Circadian integration of metabolism and energetics. Science **330**(6009), 1349–1354 (2010)
7. Dunlap, J.C., Loros, J.J., DeCoursey, P.J.: Chronobiology. Sinauer Associates, Inc. (2004)
8. Glossop, N.R., Lyons, L.C., Hardin, P.E.: Interlocked feedback loops within the drosophila circadian oscillator. Science **286**(5440), 766–768 (1999)
9. Goodwin, B.C.: Oscillatory behavior in enzymatic control processes. Adv. Enzyme Regul. **3**, 425–437 (1965)
10. Hirota, T., Lee, J.W., St. John, P.C., Sawa, M., Iwaisako, K., Noguchi, T., Pongsawakul, P.Y., Sonntag, T., Welsh, D.K., Brenner, D.A., Doyle III, F.J., Schultz, P.G., Kay, S.A.: Identification of small molecule activators of cryptochrome. Science **337**(6098), 1094–1097 (2012)
11. Huang, K.C., Meir, Y., Wingreen, N.S.: Dynamic structures in escherichia coli: Spontaneous formation of mine rings and mind polar zones. Proc. Natl. Acad. Sci. USA **100**(22), 12724–12728 (2003)
12. Ishiura, M., Kutsuna, S., Aoki, S., Iwasaki, H., Andersson, C.R., Tanabe, A., Golden, S.S., Johnson, C.H., Kondo, T.: Expression of a gene cluster kaiABC as a circadian feedback process in cyanobacteria. Science **281**(5382), 1519–1523 (1998)
13. Marquié, J.C., Tucker, P., Folkard, S., Gentil, C., Ansiau, D.: Chronic effects of shift work on cognition: findings from the visat longitudinal study. Occup. Environ. Med. **72**(4), 258–264 (2015)
14. Mirsky, H.P., Liu, A.C., Welsh, D.K., Kay, S.A., Doyle III, F.J.: A model of the cell-autonomous mammalian circadian clock. Proc. Natl. Acad. Sci. USA **106**(27), 11107–11112 (2009)
15. Mukherji, A., Kobiita, A., Damara, M., Misra, N., Meziane, H., Champy, M.F., Chambon, P.: Shifting eating to the circadian rest phase misaligns the peripheral clocks with the master scn clock and leads to a metabolic syndrome. Proc. Natl. Acad. Sci. USA **112**(48), E6691–E6698 (2015)
16. Panda, S., Antoch, M.P., Miller, B.H., Su, A.I., Schook, A.B., Straume, M., Schultz, P.G., Kay, S.A., Takahashi, J.S., Hogenesch, J.B.: Coordinated transcription of key pathways in the mouse by the circadian clock. Cell **109**(3), 307–320 (2002)
17. Ramkisoensing, A., Meijer, J.H.: Synchronization of biological clock neurons by light and peripheral feedback systems promotes circadian rhythms and health. Front. Neurol. **6**(MAY) (2015)
18. Serkh, K., Forger, D.B.: Optimal schedules of light exposure for rapidly correcting circadian misalignment. PLoS Comput. Biol. **10**(4), e1003,523 (2014)
19. Shaik, O., Sager, S., Slaby, O., Lebiedz, D.: Phase tracking and restoration of circadian rhythms by model-based optimal control. IET Syst. Biol. **2**(1), 16–23 (2008)
20. Slaby, O., Sager, S., Shaik, O.S., Kummer, U., Lebiedz, D.: Optimal control of self-organized dynamics in cellular signal transduction. Math. Comput. Model. Dyn. Syst. **13**(5), 487–502 (2007)
21. St. John, P.C., Hirota, T., Kay, S.A., Doyle III, F.J.: Spatiotemporal separation of per and cry posttranslational regulation in the mammalian circadian clock. Proc. Natl. Acad. Sci. USA **111**(5), 2040–2045 (2014)
22. St. John, P.C., Taylor, S.R., Abel, J.H., Doyle III, F.J.: Amplitude metrics for cellular circadian bioluminescence reporters. Biophys. J. **107**(11), 2712–2722 (2014)
23. Takahashi, J.S.: Transcriptional architecture of the mammalian circadian clock. Nat. Rev. Genet. (2016)
24. Taylor, S.R., Doyle III, F.J., Petzold, L.R.: Oscillator model reduction preserving the phase response: application to the circadian clock. Biophys. J. **95**(4), 1658–1673 (2008)
25. Ukai, H., Kobayashi, T.J., Nagano, M., Masumoto, K.H., Sujino, M., Kondo, T., Yagita, K., Shigeyoshi, Y., Ueda, H.R.: Melanopsin-dependent photo-perturbation reveals desynchronization underlying the singularity of mammalian circadian clocks singularity behaviour in circadian clocks. Nat. Cell Biol. **9**(11), 1327–1334 (2007)
26. Welsh, D.K., Takahashi, J.S., Kay, S.A.: Suprachiasmatic nucleus: cell autonomy and network properties. Annu. Rev. Physiol. **72**, 551–577 (2010)

27. Zhang, J., Qiao, W., Wen, J.T., Julius, A.: Light-based circadian rhythm control: entrainment and optimization. Automatica **68**, 44–55 (2016)
28. Zhang, R., Lahens, N.F., Ballance, H.I., Hughes, M.E., Hogenesch, J.B.: A circadian gene expression atlas in mammals: Implications for biology and medicine. Proc. Natl. Acad. Sci. U. S. A. **111**(45), 16219–16224 (2014)

Chapter 10
Wasserstein Geometry of Quantum States and Optimal Transport of Matrix-Valued Measures

Yongxin Chen, Tryphon T. Georgiou and Allen Tannenbaum

Abstract We overview recent results on generalizations of the Wasserstein 2-metric, originally defined on the space of scalar probability densities, to the space of Hermitian matrices and of matrix-valued distributions, as well as some extensions of the theory to vector-valued distributions and discrete spaces (weighted graphs).

10.1 Introduction

Optimal mass transport (OMT) is currently a very active area of research with applications to areas both applied and theoretical including control, transportation, econometrics, fluid dynamics, probability theory, statistical physics, shape optimization, expert systems, and meteorology; see [25, 30] for extensive lists of references. The original problem was first formulated by the civil engineer Gaspar Monge in 1781, and concerned finding the optimal way, in the sense of minimal transportation cost, of moving a pile of soil from one site to another. Much later the problem was extensively analyzed by Kantorovich [18], and is now known as the Monge–Kantorovich (MK) or optimal mass transport (OMT) problem.

This project was supported by ARO grant (W911NF-17-1-0429), AFOSR grants (FA9550-15-1-0045 and FA9550-17-1-0435), NSF (ECCS-1509387), NIH (P41-RR-013218, P41-EB-015902, 1U24CA18092401A1), and a postdoctoral fellowship at MSKCC.

Y. Chen
Department of Electrical and Computer Engineering, Iowa State University, Ames, IA 50011, USA
e-mail: yongchen@iastate.edu

T. T. Georgiou (✉)
Department of Mechanical Engineering, University of California, Irvine, CA, USA
e-mail: tryphon@uci.edu

A. Tannenbaum
Departments Computer Science and Applied Mathematics and Statistics, Stony Brook University, New York City, USA
e-mail: allen.tannenbaum@stonybrook.edu

R. Tempo et al. (eds.), *Emerging Applications of Control and Systems Theory*, Lecture Notes in Control and Information Sciences - Proceedings,
https://doi.org/10.1007/978-3-319-67068-3_10

In this paper, we present certain generalizations of OMT to matrix and vector-valued transportation. Our original motivation for this rather nontraditional viewpoint was provided by problems in Signal Analysis, more specifically, the need of a weakly continuous metric to compare (matrix-valued) power spectra of multivariate time series (see [24]). Soon afterward it became apparent Quantum Mechanics was another field that would stand to benefit from such an unusual extension of OMT. In fact, it was this latter subject that provided some of the clues of how to properly set up noncommutative OMT.

The basis of the new theory is a suitable extension of the ***Liouville (continuity) equation*** that allows flows in matrix or other spaces. To this end, in [8], we first proposed such a continuity equation and a noncommutative counterpart of OMT where probability distributions are replaced by density matrices (i.e., Hermitian positive-definite matrices with unit trace). The appropriate ***Wasserstein metric*** now corresponds to the minimal value of an action integral evaluated on flows connecting end-point density matrices. The key insight, to use such a dynamic formulation in seeking the needed generality was provided by the seminal approach of Benamou and Brenier [3]. Indeed, the Benamou–Brenier formulation recasts OMT as a stochastic control problem. The work we are reporting herein takes this idea along several different directions, and in particular to OMT between matrices, matricial distribution and vectorial distributions [8, 10]. Extensions of these results to distributions that have end-point distributions of unequal overall mass (unbalanced) are reported in [11] (and not included in the current survey).

We note that at about the same time as [8] was originally reported, closely related approaches were formulated independently and simultaneously in [6, 20]. In fact, in our work, we greatly benefited from earlier work by Carlen and Maas [7] on a fermionic Fokker–Planck equation.

10.2 Quantum Continuity Equation

The three papers [6, 8, 20] all begin with the *Lindblad equation* that describes the evolution of open quantum systems. Open quantum systems are thought of as coupled to a larger system (referred to as the environment or the ancilla) and, thereby, cannot, in general, be described by the Schrödinger equation [17]. In this case, the evolution of density operators ρ [17] is given by the Lindblad equation

$$\begin{aligned}\dot{\rho} &= -i[H,\rho] \\ &+ \sum_{k=1}^{N}(L_k\rho L_k^* - \frac{1}{2}\rho L_k^* L_k - \frac{1}{2}L_k^* L_k\rho),\end{aligned} \tag{10.1}$$

where * denotes conjugate transpose, and throughout, we assume that $\hbar = 1$. The first term on the right-hand side describes the evolution of the state under the effect of the Hamiltonian H (Schrödinger unitary evolution). The other terms on the right-hand side model diffusion and, thereby, capture the dissipation of energy–they constitute the quantum analogue of Laplace's operator Δ and are referred to as the Lindblad terms.

Our approach [8, 9] relies on a suitable continuity equation in the space of Hermitian matrices $\mathscr{H}$ (of a given dimension). To this end, we invoke suitable definitions for the gradient ∇_L and divergence ∇_L^* operators on spaces of matrices that are explained below and express the continuity equation in the familiar form

$$\dot{\rho} = \nabla_L^* J, \tag{10.2}$$

where J is a *matricial flux*, in complete analogy with the continuity equation on scalar densities.

Throughout, $\rho(t) \in \mathscr{H}$ is a positive-semidefinite matrix of trace one, i.e., a *density matrix* of quantum mechanics. Regarding notation, we let $\mathscr{H}_+$ and $\mathscr{H}_{++}$ denote the cones of nonnegative and positive-definite matrices, respectively,

$$\mathscr{D}_+ := \{\rho \in \mathscr{H}_{++} \mid \operatorname{tr}(\rho) = 1\}$$

the space of density matrices, and $\mathscr{S}$ the space of skew-Hermitian matrices (of the same dimension as $\mathscr{H}$). The flux J is taken in $\mathscr{S}^N$, i.e., a vector with matrix entries. Flux typically arises in the form

$$J = \rho \circ v \quad \text{or in the form} \quad J = \nabla_L \rho.$$

The symbol $\rho \circ v$ denotes one of several possible choices of *noncommutative multiplication*. We have considered specifically the following two choices, referred to as the *anticommutator* multiplication (i) and *Kubo-Mori* product (ii), respectively:

$$\text{(i)} \;\; \rho \circ v = \frac{1}{2}(\rho v + v\rho) \text{ and (ii) } \; \rho \circ v = \int_0^1 \rho^s v \rho^{1-s} ds,$$

where, for $\rho \in \mathscr{H}$ and $v \in \mathscr{S}^N$,

$$v\rho := \begin{bmatrix} v_1\rho \\ \vdots \\ v_N\rho \end{bmatrix}, \text{ and } \rho v := \begin{bmatrix} \rho v_1 \\ \vdots \\ \rho v_N \end{bmatrix}.$$

On the other hand, we define the gradient operator with respect to $L \in \mathscr{H}^N$ to be

$$\nabla_L : \mathscr{H} \to \mathscr{S}^N, \;\; X \mapsto \begin{bmatrix} L_1 X - X L_1 \\ \vdots \\ L_N X - X L_N \end{bmatrix}.$$

With respect to the standard Hilbert–Schmidt inner product $\langle X, Y\rangle = \operatorname{tr}(X^*Y)$ (and, for the case when X, Y are in $\mathscr{H}^N$ or $\mathscr{S}^N$, the inner product $\langle X, Y\rangle = \sum_{k=1}^N \operatorname{tr}(X_k^* Y_k)$), the divergence operator turns out to be

$$\nabla_L^* : \mathscr{S}^N \to \mathscr{H},\ Y = \begin{bmatrix} Y_1 \\ \vdots \\ Y_N \end{bmatrix} \mapsto \sum_k^N L_k Y_k - Y_k L_k,$$

and this is what is used in (10.2). We note that for technical reasons, the definition of gradient and divergence require that $L_k = L_k^*$, i.e., $L \in \mathscr{H}^N$, as above. Also, one can easily verify that ∇_L is a derivation, in that,[1]

$$\begin{aligned} \nabla_L(XY + YX) = (\nabla_L X)Y + X(\nabla_L Y) \\ + (\nabla_L Y)X + Y(\nabla_L X), \quad \forall X, Y \in \mathscr{H}. \end{aligned}$$

With these definitions in place, we define the (matricial) *Laplacian* as

$$\begin{aligned} \Delta_L X &:= -\nabla_L^* \nabla_L X \\ &= \sum_{k=1}^N (2L_k X L_k - X L_k L_k - L_k L_k X),\ X \in \mathscr{H}, \end{aligned}$$

which is exactly (after scaling by $1/2$) the diffusion term in the Lindblad equation[2] (10.1). Hence, Lindblad's equation can be rewritten as

$$\begin{aligned} \dot\rho &= -\nabla_H^*(\rho i) + \frac{1}{2}\nabla_L^*(\nabla_L \rho) \\ &= -\nabla_H(\rho i) + \frac{1}{2}\Delta_L \rho. \end{aligned}$$

10.3 Matricial Wasserstein 2-Metric

From here on we consider the *continuity equation*,

$$\dot\rho = \nabla_L^*(\rho \circ v), \tag{10.3}$$

without the diffusion term, but for a general velocity field $v \in \mathscr{S}^N$. A tacit assumption throughout is that the identity matrix I spans the null space or ∇_L; this can be ensured

[1]The domain of ∇_L is $\mathscr{H}$, hence the identity requires $XY + YX$, instead of simply XY.

[2]The Lindblad term is in the so-called *symmetric form* since the coefficients are Hermitian.

if one chooses $L_1, \ldots, L_N$ to form a basis of $\mathscr{H}$ (which is a sufficient, but not a necessary condition).

A Wasserstein distance between two density matrices can now be defined as the *least action* (minimum control problem) to steer one density matrix to another,

$$W_{2,a}(\rho_0, \rho_1)^2 := \min_{\rho, v} \int_0^1 \langle v, \rho \circ v \rangle dt, \tag{10.4a}$$

$$\dot{\rho} = \frac{1}{2} \nabla_L^* (\rho \circ v), \tag{10.4b}$$

$$\rho(0) = \rho_0, \quad \rho(1) = \rho_1. \tag{10.4c}$$

In this, ρ_0 and ρ_1 in $\mathscr{D}_+$ and the optimization is over $\rho(\cdot) \in \mathscr{D}_+$ and $v \in \mathscr{S}^N$. In fact, for $v \in \mathscr{S}^N$, (10.3) already preserves positive definiteness and trace of $\rho(\cdot)$.

The choice of the anticommutator product $\rho \circ v = \frac{1}{2}(\rho v + v \rho)$ is especially appealing since, in this case, the matricial Wasserstein metric (10.4) is readily computable. Indeed, (10.4) can be cast as a convex optimization problem in a manner analogous to that in the scalar case [3]. To see this, let $u := \rho v = [u_1^*, \ldots, u_N^*]^*$ and $u_* := [u_1, \ldots, u_N]^*$, and observe that

$$\begin{aligned} \operatorname{tr}(\rho v^* v) &= \sum_{k=1}^N \operatorname{tr}(\rho v_k^* v_k) \\ &= \sum_{k=1}^N \operatorname{tr}((\rho v_k)^* \rho^{-1} \rho v_k) = \operatorname{tr}(u^* \rho^{-1} u). \end{aligned}$$

Thus, (10.4) can be equivalently expressed as

$$W_2(\rho_0, \rho_1)^2 = \min_{\rho, u} \int_0^1 \operatorname{tr}(u^* \rho^{-1} u) dt, \tag{10.5a}$$

$$\dot{\rho} = \frac{1}{2} \nabla_L^* (u - u_*), \tag{10.5b}$$

$$\rho(0) = \rho_0, \quad \rho(1) = \rho_1. \tag{10.5c}$$

In this, it turns out that although we do not require any structural constraint on u, the optimal u satisfies $u = \rho v$ for some $v \in \mathscr{S}^N$.

The choice of the Kubo-Mori product, on the other hand, provides a matricial version of the Wasserstein metric for which the gradient flow of the von Neuman entropy $\operatorname{tr}(\rho \log(\rho))$ is the Lindblad equation [6, 8, 20]. Thus, it generalizes to the noncommutative setting, the famous result by Jordan, Kinderlehrer, and Otto [16] for the ordinary Wasserstein-2 metric on probability densities where the heat equation is the gradient flow of the entropy. However, it is interesting to note that computation of the Wasserstein metric for the Kubo-Mori product appears challenging as compared to the one based on the anticommutator product above.

To characterize the form of minimizer one can proceed to consider the dual problem, which for the case of the anticommutator product goes as follows. With $\lambda(\cdot) \in \mathscr{H}$ a smooth Lagrangian multiplier for the constraints we construct the Lagrangian

$$\mathscr{L}(\rho, v, \lambda) = \int_0^1 \left\{ \frac{1}{2}\mathrm{tr}(\rho v^* v) - \mathrm{tr}(\lambda(\dot{\rho} - \frac{1}{2}\nabla_L^*(\rho v + v\rho))) \right\} dt$$
$$= \int_0^1 \left\{ \frac{1}{2}\mathrm{tr}(\rho v^* v) + \frac{1}{2}\mathrm{tr}((\nabla_L \lambda)^*(\rho v + v\rho)) + \mathrm{tr}(\dot{\lambda}\rho) \right\} dt - \mathrm{tr}(\lambda(1)\rho_1) + \mathrm{tr}(\lambda(0)\rho_0).$$

Point-wise minimization over v yields

$$v_{opt}(t) = -\nabla_L \lambda(t)$$

and the expression for the corresponding minimum

$$\int_0^1 \left\{ -\frac{1}{2}\mathrm{tr}(\rho(\nabla_L \lambda)^*(\nabla_L \lambda)) + \mathrm{tr}(\dot{\lambda}\rho) \right\} dt$$
$$-\mathrm{tr}(\lambda(1)\rho_1) + \mathrm{tr}(\lambda(0)\rho_0),$$

from which we conclude the following sufficient condition for optimality: Suppose there exists $\lambda(\cdot) \in \mathscr{H}$ satisfying

$$\dot{\lambda} = \frac{1}{2}(\nabla_L \lambda)^*(\nabla_L \lambda) = \frac{1}{2}\sum_{k=1}^{N}(\nabla_L \lambda)_k^*(\nabla_L \lambda)_k \tag{10.6a}$$

such that the solution of

$$\dot{\rho} = -\frac{1}{2}\nabla_L^*(\rho \nabla_L \lambda + \nabla_L \lambda \rho) \tag{10.6b}$$

matches the marginals $\rho(0) = \rho_0$, $\rho(1) = \rho_1$. Then the pair (ρ, v) with $v = -\nabla_L \lambda$ solves (10.4).

The Wasserstein metric induces a Riemannian structure

$$\langle \delta_1, \delta_2 \rangle_\rho = \frac{1}{2}\mathrm{tr}(\rho \nabla \lambda_1^* \nabla \lambda_2 + \rho \nabla \lambda_2^* \nabla \lambda_1)$$

on the tangent space of Hermitian matrices with a specified trace,

$$T_\rho = \{\delta \in \mathscr{H} \mid \mathrm{tr}(\delta) = 0\}.$$

Here $\lambda_j,\ j = 1, 2$ is the solution to the *Poisson equation*

$$\delta_j = -\frac{1}{2}\nabla_L^*(\rho\nabla_L\lambda_j + \nabla_L\lambda_j\rho). \tag{10.7}$$

The proof of existence and uniqueness of the solution of (10.7) follows exactly along the same lines as in [7]; details are given in [9]. In fact, given a *tangent direction* δ, $-\nabla_L\lambda$ is the unique minimizer of $\mathrm{tr}(\rho v^* v)$ over all $v \in \mathscr{S}^N$ satisfying

$$\delta = \frac{1}{2}\nabla_L^*(\rho v + v\rho).$$

With the above definition of inner product, $W_2(\cdot,\cdot)$ indeed defines a metric on $\mathscr{D}_+$ for which $\mathscr{D}_+$ is a geodesic space, i.e., the distance between two given $\rho_0, \rho_1 \in \mathscr{D}_+$ can be rewritten as

$$W_{2,a}(\rho_0, \rho_1) = \min_\rho \int_0^1 \sqrt{\langle\dot\rho(t), \dot\rho(t)\rangle_{\rho(t)}}dt,$$

where the minimum is taken over all piecewise smooth paths on the manifold $\mathscr{D}_+$.

We finally note that, more generally, OMT can be formulated on the space of matrix-valued distributions. In this case, the mass constraint becomes $\int \mathrm{tr}\rho(x)dx = 1$, where x represents a vector of spatial coordinates and dx the volume element. Transport along spatial coordinates, e.g., with $x \in \mathbb{R}^m$, is effected by a term $\nabla_x \cdot (\rho \circ w)$ in the continuity equation, with $w \in \mathscr{H}^m$, i.e.,

$$\dot\rho = \nabla_L^*(\rho \circ v) - \nabla_x \cdot (\rho \circ w).$$

Likewise, the cost of transport is duly penalized in a corresponding problem to minimize a suitable action integral; see [8] for details.

10.4 Vector-Valued Transport on $\mathbb{R}^N$

A vector-valued density $\rho = [\rho_1, \rho_2, \cdots, \rho_\ell]^T$, on $\mathbb{R}^N$ or on a discrete space, may represent power reflected off a surface at different frequencies/colors. The "mass" of these components may transfer between different entries of the density vector (e.g., due to different angles of reflection) along time flows of the vectorial density. Thus, while the total power may be invariant (under some lighting conditions), the proportion of power at different frequencies or polarization may smoothly vary with viewing angle. As another example consider the case where the entries of ρ represent densities of different species, or particles, and allow for the possibility that mass transfers from one species to another ("mutate"), i.e., between entries of ρ. Thus, in general, we postulate that transport of vector-valued quantities captures flow across space as well as between entries of the density vector.

In this context, an OMT-inspired geometry is aimed to express a suitable continuity and to quantify transport cost for such vectorial distributions. We highlight some of the key elements in [10] for such a theory. It follows a line which is analogous to development of quantum transport that was discussed above.

We begin by considering a vector-valued density ρ on $\mathbb{R}^N$, i.e., a map from $\mathbb{R}^N$ to $\mathbb{R}_+^\ell$ such that

$$\sum_{i=1}^{\ell} \int_{\mathbb{R}^N} \rho_i(x)dx = 1,$$

and consider the entries of ρ as representing density or mass of species/particles that can mutate between one another while maintaining total mass. We denote the set of all vector-valued densities and its interior by $\mathscr{D}$ and $\mathscr{D}_+$, respectively. The dynamics are described by the following ***continuity equation***:

$$\frac{\partial \rho_i}{\partial t} + \nabla_x \cdot (\rho_i v_i) - \sum_{j \neq i} (\rho_j w_{ji} - \rho_i w_{ij}) = 0, \quad \forall i = 1, \ldots, \ell. \tag{10.8}$$

Here v_i is the velocity field of particles i and $w_{ij} \geq 0$ is the transfer rate from i to j. Equation (10.8) allows for the possibility to mutate between each pair of entries. More generally, mass transfer may only be permissible between specific types of particles and can be modeled by a graph $\mathscr{G} = (\mathscr{V}, \mathscr{E})$ (where the entries denote nodes and edges, respectively), in which case, for a subset of indices, the transfer rates w_{ji} may be restricted to be zero.

Given $\mu, \nu \in \mathscr{D}_+$, we formulate the optimal mass transport

$$\begin{aligned} W_2(\mu, \nu)^2 := \inf_{\rho, v, w} & \int_0^1 \int_{\mathbb{R}^N} \left\{ \sum_{i=1}^{\ell} \rho_i(t,x) \|v_i(t,x)\|^2 + \gamma \sum_{i,j=1}^{\ell} \rho_i w_{ij}^2(t,x) \right\} dx dt \\ & \frac{\partial \rho_i}{\partial t} + \nabla_x \cdot (\rho_i v_i) - \sum_{j \neq i} (\rho_j w_{ji} - \rho_i w_{ij}) = 0, \quad \forall i = 1, \ldots, \ell \\ & w_{ij}(t,x) \geq 0, \quad \forall i, j, t, x \\ & \rho(0, \cdot) = \mu(\cdot), \quad \rho(1, \cdot) = \nu(\cdot). \end{aligned} \tag{10.9}$$

The coefficient $\gamma > 0$ specifies the relative cost between transporting mass in space and trading mass between different types of particles. When γ is large, the solution reduces to independent OMT problems for the different entries to the degree possible. In general, W_2 is a quasi-metric in that it satisfies the triangle inequality and positivity, but may not be symmetric.

Setting $p_{ij} = \rho_i w_{ij} \geq 0$ and $u_i = \rho_i v_i$, we have $\rho_i w_{ij}^2 = \rho_i^{-1} p_{ij}^2$, and $\rho_i \|v_i\|^2 = \rho_i^{-1} \|u_i\|^2$. It follows that

$$W_2(\mu, \nu)^2 = \inf_{\rho,u,p} \int_0^1 \int_{\mathbb{R}^N} \left\{ \sum_{i=1}^{\ell} \rho_i(t,x)^{-1} \|u_i(t,x)\|^2 + \gamma \sum_{i,j=1}^{\ell} \rho_i^{-1} p_{ij}^2(t,x) \right\} dxdt$$

$$\frac{\partial \rho_i}{\partial t} + \nabla_x \cdot u_i - \sum_{j \neq i} (p_{ji} - p_{ij}) = 0, \quad \forall i = 1, \ldots, \ell$$

$$p_{ij}(t,x) \geq 0, \quad \forall i, j, t, x$$

$$\rho(0, \cdot) = \mu(\cdot), \quad \rho(1, \cdot) = \nu(\cdot) \text{ which is a convex problem.}$$

Finally, a Riemannian-like metric on $\mathscr{D}_+$ can be obtained by symmetrizing the above expression [10]. This is,

$$W_{2,\text{sym}}(\mu, \nu)^2 = \inf_{\rho,u,p} \int_0^1 \int_{\mathbb{R}^N} \left\{ \sum_{i=1}^{\ell} \rho_i(t,x)^{-1} \|u_i(t,x)\|^2 \right. \tag{10.10}$$

$$\left. + \frac{\gamma}{2} \sum_{i,j=1}^{\ell} (\rho_i^{-1} + \rho_j^{-1}) p_{ij}^2(t,x) \right\} dxdt$$

under the same constraints. This vector-valued OMT structure is further explored and developed in [10].

10.5 Vector-Valued Transport on Graphs

We conclude by highlighting elements of an OMT theory solely on graphs, cast in the setting of vector-valued densities [10]. As explained earlier, such densities may represent the distribution of multiple species/resources that are allowed to mutate between each other as they transition from node to node. The theory is aimed to capture cost of transport in such a setting.

A vector-valued mass distribution on the graph $\mathscr{G} = (\mathscr{V}, \mathscr{E})$ (with n nodes and m edges) is a ℓ-tuple $\rho = (\rho_1, \cdots, \rho_\ell)$ with each $\rho_i = (\rho_{i,1}, \cdots, \rho_{i,n})^T$ being a vector in $\mathbb{R}_+^n$ such that

$$\sum_{i=1}^{\ell} \sum_{k-1}^{n} \rho_{i,k} = 1.$$

That is, each entry ρ_i, for $i \in \{1, \ldots, \ell\}$, is a vector with nonnegative n-entries representing, e.g., color intensity for the i-th color, at the node corresponding to the respective entry. We denote the set of all nonnegative vector-valued mass distributions with $\mathscr{D}$ and its interior with $\mathscr{D}_+$. Equation (10.8) is now replaced by the following continuity equation:

$$\dot{\rho}_i - \nabla_{\mathscr{G}}^* ((D_2^T \rho_i) \circ v_i - (D_1^T \rho_i) \circ \bar{v}_i) - \sum_{j \neq i} (\rho_j \circ w_{ji} - \rho_i \circ w_{ij}) = 0, \quad \forall i = 1, \ldots, \ell, \tag{10.11}$$

since the spatial domain is now also discrete (i.e., it is $\mathscr{G}$ instead of $\mathbb{R}^N$). Here, $D = D_1 - D_2$ is the incident matrix of the graph, with D_1, D_2 are matrices with positive entries reflecting the position of sources (D_1) and sinks (D_2) by a entry equal to 1 in the corresponding place. Thus, the vector $D_1^T \rho$ represents density at the sources of an edge and, likewise, $D_2^T \rho$ represents density at the sinks. Then, also, $\nabla_{\mathscr{G}}$ represents differencing between neighboring nodes and $\nabla_{\mathscr{G}}^*$ represents its dual (i.e., negative divergence) [10]. Finally, $\circ$ represents entry-wise multiplication (Shur) between two vectors.

Now following the Benamou–Brenier [3] philosophy once again, given two marginal densities $\mu, \nu \in \mathscr{D}_+$, we define their Wasserstein distance as

$$W_2(\mu, \nu)^2 := \inf_{\rho, v, w} \int_0^1 \left\{ \sum_{i=1}^{\ell} [v_i^T ((D_2^T \rho) \circ v_i) + \bar{v}_i^T ((D_1^T \rho) \circ \bar{v}_i)] + \gamma \sum_{i,j=1}^{\ell} \sum_{k=1}^{n} \rho_{i,k} w_{ij,k}^2 \right\} dt$$

subject to (10.11) as well as $w_{ij} \geq 0, \ v_i \geq 0, \ \bar{v}_i \geq 0$ for all (or a subset) of (i, j)'s and $\rho(0) = \mu, \ \rho(1) = \nu$.

We should note that the problem of transporting vector-valued mass on a graph is quite simpler than in the case where the underlined space is continuous, since it reduces essentially to a scalar mass situation on a suitably larger graph. Indeed, we can view the vector-valued mass as a scalar mass distribution on ℓ identical layers of the graph $\mathscr{G}$ where the same nodes at different layers are connected through a complete graph. The two velocity fields v, w represent mass transfer within the same layer and between different layers, respectively.

Once again, the computation of the metric has a ***convex*** formulation by changing optimization variable from $(\rho, v_i, \bar{v}_i, w_i j, k)$ to momenta "mass" ρ and momenta $u_i = \rho_i v_i$ and $p_{ij} = \rho_i w_{ij}$, instead.

10.6 Conclusion

The basic fluid dynamical formulation of OMT can be generalized to flows on the space of matrices or vectors, that belong to a simplex of a suitable positive cone. A Wasserstein metric in these spaces can then be defined as a minimal quadratic cost for transferring between two end points. Such metrics appear natural as, in particular, for the space of quantum density matrices, render the Lindblad equation as the gradient flow of the von Neumann entropy. Our interest stems from problems in signal analysis and, more specifically, spectral and image analysis. In both of these application areas, the relevance of weakly continuous metric that can be used to quantify distances between, e.g., matrix-valued power spectra or multicolor images, is self-evident. In particular, geodesics in such spaces naturally model flows and allow morphing between spectra and images, respectively.

References

1. Ambrosio, L.: Euro Summer School Mathematical Aspects of Evolving Interfaces. Lecture Notes on Optimal Transport Theory. CIME Series of Springer Lecture Notes. Madeira, Portugal, Springer-Verlag, New York (2000)
2. Angenent, S., Haker, S., Tannenbaum, A.: Minimizing flows for the Monge-Kantorovich problem. SIAM J. Math. Anal. **35**, 61–97 (2003)
3. Benamou, J.-D., Brenier, Y.: A computational fluid mechanics solution to the Monge-Kantorovich mass transfer problem. Numerische Mathematik **84**, 375–393 (2000)
4. Benamou, J.-D.: Numerical resolution of an unbalanced mass transport problem. ESAIM. Math. Model. Numer. Anal. **37**(5), 851–868 (2010)
5. Carlier, G., Salomon, J.: A monotonic algorithm for the optimal control of the Fokker-Planck equation. In: IEEE Conference on Decision and Control (2008)
6. Carlen, E., Maas, J.: Gradient flow and entropy inequalities for quantum Markov semigroups with detailed balance (2016). https://arxiv.org/abs/1609.01254
7. Carlen, E., Maas, J.: An analog of the 2-Wasserstein metric in non-commutative probability under which the fermionic Fokker-Planck equation is gradient flow for the entropy. Commun. Math. Phys. **331**, 887–926 (2014)
8. Chen, Y., Georgiou, T.T., Tannenbaum, A.: Matrix optimal mass transport: a quantum mechanical approach (2016). https://arxiv.org/abs/1610.03041
9. Chen, Y., Gangbo, W., Georgiou, T.T., Tannenbaum, A.: On the matrix Monge-Kantorovich problem. https://arxiv.org/abs/1701.02826
10. Chen, Y., Georgiou, T.T., Tannenbaum, A.: Transport distance on graphs and vector-valued optimal mass transport (2016). https://arxiv.org/pdf/1611.09946v1.pdf
11. Chen, Y., Georgiou, T.T., Tannenbaum, A.: Interpolation of density matrices and matrix-valued measures: the unbalanced case (2017). https://arxiv.org/abs/1612.05914
12. Chen, Y., Georgiou, T.T., Pavon, M.: On the relation between optimal transport and Schrödinger bridges: a stochastic control viewpoint. J. Optim. Theory Appl. **169**(2), 671–691 (2016)
13. Chen, Y., Georgiou, T.T., Pavon, M., Tannenbaum, A.: Robust transport over networks. IEEE Trans. Autom. Control (2016). https://doi.org/10.1109/TAC.2016.2626796
14. Chen, Y., Georgiou, T.T., Pavon, M.: Entropic and displacement interpolation: a computational approach using the Hilbert metric. SIAM J. Appl. Math. **76**(6), 2375–2396 (2016)
15. Evans, L.C.: Partial Differential Equations and Monge-Kantorovich Mass Transfer, in Current Developments in Mathematics, pp. 65–126. International Press, Boston, MA (1999)
16. Jordan, R., Kinderlehrer, D., Otto, F.: The variational formulation of the Fokker-Planck equation. SIAM J. Math. Anal. **29**, 1–17 (1998)
17. Gustafson, S., Sigal, I.M.: Mathematical Concepts of Quantum Mechanics. Springer, New York (2011)
18. Kantorovich, L.V.: On a problem of Monge. Uspekhi Mat. Nauk. **3**, 225–226 (1948)
19. Kumar, A., Tannenbaum, A., Balas, G.: Optical flow: a curve evolution approach. IEEE Trans. Image Process. **5**, 598–611 (1996)
20. Mittnenzweig, M., Mielke, A.: An entropic gradient structure for Lindblad equations and GENERIC for quantum systems coupled to macroscopic models (2016). https://arxiv.org/abs/1609.05765
21. Léonard, C.: From the Schrödinger problem to the Monge-Kantorovich problem. J. Funct. Anal. **262**, 1879–1920 (2012)

22. Maas, J.: Gradient flows of the entropy for finite Markov chains. J. Funct. Anal. **261**(8), 2250–2292 (2011)
23. McCann, R.: Existence and uniqueness of monotone measure-preserving maps. Duke Math. J. **80**, 309–323 (1995)
24. Ning, L., Georgiou, T., Tannenbaum, A.: On matrix-valued Monge-Kantorovich optimal mass transport. IEEE Trans. Autom. Control **60**(2), 373–382 (2015)
25. Rachev, S., Rüschendorf, L.: Mass Transportation Problems, vol. I. Springer-Verlag, New York (1998) (Probab. Appl.)
26. Sandhu, R., Georgiou, T., Reznik, E., Zhu, L., Kolesov, I., Senbabaoglu, Y., Tannenbaum, A.: Graph curvature for differentiating cancer networks. Sci. Rep. (Nat.), **5**, 12323 (2015). https://doi.org/10.1038/srep12323
27. Sandhu, R., Georgiou, T., Tannenbaum, A.: Ricci curvature: an economic indicator for market fragility and systemic risk. Sci. Adv. **2** (2016). https://doi.org/10.1126/sciadv.1501495
28. Tannenbaum, E., Georgiou, T., Tannenbaum, A.: Optimal mass transport for problems in control, statistical estimation, and image processing. In: Dym, H., de Oliveira, M.C., Putinar, M. (eds.) Mathematical Methods in Systems, Optimization, and Control. Birkhauser, Basel (2012)
29. Tannenbaum, E., Georgiou, T., Tannenbaum, A.: Signals and control aspects of optimal mass transport and the Boltzmann entropy. In: 49th IEEE Conference on Decision and Control, pp. 1885–1890 (2010)
30. Villani, C.: Topics in Optimal Transportation, Graduate Studies in Mathematics, vol. 58. AMS, Providence, RI (2003)
31. Villani, C.: Trend to equilibirum for dissipative equations, functional inequalities and mass transportation. In: de Carvalho, M., Rodrigues, J-F. (eds.) Contemporary Mathematics: Recent Advances in the Theory and Applications of Mass Transport. American Mathematical Society Publications (2004)
32. Villani, C.: Optimal Transport, Old and New. Springer, New York (2008)
33. Wang, C., Jonckheere, E., Banirazi, R.: Wireless network capacity versus Ollivier-Ricci curvature under Heat Diffusion (HD) protocol. In: Proceedings of ACC (2013)
34. Yamamoto, K., Chen, Y., Ning, L., Georgiou, T., Tannenbaum, A.: Regularization and interpolation of positive matrices (2016). https://arxiv.org/abs/1611.07945

Chapter 11
Identification of Dynamical Networks

Michel Gevers, Alexandre S. Bazanella and Guilherme A. Pimentel

Abstract We consider the identification of networks of linear time-invariant dynamical systems whose node signals are measured and are connected by causal linear time-invariant transfer functions. The external signals at the nodes may comprise both known excitation signals and unknown stationary noise signals. The identification of such networks comprise two essentially different problems. The first is to find conditions on the external excitation signals that allow the identification of the whole network from the measured node signals and excitation signals. The second problem is the identification of a particular module (i.e., transfer function) embedded in the network. We present state of the art results for both problems.

11.1 Introduction

The identification of networks of dynamical systems has recently emerged as an active topic in the systems and control community. Attention has focused on networks in which the node signals are connected by scalar causal rational transfer functions. These node signals are excited through the network by a combination of known external excitation signals and unknown noise sources. The node signals and the known external excitation signals are assumed to be measured without error. The identification of such networks essentially contains two different questions.

M. Gevers (✉)
ICTEAM, Louvain University, Bâtiment Euler
B1348 Louvain la Neuve, Belgium
e-mail: Michel.Gevers@uclouvain.be

A. S. Bazanella · G. A. Pimentel
Department of Automation and Energy, Universidade Federal
do Rio Grande do Sul, Av. Osvaldo Aranha 99, Porto Alegre-RS, Brazil
e-mail: bazanella@ufrgs.br

G. A. Pimentel
e-mail: guilherme.pimentel@ufrgs.br

R. Tempo et al. (eds.), *Emerging Applications of Control and Systems Theory*, Lecture Notes in Control and Information Sciences - Proceedings,
https://doi.org/10.1007/978-3-319-67068-3_11

The first question is the identification of the whole network using all measured node signals and the known external excitation signals. This approach estimates all transfer functions of the network and, as a result, also delivers the topology of the network by detecting which of these transfer functions are zero, so that one can construct the directed graph describing its interconnection structure. As we shall show, there is a fundamental unidentifiability problem in the sense that it is impossible to reconstruct such dynamical networks from measurements of the nodes and of the known external excitation signals, unless some prior knowledge is available about the structure of the network. The question is thus to produce conditions on the network structure (in the form of prior knowledge) and on the external excitation signals that lead to a unique identification of the whole network. To illustrate how virgin this question was until recently, we quote from [8] published in 2010: "*Remarkably, while networks of dynamical systems have been deeply studied and analyzed in automatic control theory, the question of reconstructing an unknown dynamical network has not been formally investigated yet. Indeed, in most applicative scenarios the network is given or it is the very objective of design. However, there are also some interesting situations where the link structure is actually unknown and dynamic, such as in biological neural networks, biochemical metabolic pathways and financial markets with a high frequency trade.*"

The second question concerns the identification of a particular transfer function within the network, assuming that its interconnection structure is known. It involves questions such as which signals need to be measured, and which external excitation signals need to be applied in order to estimate the desired transfer function. A number of results on this topic have been obtained recently [2, 3, 9].

In this chapter, we present state of the art results on these two questions. We first consider the problem of global identification of a network of dynamical systems. An early result pointing to the unidentifiability problem mentioned above can be found in [6] where the authors showed that, for a strictly proper continuous time system with known inputs, the transformation from input–output form to a network form is nonunique. More recent research has focused on the modeling and identification of high-dimensional stochastic processes, where the focus has been on detecting the causal links between variables [1, 7, 11].

We examine under what conditions on the network structure and on the external signals such network can be uniquely identified from the measured node signals and the known external signals, for networks with both deterministic and stochastic inputs. Our results take the form of a range of sufficient conditions on the network structure and on the external signals that will guarantee that the network can be uniquely reconstructed from the measured data. They are close to those of [10], even though our approach takes a different route inspired by the deterministic approach of [6].

We adopt the network model structure studied in [10], and we first show that this network model can be transformed into an equivalent Multiple Input Multiple Output (MIMO) model with added noise, which can be identified in open loop. The identifiability conditions for open-loop MIMO systems are well established, and they lead to a unique Input–Output (I/O) model and a unique noise model under the usual assumptions that the system is in the model set and that the data are informative with respect to the adopted model structure. The question of whether the network model can be identified from measured data then turns into the question of whether or not the mapping from the network model to the I/O model is injective.

We first propose a definition of *network identifiability* that relates to the objective of identifying the true network from measured data. A network model structure that is able to represent the true network will be called identifiable if no other different network model structure, that is unable to represent the true network, can produce the same I/O model. By extending the results of [6] to networks with unmeasured noise signals and with transfer functions that need not be strictly proper, we then show that, generically, there is an infinity of network models that produce the same I/O model, and we provide a parametrization of all these indistinguishable network models. These indistinguishable network models may even have different interconnection structures, i.e., the zero transfer functions are in different locations, leading to different corresponding graphs. This implies that a network model structure that is able to represent the true network will be identifiable only if some adequate prior knowledge is available about its structure. Such prior knowledge can take many different forms, such as the topology of the interconnection structure between the nodes, or the topology of the external excitation structure by the known excitation signals or by the noise signals.

We present a range of sufficient conditions on the structure of the external excitation signals—reference excitation signals and noise signals—that make the network model structure identifiable. These conditions show that the known excitation signals and the unknown noise signals play the same role in terms of their capacity to make the network structure identifiable; in other words identifiability can be achieved either by the known excitation signals, or by the noise signals, or by a combination of both.

In the second part of this chapter, we consider the problem of identifying a module (i.e., a transfer function) embedded in the network. Several contributions have recently been made for this problem [2, 3, 9]. A major open problem is that of finding conditions on the external excitation signals (known or noisy) that will lead to a consistent estimate of the desired transfer function. We illustrate this problem on a 3-dimensional network. Our contribution is twofold: show which external excitation signals are required to make the data informative, and show how adding additional excitation at other nodes affects the parameter variances of the estimated transfer function.

The outline of this chapter is as follows. The network model structure is presented in Sect. 11.2. In Sect. 11.3 we present a definition of identifiability of a network which relates to the objective of identifying the true network. We then show that a network is generically unidentifable and we parametrize the set of all indistinguishable network models. Using this parametrization, we present a range of sufficient conditions on the structure of the external excitation that render the network identifiable. In Sect. 11.4 we illustrate the problem of obtaining an informative experiment for the identification of an embedded module. We conclude in Sect. 11.5.

11.2 Problem Statement

We consider a network made up of L nodes, with node signals denoted $\{w_1(t), \ldots, w_L(t)\}$. These node signals are related to each other and to external excitation signals r_j and white noise signals e_j by the following network equations, which we call the **network model** and in which the matrix G^0 will be called the **network matrix**:

$$\begin{bmatrix} w_1 \\ w_2 \\ \vdots \\ w_L \end{bmatrix} = \begin{bmatrix} 0 & G_{12} & \ldots & G_{1L} \\ G_{21} & 0 & \ddots & G_{2L} \\ \vdots & \ddots & \ddots & \vdots \\ G_{L1} & G_{L2} & \ldots & 0 \end{bmatrix} \begin{bmatrix} w_1 \\ w_2 \\ \vdots \\ w_L \end{bmatrix} + K^0(q) \begin{bmatrix} r_1 \\ r_2 \\ \vdots \\ r_L \end{bmatrix} + H^0(q) \begin{bmatrix} e_1 \\ e_2 \\ \vdots \\ e_L \end{bmatrix} \quad (11.1)$$

Or, equivalently

$$w(t) = G^0(q)w(t) + K^0(q)r(t) + H^0(q)e(t) \quad (11.2)$$

with the following properties:

- G_{ij} are proper but not necessarily strictly proper transfer functions. Some of them may be zero, indicating that there is no direct link from w_j to w_i.
- there is a delay in every loop going from one w_j to itself.
- the network is well-posed so that $(I - G^0)^{-1}$ is proper and stable.
- all node signals w_j, $j = 1, \ldots, L$ are measurable.
- r_i are external excitation signals that are available to the user in order to produce informative experiments for the identification of the G_{ij}. $K^0(q)$ reflects how the external excitation signals affect the node signals.
- $e \in \Re^L$ is a white-noise vector with a positive definite covariance matrix Σ. $H(q)$ is a $L \times L$ stable rational matrix.
- the external excitation signals r_i are assumed to be uncorrelated with all noise signals e_j, $j = 1, \ldots, L$.
- q^{-1} is the delay operator.

The network model (11.2) can be rewritten in a more traditional form as follows:

$$w(t) = T^0(q)r(t) + N^0(q)e(t) \quad (11.3)$$

where

$$T^0(q) \triangleq (I - G^0(q))^{-1}K^0(q), \quad N^0(q) \triangleq (I - G^0(q))^{-1}H^0(q). \quad (11.4)$$

The description (11.3) will be called the **input–output (I/O) description** of the network. A corresponding parametrized version $M_{io} = [T(q, \eta), N(q, \eta)]$ will be called the *input–output (I/O) model*.

11.3 Identifiability of the Whole Network

11.3.1 Definition of Network Identifiability

Consider now that our objective is to estimate the matrices $G^0(q)$, $K^0(q)$ and $H^0(q)$ of the network (11.2) using the available measurements $w(t)$ and $r(t)$. Assuming that the network is driven by sufficiently informative excitation signals $r(t)$ and white-noise signals $e(t)$, what are the conditions (in the form of required prior knowledge) on the network matrices $G^0(q)$, $K^0(q)$, $H^0(q)$ such that they can be uniquely identified from the known measured signals $w(t)$ and $r(t)$?

It is well known from the theory of identification of multi-input multi-output (MIMO) linear time-invariant (LTI) systems that from the signals $w(t)$ and $r(t)$ one can uniquely identify the matrices $T^0(q)$ and $N^0(q)$ of the input–output model (11.3) if the chosen model structure $M_{io} = [T(q,\eta), N(q,\eta)]$ is such that $[T^0(q), N^0(q)] = [T(q,\eta_0), N(q,\eta_0)]$ for some unique η_0 (this is the identifiability question), and if the signals $r(t)$ are sufficiently rich for the chosen parametrizations (this is the informativity question). The identification of (11.3) is an open loop identification problem.

The question of **network identifiability** then relates to the mapping from $[T^0(q), N^0(q)]$ to $[G^0(q), K^0(q), H^0(q)]$, namely *under what conditions (in the form of prior knowledge on the network matrices $G^0(q)$, $K^0(q)$, $H^0(q)$) can one uniquely recover the network matrices $[G^0(q), K^0(q), H^0(q)]$ from the true input–output description $[T^0(q), N^0(q)]$*? It can be formally defined as follows.

Definition 11.1 (*Identifiability of the true network model*) Consider the true network (11.2) defined by the triple $\mathscr{S} = [G^0, K^0, H^0]$ and a parametrized network model structure $\{M(\theta) = [G(\theta), K(\theta), H(\theta)], \theta \in D_\theta\}$ with the property that $M(\theta_0) = \mathscr{S} = [G^0, K^0, H^0]$ for some $\theta_0 \in D_\theta$. Let $[T^0, N^0]$ be the corresponding true I/O model defined by (11.4). Then $\mathscr{S}$ is network identifiable if there exists no other network model structure $\{\tilde{M}(\nu) = [\tilde{G}(\nu), \tilde{K}(\nu), \tilde{H}(\nu)], \nu \in D_\nu\}$ such that $(I - \tilde{G}(\nu_0))^{-1}\tilde{K}(\nu_0) = T^0$ and $(I - \tilde{G}(\nu_0))^{-1}\tilde{H}(\nu_0) = N^0$ for some ν_0, with $[\tilde{G}(\nu_0), \tilde{K}(\nu_0), \tilde{H}(\nu_0)] \neq [G^0, K^0, H^0]$.

We illustrate this definition with the following example studied in [10].

Example 11.1 Consider the following 3-node noise-free network $\mathscr{S}_1$:

$$G^0(q) = \begin{bmatrix} 0 & 0 & 0 \\ A(q) & 0 & 0 \\ 0 & B(q) & 0 \end{bmatrix}, \; K^0(q) = \begin{bmatrix} 1 & 0 & 0 \\ 0 & 1 & 0 \\ 1 & 0 & 0 \end{bmatrix} \tag{11.5}$$

i.e., $G^0_{21}(q) = A(q)$, $G^0_{32}(q) = B(q)$, where $A(q)$ and $B(q)$ are rational transfer functions, and all other G^0_{ij} are zero. The corresponding I/O description of the true network is given by (11.3) with

$$T^0(q) = \begin{bmatrix} 1 & 0 & 0 \\ A(q) & 1 & 0 \\ A(q)B(q)+1 & B(q) & 0 \end{bmatrix} \tag{11.6}$$

The following $\bar{G}$ and $\bar{K}$ yield a network $\mathscr{S}_2$ with the same I/O model T^0 as the "true" network (11.5):

$$\bar{G}(q) = \begin{bmatrix} 0 & -B(q) & 1 \\ A(q) & 0 & 0 \\ 0 & B(q) & 0 \end{bmatrix}, \bar{K}(q) = \begin{bmatrix} 0\,0\,0 \\ 0\,1\,0 \\ 1\,0\,0 \end{bmatrix} \tag{11.7}$$

Thus, the network $[\bar{G}, \bar{K}]$ is indistinguishable from the true network even though it has a different topology. It means that if the same data $\{r(t)\}$ excite the two networks (11.5) and (11.7), they will generate the same data $\{w(t)\}$, despite the fact that the graphs of these two networks are different.

11.3.2 The Set of All Indistinguishable Networks

We show in this section that there exists an infinite set of network models $M(\theta) = [G(q,\theta), K(q,\theta), H(q,\theta)] \in \mathscr{M}^*$ that produce the same I/O model $[T(q), N(q)]$, and we parametrize the set of these indistinguishable network models. This will allow us to derive conditions on prior knowledge of the true network model structure that will make this network identifiable in the sense of Definition 11.1. This parametrization is an extension to networks with noisy inputs of a result of [6] which addressed the case of a noiseless network with strictly proper transfer functions. We first introduce the notion of **admissible network matrix**.

Definition 11.2 (*Admissible network matrix*) A network matrix $G(q,\theta)$ is called admissible if the following conditions hold:

- the diagonal elements of $G(q,\theta)$ are zero;
- there is a delay in every loop going from one w_j to itself;
- all $G_{ij}(q,\theta)$ are proper
- $(I - G(q,\theta))^{-1}$ is stable

The following theorem describes the set of all network models that produce the same I/O model $[T \;\; N]$. For brevity of notations, we delete the (q,θ) dependence.

Theorem 11.1 *The set of all network models that produce an I/O model $M_{io} = [T \;\; N]$ is given by*

$$\{[\tilde{G} \; \tilde{K} \; \tilde{H}] = [\tilde{G} \;\; (I-\tilde{G})T \;\; (I-\tilde{G})N]\} \tag{11.8}$$

where $\tilde{G}$ is any admissible network matrix of size $L \times L$, in the sense of Definition 11.2.

Proof We first show that the set of network matrices defined in (11.8) produce the correct I/O model $M_{io} = [T, N]$. Indeed, the I/O transfer function matrices derived from (11.8) are

$$\tilde{T} = (I - \tilde{G})^{-1}\tilde{K} = (I - \tilde{G})^{-1}(I - \tilde{G})T = T$$
$$\tilde{N} = (I - \tilde{G})^{-1}\tilde{H} = (I - \tilde{G})^{-1}(I - \tilde{G})N = N$$

Conversely, let $[\tilde{G}, \tilde{K}, \tilde{H}]$ be any network that produces the correct T and N with $\tilde{G}$ admissible. Then, necessarily, we must have $(I - \tilde{G})^{-1}\tilde{K} = T$ and $(I - \tilde{G})^{-1}\tilde{H} = N$. Premultiplying both equations by $(I - \tilde{G})$ shows that this network has the form (11.8).

This result shows that without prior knowledge about the network structure, any admissible $\tilde{G}$ can produce the true I/O model $[T^0(q), N^0(q)]$. The choice of any particular $\tilde{G}$ fixes the corresponding $\tilde{K} = (I - \tilde{G})T$ and $\tilde{H} = (I - \tilde{G})N$, and the network $[\tilde{G}, \tilde{K}, \tilde{H}]$ is then indistinguishable from the "true" $[G^0, K^0, H^0]$. This means that if they are driven by the same $\{r(t), e(t)\}$ signals, they will generate the same $\{w(t)\}$. Thus, a network is generically not identifiable from measured data $\{w(t), r(t)\}$, unless some prior information is known about $G^0(q)$ and/or $K^0(q)$ and/or $H^0(q)$.

The following corollary, which is an extension to noisy networks of Lemma 4 of [6], will help us generate constraints that make a network identifiable.

Corollary 11.1 *Let $[G^0, K^0, H^0]$ be the transfer matrices of the "true" network. Let ΔG be any transfer function matrix of size $L \times L$ such that $\tilde{G} \triangleq G^0 + \Delta G$ is admissible in the sense of Definition 11.2. Let $\tilde{K} = K^0 + \Delta K$ and $\tilde{H} = H^0 + \Delta H$ be the corresponding matrices defined by (11.8). Then the network $[\tilde{G}, \tilde{K}, \tilde{H}]$ has the same I/O model as the true network $[G^0, K^0, H^0]$ if and only if*

$$[\Delta G \ \Delta K \ \Delta H] \begin{bmatrix} T^0 & N^0 \\ I & 0 \\ 0 & I \end{bmatrix} = [0 \ \ 0] \tag{11.9}$$

Proof The proof follows immediately from Theorem 11.1 by noting that for all these $[\tilde{G} \ \tilde{K} \ \tilde{H}]$ we have $(I - \tilde{G})^{-1}[\tilde{K} \ \ \tilde{H}] = [T^0 \ N^0]$.

11.3.3 Conditions for Network Identifiability

In this section we use the result of Corollary 11.1 to derive a range of sufficient conditions under which the true network is identifiable. These conditions take the form of prior knowledge on the structure of the excitation matrices $K^0(q)$ and $H^0(q)$. As stated in the introduction, it is a realistic situation that the way in which the external signals enter the network is known a priori. The following theorem provides a first set of sufficient conditions.

Theorem 11.2 *The network structure (11.1) is identifiable if* $L-1$ *columns of the matrix* $[K^0 \;\; H^0]$ *are known and linearly independent.*

Proof It follows from Corollary 11.1 that the network $[G^0 \;\; K^0 \;\; H^0]$ is identifiable if and only if there is no triple $[\Delta G \;\; \Delta K \;\; \Delta H]$ that satisfies

$$\Delta G(I - G^0)^{-1}[K^0 \;\; H^0] = -[\Delta K \;\; \Delta H] \tag{11.10}$$

Since ΔG has zeroes on its diagonal, it contains $L \times (L-1)$ unknown elements. Now let W denote the $L \times (L-1)$ submatrix of $[K^0 \;\; H^0]$ made up of its known and linearly independent columns. Then the corresponding columns of $[\Delta K \;\; \Delta H]$ are zero. From (11.10) we can thus extract the following subset of equations for ΔG:

$$\Delta G(I - G^0)^{-1} W = O \tag{11.11}$$

where W and O have size $L \times (L-1)$. This represents a set of $L \times (L-1)$ linearly independent equations for the $L \times (L-1)$ unknown elements of ΔG, from which it follows that $\Delta G = 0$. It then follows from (11.10) that ΔK and ΔH are also zero.

By applying this theorem to Example 11.1 we note that if K^0 is known, then the network is identifiable. A network is also identifiable when either $K^0(q)$ or $H^0(q)$ is diagonal with nonzero diagonal elements, a situation that is not covered by Theorem 11.2.

Theorem 11.3 *Consider the network structure (11.1) and assume that either* $K^0(q)$ *or* $H^0(q)$ *is diagonal and of full rank. Then the network is identifiable.*
The proof can be found in [4].

Alternative sets of sufficient conditions for identifiability of the whole network have also been derived in [10].

11.4 Identification of an Embedded Module

In this section we consider the other major problem in the identification of networks, namely the identification of a single embedded module. Without loss of generality, consider that the objective is to identify the module $G_{12}(q)$ in the network (11.1).

Historically, this problem was addressed first in [9]. In that paper, the authors proposed several solutions to this problem, based on existing closed-loop identification methods. Indeed, if the objective is to identify $G_{12}(q)$, it is easy to show that the network model (11.1) can be rewritten as a Multiple Input Single Output (MISO) closed-loop system, where w_1 acts as the single output and $[w_2\ w_3 \dots w_L]^T$ acts as the input vector of this closed-loop system. Thus, the various methods of closed-loop identification can be applied for the identification of $G_{12}(q)$.

An important problem that has not been solved so far is that of deciding which external excitation signals, measured or unmeasured, need to be applied for the identification algorithms to converge to the true G_{12}. This is the question of informativity of the identification experiment. In [2, 9] it is assumed that the vector $w(t)$ of node signals is informative, but this is an internal constraint. The difficult question is what are the requirements on the external signals, $r_i(t)$ and $e_i(t)$, that will deliver informative data for the identification of the module $G_{12}(q)$. Assuming that different choices of external signals can yield informative data, then another interesting question is how do these different choices affect the variance of the estimated $\hat{G}_{12}(q)$.

The objective of obtaining necessary and sufficient informativity conditions on the external excitation signals for the identification of a specific module, say G_{12}, is illusory, since these informativity conditions will depend on the method that is used for the identification of G_{12} and, in particular, on the signals that are used. Thus, the aim is to find sufficient conditions for informativity. The first contribution to this informativity question for an embedded module is to be found in [3], where we have analyzed a 3-node network. We have shown that, even in such simple network, different alternatives exist for the identification of G_{12} and we have proposed a framework, based on [5], for the computation of sufficient conditions for informativity depending on the identification method used.

Here we study the *direct identification* of G_{12} using the first equation of (11.12), and we extend the analysis of [3] by computing not just the informativity requirements, but also the way in which informative excitation signals affect the variance of the estimated G_{12}. In order to obtain an unbiased estimate of G_{12}, the direct method requires the identification of the vector $[G_{12}\ \ G_{13}]$. To keep the analysis simple, we shall assume that $K^0 = H^0 = I$. Thus, consider the following 3-node network:

$$\begin{bmatrix} w_1 \\ w_2 \\ w_3 \end{bmatrix} = \begin{bmatrix} 0 & G_{12} & G_{13} \\ G_{21} & 0 & G_{23} \\ G_{31} & G_{32} & 0 \end{bmatrix} \begin{bmatrix} w_1 \\ w_2 \\ w_3 \end{bmatrix} + \begin{bmatrix} r_1 \\ r_2 \\ r_3 \end{bmatrix} + \begin{bmatrix} e_1 \\ e_2 \\ e_3 \end{bmatrix} \tag{11.12}$$

where it is desired to identify G_{12}. For the purpose of analyzing the effect of different excitation scenarios on the estimates, we adopt the following model structure for the parametrization of G_{12}, G_{13}:

$$\mathscr{M} = \left\{ G_{12}(\alpha), G_{13}(\alpha, \beta),\ \theta = \left(\alpha^T\ \beta^T\right)^T \in D_\theta \subset \mathscr{R}^d \right\} \tag{11.13}$$

where $G_{12}(\alpha)$ and $G_{13}(\alpha, \beta)$ are rational transfer functions, $\theta \in \mathscr{R}^d$ is the vector of model parameters, and D_θ is a subset of admissible values for θ. Thus, α are the possibly common parameters of $G_{12}(\theta)$ and $G_{13}(\theta)$.

We shall assume that there exists some $\theta^0 = (\alpha_0^T, \beta_0^T)^T \in D_\theta$ that represents the true G_{12}^0 and G_{13}^0. The one-step ahead prediction error for $w_1(t)$ is given by

$$\varepsilon_1(t,\theta) \stackrel{\Delta}{=} w_1(t) - \hat{w}_1(t|t-1,\theta) = [w_1(t) - G_{12}(\alpha)w_2(t) - G_{13}(\alpha,\beta)w_3(t) - r_1(t)]$$

If the model structure is identifiable and the data informative, the parameter vector estimate $\hat{\theta}^N$ converges asymptotically to the true θ^0, and the per sample asymptotic covariance matrix is given by $P_\theta = [I(\theta^0)]^{-1}$ where $I(\theta)$ is the information matrix:

$$I(\theta) \stackrel{\Delta}{=} \bar{E}[\psi(t,\theta)\psi^T(t,\theta)] \tag{11.14}$$

The pseudoregressor vector $\psi(t,\theta) \stackrel{\Delta}{=} \frac{\partial \varepsilon_1(t,\theta)}{\partial \theta}$ is expressed as follows as a function of the excitation signals:

$$\psi(t,\theta) = V(q,\theta) \begin{bmatrix} r_1(t) + e_1(t) \\ r_2(t) + e_2(t) \\ r_3(t) + e_3(t) \end{bmatrix} \tag{11.15}$$

where $V(q,\theta)$ is a $d \times 3$ matrix of transfer functions obtained as follows from the partial derivatives of $G_{12}(\theta)$ and $G_{13}(\theta)$ with respect to the unknown parameters.

$$V(q,\theta) = \begin{bmatrix} V_1 & V_2 & V_3 \end{bmatrix}, \text{ where} \tag{11.16}$$

$$\begin{bmatrix} V_1 \\ V_2 \\ V_3 \end{bmatrix} = \begin{bmatrix} T_{21}^0 & T_{31}^0 \\ T_{22}^0 & T_{32}^0 \\ T_{23}^0 & T_{33}^0 \end{bmatrix} \begin{bmatrix} \nabla_1 \\ \nabla_2 \end{bmatrix}, \text{ with } \nabla_1 = \begin{bmatrix} \frac{\partial G_{12}}{\partial \alpha} \\ 0 \end{bmatrix} \text{ and } \nabla_2 = \begin{bmatrix} \frac{\partial G_{13}}{\partial \alpha} \\ \frac{\partial G_{13}}{\partial \beta} \end{bmatrix}. \tag{11.17}$$

Here the T_{ij}^0 are the elements of the second and third column of the transfer matrix $T^0 \stackrel{\Delta}{=} (I - G^0)^{-1}$ of the true network (11.12).

A data set is informative if the information matrix that it produces is nonsingular, i.e., $I(\theta) > 0$. By the above expressions, this is equivalent with the condition that there exists no vector $\mu \in \Re^d$ with $\mu \neq \mathbf{0}$ such that

$$\mu^T V(q,\theta) = \mathbf{0}. \tag{11.18}$$

We now apply this informativity analysis to the identification of $G_{12} = a_1 q^{-1} + a_2 q^{-2}$ and $G_{13} = bq^{-1}$ using the direct prediction error method based on the first equation in the following 3-node example.

$$\begin{bmatrix} w_1 \\ w_2 \\ w_3 \end{bmatrix} = \begin{bmatrix} 0 & a_1q^{-1} + a_2q^{-2} & bq^{-1} \\ q^{-1} & 0 & 0 \\ 0 & cq^{-1} & 0 \end{bmatrix} \begin{bmatrix} w_1 \\ w_2 \\ w_3 \end{bmatrix} + \begin{bmatrix} r_1 \\ r_2 \\ r_3 \end{bmatrix} + \begin{bmatrix} e_1 \\ e_2 \\ e_3 \end{bmatrix} \tag{11.19}$$

Applying expressions (11.16)–(11.17) to this example, with $\alpha = (a_1 \;\; a_2)$ and $\beta = b$, yields:

$$\begin{bmatrix} V_1 \; V_2 \; V_3 \end{bmatrix} = \frac{1}{\Delta} \begin{bmatrix} q^{-2} & q^{-1} & bq^{-3} \\ q^{-3} & q^{-2} & bq^{-4} \\ cq^{-3} & cq^{-2} & q^{-1} - a_1q^{-3} - a_2q^{-4} \end{bmatrix} \tag{11.20}$$

where $\Delta = 1 - a_1q^{-2} - (a_2 + bc)q^{-3}$. From (11.20) it is clear that $\mu^T [V_1 \;\; V_2] = \mathbf{0}$ for $\mu = [0 \;\; c \;\; -1]^T$, while $Ker(V^3) = \{\mathbf{0}\}$. This shows that applying either $r_3 \neq 0$ or $e_3 \neq 0$ is a necessary and sufficient condition for the generation of informative data, and thus for convergence of the parameters a_1, a_2, b to their true values. Additional signals at other nodes may reduce the variance of these estimates, and hence of $\hat{G}_{12}$, since $I(\theta)$ is given by (11.14), which leads to the following covariance of the estimate:

$$P_{\hat{\theta}N} = \frac{\lambda_1}{N}[I(\theta^0)]^{-1} \qquad I(\theta) = \frac{1}{2\pi}\int_0^{2\pi} \left\{ \sum_1^3 [V_i V_i^{\star} \Phi_{r_i} + V_i V_i^{\star} \lambda_i] \right\} d\omega \tag{11.21}$$

where N is the number of data used in the identification and λ_i is the variance of e_i.

We illustrate our informativity analysis and the effect of different scenarios of external excitation on parameter variance by calculating the variance from (11.21) using the following true parameters: $a_1 = -0.3$, $a_2 = 0.8$, $b = -0.5$, $c = 0.5$. We consider there different scenarii: excitation of r_3 alone, excitation of r_3 and r_1, and excitation of all inputs. In all cases, the inputs are white noise with unit variance, while a white noise with variance $\lambda_1 = 2$ is present in the first equation - the one that is used for prediction error identification. The variances of the parameter estimates are calculated for $N = 2,000$ data.

Table 11.1 below shows the different experimental scenarii and the corresponding values of the covariance matrix—note that the individual variances of each parameter correspond to the diagonal elements of this matrix. Recall that either $e_3 \neq 0$ or $r_3 \neq 0$ is necessary and sufficient for informativity. Sufficiency is confirmed in the first part of the Table, which gives a finite covariance for the first scenario where only node 3 is excited. Necessity is confirmed by simulations, which yield an infinite covariance matrix if identification is performed with $r_3 = e_3 = 0$. Notice in the Table how the variances are reduced as excitation in the other inputs is added, although this reduction is very slim in the variance of parameter b. Identical covariance matrices are obtained if $r_3 = 0$ while $e_3 \neq 0$ is applied with the same unit variance as that used for r_3.

Chapter 12
Smooth Operators Enhance Robustness

Keith Glover and Glenn Vinnicombe

Abstract The problem of synthesizing an $\mathscr{H}_\infty$ loop-shaping controller, but with a bound on its complexity, is shown to be a tractable optimization problem. Here complexity is defined in terms of the smoothness of the transfer function.

12.1 Introduction

The gap metric between a nominal plant with transfer function, $P(s)$, and a perturbed plant, $P_\Delta(s)$, given by $\delta(P, P_\Delta)$, [1] and the ν-gap, $\delta_\nu(P, P_\Delta)$, [2] have been demonstrated as very effective measures of plant uncertainty when subject to feedback control, and together with $\mathscr{H}_\infty$ loop-shaping [3] give a very straightforward and effective design methodology [4, 5]. These measures are essentially pointwise in frequency plus a closed-loop stability or winding number requirement. Hence if a plant has poles and zeros near the imaginary axis then a small change in frequency scale to give, P_Δ, can result in a large gap and perhaps the false impression that the plant and perturbed plant will be difficult to control with a single robust controller. In [6] they concluded that the gap metric could be too conservative. In [7] uncertainty in the 's' variable in $P(s)$ was analysed in detail showing that the *effective* distance between P and P_Δ could be small when the complexity/smoothness of the controller, $K(s)$, is small. In Sect. 12.2 these results are summarized. The purpose of the present paper is to consider the problem of synthesizing controllers with a bound on the complexity to meet an $\mathscr{H}_\infty$ loop-shaping specification. A motivation for revisiting this problem came in [8] where exactly these difficulties were encountered in an example of stabilizing combustion instabilities and poor stability margins were experienced with an optimal controller but a reduced order, lower complexity controller gave satisfactory

K. Glover · G. Vinnicombe (✉)
Department of Engineering, University of Cambridge, Cambridge, UK
e-mail: gv@eng.cam.ac.uk

K. Glover
e-mail: kg@eng.cam.ac.uk

R. Tempo et al. (eds.), *Emerging Applications of Control and Systems Theory*, Lecture Notes in Control and Information Sciences - Proceedings,
https://doi.org/10.1007/978-3-319-67068-3_12

results. A positive feedback convention is used so in examples the phase of $-K(j\omega)$ will be used to be consistent with the normal use of phase advance. Otherwise the notation is standard in the area[1].

12.2 Background on $\mathscr{H}_\infty$-Loop Shaping and Complexity

This paper is built on a simple premise. For a generalized plant G, and bound γ, the set of internally stabilizing controllers achieving $\|\mathscr{F}_l(G, K)\|_{\mathscr{H}_\infty} < \gamma$ is either empty or very large indeed. If there exists such a K then it will stabilize all plants of the form $\mathscr{F}_u(G, \Delta) : \Delta \in \mathscr{H}_\infty, \|\Delta\|_{\mathscr{H}_\infty} \leq 1/\gamma$. However, if you let $P_\Delta = \mathscr{F}_u(G, \Delta)$ for some $\Delta \in \mathscr{H}_\infty$ with $\bar{\sigma}(\Delta(j\omega)) > 1/\gamma$ at some frequency, then so long as there is no smaller Δ such that $P_\Delta = \mathscr{F}_u(G, \Delta)$, there will always exist an internally stabilizing controller K_{bad}, achieving $\|\mathscr{F}_l(G, K_{\text{bad}})\|_{\mathscr{H}_\infty} < \gamma$, which destabilizes P_Δ [2]. In fact, define any number of perturbed plants in the same way and, as long as there exists a disjoint set of frequencies at which each Δ exceeds the bound, there will always exist a *single* internally stabilizing controller K_{bad} achieving $\|\mathscr{F}_l(G, K_{\text{bad}})\|_{\mathscr{H}_\infty} < \gamma$ which destabilizes each and every one of them. Such a controller might be expected to have an extremely complex frequency response. Fortunately, the central solution to the $\mathscr{H}_\infty$ control problem is not usually this badly behaved, and this is particularly so for the $\mathscr{H}_\infty$ loop-shaping generalized plant, but even then can sometimes seem unnecessarily complex. One useful trick is to attempt to restrict the McMillan degree of the controller but this comes with no guarantees of added robustness. However, by restricting the complexity of K in the frequency domain, making it a smoother operator, it *is* possible to guarantee extra robustness *a priori* as the following simple example shows.

12.2.1 A Very Simple Example

The formulation of the problem in the gap framework tends to hide the simplicity of the underlying issue, so we start with a motivating example based on an additive uncertainty description. $\mathscr{H}_\infty$ control is rooted in the observation that internal stability plus

$$\left\| \frac{K}{1 - P_1 K} \right\|_{\mathscr{H}_\infty} < \gamma \tag{12.1}$$

guarantees that K will stabilize any plant P_Δ which has the same number of RHP poles as P_1 and satisfies $\|P_1 - P_\Delta\|_{\mathscr{L}_\infty} \leq 1/\gamma$. Let

[1] i.e. $P^* = (P(-s))^T$; $P(s) = D + C(sI - A)^{-1}B = \left[\begin{array}{c|c} A & B \\ \hline C & D \end{array}\right]$; $\mathscr{F}_\ell\left(\begin{bmatrix} P_{11} & P_{12} \\ P_{21} & P_{22} \end{bmatrix}, K\right) = P_{11} + P_{12}K(I - P_{22}K)^{-1}P_{21}$.

$$P_\lambda = \frac{1}{s+\lambda} \text{ so } P_1 = \frac{1}{s+1} \text{ and } P_2 = \frac{1}{s+2} = P_1 + \frac{-1}{(s+1)(s+2)}.$$

Any controller stabilizing P_1 and satisfying $\left\| \frac{K}{1-P_1K} \right\|_{\mathcal{H}_\infty} < 2$ is guaranteed to stabilize P_2. This is equivalent to requiring

$$\inf_\omega |P_1(j\omega) - 1/K(j\omega)| > 1/2.$$

Moreover, there exist controllers achieving $\inf_\omega |P_1(j\omega) - 1/K(j\omega)| = 1/2$ which destabilize P_2. An example is $K_{\text{bad}} = -\frac{s/\sqrt{2}+1}{s-1/2}$. Note that this controller has to work rather hard to destabilize P_2 and not P_1. On the other hand, it is straightforward to find controllers for P_1 which stabilize P_2. Indeed any constant controller which stabilizes P_1 (i.e. any $K < 1$) will also stabilize P_2 (which requires $K < 2$). Is there a way of capturing the notion that any sufficiently *simple* controller which stabilizes P_1 also stabilizes P_2? Note that $P_2(s) = P_1(s+1)$, and (12.1) guarantees that $|P_1(s) - 1/K(s)| > 1/\gamma$ for all $s : \Re(s) \geq 0$. So, if $1/K(s)$ and $1/K(s+1)$ are similar, that is if $1/K(s)$ doesn't change rapidly, then everything should be fine. To make this idea concrete, note that for all λ and all $\delta \geq 0$

$$\begin{aligned} |P_\lambda(j\omega) - 1/K(j\omega)| &\geq |P_1(j\omega+\delta) - 1/K(j\omega+\delta)| \\ &\quad - |P_\lambda(j\omega) - P_1(j\omega+\delta)| - |1/K(j\omega) - 1/K(j\omega+\delta)| \\ &\geq 1/\gamma - |P_\lambda(j\omega) - P_1(j\omega+\delta)| - |1/K(j\omega) - 1/K(j\omega+\delta)|. \end{aligned} \tag{12.2}$$

Taking $\delta = \lambda - 1$ for example, we have $P_\lambda(j\omega) = P_1(j\omega+\delta)$ and

$$|1/K(j\omega) - 1/K(j\omega+\delta)| \leq \delta \|\tfrac{d}{ds} 1/K(s)\|_{\mathcal{H}_\infty} \leq \|\tfrac{d}{ds} 1/K(s)\|_{\mathcal{H}_\infty} \text{ for } \lambda \in [1,2]$$

and so if $\|\frac{d}{ds} 1/K(s)\|_{\mathcal{H}_\infty} < 1/\gamma$ then $K(s)$ is guaranteed to stabilize P_λ for all $\lambda \in [1,2]$, as no closed-loop poles can cross the imaginary axis as we perturb from P_1 to P_2. Alternatively, any controller which satisfies (12.1) and yet *destabilizes* P_2 must necessarily have $\|\frac{d}{ds} 1/K(s)\|_{\mathcal{H}_\infty} \geq 1/\gamma$, and indeed stronger results are possible by optimizing (12.2) over δ. Thus the smoothness of K can be used to guarantee greater robustness. For this example, then provided $\|\frac{d}{ds} 1/K(s)\|_{\mathcal{H}_\infty} = \mu < 1$

$$|P_2(j\omega) - P_1(j\omega+\delta)| + |1/K(j\omega) - 1/K(j\omega+\delta)| \leq |P_2(j\omega) - P_1(j\omega+\delta)| + \delta\mu$$

and there will always exist a δ such that $|P_2(j\omega) - P_1(j\omega+\delta)| + \delta\mu < 0.5$, meaning that any controller which stabilizes P_1, with $\left\| \frac{K}{1-P_1K} \right\|_\infty \leq 2$, and destabilizes P_2 must necessarily satisfy $\|\frac{d}{ds} 1/K(s)\|_{\mathcal{H}_\infty} \geq 1$, whereas $\|\frac{d}{ds} 1/K_{\text{bad}}(s)\|_{\mathcal{H}_\infty} = 1 + \frac{1}{2\sqrt{2}} = 1.3536$. Note that this result is fundamentally asymmetric; to stabilize a ball around P_1 and destabilize P_2 requires a complex controller, yet it is straightforward to stabilize a ball around P_2 and destabilize P_1 with a smooth controller. For example, $K = 1$, with $\|\frac{d}{ds}1/K(s)\|_{\mathcal{H}_\infty} = 0$, achieves $\left\|\frac{K}{1-P_2K}\right\|_\infty = 2$ and destabilizes P_1.

12.2.2 A Useful Complexity Defintion

Now, the result of the previous subsection is a curiosity, but not useful in itself. The criterion chosen above is not one that would ever be used alone, and this particular approach is limited to minimum phase $K(s)$. However, in the same way, any $\mathcal{H}_\infty$ criterion will naturally induce a complexity measure on the controller by which stronger *a priori* robustness results, and "better" controllers, can be obtained. In the case of $\mathcal{H}_\infty$ loop shaping, the induced measure is particularly appealing and can be defined in such a way as to only depend on the rate of change of frequency response on the imaginary axis.

One of the great appeals of the $\mathcal{H}_\infty$ loop-shaping design procedure is that it usually results in "good" controllers, however there are cases where it doesn't, as the next example will show. First though we state a simplified version the main result from [7]. Rather than using the complex plane we measure distances as follows

$$\kappa(P_1, P_2; s_1, s_2) = \bar{\sigma}\left((I + P_1P_1^*)^{1/2}(s_1)\right)^{-1}(P_1(s_1) - P_2(s_2))\left((I + P_2^*P_2)^{1/2}(s_2)\right)^{-1}$$

where we take an antistable, with stable inverse, spectral factor in each case.[2] For $s_1 = j\omega_1$ and $s_2 = j\omega_2$ this is precisely the chordal distance between $P_1(j\omega_1)$ and $P_2(j\omega_2)$ on the Riemann sphere, or it's higher dimensional generalization. We further define an *effective* distance between P_1 and P_2 as

$$\delta_{\text{eff}}(P_1, P_2; \alpha) = \sup_{s_2:\Re(s_2)=0}\ \inf_{s_1:\Re(s_1)\geq 0} \kappa(P_1, P_2; s_1, s_2) + \alpha\,|\log s_1/s_2|$$

noticing the asymmetry here again. As before, P_1 is to be taken as the "nominal" system, and we are interested in robustness guarantees for P_2. We also define

$$V_K = \sup_{s:\Re(s)\geq 0}\lim_{\delta\to 0}\frac{\kappa(K, K; s, s+\delta)}{\log((s+\delta)/s)} = \left\|(I + KK^*)^{-1/2}\, s\frac{dK}{ds}\,(I + K^*K)^{-1/2}\right\|_{\mathcal{L}_\infty}$$

since it will attain it's maximum value on the imaginary axis, in which case any square root may be taken. State-space formulae for these objects are given in a later section. Finally, we define as usual

[2] This ensures that $\kappa(P_1, P_2; s_1, s_2)$ is analytic and bounded (by 1 in fact) for s_1, s_2 in the closed RHP, with no requirements for either P_1 or P_2 themselves to be stable or minimum phase, since RHP zeros of the inverted spectral factors cancel any unstable poles of the plants.

$$b_{P,K} = \left\| \begin{bmatrix} I \\ K \end{bmatrix} (I - PK)^{-1} \begin{bmatrix} I & P \end{bmatrix} \right\|_{\mathscr{H}_\infty}^{-1} = \left\| \mathscr{F}_\ell \left(\begin{bmatrix} I & 0 & 0 \\ 0 & 0 & I \\ I & 0 & 0 \end{bmatrix} + \begin{bmatrix} I \\ 0 \\ I \end{bmatrix} P(s) \begin{bmatrix} 0 & I & I \end{bmatrix}, K \right) \right\|_{\mathscr{H}_\infty}^{-1} \tag{12.3}$$

and note the close compatibility between δ_ν, $b_{P,K}$ and the definition of V_K that is implied by the following theorem.

Theorem 12.1 *Given* K *and* P_λ, *with* $\lambda \in [1, 2] \mapsto P_\lambda$, *continuous in the graph topology.*[3] *If* $b_{P_1,K} > \delta_{eff}(P_1, P_\lambda; V_K)$ *for all* λ *then* K *stabilizes* P_2 *and*

$$b_{P_2,K} \geq b_{P_1,K} - \delta_{eff}(P_1, P_2; V_K)$$

Note that if $P_2(s) = P_1(ks)$ then $\delta_{\text{eff}}(P_1, P_2; V_K) \leq |\log(k)| V_K$

12.2.3 Illustrative Example

In this subsection, we present a motivational example to illustrate the potential problems in $\mathscr{H}_\infty$-loop shaping when the optimal controller has high complexity. Let the nominal plant be

$$P_1(s) = \frac{s(s^2 + 1)}{(s^2/q + 0.1s + q)(s^2/q - 0.1s + q)}$$

with $q = 1.05$, and a perturbed plant be given by $P_2(s) = P_1(s/r)$ with $r = 1.1$ i.e. a 10% change in the frequency scale. The optimal controller in this case will be (using a positive feedback convention),

$$K_o(s) \approx \frac{-(1.22s^2 + 0.1282s + 1.345)}{(s^2 - 0.1048s + 1.103)}$$

giving $b_{\text{opt}}(P_1) = b_{P_1,K_o} = 0.633$, which is a good stability margin with the nominal plant. However with perturbed plant $b_{P_2,K_o} = 0.047$ which is an unsatisfactory stability margin. The log-complexity of K_o, $V_{K_o} = 19.6$, and the potential reduction in stability margin given by Theorem 12.1 of $V_{K_o} * \log(1.1) = 1.87$ gives no assurance of stability. However the simple phase lead controller, $K_1(s) = \frac{-(2.5s+1)}{(s+2.5)}$ satisfies $b_{P_1,K_1} = 0.353$ with $V_{K_1} = 0.362$ and for the perturbed plant satisfies $b_{P_2,K_1} = 0.354$, which is consistent with Theorem 12.1:

$$0.354 = b_{P_2,K_1} \geq b_{P_1,K_1} - V_{K_1}|\log(r)| = 0.353 - 0.035 = 0.318 \tag{12.4}$$

[3]This is the simplest way to state the theorem. If instead continuity requirements are placed on the implied map from s_1 to s_2 in the definition of δ_{eff} then an explicit path linking P_1 and P_2 is not required. See [7] for more details.

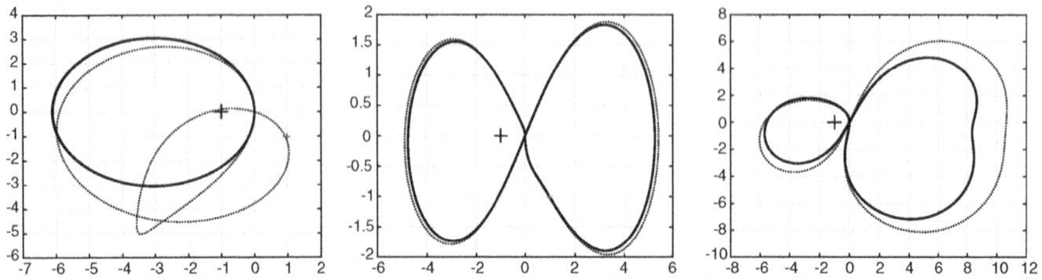

Fig. 12.1 Loop gain Nyquist diagrams for $-K_o(s)P_1(s)$ (solid) and $-K_o(s)P_2(s)$ (dotted) (for positive frequencies only) in left figure, and for $-K_1(s)P_1(s)$ (solid) and $-K_1(s)P_2(s)$ (dotted) in centre figure; and for the optimal 2nd order controller with $V_K \leq 2$ for right figure

Note that in this case K_o has two unstable poles in the neighbourhood of the poles and zeros of the plant giving both the plant and the optimal controller closely aligned and rapidly changing phase and magnitude. This alignment does not occur in the perturbed case as can be seen in the Nyquist diagrams in Fig. 12.1 (left panel). Whereas in the corresponding results for $K_1(s)$ there is very little difference between the nominal and perturbed loop gains as seen in Fig. 12.1 (centre panel). Note the the gap metric distance, $\delta_\nu(P_1, P_2) = 1$ whereas the effective distance $\delta_{\text{eff}}(P_1, P_2; V_{K_1})$ is no greater than $V_{K_1} \log(r) \approx 0.035$ (as used in (12.4)).

12.3 Controller Synthesis

In this section we will consider the synthesis of a controller, $K(s)$, that satisfies:

$$V_K \leq \alpha \text{ and } b_{P,K} \geq \beta \text{ if such exists for a given } \alpha, \beta > 0 \tag{12.5}$$

12.3.1 A State-Space Approach

Suppose a candidate controller is parameterized in the state-space by $K(s) = D(p) + C(p)(sI - A(p))^{-1}B(p)$, where p is a vector of parameters which is to be designed to meet the specification in (12.5). The specification on V_K will first be written in terms of the state space.

$$V_K := \left\| (I + KK^*)^{-1/2}\, s\frac{dK}{ds}\, (I + K^*K)^{-1/2} \right\|_{\mathscr{L}_\infty}$$

$$\text{now define } M_1 := \begin{bmatrix} -K^* \\ I \end{bmatrix} (I + KK^*)^{-1/2} \;\Rightarrow\; M_1^* M_1 = I$$

$$M_2 := (I + K^*K)^{-1/2} \begin{bmatrix} I & -K^* \end{bmatrix} \;\Rightarrow\; M_2 M_2^* = I$$

Pre and post multiplying by M_1 and M_2 gives,

$$V_K = \left\| \begin{bmatrix} -K^* \\ I \end{bmatrix} (I + KK^*)^{-1} \, s \frac{dK}{ds} \, (I + K^*K)^{-1} \begin{bmatrix} I & -K^* \end{bmatrix} \right\|_{\mathcal{L}_\infty}$$

and

$$\begin{bmatrix} -K^* \\ I \end{bmatrix} (I + KK^*)^{-1} = \mathcal{F}_\ell \left(\left[\begin{array}{c:c} I & K \\ 0 & I \\ \hdashline I & K \end{array} \right], -K^* \right)$$

$$(I + K^*K)^{-1} \begin{bmatrix} I & -K^* \end{bmatrix} = \mathcal{F}_\ell \left(\left[\begin{array}{cc:c} I & 0 & I \\ \hdashline K & I & K \end{array} \right], -K^* \right)$$

$$\begin{aligned} \text{Now } \frac{dK}{ds} &= C(sI - A)^{-1}(sI - A)^{-1}B \\ s\frac{dK}{ds} &= C(sI - A)^{-1}\left(B + A(sI - A)^{-1}B\right) \\ \Rightarrow \quad V_K &= \left\| \mathcal{F}_\ell\left(F_L, -K^*\right) \, \mathcal{F}_\ell\left(F_R, -K^*\right) \right\|_{\mathcal{L}_\infty} \end{aligned} \tag{12.6}$$

$$\text{where } F_L = \left[\begin{array}{c|cc} A & I & B \\ \hline 0 & 0 & I \\ C & 0 & D \\ C & 0 & D \end{array} \right], \quad F_R = \left[\begin{array}{c|ccc} A & B & 0 & B \\ \hline A & B & 0 & B \\ C & D & I & D \end{array} \right]$$

The state dimension for this calculation is hence just four times that of K. The specification that $b_{P,K} \geq \beta$ is a standard $\mathcal{H}_\infty$-norm condition as in (12.3). Note that the system in (12.6) has no closed-loop stability requirement.

Finding parameters, p, such that β is maximized whilst (12.6) is satisfied is a highly nonconvex problem with no guarantees of success, however software such as the `systune` toolbox in `MATLAB` can accommodate such a problem formulation and should produce a local solution. This has been implemented and in the example of Sect. 12.2.3 maximizing β with $\alpha = 2$ gave a first order controller very similar to phase lead compensator, $K_1(s)$, and with a second order controller $b_{P_1,K} = 0.486$ was achieved that also gave $b_{P_2,K} = 0.469$ as illustrated in Fig. 12.1. A further modest increase in $b_{P_1,K}$ is possible with increased controller order but no greater than 0.525 was achieved.

This section has demonstrated that adding a complexity constraint in the synthesis of a parameterized controller is a relatively straightforward addition and can avoid the occasional robustness issues that can occur with $\mathcal{H}_\infty$-loop shaping for plants with poles and zeros near the imaginary axis. Since this is not a convex problem there are no guarantees of global solutions but in the normal situation with relatively few parameters to tune good results have been demonstrated.

12.3.2 Non-parametric Optimization over Frequency in the Single-Input/Single-Output Case

The purpose of this section is to consider optimizing over $K(j\omega_i)$ for a fine grid of frequencies, ω_i, $i = 1, \dots N+1$, rather than over a state-space model as in the previous section. In the SISO case the requirement that $b_{P,K} \geq \beta$ can be written in terms of the pointwise stability margin,

$$\rho(P(j\omega), K(j\omega) := \frac{|1 - K(j\omega)\, P(j\omega)|}{\sqrt{(1 + |K(j\omega)|^2)}\sqrt{(1 + |P(j\omega)|^2)}} \tag{12.7}$$

$$\text{as } \rho(P(j\omega_i), K(j\omega_i) \geq \beta, \quad \forall\, i = 1, \dots, N+1 \tag{12.8}$$

Similarly the condition, $V_K \leq \alpha$ can be approximated by,

$$v_K(\omega_i) \;\leq\; \alpha,\ \forall\, i = 1, \dots N+1 \tag{12.9}$$

$$\text{where } v_K(\omega) := \frac{\left|\omega \frac{dK}{d\omega}\right|}{(1 + |K(j\omega)|^2)} \tag{12.10}$$

The problem will now be transferred from the imaginary axis to the unit circle via $\frac{sT}{2} = \frac{z-1}{z+1}$, giving for $z = e^{j\theta}$, $\omega = \frac{2}{T}\tan(\theta/2)$. A frequency grid on the unit circle is chosen as,

$$\theta_i = (i-1)\pi/N, \text{ and hence } \omega_i = \frac{2}{T}\tan(\theta_i/2) \text{ for } i = 1, \dots, N+1$$

Let $L(\theta) = K\left(j\frac{2}{T}\tan(\theta/2)\right)$. Then

$$\left|\omega\frac{dK}{d\omega}\right| = \left|\omega\frac{dL}{d\theta}\frac{d\theta}{d\omega}\right| = \left|\sin(\theta)\frac{dL}{d\theta}\right|$$

and (12.9, 12.10) become,

$$v_L^d(\theta_i) \leq \alpha, \forall\, i = 1, \dots N+1 \tag{12.11}$$

$$\text{where } v_L^d(\theta) := \left|\sin(\theta)\frac{dL}{d\theta}\right| \Big/ (1 + |L(\theta)|^2) \tag{12.12}$$

An approach is now to consider a coarser set of discrete frequencies, $\hat{\theta}_k = (k-1)\pi/M$ for $k = 1 \cdots M+1$ and let $L_x(\hat{\theta}_k) = \mathrm{Re}(L(\theta_k))$ be unknowns over which the optimization is performed. In order to minimize ripple between these frequencies a cubic spline is then calculated to interpolate between these frequencies at the above finer grid θ_i with N a multiple of M. The cubic spline can then also give $\frac{dL_x}{d\theta}$ at $\theta = \theta_i$. If we assume that $K(s)$ is open-loop stable then the discrete-time Hilbert transform of L_x will now give $L_y(\theta) = \text{Im}(L(\theta))$ and $\frac{dL_y}{d\theta}$ at θ_i from L_x and $\frac{dL_x}{d\theta}$ resp. at these frequencies together with $\theta = i\pi/N$ for $i = -(N-1), \ldots, -1$ with $L_x(-\theta) = L_x(\theta)$, $\frac{dL_x}{d\theta}(-\theta) = -\frac{dL_x}{d\theta}(\theta)$.

The optimization problem is therefore ((12.8) and (12.12)) maximize β over $L_x(\theta_i),\ i = 1 \cdots M+1$, subject to,

$$\begin{aligned} \rho(P(j\omega_i), L(\theta_i) &\geq \beta, \ \ \forall\, i = 1, \ldots, N+1 \\ v_L^d(\theta_i) &\leq \alpha, \ \ \forall\, i = 1, \ldots, N+1 \end{aligned}$$

This optimization problem has $M+2$ unknowns and $2N+2$ constraints which could be computationally expensive due to its scale rather than the nature of the nonlinearities. The constraints are relatively benign but unfortunately not convex.

The use of the Hilbert transform here assumes that the controller, $K(s)$ is open-loop stable. In addition, the condition is on the pointwise stability margin with no explicit test on closed-loop stability. The relevant winding number condition can be checked after the optimization and if it fails then no conclusion can be drawn. However, if the optimization starts from a stabilizing solution then it would have to cross a $b_{P,K} = 0$ boundary to violate the closed-loop stability condition.

If in addition to open-loop stability of $K(s)$ it was assumed that the controller was also minimum phase then $|L(\hat{\theta}_i)|$ could be used as the unknowns with $\angle(L(\theta_i))$ obtained from the Bode relation (i.e. the Hilbert transform of $\frac{d\log(|L(\theta)|)}{d\theta}$).

The final step is to generate $K(s)$ from $\{K(j\omega_i)\}_{i=1}^{N+1}$ or equivalently from $\{L(\theta_i)\}_{i=1}^{N+1}$. The approach adopted is to take the inverse discrete Fourier transform of $\{L(\theta_i)\}_{i=1}^{N+1}$ and form the corresponding $N \times N$ Hankel matrix and perform a truncated balanced realization. Finally the inverse bilinear transform of this discrete-time state-space system $D_d + C_d(zI - A_d)^{-1}B_d$ to continuous time is carried out, giving $K(s) = D + C(sI - A)^{-1}B$ with

$$\begin{aligned} A &= \tfrac{2}{T}(I + A_d)^{-1}(A_d - I); \quad B = \tfrac{2}{T}(I + A_d)^{-1}B_d; \\ C &= 2C_d(I + A_d)^{-1}; \quad D = D_d - C_d(I + A_d)^{-1}B_d \end{aligned}$$

12.3.2.1 Example

The example given in Sect. 12.2.3 is studied using the above non-parametric approach. The results of the optimization are given in Fig. 12.2 for $\alpha = 1$. A 10th order approximation to this frequency response gave a good match and essentially the same performance giving $\beta \approx 0.475$ which is slightly better than that obtained earlier in Sect. 12.3.1 with a 2nd-order controller but with $\alpha = 2$.

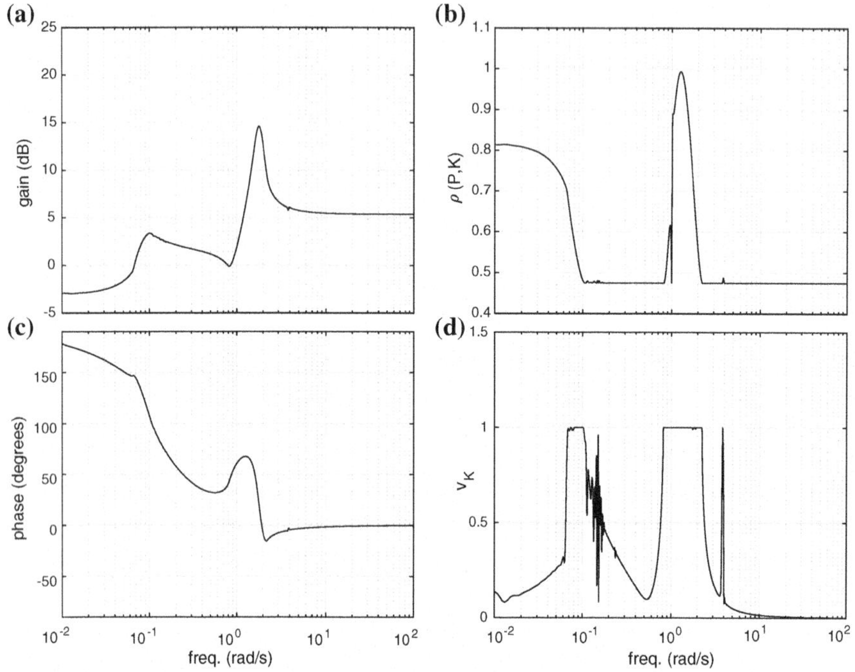

Fig. 12.2 Example with $P(s) = \frac{s(s^2+1)}{(s^2/q+0.1s+q)(s^2/q-0.1s+q)}$ with controller $K_o(s)$ maximizing b_{P,K_o} subject to $V_{K_o} \le 1$. **a** gain of $|K_o(j\omega)|$ **c** phase of $K_o(j\omega)$ **b** $\rho(P(j\omega), K_o(j\omega))$ demonstrating the maximal $\beta = 0.475$. **d** $v_{K_o}(\omega)$ demonstrating that $V_{K_o} \le 1$ as required

12.4 Observations on the Properties of Optimal Solutions

In this section the properties of the problem $\beta = \max_{K(s)}(b_{P,K})$ subject to $V_K \le \alpha$ for a given $\alpha \ge 0$ and (P, K) internally stable. Or equivalently $\max_{K(s)}(\beta)$ subject to $\rho(P(j\omega), K(j\omega)) \ge \beta$ and $v_K(\omega) \le \alpha$ for all ω. There are two extreme cases where a rational $K(s)$ gives the solution:

If $\alpha \ge V_{K_o}$ where K_o is the optimal controller maximizing $b_{P,K}$ then K_o will be optimal here since the complexity constraint is not active, and will have degree less than that of the plant.

If there exists a stabilizing constant gain controller then it will satisfy $\frac{dK}{ds} = 0$ and maximizing $b_{P,K}$ over this gain will be optimal with the constraint $V_K \le \alpha = 0$.

An interesting question is whether within these extremes the optimal solutions are typically irrational. This is a similar in style to the question addressed in [9] in considering optimal mixed $\mathscr{H}_\infty/\mathscr{H}_2$ problems and finding the solution to be typically irrational. Analytical results have not been found for our problem but some observations will be made based purely on the properties of the solution when $P(s) = \frac{1}{s^2}$ when,

$$b_{\mathrm{opt}}(P) = \frac{1}{2}\sqrt{2-\sqrt{2}} = 0.3827... = b_{P,K_o}$$
$$\text{where } K_o = \frac{(hs+1)}{(s+h)}, \quad h = 1+\sqrt{2} \tag{12.13}$$

This controller is in fact a very satisfactory controller and corresponds to classical phase advance controller. This problem has some pleasing features for example it is easily verified that since $P(s) = P^{-1}(1/s)$ then defining $J(s) = K^{-1}(1/s)$

$$\rho(P(j\omega), J(j\omega)) = \rho(P(1/j\omega), K(1/j\omega))$$
$$v_J(\omega) = v_K(1/\omega)$$

Hence the controller $J(s)$ will have identical performance to $K(s)$ except with the frequencies reversed, $\omega \leftrightarrow 1/\omega$. One might expect that the optimal solution for such a plant would satisfy $K(s) = K^{-1}(1/s)$ and the numerical evidence supports this expectation, however this conjecture has not been proved.

The numerical results for this example are presented in Fig. 12.3 and it is seen that the above expectation is confirmed for these three cases as well as with the optimal

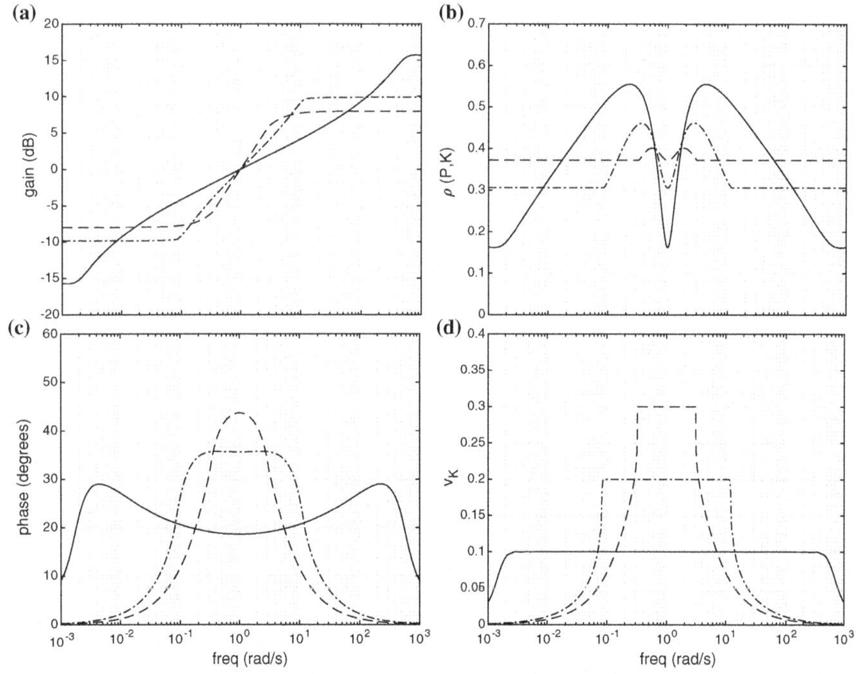

Fig. 12.3 Example with $P(s) = \frac{1}{s^2}$ with controller $K_o(s)$ maximizing b_{P,K_o} subject to $V_{K_o} \le \alpha$ with $\alpha = 0.1,\ 0.2,\ 0.3$ (solid, dot-dashed, dashed lines resp.). **a** gain of $|K(j\omega)|$ **c** phase of $K(j\omega)$ **b** $\rho(P(j\omega), K(j\omega))$ demonstrating the maximal $\beta = 0.162, 0.306, 0.372$ resp. **d** $v_K(\omega)$ demonstrating that $V_K \le 0.1,\ 0.2,\ 0.3$ as required

for $b_{opt}(P)$ in (12.13). It is noted in Fig. 12.3 that the slope of the log-magnitude plot is approximately constant around $\omega = 1$ and hence defining $K(j\omega) = r(\omega)e^{j\phi(\omega)}$, $M(u) = \log(r(e^u)$ where $u = \log(\omega)$ the Bode formula would give,

$$\phi(\omega) \approx \frac{\pi}{2}\frac{dM}{du} = \frac{\pi}{2}\frac{\omega}{r(\omega)}\frac{dr}{d\omega}$$

$$\omega\frac{dr}{d\omega}\Big/(1+r^2) \approx \frac{2r}{1+r^2}\frac{\phi(\omega)}{\pi} = \frac{\phi(1)}{\pi} \text{ at } \omega = 1 \text{ when } r = 1$$

At $\omega = 1$, $\frac{d\phi}{d\omega} = 0$ and hence $\nu_K(1) \approx \phi(1)/\pi$. It can also be shown that under these assumptions, $\rho(P(j), K(j)) = \sin(\phi(1)/2)$ and $b_{P,K} \approx \sin(V_K\pi/2)$. This approximation to the trade-off between α and β is good for $\alpha = 0.1, 0.2$, but less so when the constant slope assumption is less accurate.

In this example at each frequency at least one of the two constraints is active and there are intervals where one constraint is active for all frequencies in the interval. This would imply that the corresponding controller cannot be rational. The results for the example in Sect. 12.2.3 presented in Fig. 12.2 indicate there are frequency ranges where neither constraint is active, e.g. $\omega < 0.06$.

12.5 Conclusions

The following observations are made:

In certain situations, (although not normally), an optimal $\mathscr{H}_\infty$ loop-shaping controller can have high complexity, giving a good value for $b_{P,K}$, but nevertheless poor robustness to small perturbations in s.

In state-space controller synthesis including a side condition on the complexity, V_K, is straightforward to specify, but any optimization may have difficulty finding an optimal solution.

An optimal, $K(s)$, may well be irrational but low order approximations can be close to optimal. A non-parametric frequency domain method has been described that can give estimates of the potential reduction in complexity against which the performance of a lower order controller can be compared.

References

1. Georgiou, T.T., Smith, M.C.: Optimal robustness in the gap metric. IEEE Trans. Autom. Control **35**, 673–686 (1990). ISSN 0018-9286
2. Vinnicombe, G.: Uncertainty and feedback: $\mathscr{H}_\infty$ loop-shaping and the ν -gap metric. Imperial College Press (2001)
3. McFarlane, D.C., Glover, K.: A loop shaping design procedure using H-infinity-synthesis. IEEE Trans. Autom. Control **37**, 759–769 (1992). ISSN 0018-9286

4. Skogestad, S., Postlethwaite, I.: Multivariable Feedback Control: Analysis and Design. Wiley, Chichester (1996)
5. Hyde, R.A., Glover, K., Shanks, G.T.: VSTOL first flight of an H-infinity control law. Comput. Control Eng. J. **6**, 11–16 (1995). ISSN 0956-3385
6. Hsieh, G., Safonov, M.G.: Conservativism of the gap metric. IEEE Trans. Autom. Control **AC-38**(4):594–598 (1993)
7. Vinnicombe, G.: The robustness of feedback systems with bounded complexity controllers. IEEE Trans. Autom. Control **41**, 795–803 (1996). ISSN 0018-9286
8. Yuan, X.C., Glover, K.: Model-based control of thermoacoustic instabilities in partially premixed lean combustion-a design case study. Int. J. Control **86**, 2052–2066 (2013). ISSN 0020-7179
9. Megretski, A.: On the order of optimal controllers in the mixed H2/H-infinity control. IEEE CDC (1994)

Chapter 13
Hierarchically Decentralized Control for Networked Dynamical Systems with Global and Local Objectives

Shinji Hara, Koji Tsumura and Binh Minh Nguyen

Abstract This article deals with hierarchically decentralized control structure for large-scaled dynamical networked systems by aggregation. Our main idea to clarify the trade-off and the role-sharing of the global and the local controllers is to introduce a model set named "Global/Local Shared Model Set," which should be taken in both the global and local cites. We set up a fairly general framework and derive the global and local control problems. We then clarify the trade-off through the size of the model set and demonstrate it by a simple example.

13.1 Introduction

In recent years, systems to be treated in various fields of engineering including control have become large and complex. Typical examples include meteorological phenomena, energy network systems, traffic flow networks and biological systems. They can be regarded as hierarchical networked dynamical systems, and several new frameworks to treat such systems from the viewpoint control have been proposed (See e.g. [1–3]). Hara et al. proposed a new research area so-called "Glocal Control" meaning that both desired global and local behaviors are achieved by local actions of measurement and control [3]. The key framework is based on hierarchical networked systems with multiple-resolution in time and space, and each layer has its own objective which might be in conflict with other layers' objectives. Hence, one

S. Hara (✉) · K. Tsumura · B. M. Nguyen
Department of Information Physics and Computing,
The University of Tokyo, Tokyo, Japan
e-mail: Shinji_Hara@ipc.i.u-tokyo.ac.jp

S. Hara · K. Tsumura · B. M. Nguyen
CREST, Tokyo, Japan
K. Tsumura
e-mail: Koji_Tsumura@ipc.i.u-tokyo.ac.jp

B. M. Nguyen
e-mail: N._BinhMinh@ipc.i.u-tokyo.ac.jp

R. Tempo et al. (eds.), *Emerging Applications of Control and Systems Theory*, Lecture Notes in Control and Information Sciences - Proceedings,
https://doi.org/10.1007/978-3-319-67068-3_13

of the big issues to realize the glocal control is to establish a unified way of handling global/local objectives properly as a hierarchically decentralized control.

We consider the following situation. There are a bunch of subsystems which are slightly different each other, and we assume that each of them is equipped with a local controller which can be designed independently so that it stabilizes its own subsystem and optimizes a certain local objective. There also exists a so-called global controller which uses the average or sum of a certain quantity of the locally controlled subsystems to coordinate all the subsystems properly for optimizing a certain global objective.

There are two main reasons to consider such a situation. The first reason is from the practical viewpoint. A typical example is electric power network systems, where the global control objective is to make the balance of demand and supply of the total power of multiple generators, and each generator is locally controlled to achieve the local performance better. The second reason is from the theoretical viewpoint. As seen in [4], averaging or low-rank inter-layer interactions in general is quite effective to achieve the rapid consensus, and the property is fit to the glocal control concept based on hierarchical networked systems with multiple time/space resolutions [3].

There are two theoretical key issues to be investigated for hierarchically decentralized control by aggregation, namely (i) how to guarantee the stability of the whole system? and (ii) how to derive the global/local trade-off relation and how to compromise it?

To this end, we introduce a model set named "*Global/Local Shared Model Set*," which is defined by a standard LFT form consisting of the nominal model and norm-bounded perturbations. The nominal model is set for both the aggregated system and each locally controlled subsystem to be followed within a certain error bound. Then, each local controller is designed to make the resultant feedback loop system of the local subsystem to be in the class and simultaneously attain a local control performance. On the other hand, the global controller is designed to attain control performance for this nominal model under consideration of the errors between the nominal model and the local feedback loop systems. Thus, through changing the size of the model set we will clarify the trade-off in the hierarchically decentralized control systems and provide a simple illustrative example to show the effectiveness of the approach.

Notation: RH_∞: stable rational ring
$\mathcal{S}_c(P)$: a set of all stabilizing controllers for P

13.2 General Problem Setting

We here propose a fairly general setting, which is represented by a block diagram depicted in Fig. 13.1. The system consists of two layers, the upper layer for global control and the lower layer for local control. They are connected each other by aggregation $\frac{1}{N}\mathbf{1}^\top$ from bottom to up and distribution $\mathbf{1}$ from up to bottom, where $\mathbf{1} := [1, 1, \ldots, 1]^\top$ with size N.

$$G(s) = \begin{bmatrix} G_{yu}(s) & G_{yw}(s) & G_{yv}(s) \\ G_{zu}(s) & G_{zw}(s) & G_{zv}(s) \\ G_{ru}(s) & G_{rw}(s) & G_{rv}(s) \end{bmatrix},$$

$$L(s) = \begin{bmatrix} L_{vr}(s) & L_{vw}(s) & L_{vu}(s) \\ L_{zr}(s) & L_{zw}(s) & L_{zu}(s) \\ L_{yr}(s) & L_{yw}(s) & L_{yu}(s) \end{bmatrix},$$

$$L_{*\sharp}(s) = \mathrm{diag}\left\{L_{*\sharp}^{(i)}(s)\right\},$$

$$C_\ell(s) = \mathrm{diag}\left\{C_\ell^{(i)}(s)\right\}.$$

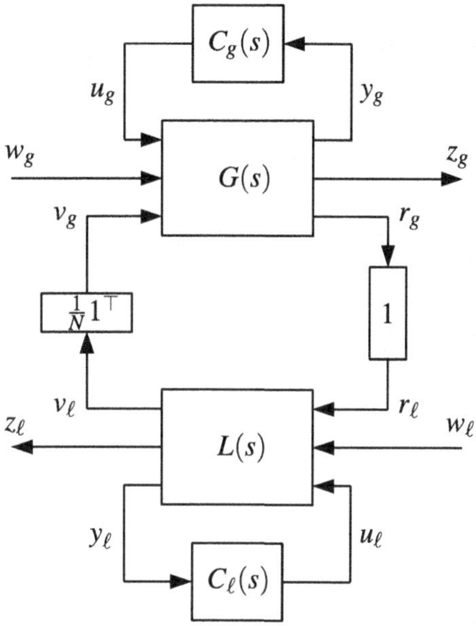

Fig. 13.1 Structure of total system

The upper and lower layer generalized plants are represented by $\boldsymbol{G}(s)$ and $\boldsymbol{L}(s)$, respectively. The block diagonal property of $\boldsymbol{L}(s)$ means that the lower layer is a collection of independent subsystems. Each subsystem thus has an independent local controller $C_\ell^{(i)}(s)$, and hence we have the block diagonal property of $\boldsymbol{C}_\ell(s)$.

The collections of inputs and outputs of the local controllers are denoted by $\boldsymbol{y}_\ell$ and $\boldsymbol{u}_\ell$, respectively. Signals $\boldsymbol{v}_\ell$ and $\boldsymbol{r}_\ell$ are N-dimensional vectors which correspond to signals to link the upper layer, i.e., $v_g = \frac{1}{N}\mathbf{1}^\top \boldsymbol{v}_\ell$ and $\boldsymbol{r}_\ell = \mathbf{1}r_g$. Signals $\boldsymbol{w}_\ell$ and $\boldsymbol{z}_\ell$ are the collections of input and output variables for representing the local objective, respectively. The element-wise representation of $\boldsymbol{u}_\ell$ is given by $\boldsymbol{u}_\ell = [u_1, u_2, \ldots, u_N]^\top$, and the same notation is used for $\boldsymbol{y}_\ell$, $\boldsymbol{v}_\ell$, $\boldsymbol{r}_\ell$, $\boldsymbol{w}_\ell$, and $\boldsymbol{z}_\ell$. The upper layer is controlled by the global controller $C_g(s)$ with input u_g and output y_g. Signals v_g and r_g are the aggregated signal from the lower layer and the reference signal to be sent out to the lower layer, respectively. Signals w_g and z_g are the input and output variables for representing the global objective, respectively.

Our main idea to achieve the two requirements (i) and (ii) mentioned above, or (i) stability of total system and (ii) global/local performance trade-off, is to introduce a set of model set named "*Global/Local Shared Model Set*," which is defined by a standard LFT form as

$$\mathscr{M}_\delta := \left\{\tilde{\boldsymbol{M}} = \mathscr{F}_\ell\left(M_o(s), \boldsymbol{\Delta}(s)\right) : \|\boldsymbol{\Delta}\|_\infty \leq \delta\right\}, \tag{13.1}$$

where

$$M_o(s) = \begin{bmatrix} M_0(s) & M_1(s) \\ M_2(s) & 0 \end{bmatrix}. \tag{13.2}$$

The set is expected to be shared by both upper and lower layers in the following sense. The upper layer expects that each local agent is controlled such that the closed-loop transfer function from r_i to v_i denoted by $\Phi_{Lvr}^{(i)}(s)$ belongs to $\mathcal{M}_\delta$, and hence the lower layer tries to optimize the local objective related to the closed-loop transfer function from w_i to z_i denoted by $\Phi_{Lzw}^{(i)}(s)$ under the requirement from the upper layer, where

$$\Phi_{Lvr}^{(i)} := \mathscr{F}_\ell\left(\begin{bmatrix} L_{vr}^{(i)} & L_{vu}^{(i)} \\ L_{yr}^{(i)} & L_{yu}^{(i)} \end{bmatrix}, C_\ell^{(i)}\right), \quad \Phi_{Lzw}^{(i)} := \mathscr{F}_\ell\left(\begin{bmatrix} L_{zw}^{(i)} & L_{zu}^{(i)} \\ L_{yw}^{(i)} & L_{yu}^{(i)} \end{bmatrix}, C_\ell^{(i)}\right).$$

Then, the upper layer designs the upper layer controller $C_g(s)$ so that the global control performance represented by the transfer function from w_g to z_g is optimized under uncertainty channel $\mathcal{M}_\delta$ connected in between r_g and v_g. Note that $\mathcal{M}_\delta$ includes the classes of additive and multiplicative perturbations and that the averaging does not change the size of uncertainty δ as will be shown in the next section.

Consequently, we can split the global/local controller design into two independent designs, *Global Controller Design* and *Local Controller Design*.

Global Controller Design:

$$\min_{C_g \in \mathscr{S}_c(G_{yu})} \left\{ \max_{\tilde{M} \in \mathcal{M}_\delta} \|\Phi_{Gzw}\|_\infty \right\}, \tag{13.3}$$

where

$$\Phi_{Gzw} := \mathscr{F}_u\left\{ \mathscr{F}_\ell\left\{ \begin{bmatrix} \begin{pmatrix} G_{yu} & G_{yw} \\ G_{zu} & G_{zw} \end{pmatrix} & \begin{pmatrix} G_{yv} \\ G_{zv} \end{pmatrix} \\ \begin{pmatrix} G_{ru} & G_{rw} \end{pmatrix} & G_{rv} \end{bmatrix}, \tilde{M} \right\}, C_g \right\}$$

Local Controller Design:

$$\min_{C_\ell^{(i)} \in \mathscr{S}_c(L_{yu}^{(i)})} \left\| \Phi_{Lzw}^{(i)} \right\|_\infty \quad \text{s.t.} \quad \Phi_{Lvr}^{(i)}(s) \in \mathcal{M}_\delta \tag{13.4}$$

The global and local problems are a robust performance problem and a 2-disk problem, respectively, and hence they are not so easy to derive the optimal controllers. However, we can investigate the global/local performance trade-off by the uncertainty level δ. We can readily see that the smaller δ leads to the better global performance and that the larger δ yields the better local performance. Note that we have a freedom of the selection of $M_o(s)$, although we assume in this paper that $M_o(s)$ is a priori selected for simplicity.

13.3 A Typical Situation with Multiplicative Perturbations

13.3.1 *Aggregation of Local Systems*

In this section, we give detailed formulations for a case of aggregation of local subsystems, where we replace notations $C_\ell^{(i)}(s)$ and $\Phi_{Lvr}^{(i)}(s)$ defined in the previous section by $C_\ell^{(i)}(s) = C_i(s)$ and $\Phi_{Lvr}^{(i)}(s) = \Phi_i(s)$ for simplicity.

We assume that the whole system consists of the following N local and SISO subsystems;

$$v_i = P_i(s)u_{oi}, \quad P_i(s) = \frac{n_i(s)}{d_i(s)}, \quad n_i(s),\ d_i(s) \in RH_\infty,$$

where $i\ (= 1,\ 2,\ \ldots,\ N)$ represents the index of a local subsystem, u_{oi} is the input, v_i is the output, $P_i(s)$ is the transfer function represented by a coprime factorization over the stable rational ring, and n_i and d_i are the numerator and the denominator. Suppose that the input u_{oi} consists of a local input u_i and a global input $r_i = r_g$ such as $u_{oi} = u_i + r_i$ and the local input u_i is generated by a local controller $C_i(s)$ as

$$u_i = C_i(s)y_i, \quad y_i = v_i + w_i \tag{13.5}$$

Suppose that $C_i(s)$ is designed to stabilize $P_i(s)$ and then it is represented by Youla parametrization such as

$$C_i(s) = -\frac{\alpha_i(s) + d_i(s)q_i(s)}{\beta_i(s) - n_i(s)q_i(s)}, \quad n_i(s)\alpha_i(s) + d_i(s)\beta_i(s) = 1, \tag{13.6}$$

where $\alpha_i(s),\ \beta_i(s),\ q_i(s) \in RH_\infty$. The local controller $C_i(s)$ is also designed to attain given local control performances.

We consider a case that the output of the group of these local subsystems is the average of their local outputs such as $v_g = \frac{1}{N}\sum_{i=1}^{N} v_i$ where we call v_g as *the aggregated output* of the group of the local subsystems. Define a transfer function $\Phi_i(s)$ from the global input $r_i = r_g$ to the local output v_i as

$$v_i = \frac{P_i(s)}{1 - P_i(s)C_i(s)} r_i =: \Phi_i(s) r_g. \tag{13.7}$$

Thus $\Phi_i(s)\ (= 1,\ 2,\ \ldots,\ N)$ can be regarded as the transfer function of the locally controlled subsystem composed of the local system $P_i(s)$ and the local controller $C_i(s)$.

Consequently, the local feedback system can be shown in Fig. 13.2, where the i-th component of $\boldsymbol{L}(s)$ is given by

$$\begin{bmatrix} P_i & 0 & P_i \\ W_{SL}P_i & W_{SL} & W_{SL}P_i \\ P_i & 1 & P_i \end{bmatrix}.$$

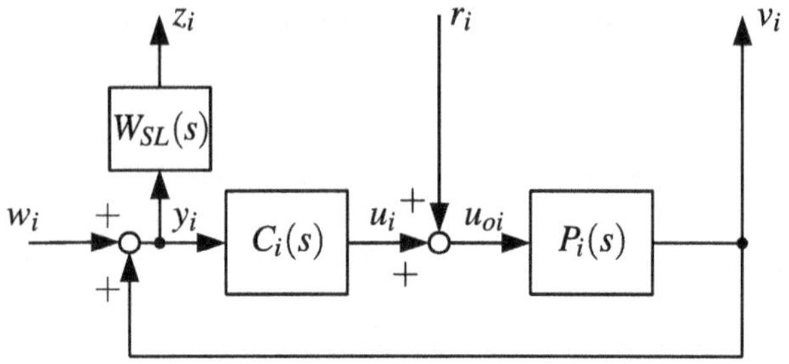

Fig. 13.2 Feedback loop of local subsystem

Also define $\boldsymbol{M}$ as the aggregated subsystems such as

$$\boldsymbol{M}(s) := \sum_{i=1}^{N} \frac{1}{N} \Phi_i(s), \tag{13.8}$$

then, the aggregated output v_g can be represented as $v_g = \frac{1}{N} \sum_{i=1}^{N} \Phi_i(s) r_i = \boldsymbol{M}(s) r_g$. We call $\boldsymbol{M}(s)$ as *the aggregated transfer function* from r_g to v_g.

Next, the global controller $C_g(s)$ is designed to generate the global input u_g from the global output y_g such as

$$u_g = C_g(s) y_g. \tag{13.9}$$

The purpose of the global controller $C_g(s)$ is basically to attain the stability of the whole system and global control performances.

The strategy of the hierarchically decentralized control system considered in this paper is divided into three layers; lower layer, upper layer, and middle layer as follows: (lower layer) design of the local controllers $C_i(s)$ in the lower layer for given local control objectives, (upper layer) design of the global controller $C_g(s)$ in the upper layer for a given global control objective, (middle layer) design of a nominal model $M_o(s)$ to which the locally controlled subsystems are designed to be close.

The actual design procedure of the whole system is as follows: At first, set a stable nominal model $M_o(s)$ appropriately in the middle layer and its information is broadcasted to each local subsystem. The control objective for a local controller $C_i(s)$ are both of to make its closed-loop system composed of $P_i(s)$ and $C_i(s)$ to be close to the nominal model $M_o(s)$ and to attain a given local control performance. On the other hand, the control objective for the global controller $C_g(s)$ in the upper layer is to attain a given global control performance with consideration of the error between $\boldsymbol{M}(s)$ and $M_o(s)$, that is, a robust control performance. When the selected nominal model $M_o(s)$ is not appropriate for both of the local control performance and the global control performance, we reset $M_o(s)$ and repeat the above procedure.

In the following subsections, we give actual formulae for the above-mentioned control strategy and clarify the relationship between controller designs in the upper layer and the lower layer, and the setting of the model set $\mathcal{M}_\delta$. Then we discuss trade-offs between the attained global control performance and the local control performance.

13.3.2 Hierarchically Decentralized Control Design

At first, in the lower layer, we consider to design a local controller $C_i(s)$ which simultaneously satisfies a given local control performance and a multiplicative perturbed model matching problem between the local closed-loop system $\Phi_i(s)$ and the nominal model $M_o(s)$ as follows:

$$\text{find } C_i(s) \text{ for each } i \text{ s.t. } \left\| W_{SL} \frac{1}{1-P_iC_i} \right\|_\infty < 1 \tag{13.10}$$

$$\left\| \frac{\Phi_i - M_o}{M_o} \right\|_\infty = \left\| \frac{1}{M_o}\left(\frac{P_i}{1-P_iC_i} - M_o \right) \right\|_\infty < \delta \tag{13.11}$$

where $W_{SL}(s)$ is a given weight function and δ represents the size of the model set $\mathcal{M}_\delta$. Note that $\Phi_i(s)$ can be represented as

$$\Phi_i(s) = \frac{P_i(s)}{1-P_i(s)C_i(s)} = \frac{n_i}{d_i}(d_i\beta_i - d_in_iq_i) = n_i(\beta_i - n_iq_i) \tag{13.12}$$

and also similarly $1/(1-P_iC_i) = d_i(\beta_i - n_iq_i)$, then, the above problem can be represented as follows:

$$\text{find } q_i(s) \in RH_\infty \text{ for each } i \text{ s.t. } \|W_{SL}d_i(\beta_i - n_iq_i)\|_\infty < 1 \tag{13.13}$$

$$\left\| M_o^{-1}(n_i(\beta_i - n_iq_i) - M_o) \right\|_\infty < \delta \tag{13.14}$$

Next, denote $\Delta_i(s)$ as the multiplicative error between $\Phi_i(s)$ and $M_o(s)$;

$$\Delta_i(s) := \frac{\Phi_i(s) - M_o(s)}{M_o(s)} = \frac{1}{M_o}\left(\frac{P_i}{1-P_iC_i} - M_o \right), \tag{13.15}$$

and the multiplicative error between the aggregated system $\boldsymbol{M}(s)$ and the nominal model $M_o(s)$ can be represented as

$$\frac{M - M_o}{M_o} = \frac{1}{M_o}\left(\sum_{i=1}^{N} \frac{1}{N}\Phi_i - M_o\right) = \frac{1}{N}\sum_{i=1}^{N} \frac{\Phi_i - M_o}{M_o} = \frac{1}{N}\sum_{i=1}^{N} \Delta_i =: \Delta. \tag{13.16}$$

Then, when the error condition (13.11) on subsystem i, that is, $\|\Delta_i\|_\infty < \delta$ is satisfied, we get

$$\|\Delta\|_\infty = \left\| \frac{1}{N}\sum_{i=1}^{N} \Delta_i \right\|_\infty < \delta. \tag{13.17}$$

On the other hand, in the upper layer, the global controller $C_g(s)$ is designed for satisfying the following global control objective under consideration of the perturbation (13.17), that is, the robust performance problem, on the aggregated system:

$$\text{find } C_g(s) \in \mathscr{S}_c(M_o) \text{ s.t. } \left\| W_S \frac{1}{1 - (1 + \tilde{\Delta})M_o C_g} \right\|_\infty < 1 \ ; \ \forall \tilde{\Delta} \text{ s.t. } \left\| \tilde{\Delta} \right\|_\infty < \delta \tag{13.18}$$

where $W_S(s)$ is an appropriate weight function. This problem can be also reduced to [5]

$$\text{find } C_g(s) \in \mathscr{S}_c(M_o) \text{ s.t. } |W_S S| + |\delta T| \le 1, \forall \omega, \tag{13.19}$$

where $S(s)$ and $T(s)$ are

$$S(s) := \frac{1}{1 - M_o(s)C_g(s)}, \quad T(s) := \frac{M_o(s)C_g(s)}{1 - M_o(s)C_g(s)}. \tag{13.20}$$

Remark 13.1 In this robust control performance problem, the detailed information on each subsystem $\Phi_i(s)$ or each $\Delta_i(s)$ is not necessary and only δ is necessary. This implies the detailed information is aggregated and the computation complexity can be restrained in the whole control system design. Such control design strategy is necessary for controlling large-scaled systems.

Note that $M_o(s)$ is assumed to be stable, then its stabilizing controller $C_g(s)$ is given as

$$C_g(s) = -\frac{q(s)}{1 - M_o(s)q(s)}, \quad q(s) \in RH_\infty, \tag{13.21}$$

and then $S(s)$ and $T(s)$ can be represented as

$$S(s) = 1 - M_o(s)q(s), \quad T(s) = M_o(s)q(s). \tag{13.22}$$

Therefore, the robust performance problem in the upper layer can be reduced into the following:

$$\text{find } q(s) \in RH_\infty \quad \text{s.t.} \quad |W_S(1 - M_o q)| + |\delta M_o q| \le 1, \ \forall s = j\omega \tag{13.23}$$

Although the nominal model $M_o(s) \in RH_\infty$ is also a design parameter, we here assume that it is set appropriately in advance. Then, we have the following method of hierarchically decentralized control design.

[Hierarchically Decentralized Control Design]

Upper layer:

$$\text{find } q(s) \in RH_\infty \quad \text{s.t.} \quad \|\,|W_S(1 - M_o q)| + |\delta M_o q|\,\|_\infty \le 1 \tag{13.24}$$

Lower layer:

$$\text{find } q_i(s) \in RH_\infty \ \text{ for each } i \ \text{ s.t.} \quad \|W_{SL} d_i(\beta_i - n_i q_i)\|_\infty < 1 \tag{13.25}$$

$$\left\| M_o^{-1}(n_i(\beta_i - n_i q_i) - M_o) \right\|_\infty < \delta \tag{13.26}$$

Remark 13.2 At first, note that $M_o(s)$ is fixed in this paper and then $q(s)$ and $q_i(s)$ are the design parameters. In (13.24), even if we set $q(s)$ in the upper layer appropriately, it is known that there exists an unavoidable trade-off between δ and the magnitude of W_{SL}. On the other hand, in order to improve the local control performance (13.25) in the lower layer, an arbitrary high control performance is attained by a setting $q_i \rightarrow \frac{\beta_i}{n_i}$ when the zeros of n_i is stable. Then, when we set $q_i = (1 + \epsilon_i)\frac{\beta_i}{n_i}$ where ϵ_i has an enough small gain, (13.26) is represented by

$$\left\| M_o^{-1}(n_i(\beta_i - n_i q_i) - M_o) \right\|_\infty = \left\| -n_i \beta_i \epsilon_i M_o^{-1} - 1 \right\|_\infty < \delta. \tag{13.27}$$

Therefore, from (13.27), it is known that the possible δ which satisfies (13.27) becomes large. In summary, there exists a trade-off between the global control performance and the local control performance and it is represented by means of the size δ of the model set $\mathscr{M}_\delta$.

13.4 An Illustrative Example

This section demonstrates the global/local performance trade-off by a very simple example defined as follows:

Local Subsystems: The plant of each local system is represented by

$$P_i(s) = \frac{k_i}{s+h_i},\ h_i,\ k_i > 0, \tag{13.28}$$
$$k_i \in [\underline{k}_p, \overline{k}_p],\ \underline{k}_p,\ \overline{k}_p > 0,\ h_i \in [\underline{h}_p, \overline{h}_p],\ \underline{h}_p > 0,\ \overline{h}_p < 1,$$

and the objective is to satisfy

$$\text{(ia)} \quad \left\| W_{SL} \frac{1}{1 - P_i C_i} \right\|_\infty < 1 \ ; \ W_{SL}(s) := \frac{\eta_\ell}{s+1},\ \eta_\ell > 0 \tag{13.29}$$

The requirement (ia) is to reduce the sensitivity and η_ℓ represents the local performance to be maximized.

Global System: The original global control objective is to satisfy

$$\text{(g)} \quad \left\| W_S \frac{1}{1 - \boldsymbol{M} C_g} \right\|_\infty < 1 \ ; \ W_S(s) = \frac{\eta_g}{\tau s + 1},\ \eta_g > 0,\ \tau > 0, \tag{13.30}$$

where $\boldsymbol{M}$ is defined by (13.8) and it represents the average of $\Phi_i(s)$. The global objective is to reduce the sensitivity, i.e., to maximize η_g.

Shared Model Set $\mathscr{M}_\delta$: The shared model set $\mathscr{M}_\delta$ is given by

$$\mathscr{M}_\delta := \left\{ \tilde{M} \mid \tilde{M} = M_o(1+\Delta),\ \|\Delta\|_\infty < \delta \right\} \ ; \ M_o(s) = \frac{b}{s+a},\ a,\ b > 0. \tag{13.31}$$

Global Controller Design: First note that (13.24) is a function of $\hat{q}(s) := M_o(s)q(s)$, which is strictly proper and stable, (13.24) can be rewritten as

$$\hat{q}(s) \in RH_\infty \ \&\ \text{strictly proper s.t.} \ \|\, |W_S(1-\hat{q})| + |\delta\hat{q}|\, \|_\infty \le 1. \tag{13.32}$$

The problem is normally difficult to solve, and hence we will consider one of standard sufficient condition of (13.32) which is represented by [5]

$$\|\, |W_S(1-\hat{q})|^2 + |\delta\hat{q}|^2\, \|_\infty \le 1/2. \tag{13.33}$$

Multiplying an appropriate inner matrix function from left yields

$$\| \begin{bmatrix} A_1(s)\hat{q}(s) - B_1(s) \\ B_2(s) \end{bmatrix} \|_{\infty}. \le \frac{1}{2}, \tag{13.34}$$

$$A_1(s) := (\sqrt{\eta_g^2 + \delta^2} - \delta s)/(1+s), \tag{13.35}$$

$$B_1(s) := \eta_g^2/(1+s)(\sqrt{\eta_g^2+\delta^2} + \delta s),\ \ B_2(s) := \eta_g \delta/(\sqrt{\eta_g^2+\delta^2} + \delta s), \tag{13.36}$$

where we assume that $\tau = 1$ without loss of generality. It is clear that $\| B_2 \|_\infty^2 \le 1/2$ is a necessary condition for the feasibility. Using this condition we can see that the necessary and sufficient condition for (13.33) is given by

$$\text{(i)} \ \ |B_1(s_z)/\hat{B}_2(s_z)| \le 1, \ \ \text{(ii)} \ \ \| B_2 \|_\infty^2 \le 1/2, \tag{13.37}$$

where

$$\hat{B}_2(s) := \frac{s + \sqrt{1 + \eta_g^2(1 - \delta^2)/\delta^2}}{\sqrt{2}(s + \sqrt{\eta_g^2 + \delta^2}/\delta}, \ \ s_z = \sqrt{1 + (\frac{\eta_g}{\delta})^2}.$$

We can also see from certain computation that condition (i) always holds if condition (ii) is satisfied, and hence we can conclude that condition (ii) is the necessary and sufficient condition for the feasibility. Consequently, we can get the achievable global performance level η_g^* as follows:

[**Global Control Performance Limitation**] $\eta_g^*(\delta)$

$$\delta > 1/\sqrt{2}: \ \eta_g^* = \delta/\sqrt{2\delta^2 - 1} \ \ (otherwise: \ \eta_g^* = +\infty). \tag{13.38}$$

Fig. 13.3 shows the plot (dotted line) of $\eta_g^*(\delta)$.

Local Controller Design:

The objective is to satisfy both of (13.11) and (13.29) for a class of $\Phi_i(s)$ given by

$$\text{(ib)} \ \ \Phi_i(s) := \frac{P_i(s)}{1 - P_i(s)C_i(s)} = \frac{k_i}{s + a_i}, \ \ a_i > 0. \tag{13.39}$$

The constraint (ib) is for specifying the class of desirable responses. By using the Youla parametrization of $C_i(s)$, any $a_i > 0$ can be attained and it is a design parameter. In order to describe the optimal $\eta_{\ell i}^*(\delta)$ of $P_i(s)$ for a given δ, we define the following functions and notations:

$$a^*(\delta) := \frac{k_i}{b(1 - \delta)} a$$

$$X^*(a^*) := -h_i^2 + \sqrt{h_i^4 + (a^*)^2(1 - h_i^2)} - h_i^2$$

$$R_\alpha(a^*) := \left(\frac{(X^*(a^*) + 1)(X^*(a^*) + (a^*)^2)}{X^*(a^*) + h_i^2} \right)^{\frac{1}{2}}, \ \ R_\beta(a^*) := \frac{a^*}{h_i}$$

$$\underline{\delta} := \frac{\overline{k}_p - \underline{k}_p}{\overline{k}_p + \underline{k}_p}, \ \ h_o := \sqrt{\frac{h_i^2}{1 - h_i^2}}$$

The following is the summary of the optimal $\eta_{\ell i}^*(\delta)$:

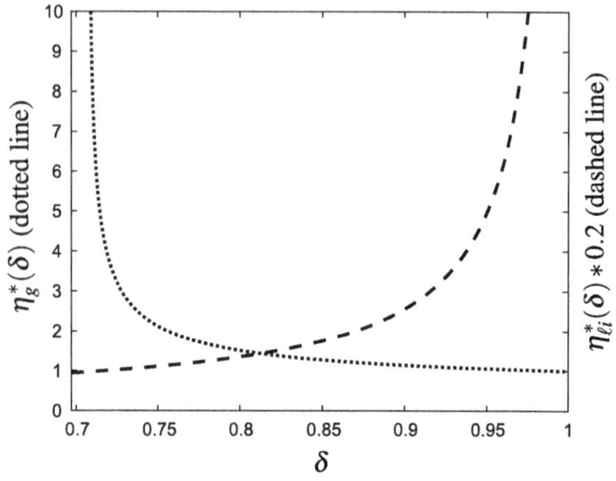

Fig. 13.3 Trade-off curves between the global control performance $\eta_g^*(\delta)$ (dotted line) and the local control performance $\eta_{\ell i}^*(\delta) * 0.2$ (dashed line)

[**Local Control Performance Limitations**] $\eta_{\ell i}^*(\delta)$

$$\text{Suppose} \quad \delta > \underline{\delta}, \ \ \overline{k}/(1+\delta) < b < \underline{k}/(1-\delta).$$

$$\text{Then} \quad \eta_{\ell i}^*(\delta) = \begin{cases} R_\alpha(a^*(\delta)), & a^*(\delta) \geq h_o \\ R_\beta(a^*(\delta)), & a^*(\delta) < h_o \end{cases} \tag{13.40}$$

Note that $\eta_{\ell i}^*(\delta)$ is an increasing function of δ. Figure 13.3 shows a numerical simulation of $\eta_g^*(\delta)$ and $\eta_{\ell i}^*(\delta)$, where $a = 1.2$, $b = 1$, $k_i = 1$, $h_i = 0.6$, $\underline{k}_p = 0.9$, and $\overline{k}_p = 1.1$. The figure shows a trade-off between $\eta_g^*(\delta)$ and $\eta_{\ell i}^*(\delta)$ by using the size δ of the shared model set $\mathcal{M}_\delta$.

13.5 Conclusion

In this article, we have proposed a fairly general formulation of hierarchically decentralized control for large-scaled systems by aggregation. We have introduced *Global/Local Shared Model Set* to split the global and local control problems and to clarify the global/local performance trade-off. The effectiveness of the method has been confirmed by a simple example. The future topics include investigating the analytical formulae of the trade-off by an optimal choice of the nominal model M_o.

References

1. Amin, M., Annaswamy, A.M., DeMarco, C., Samad, T. (eds.): Vision for smart grid control: 2030 and beyond. IEEE Standards Publication (2013)
2. Rieger C.G., Moore K.L., Baldwin, T.L.: Resilient control systems—A multi-agent dynamic systems perspective. In: Proceeding of 2013 IEEE International Conference on Electro/Information Technology, pp. 1–16 (2013)

3. Hara, S., Imura, J., Tsumura, K., Ishizaki, T., Sadamoto, T.: Glocal (global, local) control synthesis for hierarchical networked systems. In: The IEEE Control Systems Society. Multi-conference on Systems and Control, Sydney (2015)
4. Hara, S., Shimizu, H., Kim, T.-H.: Consensus in hierarchical multi-agent dynamical systems with low-rank interconnection: analysis of stability and convergence rates. In: Proceeding of ACC (2009)
5. Doyle, J.C., Francis, B.A., Tannenbaum, A.R.: Feedback control theory. Macmillan Publishing Company (1992)

Chapter 14
Bioaugmentation Approaches for Suppression of Antibiotic Resistance: Model-Based Design

Aida Ahmadzadegan, Abdullah Hamadeh, Midhun Kathanaruparambil Sukumaran and Brian Ingalls

Abstract We present a systems modelling investigation of a bioaugmentation approach to suppression of antibiotic resistance. Bioaugmentation is the manipulation of an environment by the addition of biological agents. We investigate a strategy for limiting the threat of antibiotic-resistant bacterial pathogens by delivery of engineered genetic elements to a target pathogen population. This genetic payload can either trigger cell death or suppress the expression of antibiotic resistance genes and then be passed on to other cells. We present both deterministic (ordinary differential equation) and stochastic (master equation-based) models of the proposed strategy, using the mammalian gut as a representative environment. We provide a stability analysis of the deterministic model, along with an investigation of the potential behaviours of the model through simulation over a range of parameterizations and a global sensitivity analysis. Our focus is on the role of the design parameters in achieving a successful treatment.

14.1 Introduction

The use of antibiotics to treat infectious disease marked a major advance in public health. Since the mid-twentieth century, anthropogenic production and release of antibiotics has imposed significant selection pressure on microbial populations, resulting in increased prevalence of resistance genes [1]. Antibiotic resistance has

A. Ahmadzadegan · A. Hamadeh · M. K. Sukumaran · B. Ingalls (✉)
University of Waterloo, Waterloo, Canada
e-mail: bingalls@uwaterloo.ca

A. Ahmadzadegan
e-mail: aida.ahmadzadegan@uwaterloo.ca

A. Hamadeh
e-mail: ahamadeh@uwaterloo.ca

M. K. Sukumaran
e-mail: midhun.sukumaran@uwaterloo.ca

R. Tempo et al. (eds.), *Emerging Applications of Control and Systems Theory*, Lecture Notes in Control and Information Sciences - Proceedings,
https://doi.org/10.1007/978-3-319-67068-3_14

been a rising obstacle to therapy for decades; the continually increasing prevalence of antibiotic-resistant bacterial pathogens is now recognized as a significant worldwide health concern [2, 3]. Proposed approaches for combating this threat include reductions in antibiotic use, increased surveillance of pathogen genetics, and development of new antibiotics and alternative strategies for suppression of pathogens [4, 5].

In addition to killing pathogens, alternative suppression strategies could involve targeting pathogenicity [6] or other aspects of their physiology to impact their ecology or evolution [7]. In this chapter, we apply systems theoretic modelling to explore a bioaugmentation strategy to suppress the activity of antibiotic resistance genes in microbial populations [8].

Bioaugmentation is the manipulation of environments by the addition of biological organisms. This strategy has been demonstrated to improve the metabolic function of bioremediation and waste-water treatment populations [9, 10]. In particular, it has been demonstrated that the addition of strains harbouring mobile genetic elements can result in the transfer of function to the resident microbial population [9].

Our group is investigating a bioaugmentation approach that involves transfer of genetic material to pathogens with the goal of suppressing their resistance to antibiotics. Treatment would begin with the introduction of an engineered bacterial strain. Delivery of genetic material can be achieved via plasmid conjugation (cell-to-cell transfer of extra-chromosomal DNA) or phage transduction (virus-based DNA transfer), both of which can facilitate spread of the introduced genetic element within the target population.

Suppression of antibiotic resistance pathogens by delivery of genetic material can occur through a number of mechanisms. Yosef et al. [11] demonstrated a CRISPR-based system that eliminates antibiotic resistance plasmids. Another approach is to make use of plasmid incompatibility to dilute an antibiotic resistance plasmid from the population [7, 8]. Alternatively, the introduced material could trigger cell death in the presence of active resistance genes, e.g. by activating a toehold switch [12].

Implementation of such a strategy could target reservoirs of antibiotic resistance genes in environments which are rife with gene transfer (such as waste-water treatment plants [13]) or are clinically relevant, such as hospital surfaces [14] or the human gut microbiome, from which antibiotic resistance genes have been isolated [15].

There has been much recent interest in manipulation of the gut microbiome [16, 17]. Successful deployment of a synthetic sensory gene circuit has been demonstrated in the mouse gut [18] and progress has been made in developing genetic engineering tools for species that are prevalent in the gut environment [19].

Model-based design has played a central role in the development of a wide range of synthetic biology constructs [20]. For bioaugmentation approaches, mechanistic modelling will be needed to characterize and predict aspects of the delivery of genetic material, the consequent effects on cellular function, and the population dynamics of the introduced vector and the targeted pathogen population.

Several models of propagation of conjugating plasmids have appeared in the literature, starting with [21]. Our group recently used model comparison and uncertainty analysis to confirm that this modelling framework generates accurate model predictions from quantitative time-series observations [22]. Genetic mechanisms for suppression of resistant pathogens have received less attention. We presented a model of displacement of antibiotic resistance plasmids by unilateral plasmid incompatibility [8]. Suppression by CRISPR activation has not been modelled in the context of bioaugmentation, but a related model of intracellular CRISPR activity was presented in [23].

In this chapter, we explore suppression of resistant pathogens in a model of the mammalian gut, as an instance of a promising application of the proposed bioaugmentation approach. We based our analysis on the model presented in [24], which describes the natural transfer of antibiotic resistance genes in the bovine gut. While the model used for our analysis is speculative, the qualitative results are likely to be indicative of system behaviour in a range of relevant scenarios.

14.2 Deterministic Model Development

Focusing on implementation of the proposed design in the mammalian gut, we take the ordinary differential equation model of [24] (analogous to descriptions of chemostat dynamics) as a foundation for model development. For each subpopulation, the model accounts for inflow, outflow, competition-limited growth, and gene transfer. The model describes two populations of *E. coli* (donors and recipients), each of which can flow continuously into the fixed reaction volume. Any resident microbial population that is made up of other species (which would be significant in the gut) is not represented. Thus the populations being described are presumed to occupy their own niche, within which subpopulations compete with one another.

We modified the model of [24] by including an introduced (engineered) strain, which acts as the vector for delivery of genetic material to the target population. We thus consider three subpopulations: introduced donors (abundance D), target recipient cells (T), and recipient cells that have received a delivery of the genetic payload (C, for 'cured'):

$$\begin{aligned}
\frac{d}{dt}D(t) &= \alpha_D + r_D D(t)\left(\frac{K_D - \beta_{DD}D(t) - \beta_{DT}(T(t)+C(t))}{K_D}\right) \qquad (14.1)\\
\frac{d}{dt}T(t) &= \alpha_T + r_T T(t)\left(\frac{K_T - \beta_{TT}(T(t)+C(t)) - \beta_{TD}D(t)}{K_T}\right) - (\gamma_D D(t) + \gamma_C C(t))T(t)\\
\frac{d}{dt}C(t) &= r_C C(t)\left(\frac{K_T - \beta_{TT}(T(t)+C(t)) - \beta_{TD}D(t)}{K_T}\right) + (\gamma_D D(t) + \gamma_C C(t))T(t)
\end{aligned}$$

where, for each subpopulation, α_i is the inflow rate, r_i is the growth rate, K_i is the carrying capacity, and β_{ij} characterize competition-dependent removal. The rate of delivery of genetic material from donors to recipients is characterized by γ_D. Cured cells (i.e., recipient cells that carry the genetic payload) transfer it to target cells at a rate characterized by γ_C. (Note: implementation would more likely involve pulsatile dosing of donors, rather than a constant inflow as described by α_D. The responses are similar provided the pulses are relatively frequent.) This model describes a **disarming** strategy, in which the cured cells persist, and can help transfer the genetic payload to their neighbouring recipient cells.

A simpler approach is for the genetic payload to trigger death once delivered to a target cell. We refer to this as an **elimination** strategy. In this case the C population is absent. The dynamics are described by:

$$\frac{d}{dt}D(t) = \alpha_D + r_D D(t)\left(\frac{K_D - \beta_{DD}D(t) - \beta_{DT}T(t)}{K_D}\right) \tag{14.2}$$

$$\frac{d}{dt}T(t) = \alpha_T + r_T T(t)\left(\frac{K_T - \beta_{TT}T(t) - \beta_{TD}D(t)}{K_T}\right) - \gamma_D D(t)T(t)$$

14.3 Steady States and Stability

The goal of the proposed treatment is to reduce the target (i.e., antibiotic-resistant) population. We are thus interested in steady states for which the target population is small. In the case of $\alpha_T = 0$ (i.e., no influx of the target population), we derive conditions on the parametrization that guarantee the existence of a locally stable steady state for which $T = 0$. These are necessary conditions for successful treatment with the proposed strategy.

Elimination Strategy: Considering model (14.2), and assuming no inflow of target cells (i.e., $\alpha_T = 0$), the model admits a single relevant steady state with $T = 0$, given by:

$$T = 0, \quad D = \frac{r_D + \sqrt{r_D^2 + 4\alpha_D r_D \beta_{DD}/K_D}}{2r_D\beta_{DD}/K_D}.$$

At this point the eigenvalues of the Jacobian are:

$$\lambda_1 = -\sqrt{r_D^2 + \frac{4r_D\alpha_D\beta_{DD}}{K_D}}, \tag{14.3}$$

$$\lambda_2 = r_T - \frac{K_D(r_T\beta_{TD} + K_T\gamma_D)}{2K_T\beta_{DD}}\left(1 + \sqrt{1 + \frac{4r_D\alpha_D\beta_{DD}}{K_D r_D}}\right). \tag{14.4}$$

Thus λ_1 will be negative provided the parameters are positive. The expression for λ_2 reveals that, with the other parameters fixed, stability of this $T = 0$ steady state can be achieved by increasing any of the design parameters α_D (dosage of donors cells), γ_D (transfer rate from donor cells) or K_D (carrying capacity of donor cells) sufficiently.

Stability of the $T = 0$ steady state is necessary, but not sufficient, for targets to be ultimately eliminated from the system. Model (14.2) can exhibit bistability over parameter ranges corresponding to weak competition between the populations. Bistability will not occur (at $T = 0$) provided the following condition holds.

$$\left(\frac{K_T\beta_{DT}}{\sqrt{K_D}} - \sqrt{K_D\beta_{TT}}\right)^2 < \frac{4\alpha_D\beta_{TT}(r_T\beta_{DT}\beta_{TD} - r_T\beta_{DD}\beta_{TT} + K_T\beta_{DT}\gamma_D)}{r_D r_T}. \tag{14.5}$$

This condition can be guaranteed by increasing the design parameter γ_D (delivery rate) in conjunction with increasing α_D (dosing level).

Disarming Strategy: Again focusing on the $T = 0$ case, we impose further the conservative condition that $\beta_{TD} = 0$, corresponding to the assumption that the donor population exerts no competitive effect on the resident recipient population. With this additional assumption, two steady states satisfy $T = 0$. The first is

$$C = 0, \quad D = \frac{r_D + \sqrt{r_D^2 + 4\alpha_D r_D \beta_{DD}/K_D}}{2r_D\beta_{DD}/K_D}.$$

The second steady state has $C = \frac{K_T}{\beta_{TT}}$ and D given by the positive solution to

$$0 = \alpha_D + r_D\left(1 - \frac{K_T}{K_D}\frac{\beta_{DT}}{\beta_{TT}}\right)D - r_D\frac{\beta_{DD}}{K_D}D^2$$

which exists provided $\alpha_D > 0$ and $r_D\frac{\beta_{DD}}{K_D} > 0$.

The eigenvalues of the Jacobian of (14.1) are (for $T = 0,\ \beta_{TD} = 0$):

$$\lambda_1 = -r_D\left(2\frac{\beta_{DD}}{K_D}D + \frac{\beta_{DT}}{K_D}C\right) \tag{14.6}$$

$$\lambda_2 = r_T - r_T\frac{\beta_{TT}}{K_T}C - \gamma_D D - \gamma_C C \tag{14.7}$$

$$\lambda_3 = r_C\left(1 - 2\frac{\beta_{TT}}{K_T}C\right) \tag{14.8}$$

We see that the steady state at which $C = 0$ is unstable because $\lambda_3 = r_C > 0$. In contrast, the steady state at which $C = \frac{K_T}{\beta_{TT}}$ is locally asymptotically stable since there $\lambda_1,\ \lambda_2,\ \lambda_3 < 0$.

14.4 System Performance and Sensitivity

We next carry out numeric analysis, centering our attention at a nominal parameterization, shown in Table 14.1. Simulations of system behaviour at this nominal parametrization are shown in Fig. 14.1.

Elimination Versus Disarming: Comparing the two treatment strategies, the disarming approach has the advantage that the cured cells help spread the genetic payload through the recipient population. In contrast, in the elimination strategy, a target cell is only 'cured' by being killed. However, cured cells have a secondary effect on the

Table 14.1 Nominal parametrization: The value of r_T (target growth rate) is taken from [24]. The reduced growth rate of the cured population, r_C, reflects the metabolic burden of maintaining the genetic payload. The growth rate of the introduced donors, r_D, reflects the presumed weakness of these newcomers to the environment. Population densities are given in dimensionless units relative to K_T, the carrying capacity of the recipient population. The carrying capacity of the introduced donors, K_D, is presumed smaller. The transfer rate from the donor population, γ_D, is inspired by [24]. Transfer from the cured population, γ_C, is presumed less effective. The competition terms are equal except for the donor population's effect on the resident target cells, which is presumed weak. The immigration rates are in the range corresponding to [24]

r_T	$0.16\ h^{-1}$	γ_C	$0.004\ h^{-1}$
r_C	$0.12\ h^{-1}$	β_{TT}	1
r_D	$0.016\ h^{-1}$	β_{DT}	1
K_T	1	β_{DD}	1
K_C	1	β_{TD}	0.5
K_D	0.1	α_T	$0.01\ h^{-1}$
γ_D	$0.04\ h^{-1}$	α_D	$0.4\ h^{-1}$

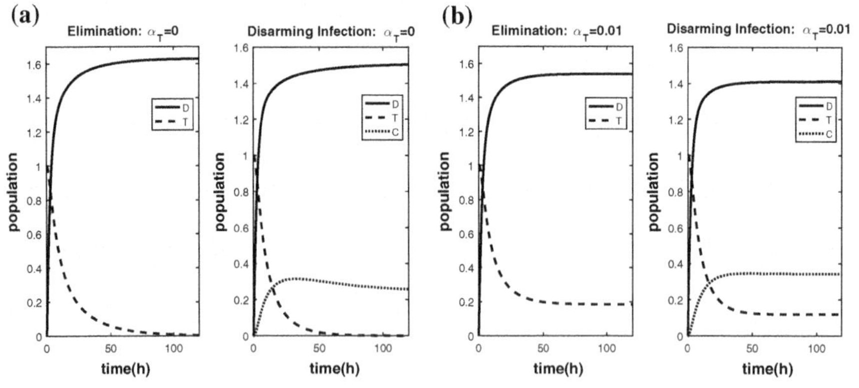

Fig. 14.1 Dynamics of the populations in the elimination (model (14.2)) and disarming (model (14.1)) scenarios. Parameter values are given in Table 14.1. For each strategy, two cases are shown: constant inflow of target cells, or no inflow of target cells. The initial target population is the steady state in the absence of the other populations

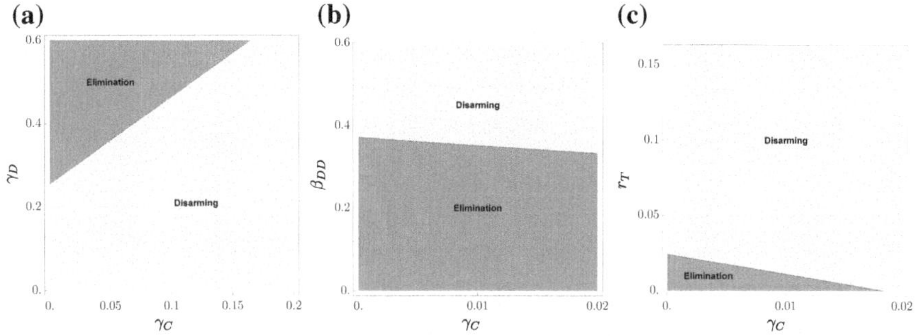

Fig. 14.2 Performance comparison of the disarming and elimination strategies. Regions in parameter space in which each strategy is more effective (as measured by the size of asymptotic target population) are indicated. Nominal parameter values in Table 14.1

treatment: they compete for resources with both the donor population (inhibiting treatment performance) and the target population (improving performance). Consequently, the elimination strategy can outperform the disarming strategy, depending on the parametrization.

To investigate the performance of the two strategies, we fixed all but two parameters to the nominal values in Table 14.1 and performed parameter scans comparing the outcome (steady state target population size) for the two strategies. The results are shown in Fig. 14.2. We focus on parameter γ_C, which characterizes the spread of the genetic payload by the cured population. For the most part, we found that in the parameter region near the nominal operating point, the disarming strategy was more effective. In particular, this approach is always preferred for sufficiently large γ_C values. In contrast, referring to Fig. 14.2, we note that for a fixed γ_C the elimination strategy is preferable for (a) sufficiently large rates of transfer γ_D, (b) sufficiently large donor populations (i.e., sufficiently small β_{DD}), or (c) sufficiently slow growth of target cells (r_T).

Sensitivity Analysis: The nominal parameterization chosen in the previous section is only loosely justified. Consequently, we explored the parameter space further through global sensitivity analysis. We focused our attention on the disarming strategy, which was found to be the more robust approach. We considered two scenarios: (i) no inflow of targets ($\alpha_T = 0$), for which we took the time at which the target population drops to 90% of its pretreatment level as the performance measure; and (ii) constant inflow of target cells ($\alpha_T = 0.01$), in which case we assigned the steady state target population as the performance measure.

The analysis was carried out with Sobol's variance-based method [25]. Given a probability distribution for one of n parameters p_i $(i = 1, \cdots, n)$, this procedure quantifies the percentage of the total variance of the performance measure $y = f(p_1, \cdots, p_n)$ (arising from the parametric uncertainty) that is attributable to any one parameter or set of parameters. The proportion of the total variance that results solely from uncertainty in the ith parameter, termed the parameter's first order Sobol index, S_i, is evaluated as

$$S_i = \frac{\mathrm{Var}(\mathscr{E}(f(p_1, \cdots, p_n | p_i)))}{\mathrm{Var}(f(p_1, \cdots, p_n))} \tag{14.9}$$

where $\mathscr{E}(\cdot)$ and $\mathrm{Var}(\cdot)$ represent the expectation and variance operators respectively. The proportion of the total variance that results from the interaction of parameter i with all other parameters is the parameter's total effects Sobol index, T_i, evaluated as

$$T_i = \frac{\mathrm{Var}(f(p_1, \cdots, p_n)) - \mathrm{Var}(\mathscr{E}(f(p_1, \cdots, p_n | p_{-i})))}{\mathrm{Var}(f(p_1, \cdots, p_n))} \tag{14.10}$$

where we define p_{-i} as the set of all parameters excluding p_i.

We calculated the first and total order indices using the procedure presented in [26], as follows. Each of the N rows of two $N \times n$ matrices, $\mathbf{A}$ and $\mathbf{B}$, is constructed by sampling each of the n parameters once from its respective distribution. The performance measure is evaluated once for each row of parameters, for each matrix, to form the column vectors $\hat{f}(\mathbf{A})$, $\hat{f}(\mathbf{B}) \in \mathbb{R}^N$. In addition, the matrix $\mathbf{B}_{\mathbf{A}_i}$ is formed by replacing the column of samples of parameter i in matrix $\mathbf{B}$ with its corresponding column in matrix $\mathbf{A}$. The performance measure is then evaluated at each row of $\mathbf{B}_{\mathbf{A}_i}$ to form the column vector $\hat{f}(\mathbf{B}_{\mathbf{A}_i})$. Finally, for N large, we obtain the variance estimates

$$\begin{aligned}
\mathrm{Var}(f(p_1, \cdots, p_n)) &\approx \frac{1}{N}\hat{f}(\mathbf{A})^T\hat{f}(\mathbf{A}) - \left(\frac{1}{N}\mathbf{1}^T\hat{f}(\mathbf{A})\right)^2 \\
\mathrm{Var}(\mathscr{E}(f(p_1, \cdots, p_n | p_i))) &\approx \frac{1}{N}\hat{f}(\mathbf{A})^T\hat{f}(\mathbf{B}_{\mathbf{A}_i}) - \left(\frac{1}{N}\mathbf{1}^T\hat{f}(\mathbf{A})\right)^2 \\
\mathrm{Var}(\mathscr{E}(f(p_1, \cdots, p_n | p_{-i}))) &\approx \frac{1}{N}\hat{f}(\mathbf{B})^T\hat{f}(\mathbf{B}_{\mathbf{A}_i}) - \left(\frac{1}{N}\mathbf{1}^T\hat{f}(\mathbf{A})\right)^2
\end{aligned} \tag{14.11}$$

where $\mathbf{1}$ is the column vector of ones in $\mathbb{R}^N$.

From Figs. 14.3a, b, we note that the design parameters α_D (the dose), K_D (the donor carrying capacity), and γ_D (rate of delivery) have significant impacts on the rate of loss of target cells. In contrast, the growth rate of cured cells r_C, and the competitive effect of donors on targets, β_{TD}, have little impact. From Figs. 14.3c and d, we see that when considering the asymptotic performance, the parameters α_D and α_T (inflow), and β_{DD} and β_{TD} (competitive effects of the donor population) have significant impact, while R_T, β_{TT}, and K_T, which characterize the target population's growth dynamics, have little effect. Overall, the finding that the design parameters (i.e., features of the donor strain) have significant impact on the performance is encouraging.

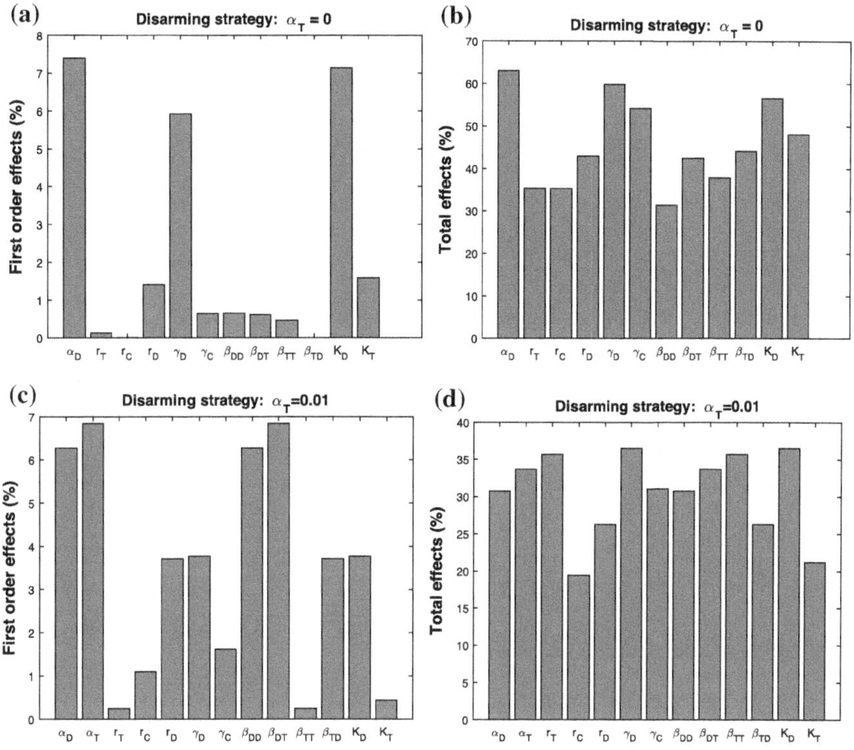

Fig. 14.3 Global sensitivity analysis. First and total order effects of parameter variation. For model (14.1), each parameter was varied two orders of magnitude above and below the nominal value (Table 14.1), over 5000 samples

14.5 Stochastic Modelling

Environmental microbial populations are generally large, and so deterministic models are sometimes sufficient to capture population-averaged behaviours. Of the antibiotic resistance suppression approaches described above, one such strategy—displacement of resistance-conferring plasmids by incompatibility—relies on intrinsic variability in plasmid copy numbers, and so cannot be accurately described by deterministic models. In [8], we presented a simple stochastic model of this process. Here, we present an improved model formulation, which incorporates appropriate descriptions of the cell population dynamics as described in the previous sections.

The model describes a population of cells in a reaction vessel of fixed volume. Two cell types are considered: donor cells and recipient cells. Donor cells can contain any number of engineered (self-conjugative) plasmids (hereafter E plasmids); recipient cells can contain any number of E plasmids and of antibiotic resistance plasmids (A plasmids). In the ODE model described above, recipient cells are presumed to be cured of their resistance trait the instant they receive the engineered plasmid.

References

1. Davies, J., Davies, D.: Microbiol. Mol. Biol. Rev. **74**(3), 417 (2010)
2. Ventola, C.L.: Pharma. Ther. **40**(4), 277 (2015)
3. Ventola, C.L.: Pharma. Ther. **40**(5), 344 (2015)
4. Levy, S.B.: J. Antimicrob. Chemother. **49**(1), 25 (2002)
5. Shlaes, D.M., Sahm, D., Opiela, C., Spellberg, B.: Antimicrob. Agents Chemother. **57**(10), 4605 (2013)
6. Balaban, N., Goldkorn, T., Nhan, R.T., Dang, L.B., Scott, S., Ridgley, R.M., Rasooly, A., Wright, S.C., Larrick, J.W., Rasooly, R., et al.: Science **280**(5362), 438 (1998)
7. Baquero, F., Coque, T.M., de la Cruz, F.: Antimicrob. Agents Chemother. **55**(8), 3649 (2011)
8. Gooding-Townsend, R., Ten Holder, S., Ingalls, B.: IEEE Life Sci. Lett. **1**(1), 19 (2015)
9. Top, E.M., Springael, D., Boon, N., Microbiol, F.E.M.S.: Ecol. **42**(2), 199 (2002)
10. Cheng, Z., Chen, M., Xie, L., Peng, L., Yang, M., Li, M.: Biotechnol. Lett. **37**(2), 367 (2015)
11. Yosef, I., Manor, M., Kiro, R., Qimron, U.: Proc. Natl. Acad. Sci. U.S.A **112**(23), 7267 (2015)
12. Green, A.A., Silver, P.A., Collins, J.J., Yin, P.: Cell **159**(4), 925 (2014)
13. Rizzo, L., Manaia, C., Merlin, C., Schwartz, T., Dagot, C., Ploy, M., Michael, I., Fatta-Kassinos, D.: Sci. Total Environ. **447**, 345 (2013)
14. Lax, S., Gilbert, J.A.: Trends Mol. Med. **21**(7), 427 (2015)
15. Hu, Y., Yang, X., Qin, J., Lu, N., Cheng, G., Wu, N., Pan, Y., Li, J., Zhu, L., Wang, X. et al.: Nat. Commun. **4** (2013)
16. Sonnenburg, J.L., Fischbach, M.A.: Sci. Transl. Med. **3**(78), 78ps12 (2011)
17. Claesen, J., Fischbach, M.A., Synth, A.C.S.: Biol. **4**(4), 358 (2014)
18. Kotula, J.W., Kerns, S.J., Shaket, L.A., Siraj, L., Collins, J.J., Way, J.C., Silver, P.A.: Proc. Natl. Acad. Sci. U.S.A **111**(13), 4838 (2014)
19. Mimee, M., Tucker, A.C., Voigt, C.A., Lu, T.K.: Cell Syst. **1**(1), 62 (2015)
20. Heinemann, M., Panke, S.: Bioinformatics **22**(22), 2790 (2006)
21. Levin, B.R., Stewart, F.M., Rice, V.A.: Plasmid **2**(2), 247 (1979)
22. Malwade, A., Nguyen, A., Sadat Mousavi, P., Ingalls, B.P.: Front. Microbiol. **8**, 461 (2017)
23. Djordjevic, M., Djordjevic, M., Severinov, K.: Biol. Direct. **7**(1), 1 (2012)
24. Volkova, V.V., Lanzas, C., Lu, Z., Gröhn, Y.T.: PLoS ONE **7**(5), e36738 (2012)
25. Sobol, I.: Math. Mod. Comp. Exp. **1**(4), 407414 (1993)
26. Homma, T., Saltelli, A.: Reliab. Eng. Syst. Safe. **52**, 1 (1996)
27. Gillespie, D.T.: J. Phys. Chem. **81**(25), 2340 (1977)
28. Paulsson, J., Ehrenberg, M., Rev, Q.: Biophys. **34**(01), 1 (2001)

Chapter 15
Grid Integration of Renewable Electricity and Distributed Control

Pratyush Chakraborty, Enrique Baeyens and Pramod P. Khargonekar

Abstract Motivated by climate change and sustainability, and the resulting need to decarbonize the electricity sector, there is a major global movement toward large-scale integration of renewable energy, i.e., wind and solar, into the existing power grid. The inherent variability of wind and solar energy production poses a major challenge in achieving these goals. The problem becomes more challenging as we consider issues of competitive markets, low cost and high reliability. In the last few years, we have been working on new systems and control problems that arise from these considerations. In this paper, we will present some highlights of our work on developing demand response methods using distributed control and bounding the loss of efficiency in these methods.

15.1 Introduction

Carbon emissions leading to climate change and sustainability are some of the major reasons motivating adoption of renewable energy sources such as wind and solar into the electric energy system. Large-scale integration of wind and solar electric energy poses significant technological challenges. These energy sources are inherently uncertain (power generation not known in advance), intermittent (large fluctuations and

This work was supported by NSF Grants ECCS-1723849 (previously ECCS-1129061) and CNS-1723856 (previously CNS-1239274).

P. Chakraborty (✉)
University of California, Berkeley, CA, USA
e-mail: pchakraborty@berkeley.edu

E. Baeyens
University of Valladolid, Valladolid, Spain
e-mail: enrbae@eis.uva.es

P. P. Khargonekar
University of California, Irvine, CA, USA
e-mail: pramod.khargonekar@uci.edu

R. Tempo et al. (eds.), *Emerging Applications of Control and Systems Theory*, Lecture Notes in Control and Information Sciences - Proceedings,
https://doi.org/10.1007/978-3-319-67068-3_15

ramps) and non-dispatchable (unable to follow a command). The term *variability* is used to represent these three characteristics [17] and is a significant hurdle in the large-scale integration of renewables. A promising solution to address the variability is to deploy *demand side management* (DSM) or *demand response* (DR) programs that adjust the consumption to match the predicted generation.

A paradigm shift in the power system operations is underway where consumers will be incentivized to manage their demand by leveraging the flexibility of their loads such as electric vehicles (EV), air conditioning, heat pumps, water heaters, etc. [2, 18]. DSM or DR programs in power systems operations exploit this flexibility in power consumption loads. Distributed control has been used as a major tool to solve problems where a central authority sends a control signal, e.g., price of electricity, and consumers decide their consumption schedules according to some private utility function [13]. A strategy for assigning quantities in a distributed price-based framework is the *proportional allocation mechanism* where the central authority calculates a price for all the consumers in such a way that the assigned quantity to each agent is proportional to the monetary value that the agent is willing to pay [12]. This mechanism has been used to formulate EV charging problems [10, 22]. However, it has not been examined in broader smart grid settings.

Consumer behavior plays a critical role in the implementation of demand response programs in distributed mode and the assumption of price-taking consumers might not always be true. Selfish behavior of agents in a non-cooperative game leads to inefficiency with respect to the solution that maximizes system welfare. Consequently, it is crucial to design distributed control systems in such a way that the efficiency loss due to selfish behavior is bounded. The term *price of anarchy* (PoA) has been coined as a measure of efficiency of distributed control as compared with centralized optimal solution. Bounds on the PoA for various cost-sharing games, congestion games and payoff maximization games have been derived in [11, 16, 20].

In the smart grid scenario, non-cooperative game theoretic methods have been used to model problems [14, 15, 23], however the loss of efficiency by selfish behavior has not been widely investigated. The Nash equilibrium has been shown to be efficient in an infinite population game when the charging rates of all the EVs are equal [14] and in a DR problem with different consumers [15], however in the first case the assumptions are rather impractical and in the second one, the consumers' utility functions were ignored. Regarding wind variability, a game has been formulated among various power consumers in [23] where the PoA bound has been calculated for an example case.

During the last few years, we have been addressing various challenging problems involving both technical and economic issues of smart grid [3–9]. We present here a few salient results on two demand response methods. In the first one, the available power is limited and the control authority designs a price signal aiming to maximize social welfare subject to supply-demand balancing. We show that if proportional allocation is used to design the price signal, then the lower bound of the price of anarchy is 75%. We also develop some strategies for improving efficiency further. In the second case, we consider a demand response problem where the available power is not limited and the price signal is set by the load consumption. Under some conditions of the utility functions of the consumers with respect to the price, we obtain a robust lower bound of the price of anarchy of 50%.

15.2 A Demand Response Program Using Proportional Allocation Mechanism with Tight PoA Bound

In this section, we develop a distributed method for controlling the consumers' flexible demand with intra-day supply forecasts. Flexible consumers are modeled as individually rational agents that maximize their net utilities in presence of load consumption constraints. The consumers bid the monetary value they are willing to pay for each time interval and the central authority obtains a price signal based on a proportional allocation mechanism. Two scenarios are considered, *price taking* and *price anticipating* consumers. In the first case, the proportional allocation method provides a competitive equilibrium that maximizes the system welfare. In the second case, the consumers' selfish behavior is modeled using a non-cooperative game. A Nash equilibrium always exists for this game but it is not efficient. We are able to obtain a lower bound on the PoA of 75% and we develop some strategies to improve the game efficiency.

15.2.1 Problem Formulation

Let us consider a residential area where electric power is supplied by thermal and renewable generators. The power consumption in the area is controlled by a central *control authority*. Two types of residential consumers are considered: *fixed consumers* and *flexible consumers*. Only flexible consumers are willing to adjust their consumption schedules in response to some signal from the authority.

Let us consider a set $\mathcal{N} := \{1, 2, \ldots, N\}$ of flexible consumers. Each flexible consumer possesses a smart energy scheduling device with two-way communication capability. We assume that the supply is initially scheduled in a traditional day ahead market, based on demand predictions. The time interval of interest $[t_0, t_f]$, corresponding to the intra-day horizon, is divided into T slots of length $\Delta t = (t_f - t_0)/T$. The set of time slots is $\mathcal{T} := \{1, 2, \ldots, T\}$, and we consider the following variables at time slot $t \in \mathcal{T}$: $q_i(t) \in \mathbb{R}_+$ is the power consumption of all the flexible loads of the i-th flexible consumer (no power transfer from the consumers to the grid is allowed), $c(t) \in \mathbb{R}_+$ is the total scheduled power generation of all the thermal power plants, $\widehat{w}(t) \in \mathbb{R}_+$ is the estimate of the power generation of all the renewable sources, $\widehat{n}(t) \in \mathbb{R}_+$ is the estimate of the total power consumption of all fixed loads of both fixed and flexible consumers. Let $v(t)$ denote the estimated net generation available for flexible demand at time slot $t \in \mathcal{T}$, i.e., $v(t) := c(t) + \widehat{w}(t) - \widehat{n}(t)$.

The control authority obtains a forecast of renewable generation and balances the estimated demand with supply for each time slot of the operating day, i.e.,

$$v(t) = \sum_{i \in \mathscr{N}} q_i(t), \; t \in \mathscr{T}. \tag{15.1}$$

Assuming that net power supply is always sufficient to meet the fixed loads' demand, i.e., $v(t) > 0$ for any $t \in \mathscr{T}$, the supply-demand balancing is accomplished by adjusting the power consumption of the flexible loads. There is always an inevitable mismatch between the estimated power generation and consumptions and their realized values. Ancillary services are implemented to handle this real-time mismatch. However, the use of the intra-day flexible load control mechanism proposed here will reduce the need for ancillary services while making large-scale renewable integration less burdensome.

Let $\mathbf{q}_i, \mathbf{v} \in \mathbb{R}_+^T$ denote vectors of dimension T that collect the consumption of flexible consumer $i \in \mathscr{N}$ and the net power generation available for flexible consumers, respectively, for every time slot $t \in \mathscr{T}$. The output of flexible consumption $i \in \mathscr{N}$ is represented in monetary units by the utility function $U_i(\mathbf{q}_i) : \mathbb{R}^T \to \mathbb{R}$, which is assumed to be non-negative, concave and continuously differentiable. In addition, it is also assumed to be a strictly increasing, i.e., $\nabla U_i(\mathbf{q}_i) > 0$, where $\nabla U_i : \mathbb{R}^T \to \mathbb{R}^T$ denotes the gradient of U_i. The operational constraints of the flexible consumers depending upon the type of loads can be expressed by a set of linear inequalities:

$$\mathbf{H}_i \mathbf{q}_i \leq \mathbf{b}_i, \; i \in \mathscr{N}, \tag{15.2}$$

where $\mathbf{H}_i \in \mathbb{R}^{M \times T}$, $\mathbf{b}_i \in \mathbb{R}^M$ and M is the number of constraints. Let $\mathscr{Q}_i := \{\mathbf{q} \in \mathbb{R}^T : \mathbf{b}_i - \mathbf{H}_i \mathbf{q} \geq \mathbf{0}\}$ denote the set of consumption vectors satisfying the operational constraints for $i \in \mathscr{N}$ and $\mathscr{S} := \{\mathbf{q}_i \in \mathscr{Q}_i : \mathbf{v} - \sum_{i \in \mathscr{N}} \mathbf{q}_i = \mathbf{0}, \; i \in \mathscr{N}\}$ the *feasibility set* of the consumption vectors satisfying both the supply-demand power balance constraint and the operational constraints. We assume that the feasibility set $\mathscr{S}$ is nonempty.

15.2.2 Centralized Control

In this idealized scenario, the central control authority dictates how much power is assigned to each flexible consumer by maximizing the social welfare. The centralized control problem is

$$\max_{\mathbf{q}_i} \left\{ \sum_{i \in \mathscr{N}} U_i(\mathbf{q}_i) : \mathbf{q}_i \in \mathscr{S} \right\} \tag{15.3}$$

The existence of a maximum is guaranteed because the objective function is concave and the search space is a nonempty compact convex set. A solution of (15.3) maximizes the social welfare and is referred to as the *centralized optimal solution*. But the consumers may want to control their loads on their own and the central authority may not have computational capability to solve the optimization problem for a large number of residential consumers. A feasible alternative is a distributed control approach.

15.2.3 Distributed Control with Price-Taking Consumers

In a distributed control model, the behavior of the consumers is an important aspect to consider. We begin by considering individually rational flexible consumers that behave as price takers. Let $\mathbf{k}_i \in \mathbb{R}_+$ denote the amount of money the consumer $i \in \mathcal{N}$ is willing to pay for the energy $\mathbf{q}_i$. The consumers bid the monetary values or expenditures $\mathbf{k}_i$ to the control authority. In this scenario, the control authority obtains the value of the available net supply $v(t)$ for every $t \in \mathcal{T}$ and computes a system price $p(t)$ according to the following proportional allocation mechanism.

Definition 15.1 (*The proportional allocation mechanism*) The proportional allocation of the energy consumption at time slot $t \in \mathcal{T}$ is given by:

$$q_i(t) = \frac{k_i(t)}{p(t)}, \quad i \in \mathcal{N}, \tag{15.4}$$

where $p(t) > 0$ is the price of electricity at time $t \in \mathcal{T}$, obtained by

$$p(t) = \frac{\sum_{i \in \mathcal{N}} k_i(t)}{v(t)}, \quad t \in \mathcal{T}. \tag{15.5}$$

Since $v(t)$ is always positive, the system price $p(t)$ is well defined for every time slot $t \in \mathcal{T}$ and guarantees $v(t) = \sum_{i \in \mathcal{N}} q_i(t)$ for all t. Each consumer (flexible or fixed) is charged at the system price. Let the net utility of a consumer be defined as the total utility minus the expenditure. The flexible consumers maximize their own net utility function by a suitable selection of their consumptions $\mathbf{q}_i$. Let $\mathbf{p} \in \mathbb{R}^T$ denote the vector that collects the system prices for every time slot $t \in \mathcal{T}$. The distributed control problem for price takers is formulated as follows:

$$\max_{\mathbf{q}_i} \left\{ U_i(\mathbf{q}_i) - \mathbf{p}^\top \mathbf{q}_i : \mathbf{q}_i \in \mathcal{S}_i^{pt} \right\}, \; i \in \mathcal{N}, \tag{15.6}$$

where the set of feasible power consumptions is $\mathcal{S}_i^{pt} := \{\mathbf{q}_i : \mathbf{b}_i - \mathbf{H}_i \mathbf{q}_i \geq 0\}$.

The solution concept for the distributed control problem with price taking flexible consumers is the competitive equilibrium.

Definition 15.2 The set $\{(\mathbf{q}_i^E, \mathbf{p}^E) : i \in \mathcal{N}\}$ is a *competitive equilibrium* if each consumer selects its consumption vector $\mathbf{q}_i^E$ by solving the optimization problem (15.6) and the control authority obtains the price vector $\mathbf{p}^E$ using the proportional allocation mechanism (15.4)–(15.5).

The competitive equilibrium always exists if the feasibility set $\mathcal{S}$ is nonempty. Moreover, in such a case a competitive equilibrium is equivalent to a solution of the centralized control problem and maximizes the social welfare.

Theorem 15.1 *The set* $\{(\mathbf{q}_i^E, \mathbf{p}^E) : i \in \mathcal{N}\}$ *is a competitive equilibrium if and only if the set of consumptions* $\{\mathbf{q}_i^E : i \in \mathcal{N}\}$ *is a solution to the centralized control problem.*

15.2.4 Distributed Control with Price Anticipating Consumers

If the consumers can predict the mechanism that the control authority uses to set the price vector $\mathbf{p}$, they adjust their consumption decisions according to their impact on the price, and we say that they behave as price anticipators. By using as decision variables the monetary value vectors $\mathbf{k}_i$, where $k_i(t) = p(t)q(t)$ for $t \in \mathcal{T}$, the consumers can obtain the price vector $\mathbf{p}$ as a function of $\sum_{i\in\mathcal{N}} \mathbf{k}_i$, because we assume they know that $\mathbf{p}$ is decided by the formula $p(t) = \sum_{i\in\mathcal{N}} k_i(t)/v(t)$. Each consumer's monetary value depends on the sum of all the consumers' expenditures and the consumption assignment can be modeled as a non-cooperative game where the players are the flexible consumers.

The problem can be formulated in terms of the monetary expenditures by eliminating the price and the consumptions variables. Let $\mathbf{k}_{-i} = \{\mathbf{k}_j : j \in \mathcal{N} \setminus \{i\}\}$ denote the collection of monetary value vectors of all flexible consumers other than the consumer i. Note that $\mathbf{p}$ and $\mathbf{q}_i$ can be expressed as functions of $\mathbf{k}_i$ as $\mathbf{p}(\mathbf{k}_i; \mathbf{k}_{-i}) = \mathbf{D}^{-1}(\mathbf{v}) \sum_{j\in\mathcal{N}} \mathbf{k}_j$ and $\mathbf{q}_i(\mathbf{k}_i; \mathbf{k}_{-i}) = \mathbf{D}^{-1}(\mathbf{p}(\mathbf{k}_i; \mathbf{k}_{-i}))\mathbf{k}_i = \mathbf{D}^{-1}(\sum_{i\in\mathcal{N}} \mathbf{k}_i)\mathbf{D}(\mathbf{v})\mathbf{k}_i$ where $\mathbf{D}(\mathbf{x})$ denotes a diagonal square matrix whose main diagonal has the components of vector $\mathbf{x}$. Considering $\mathcal{S}_i^{pa}(\mathbf{k}_{-i}) := \{\mathbf{k}_i : \mathbf{b}_i - \mathbf{H}_i\mathbf{D}^{-1}\}$ $\{(\sum_{i\in\mathcal{N}} \mathbf{k}_i)\mathbf{D}(\mathbf{v})$ $\mathbf{k}_i \geq \mathbf{0}\}$, the distributed control problem for price anticipators is given by

$$\max_{k_i} \left\{ U_i(\mathbf{D}^{-1}(\sum\nolimits_{j\in\mathcal{N}} \mathbf{k}_j)\mathbf{D}(\mathbf{v})\mathbf{k}_i) - \mathbf{1}^\top \mathbf{k}_i : \mathbf{k}_i \in \mathcal{S}_i^{pa}(\mathbf{k}_{-i}) \right\}, \ i \in \mathcal{N}. \quad (15.7)$$

Each consumer will try to maximize her own net utility, assuming that all other consumers' expenditures are fixed. This is called the *best response strategy* and the solution is called a Nash equilibrium. In a Nash equilibrium, no player has an incentive to deviate unilaterally of the equilibrium [20]. The Nash equilibrium for the distributed control problem with price anticipators is the set of expenditures $\{\mathbf{k}_i^G : i \in \mathcal{N}\}$ such that

$$U_i(\mathbf{q}_i(\mathbf{k}_i^G, \mathbf{k}_{-i}^G)) - \mathbf{1}^\top \mathbf{k}_i^G \geq U_i(\mathbf{q}_i(\mathbf{k}_i, \mathbf{k}_{-i}^G)) - \mathbf{1}^\top \mathbf{k}_i, \ \mathbf{k}_i \in \mathscr{S}_i^{pa}(\mathbf{k}_{-1}^G), \ i \in \mathscr{N}. \tag{15.8}$$

It can be proved that a Nash equilibrium always exists if the feasibility set $\mathscr{S}$ is nonempty.

Theorem 15.2 (Existence of Nash equilibrium) *The non-cooperative game described by Eq. (15.8) has a Nash equilibrium if the space $\mathscr{S}$ is nonempty.*

15.2.5 Price of Anarchy and Efficiency Improvement

The selfish behavior of agents in a non-cooperative game theoretic setting renders lower performance as compared to the optimal centralized control. Price of anarchy (PoA) is a measure to quantify the loss of efficiency in using game theoretic control over centralized control. PoA is defined as the worst-case ratio of the objective function value of a Nash equilibrium of a game and that of a centralized optimal solution [20]. The quantity $1 - \text{PoA}$ is a worst-case estimate of the loss of performance due to price anticipating behavior of agents.

In our energy assignment problem for flexible consumers, $\{\mathbf{q}_i^C : i \in \mathscr{N}\}$ denotes a solution of the centralized problem (15.3) and $\{\mathbf{q}_i^G : i \in \mathscr{N}\}$ denotes a Nash equilibrium for the distributed control problem with price anticipating consumers. The PoA is defined as follows:

$$\text{PoA} := \frac{\sum_{i \in \mathscr{N}} U_i(\mathbf{q}_i^G)}{\sum_{i \in \mathscr{N}} U_i(\mathbf{q}_i^C)}. \tag{15.9}$$

Theorem 15.3 *The tight lower bound of* PoA *of the Nash equilibrium solution for the distributed consumption assignment with flexible consumers that behave as price anticipators is* 0.75.

The worst-case loss of efficiency corresponds to the case where one agent consumes half of the total power consumed by all the agents at each time slot. Thus, the market power of a consumer plays a key role in the efficiency of the game. The theoretical worst case could only be attained under a particular setting. We are interested in developing strategies to improve efficiency. The following corollaries show two different ways to improve efficiency.

Corollary 15.1 *If all the consumers have same utility function, i.e., $U_i = U$, there is no efficiency loss at Nash equilibrium solution, i.e., PoA is 1.*

Corollary 15.2 *Suppose $\{\mathbf{q}_i = \mathbf{0} : i \in \mathscr{N}\}$ belongs to the set of load operational constraints, then the PoA approaches 1 as the number N of flexible consumers goes to infinity.*

Efficiency can be improved by recruiting consumers with similar utility functions, or by classifying them into groups of similar utility and designing specific programs for each group. The distributed control approach will have better efficiency if consumers share their utility functions with the central control authority. Another option is to reduce individual market power by increasing the number of consumers.

15.3 A Demand Response Program with Robust PoA Bound

In this section, we consider a different model for demand response. Unlike in Sect. 15.2, the price here is decided by desired energy consumption. Here, we formulate a decentralized control model assuming that the consumers are price anticipators and quantify that, in the worst case loss of efficiency of this problem is never greater than 50%.

We introduce some additional notation in this section. Let $\{q_i \in \mathbb{R}^T : i \in \mathcal{N}\}$ denote the set of power demand vectors for each consumer in the system. The vector of aggregated power demand in the system is $\mathbf{q}_{\mathcal{N}} = \sum_{i \in \mathcal{N}} \mathbf{q}_i$ where the entry t is denoted by $q_{\mathcal{N}}(t)$ and corresponds to the aggregated consumption at time slot $t \in \mathcal{T}$. The price of electricity in the system at time $t \in \mathcal{T}$ is a function of the aggregated consumption at that time and is denoted by $p(t) = p(q_{\mathcal{N}}(t))$. We assume that the price function is a convex, continuously differentiable and monotonically increasing function.

15.3.1 Centralized Control

We assume that the authority has full information about the supply function $p(q_{\mathcal{N}}(t))$ for that system. Let $\mathbf{p}(\mathbf{q}_{\mathcal{N}}) \in \mathbb{R}^T$ denote the vector of system prices for all time slots $t \in \mathcal{T}$. The authority aims to maximize the consumer's aggregated net utility subject to their operational constraints. For any feasible set of consumptions $\{\mathbf{q}_i \in \mathcal{Q}_i : i \in \mathcal{N}\}$, the objective is

$$\max \left\{ \sum_{i=1}^{N} U_i(\mathbf{q}_i) - \mathbf{p}^{\top}(\mathbf{q}_{\mathcal{N}})\mathbf{q}_{\mathcal{N}} \ : \mathbf{q}_i \in \mathcal{Q}_i,\ i \in \mathcal{N} \right\}. \tag{15.10}$$

Since the objective function is concave, the non-emptiness of the convex sets $\{\mathcal{Q}_i : i \in \mathcal{N}\}$ defined by the operational constraints ensure that a global maxima always exist [1]. But if the consumers want to control the power consumption of their loads on their own with the help of available information on their smart meters, then centralized control will not work. If consumers are price takers, like the earlier problem, it is easy to show that the distributed control has a competitive equilibrium where the solution is same as the centralized solution. We are more interested in the price anticipatory case. Thus we model this scenario as a game problem in the next subsection.

15.3.2 Decentralized Control with Price Anticipating Consumers

The consumers know the price function and they optimize their consumption schedules accordingly. As the price is a function of power consumption of all the consumers, we model the resulting situation as a non-cooperative game.

Definition 15.3 (*Demand response game*) The *demand response game* is defined by the triple $(\mathcal{N}, \mathcal{E}, V)$ where $\mathcal{N}$ is the set of players, $\mathcal{E} = \cup_{i \in \mathcal{N}} \mathcal{Q}_i$ is the set of feasible strategies, and $V : 2^{\mathcal{N}} \times \mathcal{E} \to \mathbb{R}$ is the welfare function for a subset of players $\mathcal{S} \in 2^{\mathcal{N}}$ and a strategy set $\{\mathbf{q}_i : i \in \mathcal{N}\} \in \mathcal{E}$.

Each consumer is individually rational and maximizes her individual welfare, assuming that the remaining players' strategies are fixed. Denoting the strategies of other players by $\mathbf{q}_{-i} = \{\mathbf{q}_j : j \in \mathcal{N} \setminus \{i\}\}$, the individual welfare of player $i \in \mathcal{N}$ is $V(\{i\}, \{\mathbf{q}_i, i \in \mathcal{N}\})$ and can be expressed as a function of the strategy of the player i and the strategies of the other players as follows:

$$L_i(\mathbf{q}_i, \mathbf{q}_{-i}) := V(\{i\}, \{\mathbf{q}_i, i \in \mathcal{N}\}) = U_i(\mathbf{q}_i) - \mathbf{p}^\top(\mathbf{q}_i, \mathbf{q}_{-i})\mathbf{q}_i \qquad (15.11)$$

The Nash equilibrium for the demand response game is the set of all players' strategies such that no player has an incentive to deviate unilaterally. Mathematically, Nash equilibrium is defined by the set of strategies $\{\mathbf{q}_i^* \in \mathcal{Q}_i : i \in \mathcal{N}\}$ such that $L_i(\mathbf{q}_i^*, \mathbf{q}_{-i}^*) \geq L_i(\mathbf{q}_i, \mathbf{q}_{-i}^*)$ for all $\mathbf{q}_i \in \mathcal{Q}_i$, $i \in \mathcal{N}$. Since each consumer's objective function is concave and the strategies set is convex and compact, a Nash equilibrium solution exists according to Rosen's theorem [19].

The demand response game as defined by Definition 15.3 belongs to the class of valid monotone utility games and this will allow us to bound its efficiency. Let us begin by characterizing this class of games. Consider a payoff maximization game given by the triple $(\mathcal{N}, \mathcal{E}, V)$ where $\mathcal{N}$ is the set of players, $\mathcal{E} = \cup_{i \in \mathcal{N}} \mathcal{E}_i$ is the set of feasible strategies for each player and $V : 2^{\mathcal{N}} \times \mathcal{E} \to \mathbb{R}$ is a function that provides the welfare associated with a subset of players for a given strategy.

Definition 15.4 (*Valid Utility Game* [21]) The payoff maximization game $(\mathcal{N}, \mathcal{E}, V)$ is a *valid utility game* if it satisfies the following three properties:

(i) V is submodular, i.e., for any $\mathcal{S} \subseteq \mathcal{S}' \subseteq \mathcal{N}$ and any player $i \in \mathcal{N} \setminus \mathcal{S}'$

$$V(\mathcal{S} \cup \{i\}, e) - V(\mathcal{S}, e) \geq V(\mathcal{S}' \cup \{i\}, e) - V(\mathcal{S}', e), \ \forall e \in \mathcal{E} \qquad (15.12)$$

(ii) The welfare of a player is never less than the value added to the social welfare, i.e., for any $\mathcal{S} \subseteq \mathcal{N}$ and any $i \in \mathcal{S}$,

$$V(\{i\}, e) \geq V(\mathscr{S}, e) - V(\mathscr{S} \setminus \{i\}, e), \ \forall e \in \mathscr{E} \quad (15.13)$$

(iii) The aggregated value of the individual welfare of a group of players is never greater than the social welfare of the group, i.e., for any $\mathscr{S} \subseteq \mathscr{N}$

$$\sum_{i \in \mathscr{S}} V(\{i\}, e) \leq V(\mathscr{S}, e), \forall e \in \mathscr{E} \quad (15.14)$$

Definition 15.5 (*Monotone Game* [21]) The payoff maximization game $(\mathscr{N}, \mathscr{E}, V)$ is a *monotone game* if for any $\mathscr{S} \subseteq \mathscr{S}' \subseteq \mathscr{N}$, $V(\mathscr{S}, e) \leq V(\mathscr{S}', e)$.

Definition 15.6 (*Valid Monotone Utility Game*) The payoff maximization game $(\mathscr{N}, \mathscr{E}, V)$ is a *valid monotone utility game* if it is simultaneously a valid utility and a monotone game.

For the demand response game of Definition 15.3, we assume that each consumer's utility function satisfies the following condition.

Assumption 15.1 The utility function of any consumer $i \in \mathscr{N}$ is such that

$$U_i(\mathbf{q}_i) \geq \sum_{t=1}^{T} p(\tilde{q}_{-i} + q_i(t))(\tilde{q}_{-i} + q_i(t)) - p(\tilde{q}_{-i})\tilde{q}_{-i}$$

where $\tilde{q}_{-i} = \sum_{j \in \mathscr{N} \setminus \{i\}} q_j^{\max}$.

The above condition implies that the value of the utility function of a consumer $i \in \mathscr{N}$ for any feasible demand $\mathbf{q}_i$ will be greater than the increase in the maximum cost of power consumption in the system due to the addition of that demand $\mathbf{q}_i$. The central authority can broadcast this requirement to all the consumers as a prerequisite for participation in the demand response program. We also assume that the authority have an estimate of the upper-bound of $\sum_{i \in \mathscr{N}} q_i^{\max}$ which it also broadcasts to all the consumers. Under Assumption 15.1, the demand response game is a valid monotone utility game.

Theorem 15.4 *The demand response game as defined by Definition 15.3 is a valid monotone utility game if Assumption 15.1 is satisfied.*

15.3.3 Price of Anarchy

A payoff maximization game which satisfies (15.14) for any solution set $e \in \mathscr{E}$ is said to be a (λ, μ) smooth game if it satisfy

$$\sum_{i \in \mathscr{N}} V(\{i\}, e^*) \geq \lambda V(\{\mathscr{N}\}, e') - \mu V(\{\mathscr{N}\}, e^*) \quad (15.15)$$

where $e^*, e' \in \mathscr{E}$ are any two solution strategies of the game.

The (1, 1)-smooth games have the property that a lower bound for its price of anarchy is 0.5. Since any valid monotone game is (1, 1)-smooth [20], we have the following result.

Corollary 15.3 *The demand response game as defined by Definition 15.3 is a (1, 1)-smooth game. Moreover, the lower bound of the price of anarchy of any pure Nash equilibrium is at least 0.5.*

We have shown that Nash equilibrium for our game exists, but there can be a number of reasons for which the players may not reach an equilibrium [20]. So, we can consider a weaker notion of equilibria i.e., coarse correlated equilibria for a game for which Nash equilibria does not exist or exists but can not be reached. Since the demand response game is a (1, 1) smooth game, the bound derived via smoothness argument extends automatically, with no quantitative degradation to other weaker equilibria notions [20]. This is called intrinsic robustness property of the price of anarchy. So, lower bound of price of anarchy is 0.5 even in case of coarse correlated equilibrium solution.

15.4 Conclusions

Variability of renewable resources can be accommodated by shaping demand. Effective solutions to this problem will necessarily require distributed control. In this paper, we have discussed our initial research on bounding loss of efficiency by using distributed control. These ideas can be applied to other distributed control problems like supply side market with deep renewable penetration, energy resource aggregation, and scheduling, energy trading between microgrids, etc.

References

1. Boyd, S., Vandenberghe, L.: Convex Optimization. Cambridge University Press, Cambridge, UK (2009)
2. Callaway, D., Hiskens, I.: Achieving controllability of electric loads. Proc. IEEE **99**(1), 184–199 (2011)
3. Chakraborty, P.: Optimization and control of flexible demand and renewable supply in a smart power grid. Ph.D. thesis, University of Florida. Department of Electrical and Computer Engineering (2016)
4. Chakraborty, P., Khargonekar, P.P.: Flexible loads and renewable integration: distributed control and price of anarchy. In: Proceedings of the IEEE 52nd Annual Conference on Decision and Control (CDC), pp. 2306–2312. Firenze, Italy (2013)
5. Chakraborty, P., Khargonekar, P.P.: A demand response game and its robust price of anarchy. In: Proceedings of the 2014 IEEE Control Conference on Smart Grid Communications (SmartGridComm), pp. 644–649. Venice, Italy (2014a)

6. Chakraborty, P., Khargonekar, P.P.: Impact of irrational consumers on rational consumers in a smart grid. In: Proceedings of the American Control Conference (ACC), pp. 58–64. Portland, OR (2014b)
7. Chakraborty, P., Baeyens, E., Khargonekar, P.P, Poolla, K.: A cooperative game for the realized profit of an aggregation of renewable energy producers. In: Proceedings of 55th IEEE Conference on Decision and Control, pp. 5805–5812. Las Vegas, NV (2016)
8. Chakraborty, P., Baeyens, E., Khargonekar, P.P.: Cost causation based allocation of costs for market integration of renewable energy. IEEE Trans. Power Syst. (2017). (to appear)
9. Chakraborty, P., Baeyens, E., Khargonekar, P.P.: Distributed control of flexible demand using proportional allocation mechanism in a smart grid: game theoretic interaction and price of anarchy. J. Sustainable Energy Grids Netw. **12**, 30–39 (2017)
10. Fan, Z.: Distributed charging of PHEVs in a smart grid. In: Proceedings of the IEEE International Conference on Smart Grid Communications, Brussels, pp. 255–260 (2011)
11. Johari, R.: The price of anarchy and the design of scalable resource allocation mechanisms. In: Nisan, N., Roughgarden, T., Tardos, E., Vazirani, V.V. (eds.) Algorithmic Game Theory. Cambridge University Press, Avenue of the Americas, NY (2007)
12. Kelly, F.P.: Charging and rate control for elastic traffic. Eur. Trans. Telecommun. **8**(1), 33–37 (1997)
13. Li, N., Chen, L., Low, S.H.: Optimal demand response based on utility maximization in power networks. In: Proceedings of IEEE Power Engineering Society General Meeting, pp. 1–8. San Diego, CA (2011)
14. Ma, Z., Callaway, D., Hiskens, I.: Decentralized charging control of large populations of plug-in electric vehicles. IEEE Trans. Control Syst. Technol. **21**(1), 67–78 (2013)
15. Mohsenian-Rad, A., Wong, V., Jatskevich, J., Schober, R., Garcia, A.L.: Autonomous demand-side management based on game-theoretic energy consumption scheduling for the future smart grid. IEEE Trans. Smart Grid **1**(3), 320–331 (2010)
16. Moulin, H.: The price of anarchy of serial, average and incremental cost sharing. Econ. Theor. **36**(3), 379–405 (2008)
17. NERC Special Report: Accomodation of high levels of variable generation. North American Electric Reliability Corporation (NERC), Technical report (2009)
18. Rehman, S., Shrestha, G.: An investigation into the impact of electric vehicle load on the electric utility distribution system. IEEE Trans. Power Delivery **8**(2), 591–597 (1993)
19. Rosen, J.B.: Existence and uniqueness of equilibrium points for concave n-person games. Econometrica **33**(3), 520–534 (1965)
20. Roughgarden, T.: Intrinsic robustness of the price of anarchy. Commun. ACM **55**(7), 116–123 (2012)
21. Tardos, E., Wexler, T.: Network formation games and the potential function method. In: Nisan, N., Roughgarden, T., Tardos, E., Vazirani, V.V. (eds.) Algorithmic Game Theory. Cambridge University Press, Avenue of the Americas, NY (2007)
22. Vasirani, M., Ossowski, S.: A proportional share allocation mechanism for co-ordination of plug-in electric vehicle charging. Eng. Appl. Artif. Intell. **26**(3), 1185–1197 (2013)
23. Wu, C., Mohsenian-Rad, H., Huang, J.: Wind power integration via aggregator-consumer coordination: a game theoretic approach. In: Proceedings of IEEE PES Innovative Smart Grid Technologies, Washington, DC, pp. 1–6 (2012)

Chapter 16
Control Systems Under Attack: The Securable and Unsecurable Subspaces of a Linear Stochastic System

Bharadwaj Satchidanandan and P. R. Kumar

Abstract The ideas of controllable and unobservable subspaces of a linear dynamical system introduced by Kalman play a central role in the theory of control systems. They determine, among other aspects, the existence of solutions to many control problems of interest. Analogous to the notions of controllable and unobservable subspaces are the notions of *"securable"* and *"unsecurable"* subspace of a linear dynamical system, which have operational significance in the context of secure control. We examine what guarantees can be provided with respect to securable subspace, especially in the case when there is process noise in the system.

16.1 Introduction

The ideas of controllable and unobservable subspaces of a linear dynamical system, introduced by Kalman in [1], play a central role in the theory of control systems. They provide, for example, necessary and sufficient conditions for the existence of a stabilizing control law for any linear dynamical system of interest. Analogous to the notions of controllable and unobservable subspaces, we examine, in this paper, the notions of *"securable"* and *"unsecurable"* subspaces of a linear dynamical system, which we show have operational significance in the context of secure control.

Consider a multiple-input, multiple-output, discrete-time linear dynamical system, an arbitrary subset of whose sensors and actuators may be "malicious." The

This material is based upon work partially supported by NSF under Contract Nos. ECCS-1646449, ECCS-1547075, CCF-1619085 and Science & Technology Center Grant CCF-0939370, the U.S. Army Research Office under Contract No. W911NF-15-1-0279, the Power Systems Engineering Research Center (PSERC), and NPRP grant NPRP 8-1531-2-651 from the Qatar National Research Fund, a member of Qatar Foundation.

B. Satchidanandan (✉) · P. R. Kumar
Texas A&M University, College Station, Texas, USA
e-mail: bharadwaj.s1990@tamu.edu

P. R. Kumar
e-mail: prk.tamu@gmail.com

R. Tempo et al. (eds.), *Emerging Applications of Control and Systems Theory*, Lecture Notes in Control and Information Sciences - Proceedings,
https://doi.org/10.1007/978-3-319-67068-3_16

malicious sensors may not truthfully report the measurements that they observe, and the malicious actuators may not apply their control inputs as per the specified control law. In such a setting, even if the system is controllable and observable, the desired control objective may not be achievable. The honest nodes in the system may believe the state trajectory to be a certain sequence $\{\mathbf{x}[0], \mathbf{x}[1], \ldots\}$ whereas the actual state trajectory of the system may be very different. It is against this backdrop that we define the notions of *securable* and *unsecurable* subspaces of a linear dynamical system. The unsecurable subspace is defined, roughly, as the set of states that the system could actually be in, or ever reach, without the honest sensors ever being able to detect, based on their measurements, that the system had visited that state, or that there was any malicious activity in the system. Theorems 16.1 and 16.2 in the paper characterize the securable and unsecurable subspaces of a linear system. These results are analogous to those reported in [5], and in [2–4] for continuous-time linear dynamical systems, which examine what sorts of attacks are possible on control systems while remaining undetected. We formalize the results as characterizations of securable and unsecurable subspaces. They may be regarded as the analogs of the controllable and unobservable subspaces reexamined in an era where there is intense interest in cybersecurity of control systems. We then turn to the case of systems with noise, i.e., linear stochastic dynamical systems. We show that the securable and unsecurable subspaces defined in the context of deterministic systems also have operational meaning in the context of stochastic systems.

One way to view these results is as negative or impossibility results which state that given a linear control system with certain malicious sensors and actuators, it is impossible for the honest sensors to distinguish certain state trajectories from others. Consequently, it may be impossible to guarantee that the system does not reach certain states that are considered "unsafe." An alternate viewpoint is to look at these results from a system designer's perspective. These results could be regarded as providing guidelines for designing secure control systems. For example, for a specified amount of resilience required of the control system, typically quantified by the number of Byzantine nodes that the system should tolerate, or for a specification that the system should not visit certain "unsafe" states, the results can be translated into conditions that the securable and unsecurable subspaces should satisfy in order to meet the security specifications. This can potentially constitute a principled approach to design systems that are secure by construction, as opposed to designing systems to maximize a performance metric, and only subsequently installing ad-hoc security measures as an afterthought.

As mentioned before, many of the results in this paper pertaining to deterministic linear dynamical systems are mathematically isomorphic to some of the results contained in [2–5]. In addition, we report preliminary results on the extension of the above results to the context of stochastic linear dynamical systems where only noisy measurements of states are available.

16.2 Problem Formulation

Consider a *pth* order discrete-time linear dynamical system with m inputs and n outputs described by

$$\begin{aligned}\bar{\mathbf{x}}[t+1] &= A\bar{\mathbf{x}}[t] + B\bar{\mathbf{u}}[t],\\ \bar{\mathbf{y}}[t+1] &= C\bar{\mathbf{x}}[t+1],\\ \bar{\mathbf{x}}[0] &= \mathbf{x}_0.\end{aligned} \tag{16.1}$$

where $\bar{\mathbf{x}}[t] \in \mathbb{R}^p$ denotes the state of the system at time t, $\bar{\mathbf{u}}[t] \in \mathbb{R}^m$ denotes the input applied to the system at time t, $\bar{\mathbf{y}}[t] \in \mathbb{R}^n$ denotes the output of the system at time t, and A, B, and C are real matrices of appropriate dimensions.

We denote by $\mathbf{z}[t]$ the values reported by the sensors at time t. If sensor i, $i \in \{1, 2, ..., n\}$, is honest, then $z_i[t] = \bar{y}_i[t]$ for all t. We assume that an arbitrary, known, possibly history-dependent control policy $g = \{g_1, g_2, ...\}$ is in place, and denote by $\bar{\mathbf{u}}^g[t]$ the control policy-specified input at time t, so that $\bar{\mathbf{u}}^g[t] = g_t(\mathbf{z}^t)$, where $\mathbf{z}^t := [\mathbf{z}^T[0] \;\; \mathbf{z}^T[1] \;\; \ldots \;\; \mathbf{z}^T[t]]^T$. If actuator i is honest, then $\bar{u}_i[t] = \bar{u}_i^g[t]$ for all t.

We assume the adversarial nodes in the system to be near-omniscient, in the sense that at time $t = 0$, they have perfect knowledge of the initial state $\mathbf{x}_0$ of the system. On the other hand, the honest nodes in the system, at any time t, have access only to the measurements $\mathbf{z}^t$ that are reported until that time. Clearly, this assumption represents a worst-case scenario from the point of view of the honest nodes in the system. Consequently, the results presented in this paper serve as fundamental bounds that apply regardless of the capabilities of the attacker, and in particular, even for systems where the adversary's knowledge may be more limited.

Note that if all the nodes in the system are honest, and if the pair (A, C) is observable, then the nodes can correctly estimate the initial state $\mathbf{x}_0$ of the system by time $p - 1$. Consequently, they can correctly estimate the state $\bar{\mathbf{x}}[t]$ of the system at any time t. However, when there are malicious sensors and/or actuators present in the system, this need not be the case. Specifically, the honest nodes in the system could be under the impression that the state of the system at some time t is $\widehat{\mathbf{x}}[t]$, while in reality, the system could be in state $\bar{\mathbf{x}}[t] \neq \widehat{\mathbf{x}}[t]$. This brings us to the central question that is addressed in this paper: *Suppose that there are malicious nodes present in the system and that they act in a fashion that keeps them undetected. Suppose also that the honest nodes believe the system's state evolution to be* $\{\widehat{\mathbf{x}}[0], \widehat{\mathbf{x}}[1], \widehat{\mathbf{x}}[2], \ldots\}$. *Under these conditions, what are the set of states that the system can actually be in, or ever reach? This set essentially contains the set of states that the malicious nodes can steer the system to.* For this reason, we term this set as the *"unsecurable"* subspace of the system (A, B, C) for state $\widehat{\mathbf{x}}[0]$. The orthogonal complement of this is called the *"securable"* subspace. The projection of the uncertain state on this subspace is actually what the honest sensors and actuators believe it is, whether the system is not under attack or is under a stealthy attack. It is the largest such subspace. A formal definition of securable and unsecurable subspaces is presented in the next section.

16.3 Securable and Unsecurable Subspaces of Linear Control Systems

In order to determine if malicious nodes are present in the system or not, each honest sensor i subjects the reported measurement sequence $\{\mathbf{z}\}$ to the following test. If and only if the test fails (at any time t) does the sensor declare that malicious nodes are present in the system.

The rest of the paper follows the notation specified in the appendix of the paper.

Test: At each time t, check if the reported sequence of measurements up to that time $\mathbf{z}^t$ satisfies the following condition: $\exists \widehat{\mathbf{x}}_0 \in \mathbb{R}^p$ such that,

$$\mathbf{z}^t - F[t-1]\bar{\mathbf{u}}^{g^{t-1}} = \Gamma[t]\widehat{\mathbf{x}}_0. \tag{16.2}$$

Proposition 16.1 *If all the nodes in the system are honest, the reported measurements* $\{\mathbf{z}\}$ *pass (16.2) at each time t. Conversely, if the reported measurements* $\{\mathbf{z}\}$ *pass (16.2) at each time t, then there exists an initial state* $\mathbf{x}[0]$ *such that* $\mathbf{z}[t]$ *is the output of the system at time t under control* $\{\bar{\mathbf{u}}^g\}$*, and so, there is no definitive reason for the honest sensor to declare that malicious nodes are present in the system.*

Proof Omitted. ■

In what follows, we assume that the measurements reported by the malicious sensors pass the above test, and examine the limits of what the malicious nodes can do under this constraint.

Since the reported measurements $\{\mathbf{z}\}$ pass (16.2), it follows in particular that $\exists \widehat{\mathbf{x}}_0 \in \mathbb{R}^p$ such that $\forall t$,

$$\mathbf{z}^{t-1} - F[t-2]\bar{\mathbf{u}}^{g^{t-2}} = \Gamma[t-1]\widehat{\mathbf{x}}_0, \tag{16.3}$$

$$\bar{\mathbf{y}}_H[t] - \sum_{i=0}^{t-1} C_H A^i B\bar{\mathbf{u}}^g[t-1-i] = C_H A^t \widehat{\mathbf{x}}_0. \tag{16.4}$$

The following proposition is a (partial) converse of the above statement.

Proposition 16.2 *Suppose that there exist* $\widehat{\mathbf{x}}_0$, $\mathbf{z}_M^{\tau-1}$, *and* $\bar{\mathbf{d}}^{\tau-1}$ *such that (16.3) and (16.4) hold for* $t = \tau$. *Then, there is a vector* $\mathbf{z}_M[\tau]$ *that satisfies Test (16.2) at time* τ.

Proof Consider $\mathbf{z}_M[\tau] = C_M A^\tau \widehat{\mathbf{x}}_0 + \sum_{i=0}^{\tau-1} C_M A^i B\bar{\mathbf{u}}^g[\tau-1-i]$. It is straightforward to verify that it satisfies (16.2). ■

The above proposition states that it is sufficient for the malicious nodes to consider strategies that only ensure "consistency" at the outputs of the honest sensors. The outputs to be reported by the malicious sensors can be fabricated accordingly.

The next proposition, along with Theorem 16.2, shows that one can consider a simpler system consisting of only malicious actuators, honest sensors, and a control policy that is identically zero, and translate the conclusion obtained from the analysis of such a system to the more general system (16.1). In other words, one can dispense with the honest actuators and malicious sensors. There is no loss of generality in assuming that the control policy is identically equal to zero, and that the system has only honest sensors and malicious actuators.

Given the system described by (16.1), consisting of honest and malicious nodes as described before, consider the following reduction of the system where all sensors are honest, all actuators are malicious, and the control policy is identically equal to zero:

$$\begin{aligned} \mathbf{x}[t+1] &= A\mathbf{x}[t] + B_M\mathbf{d}[t], \\ \mathbf{y}_H[t+1] &= C_H\mathbf{x}[t+1], \\ \mathbf{x}[0] &= \mathbf{x}_0. \end{aligned} \tag{16.5}$$

where $\mathbf{y}_H[t]$ are the measurements observed by the (honest) sensors at time t, $\mathbf{d}[t]$ are the inputs applied by the (malicious) actuators at time t. We will refer to system (16.5) as the "reduced system" of system (16.1), or simply the "reduced system" when there is no ambiguity. Note that the reduced system has the same state space as its parent system (16.1), and is also initialized with the same state as its parent. It is only the inputs and the outputs of the systems that are different. As before, the malicious actuators are assumed to be near-omniscient so that they have perfect knowledge of the initial state $\mathbf{x}_0$. For the reduced system, Test (16.2) reduces to the following, and is performed by the (honest) sensors.

Test for the reduced system: Check if $\exists\tilde{\mathbf{x}}_0 \in \mathbb{R}^p$ such that for all t,

$$\mathbf{y}_H^t = \Gamma_H[t]\tilde{\mathbf{x}}_0. \tag{16.6}$$

Proposition 16.3 *Suppose that there exists a sequence* $\{\mathbf{d}\}$ *for the reduced system satisfying test (16.6). Then, if the malicious actuators in the parent system (16.1) inject* $\{\bar{\mathbf{d}}\} \equiv \{\mathbf{d}\}$, *there exist fabricated measurements* $\{\mathbf{z}_M\}$ *that can be reported by the malicious sensors in the parent system that pass Test (16.2) with* $\hat{\mathbf{x}}_0 = \tilde{\mathbf{x}}_0$.

Proof For the reduced system, we have

$$\mathbf{y}_H[t] = C_H A^t \mathbf{x}_0 + \sum_{i=0}^{t-1} C_H A^i B_M \mathbf{d}[t-1-i]. \tag{16.7}$$

Now, suppose for induction that there exist measurements $\mathbf{z}_M[0], \mathbf{z}_M[1], \ldots, \mathbf{z}_M[t-1]$ that the malicious sensors can report for system (16.1) when the malicious actuators inject $\bar{\mathbf{d}}[i] = \mathbf{d}[i]$, $i = 0, 2, ..., t-2$, such that the reported measurements pass test (16.2) up to time $t-1$ with $\hat{\mathbf{x}}_0 = \tilde{\mathbf{x}}_0$. The base case of $t = 1$ holds since the malicious sensors in the parent system can report $\mathbf{z}_M[0] = C_M\tilde{\mathbf{x}}_0$. This amounts to assuming that (16.3) holds with $\hat{\mathbf{x}}_0 = \tilde{\mathbf{x}}_0$. Now, if the malicious actuators in the parent system inject, at time $t-1$, $\bar{\mathbf{d}}[t-1] = \mathbf{d}[t-1]$, then,

$$\bar{\mathbf{y}}_H[t] = C_H A^t \mathbf{x}_0 + \sum_{i=0}^{t-1} C_H A^i B \bar{\mathbf{u}}^g[t-1-i] + \sum_{i=0}^{t-1} C_H A^i B_M \mathbf{d}[t-1-i].$$

Substituting (16.7) in the above gives

$$\bar{\mathbf{y}}_H[t] = \mathbf{y}_H[t] + \sum_{i=0}^{t-1} C_H A^i \bar{\mathbf{u}}^g[t-1-i].$$

Since the output of the reduced system satisfies (16.6), we have $\mathbf{y}_H[t] = C_H A^t \widetilde{\mathbf{x}}_0$. Substituting this into the above equation gives $\bar{\mathbf{y}}_H[t] - \sum_{i=0}^{t-1} C_H A^i \bar{\mathbf{u}}^g[t-1-i] = C_H A^t \widetilde{\mathbf{x}}_0$, which satisfies (16.4) for $\widehat{\mathbf{x}}_0 = \widetilde{\mathbf{x}}_0$. The desired result follows from Proposition 16.2. ■

The following definition is of central importance.

Definition 16.1 Consider a system (A, B, C) of the form (16.1) with initial state $\mathbf{x}_0$. The *unsecurable subspace for state* $\mathbf{s}_0$ of the system is the maximal set of states $V(\mathbf{s}_0)$ such that for each $\mathbf{v} \in V(\mathbf{s}_0)$, there exist t, $\{\bar{\mathbf{d}}\}$, $\{\mathbf{z}_M\}$ such that $\bar{\mathbf{x}}[t] = \mathbf{v}$ and (16.2) holds for $\widehat{\mathbf{x}}_0 = \mathbf{s}_0$.

In particular, for the reduced system (A, B_M, C_H), the unsecurable subspace for state $\mathbf{s}_0$ is the maximal set of states $V_R(\mathbf{s}_0)$ such that for each $\mathbf{v} \in V_R(\mathbf{s}_0)$, there exist t, $\{\mathbf{d}\}$ such that $\mathbf{x}[t] = \mathbf{v}$ and (16.6) holds for $\widetilde{\mathbf{x}}_0 = \mathbf{s}_0$.

In other words, the unsecurable space for $\mathbf{s}_0$ is the set of states that the system can be in if the honest nodes are deceived into inferring the initial state as $\mathbf{s}_0$. If the unsecurable subspace is of dimension greater than zero, it (i) states that the malicious nodes cannot distort certain linear combinations of the state without being detected, and (ii) specifies those linear combinations that are "intact."

The following theorem characterizes the unsecurable subspace and suggests an algorithm to compute it.

Theorem 16.1 *Consider a reduced system (A, B_M, C_H) of the form (16.5). For such a system,*

(i) The unsecurable subspace $V_R(0)$ for state 0 is the maximal set $W \subseteq \mathbb{R}^p$ such that $\forall \mathbf{w} \in W$,

  a. $C_H \mathbf{w} = 0$, and
  b. $\exists \mathbf{d}$ such that $A\mathbf{w} + B_M \mathbf{d} \in W$.

(ii) The unsecurable subspace for state $\mathbf{s}_0$, $V_R(\mathbf{s}_0)$, is

$$V_R(\mathbf{s}_0) = \{\mathbf{s}_0 + \mathbf{w} : \mathbf{w} \in V_R(0)\}. \tag{16.8}$$

Proof **Lemma 16.1** *The set W is a subspace.*

Proof Omitted. ■

We now show that W is equal to $V_R(0)$. The crux of the argument is that W is an invariant subspace in the following sense.

Lemma 16.2 *If the system's state visits W at any time t, then the malicious actuators can synthesize control actions that keep the state in W at all subsequent times.*

Proof We show this via induction. Let $\mathbf{w} \in W$, and let $\mathbf{x}[t] = \mathbf{w}$, which also serves as the base case for induction. Assume for induction that $\mathbf{x}[\tau] \in W$, where $\tau \geq t$ is a fixed time. Then, $\mathbf{x}[\tau + 1] = A\mathbf{x}[\tau] + B_M\mathbf{d}[\tau]$. Since $\mathbf{x}[\tau] \in W$, it follows from the definition of W that there exists a control choice $\mathbf{d}$ for $\mathbf{d}[\tau]$ such that $A\mathbf{x}[\tau] + B_M\mathbf{d}[\tau] \in W$, implying that $\mathbf{x}[\tau + 1] \in W$. ∎

Remark: Owing to the above Lemma, W is called the controlled invariant subspace in linear system theory [6].

Now, suppose that $\mathbf{x}[0] = \mathbf{w}$ and $\mathbf{w} \in W$. We then have from Lemma 16.2 that there exists a sequence $\{\mathbf{d}\}$ that the malicious actuators can apply as inputs such that $\mathbf{x}[t] \in W$ for all t. Since $W \subseteq \mathrm{N}(C_H)$ by definition, we have $\mathbf{y}_H[t] = C_H\mathbf{x}[t] = 0$ for all t. Consequently, (16.6) holds for $\tilde{\mathbf{x}}_0 = 0$, and it follows from Definition 16.1 that $\mathbf{w} \in V_R(0)$. Hence, $W \subseteq V_R(0)$.

Now suppose that $\mathbf{v} \in V_R(0)$. We then have from Definition 16.1 that $\exists\{\mathbf{d}\}$ that the malicious actuators can apply as inputs to the system such that $\mathbf{x}[t] = \mathbf{v}$ for some t and (16.6) holds for $\tilde{\mathbf{x}}_0 = 0$. This implies that $\mathbf{y}_H[t] = 0$ for all t. Since $0 = \mathbf{y}_H[t] = C_H\mathbf{x}[t] = C_H\mathbf{v}$, we have that $\mathbf{v} \in \mathrm{N}(C_H)$, satisfying the first condition to be an element of W. Since $\{\mathbf{d}[t], \mathbf{d}[t+1], ...\}$ is a sequence that the malicious actuators can apply such that $\mathbf{x}[t'] \in \mathrm{N}(C_H)$ for all $t' \geq t$, $\mathbf{v}$ satisfies the second condition to be an element of W. Therefore, $\mathbf{v} \in W$, and $V_R(0) \subseteq W$. Combining the two results, we have $W = V_R(0)$.

(ii) Let $\mathbf{v} \in \{\mathbf{s}_0 + \mathbf{w} : \mathbf{w} \in V_R(0)\}$, and let $\mathbf{x}[0] = \mathbf{v}$. Then, $\mathbf{y}_H^t = \Gamma_H[t]\mathbf{v} + H_M[t-1]\mathbf{d}^{t-1} = \Gamma_H[t]\mathbf{s}_0 + \Gamma_H[t]\mathbf{w} + H_M[t-1]\mathbf{d}^{t-1}$ for all t. Since $\mathbf{w} \in V_R(0)$, it follows from the definition of $V_R(0)$ that there exists sequence $\{\mathbf{d}\}$ so that $\Gamma_H[t]\mathbf{w} + H_M[t-1]\mathbf{d}^{t-1} = 0$ for all t. Therefore, if the actuators inject such a sequence $\{\mathbf{d}\}$, then, $\mathbf{y}_H^t$ reduces to $\mathbf{y}_H^t = \Gamma_H[t]\mathbf{s}_0$. Therefore, (16.6) holds with $\tilde{\mathbf{x}}_0 = \mathbf{s}_0$, and so, $\{\mathbf{s}_0 + \mathbf{w} : \mathbf{w} \in V(0)\} \subseteq V_R(s_0)$.

Next, let $\mathbf{v} \in V_R(s_0)$. Then, we have from the definition of $V_R(\mathbf{s}_0)$ that $\exists\{\mathbf{d}'\}, \tau$ such that $\mathbf{x}[\tau] = \mathbf{v}$ and $\Gamma_H[t]\mathbf{s}_0 = \mathbf{y}_H^t$ for all t. This in turn implies that $\exists\{\mathbf{d}\}$ such that $\mathbf{x}_0 = \mathbf{v}$ and $\Gamma_H[t]\mathbf{s}_0 = \mathbf{y}_H^t$ for all t. Also, when $\mathbf{x}_0 = \mathbf{v}$, we have for all t, $\mathbf{y}_H^t = \Gamma_H[t]\mathbf{v} + H_M[t-1]\mathbf{d}^{t-1}$. Combining the two, we have that there exists $\{\mathbf{d}\}$ such that $\Gamma_H[t]\mathbf{s}_0 = \Gamma_H[t]\mathbf{v} + H_M[t-1]\mathbf{d}^{t-1}$ for all t. This means that $\mathbf{v}$ solves, for all t,

$$[\Gamma_H[t] \;\; H_M[t-1]]\begin{bmatrix}\mathbf{v}\\ \mathbf{d}^{t-1}\end{bmatrix} = [\Gamma_H[t] \;\; H_M[t-1]]\begin{bmatrix}\mathbf{s}_0\\ 0\end{bmatrix},$$

so that for all t,

$$\begin{bmatrix} \mathbf{v} \\ \mathbf{d}^{t-1} \end{bmatrix} = \begin{bmatrix} \mathbf{s}_0 \\ 0 \end{bmatrix} + \widetilde{\mathbf{w}},$$

where $\widetilde{\mathbf{w}} \in \mathrm{N}([\Gamma_H[t] \;\; H_M[t-1]])$. Denote by $\mathbf{w}$ the first p entries of $\widetilde{\mathbf{w}}$, and it follows from the definition of $V_R(0)$ that $\widetilde{\mathbf{w}} \in V_R(0)$. Hence, $\mathbf{v}$ must be of the form $\mathbf{s}_0 + \mathbf{w}$, $\mathbf{w} \in V_R(0)$, and hence, $V_R(s_0) \subseteq \{\mathbf{s}_0 + \mathbf{w} : \mathbf{w} \in V(0)\}$.

Combining the two results, we have $V_R(\mathbf{s}_0) = \{\mathbf{s}_0 + \mathbf{w} : \mathbf{w} \in V(0)\}$. ■

The following theorem translates the above conclusions obtained from the reduced system (16.5) to the original system (16.1) that is of interest.

Theorem 16.2 *The unsecurable subspace $V(\mathbf{s}_0)$ for the system (A, B, C) is the same as the unsecurable subspace $V_R(\mathbf{s}_0)$ for its reduction (A, B_M, C_H).*

Proof Let $\mathbf{v} \in V_R(\mathbf{s}_0)$. Then, it follows form the definition of $V_R(\mathbf{s}_0)$ that for the reduced system (16.5), there exists $\{\mathbf{d}\}$ that can be applied by the actuators so that (16.6) is satisfied for $\widetilde{\mathbf{x}}_0 = \mathbf{s}_0$ when $\mathbf{x}_0 = \mathbf{v}$. Therefore, by Proposition 16.3, $\mathbf{v} \in V(\mathbf{s}_0)$, and so, $V_R(\mathbf{s}_0) = V(\mathbf{s}_0)$.

Next, let $\mathbf{v} \in V(\mathbf{s}_0)$. Then, from the definition of $V(\mathbf{s}_0)$, we have for system (16.1) that $\exists\{\mathbf{d}\}, \{\mathbf{z}_M\}$ such that for all t, $\mathbf{z}^t = \Gamma[t]\mathbf{s}_0 + F[t-1]\bar{\mathbf{u}}^{g^{t-1}}$ when $\mathbf{x}_0 = \mathbf{v}$. This implies that $\bar{\mathbf{y}}^t_H = \Gamma_H[t]\mathbf{s}_0 + H[t-1]\bar{\mathbf{u}}^{g^{t-1}}$. Since we also have $\bar{\mathbf{y}}^t_H = \Gamma_H[t]\mathbf{v} + H[t-1]\bar{\mathbf{u}}^{g^{t-1}} + H_M[t-1]\mathbf{d}^{t-1}$, substituting this in the previous equation gives

$$\Gamma_H[t]\mathbf{v} + H_M[t-1]\mathbf{d}^{t-1} = \Gamma_H[t]\mathbf{s}_0. \tag{16.9}$$

Now, if the actuators apply the above sequence $\{\mathbf{d}\}$ to the reduced system (with initial state $\mathbf{v}$), we have for each t, $\mathbf{y}^t_H = \Gamma_H[t]\mathbf{v} + H_M[t-1]\mathbf{d}^{t-1} = \Gamma_H[t]\mathbf{s}_0$, where the last equality follows from the (16.9). Hence, (16.6) is satisfied with $\widetilde{\mathbf{x}}_0 = \mathbf{s}_0$, and hence, $V(\mathbf{s}_0) \subseteq V_R(\mathbf{s}_0)$.

Combining the two results, we have $V(\mathbf{s}_0) = V_R(\mathbf{s}_0)$. ■

The characterization of $V_R(0)$ given in Theorem 16.1 allows one to use standard algorithms that compute $(A, \mathscr{R}(B_M))$−controlled invariant subspaces of a linear dynamical system for computing its unsecurable subspace.

Definition 16.2 The *securable subspace* S of a discrete-time linear dynamical system of the form (16.1) is the orthogonal complement of $V(0)$, the unsecurable subspace of the zero state.

The securable subspace has the interpretation of the maximal set of states that the malicious nodes cannot steer the system to without leaving a nonzero trace at the output of the honest sensors. The following section examines the performance of a stochastic linear dynamical system in the securable subspace, which provides further operational meaning to it.

16.4 Performance in the Securable Subspace in the Context of Stochastic Systems

The previous section contained results analogous to some of those developed for continuous-time, deterministic, linear dynamical systems in [2, 3], and also those reported in [5]. In this section, we report preliminary results of an ongoing work which show that the notion of a securable subspace, as defined in the previous section, could also have operational significance in the context of stochastic systems. While we show this in the context of a simple class of stochastic systems in which the process noise and the initial state are the only sources of uncertainty, similar ideas and proof technique could be applied for the more general model with partial and noisy state observations.

Consider a multiple-input, multiple-output stochastic linear dynamical system described by

$$\mathbf{x}[t+1] = A\mathbf{x}[t] + B\mathbf{u}[t] + \mathbf{w}[t+1], \tag{16.10}$$

$$\mathbf{y}[t+1] = \mathbf{x}[t+1], \tag{16.11}$$

where $\mathbf{x}[t] \in \mathbb{R}^p$, $\mathbf{u}[t] \in \mathbb{R}^m$, $\mathbf{w}[t+1]$ has a known covariance Σ_w, and is independent and identically distributed across time,[1] and A and B are known real matrices of appropriate dimensions. As before, let $\mathbf{u}^g[t] = g_t(z^t)$ denote the control policy-specified input at time t, where $\mathbf{z}[t]$ is the measurement vector reported at time t. Let $\mathbf{d}[t] := \mathbf{u}_M[t] - \mathbf{u}_M^g[t]$, where the subscript 'M,' as usual, denotes the indices of the malicious actuators. Note that without loss of generality, we can assume the honest sensors to be indexed from 1 to h_s, and the honest actuators from 1 to h_a (since the rows and columns of $\mathbf{x}$, A, and B can be reordered accordingly). The system evolves in closed loop as

$$\mathbf{x}[t+1] = A\mathbf{x}[t] + B\mathbf{u}^g[t] + B_M\mathbf{d}[t] + \mathbf{w}[t+1], \tag{16.12}$$

$$\mathbf{y}[t+1] = \mathbf{x}[t+1]. \tag{16.13}$$

The honest nodes in the system perform the following test to determine the presence of malicious nodes in the system.

Test: A honest node checks if

$$\lim_{T\to\infty} \frac{1}{T} \sum_{k=0}^{T-1} (\mathbf{z}[k+1] - A\mathbf{z}[k] - B\mathbf{u}^g[k]) \, (\mathbf{z}[k+1] - A\mathbf{z}[k] - B\mathbf{u}^g[k])^T = \Sigma_w. \tag{16.14}$$

[1] More generally, this could be generalized to a martingale difference sequence with a one-step ahead conditional covariance that is uniformly bounded below by a positive definite matrix.

Note that (i) the above test is based only on the information available to the honest nodes in the system, and (ii) if all the nodes in the system are honest, the reported measurements would pass the above test almost surely. The following theorem gives an operational meaning to the securable subspace in the context of stochastic systems.

Theorem 16.3 *Let* $\mathbf{m}[t] := \mathbf{z}[t] - \mathbf{x}[t]$ *be the distortion in the state estimate of the honest nodes. Assume that* $Dim(S \cap \mathrm{N}(C_H)) = 1$. *If the reported sequence of measurements* $\{\mathbf{z}\}$ *passes test (16.14), then,*

$$\lim_{T\to\infty} \frac{1}{T} \sum_{k=0}^{T-1} ||\mathbf{m}_S[k]||^2 = 0. \tag{16.15}$$

In other words, the state estimation error caused by malicious sensors and actuators can only be of zero power in the securable subspace.

Proof We define $\gamma[t+1] := \mathbf{m}[t+1] - A\mathbf{m}[t] + B_M\mathbf{d}[t]$, so that test (16.14) reduces to $\lim_{T\to\infty} \frac{1}{T}\sum_{k=0}^{T-1}(\mathbf{w}[k+1]+\gamma[k+1])\,(\mathbf{w}[k+1]+\gamma[k+1])^T = \Sigma_w$. In particular, we have $\lim_{T\to\infty} \frac{1}{T}\sum_{k=0}^{T-1}(\mathbf{w}_H[k+1]+\gamma_H[k+1])\,(\mathbf{w}_H[k+1]+\gamma_H[k+1])^T = \Sigma_{w,H}$, where $\Sigma_{w,H}$ denotes the top-left $h_s \times h_s$ matrix of Σ_w. Since $\mathbf{m}_H[t+1] = 0$ for all t, it follows from the definition of $\gamma[t+1]$ that $\gamma_H[t+1] \in \sigma(\mathbf{m}^t, \mathbf{d}^t)$, where $\gamma_H[t]$ and $\mathbf{m}_H[t]$ are defined in the usual manner. Since $\mathbf{w}_H[k+1]$ is independent of $\sigma(\mathbf{m}^t, \mathbf{d}^t)$, the above equality yields

$$\lim_{T\to\infty} \frac{1}{T} \sum_{k=0}^{T-1} \gamma_H[k+1]\,\gamma_H^T[k+1] = 0. \tag{16.16}$$

Now, from the definition of $\gamma[t+1]$, we have

$$\begin{aligned}\mathbf{m}_V[t+1] + \mathbf{m}_S[t+1] = {} & A\mathbf{m}_V[t] + (A\mathbf{m}_S[t])_V + (A\mathbf{m}_S[t])_S \\ & + B_M\mathbf{d}_C[t] + B_M\mathbf{d}_U[t] + \gamma_V[t+1] + \gamma_S[t+1],\end{aligned} \tag{16.17}$$

where $\mathbf{m}_V[t]$ denotes the projection of $\mathbf{m}[t]$ on the unsecurable subspace $V := V(0)$ as in Definition 16.1, $\mathbf{m}_S[t], \gamma_V[t], \gamma_S[t], (A\mathbf{m}_S[t])_V$, and $(A\mathbf{m}_S[t])_S$ are defined likewise, $\mathbf{d}_U[t] := \mathbf{d}[t] - \mathbf{d}_C[t]$, and $\mathbf{d}_C[t]$ is a vector such that $A\mathbf{m}_V[t] + B_M\mathbf{d}_C[t] \in V$, which is guaranteed to exist from the characterization of V given in Theorem 16.1(i).

Now, define $\mathscr{H} := \mathrm{Span}(\mathbf{e}_1, \mathbf{e}_2, \ldots, \mathbf{e}_{h_s})$, where $\mathbf{e}_i \in \mathbb{R}^p$ is a vector all of whose components are zeros except for the *ith* component, which is unity. Then, we have $\gamma_{\mathscr{H}}[t+1] = \mathbf{m}_{\mathscr{H}}[t+1] - ((A\mathbf{m}_S[t])_S + B_M\mathbf{d}_U[t])_{\mathscr{H}}$. Since $\mathbf{m}_H[t] \equiv 0$, we have $\mathbf{m}_{\mathscr{H}}[t] \equiv 0$. It follows from the above that

$$\lim_{T\to\infty} \frac{1}{T} \sum_{k=0}^{T-1} ||((A\mathbf{m}_S[k])_S + B_M\mathbf{d}_U[k])_{\mathscr{H}}||^2 = \lim_{T\to\infty} \frac{1}{T} \sum_{k=0}^{T-1} ||\gamma_H[k+1]||^2 = 0, \tag{16.18}$$

where the last equality follows by equating the trace of (16.16).

Now, suppose for contradiction that $\lim_{T\to\infty}\frac{1}{T}\sum_{k=0}^{T-1}||\mathbf{m}_S[k]||^2=\epsilon$ for some $\epsilon>0$. Since $Dim(S\cap \mathrm{N}(C_H))=1$, it follows that if $\mathbf{m}_S[k]\neq 0$, then for no choice of $\mathbf{d}_U[k]$ will $((A\mathbf{m}_S[k])_S+B_M\mathbf{d}_U[k])\in \mathscr{H}^C$. Therefore, if $\lim_{T\to\infty}\frac{1}{T}\sum_{k=0}^{T-1}$ $||\mathbf{m}_S[k]||^2=\epsilon$ for any $\epsilon>0$, then $\lim_{T\to\infty}\frac{1}{T}\sum_{k=0}^{T-1}||((A\mathbf{m}_S[k])_S+B_M\mathbf{d}_U[k])_{\mathscr{H}}||^2$ $>\epsilon\delta$ for some $\delta>0$, contradicting (16.18). ■

16.5 Conclusion

We consider the problem of securing control systems from malicious sensors and actuators. Towards this, we formalize the notion of the securable and unsecurable subspace of a linear dynamical system. The unsecurable subspace has the interpretation as a set of states that the system could actually be in, or ever reach, as a consequence of malicious actions of the adversarial nodes, without the honest sensors in the system ever detecting definitively any malicious activity. This is an invariant subspace in the sense that once the state of the system enters this space, the malicious sensor and actuator nodes in the system can collude to keep the system in this space forever without the honest sensors ever being able to confirm any malicious activity based on their own observations or the ones being reported to them. The orthogonal complement of this subspace, the securable subspace, has the interpretation in the context of deterministic systems as the set of states that the malicious nodes cannot steer the system to without leaving a nonzero trace at the output of the honest sensors. These subspaces also have relevance to the case where the system is noisy. We have presented some preliminary results to show that the notion of a securable subspace has operational significance in the broader context of linear stochastic systems. Specifically, in the context of stochastic systems, the securable subspace has the interpretation as the subspace along which the state estimation error of the honest nodes in the system is what it would have been had there been no malicious nodes in the system, in spite of arbitrary attack strategies of malicious sensors and actuators that are actually present in the system. A characterization of these subspaces, and an algorithm to compute them for any linear system and any combination of malicious sensors and actuators is obtained by a standard recourse to geometric control methods.

Notation

The following notation is used throughout the paper:

1. Let $s_1<s_2<\ldots<s_{h_s}$ denote the indices of the honest sensors, $\psi_1,\psi_2,\ldots,\psi_{m_s}$ denote those of the malicious sensors, and $a_1<a_2<\ldots<a_{m_a}$ denote those of the malicious actuators. Then,
   - C_H is the $h_s\times p$ matrix whose *ith* row is the s_i^{th} row of C, $i=1,2,\ldots,h_s$,
   - B_M is the $p\times m_a$ matrix whose *ith* column is the a_i^{th} column of B, $i=1,2,\ldots,m_a$,

- $\bar{\mathbf{y}}_H[t]$ is the $h_s \times 1$ vector whose *ith* component is the s_i^{th} entry of $\bar{\mathbf{y}}[t]$, $i = 1, 2, ..., h_s$,
- $\mathbf{z}_M[t]$ is the $m_s \times 1$ vector whose *ith* entry is the ψ_i^{th} entry of $\mathbf{z}[t]$, $i = 1, 2, ..., m_s$,
- $\bar{\mathbf{d}}[t]$ is the $m_a \times 1$ vector whose *ith* component is $\bar{d}_i[t] := \bar{u}_{a_i}[t] - \bar{u}^g_{a_i}[t]$, $i = 1, 2, ..., m_a$.

2. $\mathbf{x}^t$ denotes $[\mathbf{x}^T[0] \;\; \mathbf{x}^T[1] \;\; \ldots \;\; \mathbf{x}^T[t]]^T$.
3. $\Gamma[t] := [C^T \;\; (CA)^T \;\; (CA^2)^T \;\; \ldots \;\; (CA^t)^T]^T$.
4. $\Gamma_H[t] := [(C_H)^T \;\; (C_H A)^T \;\; \ldots \;\; (C_H A^t)^T]^T$.
5.

$$F[t] := \begin{bmatrix} 0 & 0 & \ldots & 0 \\ CB & 0 & \ldots & 0 \\ CAB & CB & \ldots & 0 \\ \vdots & \vdots & \ddots & \vdots \\ CA^tB & CA^{t-1}B & \ldots & CB \end{bmatrix},$$

6.

$$H[t] := \begin{bmatrix} 0 & 0 & \ldots & 0 \\ C_H B & 0 & \ldots & 0 \\ C_H AB & C_H B & \ldots & 0 \\ \vdots & \vdots & \ddots & \vdots \\ C_H A^t B & C_H A^{t-1} B & \ldots & C_H B \end{bmatrix}, \quad H_M[t] := \begin{bmatrix} 0 & 0 & \ldots & 0 \\ C_H B_M & 0 & \ldots & 0 \\ C_H AB_M & C_H B_M & \ldots & 0 \\ \vdots & \vdots & \ddots & \vdots \\ C_H A^t B_M & C_H A^{t-1} B_M & \ldots & C_H B_M \end{bmatrix}.$$

7. $\mathrm{N}(\cdot)$ denotes the null space of a matrix, and $\mathscr{R}(\cdot)$ denotes the range space of a matrix.

References

1. Kalman, Rudolf: On the general theory of control systems. IRE Trans. Automatic Control **4**(3), 110–110 (1959)
2. Pasqualetti, Fabio, Dorfler, Florian, Bullo, Francesco: Control-theoretic methods for cyberphysical security: geometric principles for optimal cross-layer resilient control systems. IEEE Control Syst. **35**(1), 110–127 (2015)
3. Pasqualetti, Fabio, Drfler, Florian, Bullo, Francesco: Attack detection and identification in cyber-physical systems. IEEE Trans. Autom. Control **58**(11), 2715–2729 (2013)
4. Pasqualetti, F., Dorfler, F., Francesco, B.: Cyber-physical security via geometric control: distributed monitoring and malicious attacks. In: 2012 IEEE 51st Annual Conference on Decision and Control (CDC), pp. 3418–3425. IEEE (2012)
5. Teixeira, A., Shames, I., Henrik, S., Johansson, K.H.: Revealing stealthy attacks in control systems. In: 2012 50th Annual Allerton Conference on Communication, Control, and Computing (Allerton), pp. 1806–1813. IEEE (2012)
6. Basile, G., Marro, G.: Controlled and conditioned invariant subspaces in linear system theory. J Optimization Theor. Appl. 306–315 (1969)

Chapter 17
System Completion Problem: Theory and Applications

Pradeep Misra

Abstract This chapter addresses the following problems: (a) given a system with n states, m inputs and p outputs with $m \neq p$, is it possible to introduce additional inputs or outputs that make the system square (complete the non-square system to a square system) while assigning finite transmission zeros of the resulting squared system to desired locations, equivalently *squaring-up or partial completion* problem and (b) if only the state matrix and either the input or the output matrix has been given, is it possible to find, respectively, outputs or inputs such the resulting square system has its finite transmission zeros at desired locations, i.e. *system completion* problem.

17.1 Introduction

We consider the standard state-space representation of linear time-invariant systems described by

$$\begin{aligned} \mathscr{D}\mathbf{x}(t) &= A\mathbf{x}(t) + B\mathbf{u}(t) \\ \mathbf{y}(t) &= C\mathbf{x}(t) + D\mathbf{u}(t) \end{aligned} \tag{17.1}$$

with n states, m inputs and p outputs, $A \in \mathbb{R}^{n\times n}$, $B \in \mathbb{R}^{n\times m}$, $C \in \mathbb{R}^{p\times n}$, $D \in \mathbb{R}^{p\times m}$ and $\mathscr{D}$ represents differential operator for continuous time systems. If $m > p$, $m < p$ and $m = p$, we will refer to the system as *wide*, *tall* and *square*, respectively. While the results in the rest of the chapter are developed for continuous time system, they apply, *mutatis mutandis*, to discrete time systems as well.

P. Misra (✉)
Department of Electrical Engineering, Wright State University, Dayton 45435, USA
e-mail: pradeep.misra@wright.edu

R. Tempo et al. (eds.), *Emerging Applications of Control and Systems Theory*, Lecture Notes in Control and Information Sciences - Proceedings,
https://doi.org/10.1007/978-3-319-67068-3_17

It is well known that for a *non-degenerate*, controllable and observable system described as above with the 4-tuple $\Sigma(A, B, C, D)$, transmission zeros are defined by $\lambda \in \mathbb{C}$ such that

$$\rho[\mathscr{R}(\lambda)] = \rho \left[\begin{array}{c|c} A - \lambda I & B \\ \hline C & D \end{array} \right] < n + \min(m, p) \qquad (17.2)$$

where ρ represents the rank of the matrix pencil $\mathscr{R}(\lambda)$ using the Rosenbrock system representation [1]. Further, it is also well known that transmission zeros are invariant under state or output feedback. Therefore, once the matrices in the 4-tuple $\Sigma(A, B, C, D)$ have been specified, locations of transmissions zeros are governed by the rank condition in (17.2).

This chapter explores the following two closely related problems:

1. Given a tall ($m < p$) or wide ($m > p$), system $\Sigma(A, B, C, D)$, under what conditions is it possible to find additional columns in B and D or additional rows in C and D such that the resulting squared-up system has a desired set of transmission zeros. We will refer to this as *squaring-up* or *partial system completion* problem.
2. If only the pair (A, B) or (A, C) has been completely specified, with the pair (A, B) controllable or (A, C) observable, respectively, is it possible to find the two remaining matrices in each case such that the resulting square system has the desired set of transmission zeros. We will refer to this as *system completion* problem.

It will be demonstrated that the system completion problem can be transformed into an equivalent state feedback compensation problem for which necessary and sufficient conditions are well known. In each case, the mathematical analysis will yield a constructive procedure to compute the desired matrices.

It should be mentioned that an alternate method to square a given non-square system would be to find a transformation such that the resulting systems are squared *down* instead of squared *up*. However, squaring down does not provide the same degree of freedom as squaring-up does, in particular, it presents considerable difficulty in ensuring that the resulting squared down system has its transmission zeros at desired locations. It has been shown that finding such a transformation is equivalent to the solution of a static or dynamic output feedback compensation problem. The reader is referred to [2, 3].

17.2 Squaring-up or Partial System Completion

The following two possibilities may arise: (a) we are given the triple (A, B, C) and $m > p$ (wide system), augment C with $(m - p)$ additional rows and assign the resulting transmission zeros to desired locations, (b) we are given the 4-tuple (A, B, C, D) and $m > p$, augment C and D with $(m - p)$ additional rows and simultaneously assign the transmission zeros to desired locations [4, 5]. In light of subsequent findings in [6], for now we will assume that the given non-square system does not possess any finite transmission zeros.

17.2.1 Input–Output Interaction Matrix $D = O$

The objective here is, given a non-degenerate system (A, B, C, D), $D(= O), m > p$, to find a matrix $\hat{C} \in R^{(m-p)\times n}$ such that the resulting squared-up system has its transmission zeros at desired locations in the left half plane.

It is assumed that

- (A, B, C) is controllable and observable, and $\rho(B) = m$, $\rho(C) = p$
- $\rho(CB) = p$
- the system has minimal order, i.e. it has no uncontrollable or uncontrollable modes.

Remark 1.1: If the system has uncontrollable or uncontrollable modes, they can be removed to work with least order representation of the system.

Remark 1.2: Note that generically non-square systems do not possess any finite transmission zeros.

Under above assumptions on the system, there exist orthogonal state transformations such that matrix pencil may be transformed to

$$\left[\begin{array}{cc|c} A_{11} - \lambda I_m & A_{12} & B_1 \\ A_{21} & A_{22} - \lambda I_{n-m} & O \\ \hline C_{11} & C_{12} & O \end{array}\right]$$

such that $\rho(C_{11}) = p$.

Defining pseudo-output matrix as $[C_{21} \quad C_{22}]$, the augmented system is given by

$$\mathscr{R}(\lambda) \triangleq \left[\begin{array}{cc|c} A_{11} - \lambda I_m & A_{12} & B_1 \\ A_{21} & A_{22} - \lambda I_{n-m} & O \\ \hline C_{11} & C_{12} & O \\ C_{21} & C_{22} & O \end{array}\right] \tag{17.3}$$

With $[C_1 \quad C_2] \triangleq \begin{bmatrix} C_{11} & C_{12} \\ C_{21} & C_{22} \end{bmatrix}$ and selecting C_{21} such that $\rho(C_1) = m$, we define pseudo-output matrix as $[C_{21} \quad C_{22}]$, such that the rank of the augmented system is given by

$$\rho(\mathscr{R}(\lambda)) = \rho \left[\begin{array}{cc|c} A_{11} - \lambda I_m & A_{12} & B_1 \\ A_{21} & A_{22} - \lambda I_{n-m} & O \\ \hline C_{11} & C_{12} & O \\ C_{21} & C_{22} & O \end{array}\right] \tag{17.4}$$

$$\begin{aligned} &= \rho \left(\mathscr{R}(\lambda) \left[\begin{array}{cc|c} I_m & C_1^{-1}C_2 & O \\ O & I_{n-m} & O \\ \hline O & O & I_m \end{array}\right] \right) \\ &= \rho \left[\begin{array}{cc|c} A_{11} - \lambda I_m & X & B_1 \\ A_{21} & A_{22} - A_{21}C_1^{-1}C_2 - \lambda I_{n-m} & O \\ \hline C_1 & O & O \end{array}\right] \end{aligned} \tag{17.5}$$

where matrix X is irrelavant. Note that $\rho(B_1) = m$ by assumption and $\rho(C_1) = m$ by construction, clearly

$$\rho(\mathscr{R}(\lambda)) = 2m + \rho(A_{22} - A_{21}C_1^{-1}C_2 - \lambda I_{n-m}).$$

Further, $\rho(\mathscr{R}(\lambda)) < n + m$ at all eigenvalues of the matrix $A_{22} - A_{21}C_1^{-1}C_2$.

Since, matrices A_{22} and $A_{21}C_1$ are known, C_2 can be computed such that the matrix $A_{22} - A_{21}C_1^{-1}C_2$ has its eigenvalue at desired locations. Given that C_1 in invertible, this can be done provided that (A_{22}, A_{21}) is a controllable pair.

Computation of C_1 and C_2

Recall that $[C_1 \quad C_2] = \begin{bmatrix} C_{11} & C_{12} \\ C_{21} & C_{22} \end{bmatrix}$, $\rho(C_{11}) = p$. Therefore, computation of C_{21} in C_1 is trivial as it can be any matrix in the row null-space of C_{11}.

To compute C_2, let $C_2 \triangleq [\tilde{C}_2 + \hat{C}_2]$ where

$$\tilde{C}_2 = \begin{bmatrix} C_{12} \\ O_{(m-p)\times(n-m)} \end{bmatrix}, \quad \hat{C}_2 = \begin{bmatrix} O_{p\times(n-m)} \\ C_{22} \end{bmatrix} \tag{17.6}$$

Let $\tilde{A}_{22} \triangleq A_{22} - A_{21}C_1^{-1}\tilde{C}_2$. Then the problem of computing $\hat{C}_2$ reduces to a *state feedback design problem* $(\tilde{A}_{22} - A_{21}C_1^{-1}\hat{C}_2)$, where C_2 is the unknown 'state feedback' matrix.

We illustrate the above by a simple example. Consider a systems $\Sigma(A, B, C, D)$ with various matrices given as

$$\left[\begin{array}{c|c} A & B \\ \hline C & D \end{array}\right] = \left[\begin{array}{cccccc|ccc} 3&4&1&3&0&3&1&4&4 \\ 3&5&2&1&1&1&2&3&1 \\ 3&4&2&0&2&2&2&0&5 \\ 3&2&0&4&4&5&3&5&3 \\ 2&1&3&0&5&4&4&1&1 \\ 4&5&4&4&1&4&2&0&3 \\ \hline 4&1&2&5&1&3&0&0&0 \\ 4&4&1&4&4&5&0&0&0 \end{array}\right].$$

The given system has no transmission zeros. After squaring-up, the system will have three transmission zeros, which we assign arbitrarily at $\{-1, -2, -3\}$.

The augmented system with the output matrix given below assigns the three transmission zeros as desired .

$$C = \begin{bmatrix} 4 & 1 & 2 & 5 & 1 & 3 \\ 4 & 4 & 1 & 4 & 4 & 5 \\ 0.354 & -0.46 & 0.034 & 1.174 & 0.21 & 0.76 \end{bmatrix}.$$

The locations of the assigned transmission zeros of augmented system are $\{-0.96, -2.00, -3.07\}$ after rounding off the calculated elements of C to three decimal places.

17.2.2 Input–Output Interaction Matrix $D \neq O$

Depending on the given 4-tuple (A, B, C, D) and desired outcome of squaring-up process, the following cases may arise:

1. Given input–output interaction matrix $D = O$
   a. Augmented D should remain O. This was the case discussed in Sect. 17.2.1.
   b. Augmented D should have rank $m - p$
2. Rank of the given input–output interaction matrix $D = p$
   a. Augmented D should have full rank m
   b. Augmented D should have rank p
3. Rank of the given input–output interaction matrix $D = r(< p)$
   a. Augmented D should have rank $r + m - p$
   b. Augmented D should have rank r

Of course, there are also other possibilities such as given D has rank p, but the augmented D should have rank r such that $p < r < m$. Aside from some tedious notation, it does not pose any technical challenge to address these cases.

17.2.2.1 Given $\rho(D) = 0$, Desired Rank of Augmented $D = m - p$

Next, we consider various cases where the resulting input–output interaction matrix D must also be modified to meet the squaring-up or partial completion problem. It is important to note that as we change the rank condition on D, the number of finite transmission zeros to be assigned also changes accordingly.

Let $\hat{D}_{22}$ denote a matrix of rank $m - p$. Partitioning the system conformably to $\hat{D}_{22}$, we have

$$\mathscr{R}(\lambda) := U\mathscr{R}(\lambda)V = \left[\begin{array}{cc|cc} A_{11} - \lambda I_p & A_{12} & B_{11} & B_{12} \\ A_{21} & A_{22} - \lambda I_{n-p} & O & B_{22} \\ \hline C_{11} & C_{12} & O & O \\ \hat{C}_{21} & \hat{C}_{22} & O & \hat{D}_{22} \end{array}\right].$$

where U and V are orthogonal matrices used for various state coordinate transformations. Matrices $\hat{C}_{21}$, $\hat{C}_{22}$ and $\hat{D}_{22}$ need to be determined. It has been assumed for convenience that $\rho(B_{11}) = \rho(C_{11}) = p$, if that is not the case, it may be necessary to permute columns of B to meet this requirement.

Using some straightforward block matrix transformations, it can be shown that

$$\rho(\mathscr{R}(\lambda)) = 2p + \rho\left[\begin{array}{c|c} A_{22} - A_{21}C_{11}^{-1}C_{12} - \lambda I_{n-p} & B_{22} \\ \hline \hat{C}_{22} & \hat{D}_{22} \end{array}\right].$$

Then with $\hat{D}_{22}$ invertible, so long as $(\hat{A}, \hat{B})$ ($\triangleq (A_{22} - A_{21}C_{11}^{-1}C_{12}, B_{22})$) forms a controllable pair, $\hat{C}_{22}$ can be found by solving the state feedback problem $\hat{A} - (\hat{B}\hat{D}_{22}^{-1})\hat{C}_{22}$.

Consider the same system as before, except, now rank of augmented D must be $m - p = 1$.

$$\left[\begin{array}{c|c} A & B \\ \hline C & D \end{array}\right] = \left[\begin{array}{cccccc|ccc} 3 & 4 & 1 & 3 & 0 & 3 & 1 & 4 & 4 \\ 3 & 5 & 2 & 1 & 1 & 1 & 2 & 3 & 1 \\ 3 & 4 & 2 & 0 & 2 & 2 & 2 & 0 & 5 \\ 3 & 2 & 0 & 4 & 4 & 5 & 3 & 5 & 3 \\ 2 & 1 & 3 & 0 & 5 & 4 & 4 & 1 & 1 \\ 4 & 5 & 4 & 4 & 1 & 4 & 2 & 0 & 3 \\ \hline 4 & 1 & 2 & 5 & 1 & 3 & 0 & 0 & 0 \\ 4 & 4 & 1 & 4 & 4 & 5 & 0 & 0 & 0 \\ \times & \times & \times & \times & \times & \times & 0 & 0 & 1 \end{array}\right].$$

The resulting squared-up the system will have four transmission zeros. We assign these finite transmission zeros at $\{-1, -2, -3, -4\}$. For convenience, we have chosen $\hat{D}_{22} = 1$.

Using the method described earlier, the desired set of transmission zeros were assigned with

$$C = \begin{bmatrix} 4 & 1 & 2 & 5 & 1 & 3 \\ 4 & 4 & 1 & 4 & 4 & 5 \\ -3.6037 & -24.073 & -4.3615 & 17.759 & -2.1623 & 7.922 \end{bmatrix}.$$

The resulting assigned transmission zeros are located at −1.0040, −1.9900, −3.0048 and −4.0014.

17.2.2.2 Given $\rho(D) = p$, Desired Rank of Augmented $D = m$

This case is quite straightforward.

- Find $m - p$ rows in the row null-space of D to get a full rank matrix D.
- Solve the state feedback problem $(A, BD^{-1}C)$ with first p rows of C already known.
- Following the steps in the first case by settings $C = [\hat{C} + \tilde{C}]$, the problem can be solved. Note that in this case, $\hat{C}$ would be the given output matrix with a zero block of size $(m - p) \times n$ at the bottom, and $\tilde{C}$ will have a zero block of size $p \times n$ at the top.
- The resulting augmented system will have n finite transmission zeros.

17.2.2.3 Given $\rho(D) = p$, Rank of Augmented $D = p$

Perform column compression on D and partition as the system as

$$\mathscr{R}(\lambda) := U\mathscr{R}(\lambda)V = \left[\begin{array}{c|cc} A - \lambda I & B_1 & B_2 \\ \hline C & D_{11} & O \\ \hat{C} & O & O \end{array}\right].$$

Using $\hat{D}_{11}$ as pivot, it is easy to see that

$$\rho(\mathscr{R}(\lambda)) = p + \rho \left[\begin{array}{c|c} A - \lambda I & B_2 \\ \hline \hat{C} & O \end{array}\right].$$

The problem now reduces to that of $D = O$ case and number of inputs and outputs for the transformed system being $(m - p)$.

On row compression of B_2, the system can be partitioned as

$$\mathscr{R}(\lambda) := U\mathscr{R}(\lambda)V = \left[\begin{array}{cc|c} A_{11} - \lambda I_{m-p} & A_{12} & B_{21} \\ A_{21} & A_{22} - \lambda I_{n-(m-p)} & O \\ \hline \hat{C}_{21} & \hat{C}_{22} & O \end{array}\right].$$

By selecting an invertible $\hat{C}_{21}$, and solving the state feedback problem $(A_{22} - (A_{21}\hat{C}_{21}^{-1})\hat{C}_{22})$, $(n - (m - p))$ transmission zeros can be assigned at desired locations.

17.2.2.4 Given $\rho(D) = r, r < p$ and Desired Rank of Augmented $D = r + (m - p)$

Using suitable orthogonal transformations, the system in this can be repartitioned as

$$\mathscr{R}(\lambda) := U\mathscr{R}(\lambda)V = \left[\begin{array}{cc|ccc} A_{11} - \lambda I_r & A_{12} & B_{11} & B_{12} & B_{13} \\ A_{21} & A_{22} - \lambda I_{n-r} & O & B_{22} & B_{23} \\ \hline C_{11} & C_{12} & D_{11} & O & O \\ C_{21} & C_{22} & O & O & O \\ \hat{C}_{31} & \hat{C}_{32} & O & O & \hat{D}_{33} \end{array}\right].$$

Using the rank r matrix D_{11} as a pivot and performing appropriate block matrix transformation, the rank condition above system matrix becomes

$$\rho(\mathscr{R}(\lambda)) = r + \rho \left[\begin{array}{cc|cc} \tilde{A}_{11} - \lambda I_r & \tilde{A}_{12} & B_{12} & B_{13} \\ A_{21} & A_{22} - \lambda I_{n-r} & B_{22} & B_{23} \\ \hline C_{21} & C_{22} & O & O \\ \hat{C}_{31} & \hat{C}_{32} & O & \hat{D}_{33} \end{array}\right],$$

where $\tilde{A}_{11} = A_{11} - B_{11}D_{11}^{-1}C_{11}$ and $\tilde{A}_{12} = A_{12} - B_{11}D_{11}^{-1}C_{11}$.

This is similar to Case 1.b discussed earlier in Sect. 17.2.2.1. Specifically, provided that $\rho[C_{21} \quad C_{22}] \times [B_{12} \quad B_{22}]^T$ is full $(= p - r)$, we can assign $(n - (m - r))$ transmission zeros at the desired locations.

17.2.2.5 Given $\rho(D) = r, r < p$ and Desired Rank of Augmented $D = r$

Using suitable orthogonal transformations, the system, in this case, can be easily repartitioned as

$$\mathscr{R}(\lambda) := U\mathscr{R}(\lambda)V = \left[\begin{array}{cc|ccc} A_{11} - \lambda I_r & A_{12} & B_{11} & B_{12} & B_{13} \\ A_{21} & A_{22} - \lambda I_{n-r} & O & B_{22} & B_{23} \\ \hline C_{11} & C_{12} & D_{11} & O & O \\ C_{21} & C_{22} & O & O & O \\ \hat{C}_{31} & \hat{C}_{32} & O & O & O \end{array}\right],$$

$$\rho(\mathscr{R}(\lambda)) = r + \rho \left[\begin{array}{cc|cc} \tilde{A}_{11} - \lambda I_r & \tilde{A}_{12} & B_{12} & B_{13} \\ A_{21} & A_{22} - \lambda I_{n-r} & B_{22} & B_{23} \\ \hline C_{21} & C_{22} & O & O \\ \hat{C}_{31} & \hat{C}_{32} & O & O \end{array}\right]$$

where $\tilde{A}_{11} = A_{11} - B_{11} D_{11}^{-1} C_{11}$, $\tilde{A}_{12} = A_{12} - B_{11} D_{11}^{-1} C_{11}$. It can be solved along the lines of Case 1.a.

17.2.3 Squaring-up in Presence of Existing Transmission Zeros

In references [4, 5], no attention was paid to the case where the given non-square systems may have finite transmission zeros, hence reported results did not account for that case. In Sects. 17.2.1 and 17.2.2, it has been assumed that the given non-square system does not have any finite transmission zeros. Only recently, it was reported in [6] that if the given plant possesses a finite transmission zero, then the process of squaring-up will neither eliminate, nor be able to reassign any existing transmission zero. This additional result provides further insight into the squaring-up problem and, in fact, solves the problem completely. Private communications with the authors of [6] is gratefully acknowledged.

For ease of presentation, consider the simplest case of the reduced system in Eq. (17.5), where after suitable state transformations, the rank condition becomes

$$\rho(\mathscr{R}(\lambda)) = \rho \left[\begin{array}{cc|c} A_{11} - \lambda I_m & X & B_1 \\ A_{21} & A_{22} - A_{21} C_1^{-1} C_2 - \lambda I_{n-m} & O \\ \hline C_1 & O & O \end{array}\right],$$

such that $\rho(\mathscr{R}(\lambda)) = 2m + \rho(A_{22} - A_{21} C_1^{-1} C_2 - \lambda I_{n-m})$. It is quite straightforward to see that $\rho(A_{22} - A_{21} C_1^{-1} C_2 - \lambda I_{n-m}) < (n - m)$ if and only if $(A_{22}, A_{21} C_1^{-1})$ is an uncontrollable pair. Similar results can be easily derived for other cases explored in Sect. 17.2.2.

Therefore, if λ_0 is an uncontrollable mode of the pair $(A_{22}, A_{21} C_1^{-1})$, then $\rho(\mathscr{R}(\lambda_0)) < (n + m)$, implying that the given non-square system has a transmission zero at λ_0. Further, since C_2 represents a state feedback matrix, only the eigenvalues of the controllable subsystem of $(A_{22}, A_{21} C_1^{-1})$ may be reassigned, eigenvalues of uncontrollable subsystem will remain unaffected. The reader is referred to [6] for additional details.

In light of the findings of [6] and earlier results presented in [4, 5], the squaring-up results may be summarized as

Result 17.2.1: Given a least order, non-degenerate, wide non-square $(m > p)$ system $\Sigma(A, B, C, D)$, it is always possible to find pseudo-outputs such that all the finite transmission zeros of the squared-up system are assigned at desired locations, provided that $\rho(BC) = p$ and the given system does not have any finite transmission zeros.

Result 17.2.2: Given a least order, non-degenerate, non-square $(m > p)$ *minimum phase* system $\Sigma(A, B, C, D)$, it is always possible to find pseudo outputs such that the resulting squared-up system remains minimum phase, provided that $\rho(BC) = p$ and any finite transmission zeros of the given system are in the left half plane.

17.3 System Completion

The problem addressed in this section is a variation of the partial system completion problem discussed in Sect. 17.2. Specifically, given a controllable pair (A, B), where B has *full column rank*, we would like to find a matrix C and optionally D such that the resulting 'completed system' has its transmission zeros at desired locations.

Remark 17.3.1: It is worth noting that unlike the squaring-up problem, here one need not be concerned about existing transmission zeros since matrices C and D are completely unspecified.

Remark 17.3.2: While typically one would expect to make the resulting system square, the results described below are easily extended to the case where $m \neq p$. However, due to space limitations, they have not been discussed here.

Remark 17.3.3: The case when an observable pair (A, C) is given, then using duality, finding matrices B, D is straightforward. Therefore, we will only consider the case when a controllable pair (A, B) is given and matrices C and D need to be computed.

As in Sect. 17.2, there are many possible cases in the choice of C and D, with $\rho(D)$ having a desired value.

17.3.1 $p = m$ and $\rho(D) = m$

$$\rho(\mathscr{R}(\lambda)) = \rho\left[\begin{array}{c|c} A - \lambda I_n & B \\ \hline C & D \end{array}\right] = \rho\left[\begin{array}{c|c} A - BD^{-1}C - \lambda I_n & O \\ \hline D^{-1}C & I_m \end{array}\right]$$

with (A, B) a controllable pair, $\rho(\mathscr{R}(\lambda)) = m + \rho(A - BD^{-1}C - \lambda I_n)$. Therefore, all n transmission zeros can be assigned to desired locations, using state feedback techniques to assign eigenvalues of the controllable pair (A, BD^{-1}) with C serving as the state feedback matrix.

17.3.2 $p = m$ and $\rho(D) = 0$

Since $\rho(B) = m$, using orthogonal transformations, the system can be transformed to and repartitioned as

$$\mathscr{R}(\lambda) := U\mathscr{R}(\lambda)V = \left[\begin{array}{cc|c} A_{11} - \lambda I_m & A_{12} & B_1 \\ A_{21} & A_{22} - \lambda I_{n-m} & O \\ \hline C_1 & C_2 & O \end{array}\right],$$

where $B_1 \in \mathbb{R}^{m\times m}$ with full rank and both C_1 and C_2 are unknown matrices that can be selected arbitrarily. Then,

$$\rho(\mathscr{R}(\lambda)) = m + \rho\left[\begin{array}{c|c} A_{21} & A_{22} - A_{21}C_1^{-1}C_2 - \lambda I_{n-m} \\ \hline C_1 & O \end{array}\right].$$

by selecting C_1 to be any invertible $(m \times m)$ matrix, e.g. $C_1 = I_m$. Having selected C_1, we can reformulate computation of C_2 into a state feedback problem. Note that due to the controllability assumption on (A, B), controllability of $(A_{22}, A_{21}C_1^{-1})$ is trivially guaranteed and, therefore, a C_2 can always be found such that the $(n - m)$ finite transmission zeros of the resulting square system $\Sigma(A, B, C, D)$ with $D = O$ can be placed at the desired locations.

17.3.3 $p = m$ and $\rho(D) = r$, $r < m$

Without the loss of generality, it can be assumed that $D = \left[\begin{smallmatrix} D_{11} & O \\ O & O \end{smallmatrix}\right]$, with $\rho(D_{11}) = r$. Then, partitioning matrices A, B, C conformably to $D_{11}$and row compressing $[(r + 1) : m]$ columns of B, knowing that $\rho(B) = m$, the resulting system matrix becomes

$$\mathscr{R}(\lambda) = \left[\begin{array}{cc|cc} A_{11} - \lambda I_{m-r} & A_{12} & B_{11} & B_{12} \\ A_{21} & A_{22} - \lambda I_{n-m+r} & B_{21} & O \\ \hline C_{11} & C_{12} & D_{11} & O \\ C_{21} & C_{22} & O & O \end{array}\right].$$

Therefore,

$$\rho(\mathscr{R}(\lambda)) = (m - r) + \rho\left[\begin{array}{cc|cc} A_{21} & A_{22} - \lambda I_{n-m+r} & B_{21} & O \\ \hline C_{11} & C_{12} & D_{11} & O \\ C_{21} & C_{22} & O & O \end{array}\right].$$

Since we have complete freedom in the choice of C, we can set $C_{22} = O$ and C_{21} as an invertible matrix of dimension $(m - r)$. Again, without any loss of generality, we set $C_{21} = I_{m-r}$. This leads to

$$\rho(\mathscr{R}(\lambda)) = 2(m - r) + \rho \left[\begin{array}{c|c} A_{22} - \lambda I_{n-m+r} & B_{21} \\ \hline C_{12} & D_{11} \end{array} \right].$$

The reduced order system matrix has the same structure as the system matrix considered in Sect. 17.3.1. Hence, the $(n - m + r)$ finite transmission zeros can be assigned by computing C_{12} using state feedback techniques, such that the controllable pair $(A_{22}, B_{21} D_{11}^{-1})$ has all its eigenvalues at the desired locations.

It should be pointed out that similar results can be derived for cases where $p \neq m$, but they are not so interesting from a practical viewpoint. However, interested reader is referred to [5].

17.3.4 System Completion with $CA^{i-1}B = O$

A number of design techniques such as sliding mode controls [7] (single input, single output) and multivariable PID control [8] (multiple inputs, multiple outputs), require that given a controllable pair (A, B) with B having full column rank, we find a C such that $CA^{(i-1)}B = O$, $1 \leq i < (\ell - 1)$ and $CA^{(\ell-1)}B \neq O$. At the same time, the transmission zeros of the resulting square system lie at desired locations (in the left half plane). While for the single input, single output systems, this implies that ℓ is the relative degree of the system, for multiple input, multiple outputs systems, the notion of relative degree is not so straightforward, instead a *vector* relative degree is used [9].

Instead of delving into relative degree, we will limit our discussion to the computation of C to meet the conditions $CA^{(i-1)}B = O$, $1 \leq i < (k - 1)$ and $CA^{(k-1)}B \neq O$. To this end, assume that the controllable pair (A, B) has been transformed to a block *upper Hessenberg* form [10], with the resulting system described by

$$\left[\begin{array}{c|c} A & B \\ \hline C & O \end{array} \right] = \left[\begin{array}{cccccc|c} A_{11} & A_{12} & A_{13} & \cdots & A_{1,k-1} & A_{1k} & B_1 \\ A_{21} & A_{22} & A_{23} & \cdots & A_{2,k-1} & A_{2k} & O \\ O & A_{32} & A_{33} & \cdots & A_{3,k-1} & A_{3k} & O \\ \vdots & \vdots & \vdots & \ddots & \vdots & \vdots & \vdots \\ O & O & O & \cdots & A_{k-1,k-1} & A_{k-1,k} & O \\ O & O & O & \cdots & A_{k,k-1} & A_{k,k} & O \\ \hline C_1 & C_2 & C_3 & \cdots & C_{k-1} & C_k & O \end{array} \right], \tag{17.7}$$

where the block sub-diagonal matrices $A_{r+1,r}$, $r = 1, \ldots, (k-1)$ have full row rank. To enforce $CA^{i-1}B = O$, it suffices to set $C_r = O, r = 1, \ldots, i$. The corresponding system pencil is given by

$$\mathscr{R}(\lambda) = \left[\begin{array}{ccccc|c} A_{11} - \lambda I_1 & A_{12} & \cdots & A_{1,k-1} & A_{1k} & B_1 \\ A_{21} & A_{22} - \lambda I_2 & \cdots & A_{2,k-1} & A_{2k} & O \\ O & A_{32} & \cdots & A_{3,k-1} & A_{3k} & O \\ \vdots & \vdots & \ddots & \vdots & \vdots & \vdots \\ O & O & \cdots & A_{k,k-1} & A_{k,k} - \lambda I_k & O \\ \hline C_1 & C_2 & \cdots & C_{k-1} & C_k & O \end{array}\right],$$

Case: $CB = O$ **and** $CAB \neq O$

For illustration, let us consider the case when $i = 1$. By controllability of (A, B), $A_{21} \neq O$ and B_1 is a full rank $(m \times m)$ matrix. Selecting $C_1 = O$ ensures that $CB = O$ and $CAB = C_2 A_{21} B_1$. Since A_{21} has full row rank, C_2 may be chosen such that $C_2 A_{21} B_1 \neq O$. Further,

$$\rho(\mathscr{R}(\lambda)) = m + \rho \left[\begin{array}{c|cccc} A_{21} & A_{22} - \lambda I_2 & \cdots & A_{2,k-1} & A_{2k} \\ O & A_{32} & \cdots & A_{3,k-1} & A_{3k} \\ \vdots & \vdots & \ddots & \vdots & \vdots \\ O & O & \cdots & A_{k,k-1} & A_{k,k} - \lambda I_k \\ \hline O & C_2 & \cdots & C_{k-1} & C_k \end{array}\right], \quad (17.8)$$

For a scalar (single input, single output) system, it is easily seen that $k = n$ and since scalar elements $A_{i+1,i} \neq 0, i = 1, \ldots, (n-1)$ the rank condition may be rewritten as the rank condition of the system matrix in Sect. 17.3.2, thereby meeting the conditions to assign all finite transmission zeros at desired location.

However, when $m, p > 1$, the resulting A_{21} may or may not have full column rank m. If it has full column rank, then the resulting $(n - 2m)$ finite transmission zeros may be assigned using the results presented in Sect. 17.3.2. On the other hand, if $\rho(A_{21}) < m$, then on rearranging and simplifying the pencil notation, we have the following rank condition on the system matrix:

$$\rho(\mathscr{R}(\lambda)) = m + \rho \left[\begin{array}{c|c} \hat{A} - \lambda I_{n-m} & \hat{B} \\ \hline \hat{C} & O \end{array}\right], \quad (17.9)$$

where $\hat{A} \in \mathbb{R}^{(n-m)\times(n-m)}$, $\hat{C} \in \mathbb{R}^{m\times(n-m)}$ and $\hat{B} \in \mathbb{R}^{(n-m)\times m}$ are appropriate matrices in (17.8), with $\rho(\hat{B}) < m$ $(= r$, say) and $(\hat{A}, \hat{B})$ is a controllable pair by virtue of controllability assumption on the given (A, B). Using rank revealing transformation on $\hat{B}$, Eq. (17.9) may be rewritten as

$$\rho(\mathscr{R}(\lambda)) = m + \rho \left[\begin{array}{cc|cc} \hat{A}_{11} - \lambda I_r & \hat{A}_{12} & \hat{B}_{11} & O \\ \hat{A}_{21} & \hat{A}_{22} - \lambda I_{(n-(m+r))} & O & O \\ \hline \hat{C}_{11} & \hat{C}_{12} & O & O \\ \hat{C}_{21} & \hat{C}_{22} & O & O \end{array}\right]$$

$$= m + r + \rho \left[\begin{array}{c|c} \hat{A}_{21} & \hat{A}_{22} - \lambda I_{(n-(m+r))} \\ \hline \hat{C}_{11} & \hat{C}_{12} \end{array}\right]$$

upon setting $\hat{C}_{21}$ and $\hat{C}_{22} = O$ and letting $\hat{C}_{11}$ be an invertible ($r \times r$ matrix, all finite zeros of the system can be assigned at the eigenvalues of $\hat{A}_{22} - (\hat{A}_{21}\hat{C}_{11}^{-1})\hat{C}_{12}$ by appropriate choice of the unknown 'state feedback' matrix $\hat{C}_{12}$.

By taking advantage of block upper Hessenberg structure, these results can be easily extended to $CA^{(i-1)}B = O$, $1 \leq i < (\ell - 1)$ and $CA^{(\ell-1)}B \neq O$ for $\ell > 1$.

17.4 Concluding Remarks

This chapter investigated a system completion problem, whereby missing inputs or outputs can be found to create a square or squared-up system with transmission zeros at desired locations. A number of classical results from scalar systems are easier to interpret when the multivariable system is a square one. Using the squaring process to embed a nonsquare system into a square one as an intermediate step to eventual design may be a useful way to extend a number of classical results to multivariable settings. For an application in adaptive control, the reader is referred to [11].

References

1. Rosenbrock, H.H.: State Space and Multivariable Theory. Nelson, London (1970)
2. Saberi, A., Sannuti, P.: Squaring down by static and dynamic compensators. IEEE Trans. Aut. Contr. **33**, 358–365 (1988)
3. Kouvaritakis, B., MacFarlane, A.G.J.: Geometric approach to analysis and synthesis of system zeros. II Non-square systems. Int. J. Control **23**, 167–181 (1976)
4. Misra, P.: A computational algorithm for squaring-up. Part I - Zero input output matrix. In: Proceedings of 1992 31st IEEE Conference on Decision and Control, pp. 149–150. IEEE, New York, NY, USA (1992)
5. Misra, P.: Numerical algorithms for squaring-up non-square systems. Part II: general case. In: Proceedings of the 1993 American Control Conference, pp. 1573–1577, June 1993
6. Qu, Z., Wiese, D., Annaswamy, A.M.: Squaring-up method in the presence of transmission zeros. In: Proceedings of the 19th IFAC World Congress, pp. 4164-4169. Cape Town, South Africa, August 2014
7. Hernandez, D., Castnnos, F., Fridman, L.: Pole-placement in higher-order sliding-mode control. In: Proceedings of the 19th IFAC World Congress, pp. 1386–1391. Cape Town, South Africa, August 2014

8. Anderson, B.D.O., Moore, J.B.: Optimal Control: Linear Quadratic Methods. Prentice Hall, Upper Saddle River (2007)
9. Isidori, A.: Nonlinear Control System, 3rd edn, p. 1994. Springer-Verlag, New York (1994)
10. Patel, R.V.: Computation of minimal order state space realizations and observability indices using orthogonal transformations. Int. J. Cont. **33**, 227–246 (1981)
11. Lavretsky, E., Wise, K.: Robust and Adaptive Control: With Aerospace Applications. Advanced Textbooks in Control and Signal Processing. Springer, London (2012)

Chapter 18
The Role of Sensor and Actuator Models in Control of Distributed Parameter Systems

Kirsten Morris

Abstract Many systems are modelled by partial differential equations. The boundary conditions are important and affect the dynamics. Also, the modelling of actuation and sensing is not straightforward. The modelling of the actuators and sensors, as well as their locations, can affect control and estimation performance and design.

18.1 Introduction

Many systems have dynamics that depend on space as well as on time. Examples include transmission lines, acoustic noise, structural vibrations. These systems are known as distributed parameter systems (DPS). This is different from lumped parameter systems, such as circuits, where the dynamics only depend on time. Physics-based models for DPS involve partial differential equations.

There are a number of issues in modelling DPS that do not arise in lumped parameter systems. The handling of the boundary conditions in partial differential equation (PDE) models affects the dynamics of the system and other control-related properties. The well-posedness of the model can be affected. Also, the modelling and location of the actuators and sensors can affect the achievable performance of the controller and estimator. These issues are explored in this paper through a number of examples.

The research described here was supported by an NSERC Discovery Grant and by AFOSR Grant FA9550-16-1-0061.

K. Morris (✉)
Department of Applied Mathematics, University of Waterloo,
Waterloo, ON N2L 3G1, Canada
e-mail: kmorris@uwaterloo.ca

R. Tempo et al. (eds.), *Emerging Applications of Control and Systems Theory*, Lecture Notes in Control and Information Sciences - Proceedings,
https://doi.org/10.1007/978-3-319-67068-3_18

18.2 Dynamics

Systems modelled by partial differential equations (PDEs) can be written in state-space form similar to that for linear ordinary differential equation (ODE) models; that is

$$\dot{z}(t) = Az(t) + Bu(t), \quad z(0) = z_0 \tag{18.1}$$

For ODE's $z(t)$ is a vector and A, B are matrices. The main difference between ODEs and PDEs is that the matrix A becomes an operator on an infinite-dimensional Hilbert space, $\mathscr{Z}$. Similarly, B is an operator mapping the input space into $\mathscr{Z}$. See [6] for more detail on the systems theory described briefly here.

Consider first an uncontrolled system. A unique solution needs to exist for all initial conditions and small changes in the initial condition should lead to small changes in the solution. This means that the solution operator $S(t) : x(0) \to x(t)$ exists and is a continuous operator in the Hilbert space norm. Also, at time 0, the initial condition should be recovered, and the solution at time $t + s$ with initial condition $x(0) = x_0$ should be the same as that at time t with initial condition $S(s)x_0$. This motivates the following definition. Let $\mathscr{L}(\mathscr{X}_1, \mathscr{X}_2)$ indicate bounded linear operators from a Hilbert space $\mathscr{X}_1$ to a Hilbert space $\mathscr{X}_2$.

Definition 18.1 A strongly continuous (C_0-) semigroup $S(t)$ on Hilbert space $\mathscr{Z}$ is a family of operators $S(t) \in \mathscr{L}(\mathscr{Z}, \mathscr{Z})$, $t \geq 0$, such that

1. $S(0) = I$,
2. $\lim_{t\downarrow 0} S(t)z = z$, for all $z \in \mathscr{Z}$,
3. $S(t)S(s) = S(t+s)$, for all s, t ≥ 0.

Definition 18.2 The infinitesimal generator A of a C_0-semigroup on $\mathscr{Z}$ is

$$Az = \lim_{t\downarrow 0} \frac{1}{t}(S(t)z - z) \quad \text{for all } z_0 \in \text{D(A)},$$

with $D(A)$ the set of elements $z \in \mathscr{Z}$ for which the limit exists.

Provided that A is the generator of a C_0-semigroup $S(t)$ on a Hilbert space $\mathscr{Z}$,

$$\frac{d(S(t)z_0)}{dt} = AS(t)z_0 = S(t)Az_0 \quad \text{for all } z_0 \in \text{D(A)}.$$

Then the differential equation on $\mathscr{Z}$

$$\frac{dz(t)}{dt} = Az(t), \qquad z(0) = z_0 \tag{18.2}$$

has the solution $z(t) = S(t)z_0$. Due to the properties of a C_0-semigroup, this solution is unique, and depends continuously on the initial data z_0.

The matrix exponential is an example of a strongly continuous (C_0) semigroup; in this case, the state space is $\mathbb{R}^n$. An example of a PDE is described below.

Example 18.1 Heat Conduction. The temperature $z(x, t)$ at time t at position x from the left-hand end in a long thin bar of length L with constant thermal conductivity K_0, mass density ρ and specific heat capacity C_p is modelled by [11, Chap. 1]

$$C_p\rho\frac{\partial z(x,t)}{\partial t} = K_0\frac{\partial^2 z(x,t)}{\partial x^2}, \quad z(x,0) = z_0(x) \qquad x \in (0, L), \qquad t \geq 0. \quad (18.3)$$

The boundary conditions at each end need to be specified. Suppose the temperature at both ends is fixed. Setting the temperature of the immersing medium to 0,

$$z(0,t) = 0, \qquad z(L,t) = 0. \quad (18.4)$$

The PDE (18.3) can be rewritten in state-space form (18.2) with state space $\mathscr{L}^2(0, L)$. by defining

$$Az = \frac{\partial^2 z}{\partial x^2}, \quad D(A) = \{z \in \mathscr{L}^2(0,L);\ z', z'' \in \mathscr{L}^2(0,L);\ z(0) = z(L) = 0\}. \quad (18.5)$$

Letting $\phi_n(x) = \sqrt{\frac{2}{L}}\sin(n\pi\frac{x}{L})$, $\lambda_n = -n^2\pi^2$, $n = 1, 2, \ldots$ the solution is the well-known Fourier series,

$$S(t)z_0 = \sum_{n=1}^{\infty}\langle z_0, \phi_n\rangle\phi_n(x)e^{\lambda_n t}. \quad (18.6)$$

For all initial conditions in $\mathscr{L}^2(0, L)$ the solutions decays to zero: the system is asymptotically stable.

Definition 18.1(2) states strong convergence of $S(t)$ to the identity I as $t \to 0$. In (18.6) consider an initial condition $z_0 = \phi_n$.

$$\lim_{t\downarrow 0}\|S(t)\phi_n - \phi_n\| = \lim_{t\downarrow 0} 1 - e^{-n^2\pi^2 t} = 0$$

as required. But no single value of t will give a uniformly small error for all z_0. In fact, uniform convergence implies that the generator is a bounded operator defined on the whole space, for example, a matrix. However, for partial differential equations, the generator A is a differential operator and not bounded. Only strong convergence to the initial condition is possible [19].

The boundary conditions of partial differential equations are an important part of the model and affect the dynamics.

(Example 18.1 *cont.) Heat Conduction.* Suppose that instead of (18.4) the ends are insulated:

$$\frac{\partial z}{\partial x}(0,t) = 0, \qquad \frac{\partial z}{\partial x}(L,t) = 0. \quad (18.7)$$

The state space is still $\mathscr{L}^2(0, L)$ but the generator is now

$$Az = \frac{\partial^2 z}{\partial x^2}, \quad D(A) = \{z, z', z'' \in \mathscr{L}^2(0, L);\ z'(0) = z'(L) = 0\}.$$

The different boundary conditions lead to a different domain, and so a different operator. The eigenfunctions are $\psi_n(x) = \sqrt{\frac{2}{L}}\cos(n\pi\frac{x}{L})$, with eigenvalues $\lambda_n = -n^2\pi^2$, $n = 0, 1, \ldots$ The solution is

$$S(t)z_0 = \langle z_0, \psi_0\rangle\psi_0(x) + \sum_{n=1}^{\infty}\langle z_0, \psi_n\rangle\psi_n(x)e^{\lambda_n t}.$$

Because of the 0 eigenvalue, the solution no longer decays to zero for all initial conditions. The system is not asymptotically stable.

Example 18.2 Acoustic Noise. Consider acoustic noise in a duct of length L with radius $a << L$. Letting ρ_0 indicate density, c the speed of sound, $v(x, t)$ wave velocity, and $p(x, t)$ acoustic pressure, wave propagation can be modelled by [20, e.g]

$$\frac{1}{c^2}\frac{\partial p(x,t)}{\partial t} = -\rho_0\frac{\partial v(x,t)}{\partial x}, \tag{18.8}$$

$$\rho_0\frac{\partial v(x,t)}{\partial t} = -\frac{\partial p(x,t)}{\partial x}. \tag{18.9}$$

Suppose the end the end $x = 0$ is plugged and the other end is open and set

$$v(0, t) = 0, \qquad p(L, t) = 0.$$

The model is well-posed with states (p, v) on state space $(\mathscr{L}^2(0, L))^2$ with norm proportional to the sum of potential and kinetic energies. The eigenvalues are $\lambda_n = j\frac{c}{2L}(2n+1)\pi, \quad n = 1, 2\ldots.$ All eigenvalues are imaginary. This model predicts reflection of all pressure waves and any initial condition will lead to a wave that oscillates indefinitely. A more accurate boundary condition is [29]

$$p(L, t) - \beta v(L, t) = 0, \quad 0 < \beta < \rho_0 c.$$

The eigenvalues are now

$$\lambda_n = -\frac{c}{2L}\ln\left(\left|\frac{\rho_0 c + \beta}{\rho_0 c - \beta}\right|\right) + j\frac{c}{L}(n + \frac{1}{2})\pi$$

and $\text{Re}\lambda_n < 0$ which indicates the dissipation present. This model is asymptotically stable. The impedance β is frequency dependent and modelled by a more complex frequency dependent boundary condition [29]. This model has eigenvalues with negative real parts that asymptote to a vertical line in the left-hand plane [4, 29].

Thus, the handling of the boundary conditions is a significant aspect of the model. In particular, the location of the eigenvalues and stability are dependent on the boundary conditions.

18.3 Effect of Actuator/Sensor Model on Dynamics

The modelling of an actuator affects the maps from the input to the state and from the input to the output. In the state-space formulation, the nature of the operator B in (18.1) is affected by the type of actuation and its model.

(Example 18.1 *cont.) Heat Conduction.* Consider Dirichlet control:

$$z(0,t)=0, \qquad z(L,t)=u(t). \tag{18.10}$$

The transfer function can be derived by taking Laplace transforms of the PDE (18.3) with respect to the time variable t, assuming an initial condition of zero and then using the boundary conditions. (For the mathematical justification of this procedure for partial differential equations see [3].) Denoting the Laplace transforms by $\hat{z}$, $\hat{u}$, the resulting boundary value problem is

$$K_0 \frac{d^2\hat{z}(x,s)}{dx^2} = C_p \rho s \hat{z}(x,s), \tag{18.11}$$

$$z(0,s)=0, \qquad z(L,s)=\hat{u}(s). \tag{18.12}$$

Measuring the temperature at point x_0, $0 < x_0 \leq L$, yields observation

$$y(t) = z(x_0,t). \tag{18.13}$$

The transfer function is, defining $\alpha^2 = \frac{K_0}{C_p \rho}$,

$$G_{heat0}(s) = \frac{\sinh\left(\frac{\sqrt{s}x_0}{\alpha}\right)}{\sinh\left(\frac{\sqrt{s}L}{\alpha}\right)}. \tag{18.14}$$

This function is in $\mathcal{H}_\infty$ and the control system is $\mathcal{L}_2$-stable.

This model is written in state-space form with state-space $\mathcal{L}^2(0,L)$ as, letting $\delta'(x-L)$ indicate the derivative of the impulse distribution at $x=L$,

$$\dot{z}(t) = Az(t) + \delta'(x-L)u(t)$$

where A is defined in (18.5). Because the control operator $B = \delta'$ this differential equation needs to be understood as a differential equation not on the natural state space $\mathscr{X} = \mathscr{L}^2(0, L)$ but on a larger space of distributions $\mathscr{V} = [\mathscr{D}(A)]'$.

The map from the control $u \in \mathscr{L}^2(0, T)$ to the state $z(T) \in \mathscr{X}$ is not bounded [5]. This implies that small changes in the control variable could lead to large changes in the value of the state. This reflects the unphysical nature of temperature $z(L, t)$ being instantaneously changed by the control.

A more realistic model for the control than (18.12) is, for $h > 0$

$$K_0 \frac{\partial z}{\partial x}(L, t) = h(u(t) - z(L, t))$$

With the same observation (18.13), this leads to the transfer function

$$G_{heat1}(s) = \frac{h\alpha \sinh\left(\frac{\sqrt{s}x_0}{\alpha}\right)}{K_0\sqrt{s}\cosh\left(\frac{\sqrt{s}L}{\alpha}\right) + h\alpha \sinh(\frac{\sqrt{s}L}{\alpha})}. \tag{18.15}$$

As the conductivity $K_0 \to 0$, the transfer function (18.14) with Dirichlet boundary control (18.10) is recovered. But with $K_0 \neq 0$, the poles (and the eigenvalues of the generator A) are different from those with the original boundary conditions and reflect different dynamics. The state space is still $\mathscr{L}^2(0, L)$ and defining

$$Az = \frac{\partial^2 z}{\partial x^2}, \quad D(A) = \{z \in \mathscr{L}^2(0, L);\ z', z'' \in \mathscr{L}^2(0, L);\ z(0) = 0,\ K_0 z'(L) + hz(L) = 0\},$$

the state-space representation is

$$\begin{aligned} \dot{z}(t) &= Az(t) + \delta(x - L)u(t) \\ y(t) &= z(x_0, t) \end{aligned}$$

The map from the control to the state is now bounded [5].

(Example 18.2 *cont.) Acoustic Noise.* [29] Plugging the end $x = 0$ with a loudspeaker as a noise source can be modelled by

$$v(0, t) = u(t).$$

This implies that if $u = 0$ the end is rigid. But an undriven loudspeaker has nonzero compliance. Also, the control variable is not velocity at $x = 0$ but voltage to the speaker $V(t)$. Let d indicate the speaker cone displacement and a, A_s, m, ξ, k_s, B_s various physical parameters. A more accurate boundary condition is

$$A_s \dot{d}(t) = \pi a^2 v(0, t), \tag{18.16}$$

$$m\ddot{d}(t) + \xi \dot{d}(t) + k_s d(t) = B_s V(t) - A_s p(0, t). \tag{18.17}$$

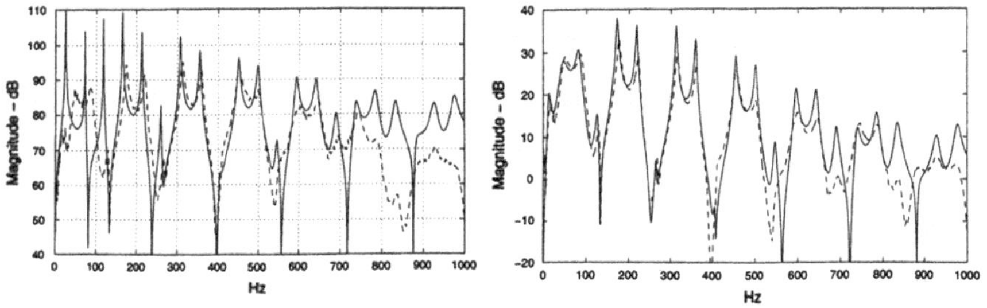

(a) v(0,t)=0 (b) Dynamic model ($V(t) = 0$)

Fig. 18.1 Comparison of experimental and calculated frequency response in a duct of length 3.54 with different models for the boundary condition at $x = 0$. Figures **a** and **b** show the frequency response of the pressure at $x = 1.095$ compared to the velocity and voltage respectively of a loudspeaker at $x = 2.32$ Including a dynamic model for the behaviour at $x = 0$ increases the accuracy considerably. The error is very small until about 400 Hz, which is where the one space dimension assumption becomes inaccurate. Experiment (- - -) Simulation (—). Reproduced with permission from © 2003 ASME [29]

The coupled differential Equations (18.8, 18.9, 18.17) with boundary condition (18.16) at $x = 0$, and an appropriate boundary condition at the open end $x = L$ (as discussed earlier), lead to a model with state $\left(p, v, d, \dot{d}\right)$ that is well-posed on the state space $(\mathscr{L}^2(0, L))^2 \times \mathbb{R}^2$. The control operator B is now bounded into the state space. Most importantly, this model is considerably more accurate than the simple boundary condition $v(0, t) = u(t)$; see the experimental results in Fig. 18.1.

Example 18.3 Beam Vibrations. [1, 3] The simplest example of transverse vibrations in a structure is a beam, where the vibrations can be considered to occur only in one dimension. Consider a homogeneous beam of length L with only small transverse vibrations. The classic Euler-Bernoulli beam model for the deflection $w(x, t)$ is

$$\frac{\partial^2 w(x,t)}{\partial t^2} + EI\frac{\partial^4 w(x,t)}{\partial x^4} = 0,$$

where E, I are material constants [11, Chap. 6]. This simple model does not include any damping, and predicts that a beam, once disturbed, would vibrate forever. To model more realistic behaviour, damping should be included. The most common model of damping is Kelvin-Voigt damping, which leads to the PDE

$$\frac{\partial^2 w}{\partial t^2} + \frac{\partial^2}{\partial x^2}\left(EI\frac{\partial^2 w(x,t)}{\partial x^2} + c_d I\frac{\partial^3 w(x,t)}{\partial x^2 \partial t}\right) = 0, \qquad (18.18)$$

where c_d is the damping constant. Assume that the beam is clamped at $x = 0$ and free at the tip $x = L$, with control of the shear force at the tip. Letting $u(t)$ be the applied force,

$$w(0,t) = 0, \qquad \frac{\partial w}{\partial x}(0,t) = 0, \tag{18.19}$$

$$\frac{\partial^2 w}{\partial x^2}(L,t) = 0, \qquad EI\frac{\partial^3 w}{\partial x^3}(L,t) = u(t), \qquad t \geq 0. \tag{18.20}$$

With measurement of the tip velocity, the output is

$$y(t) = \frac{\partial w}{\partial t}(L,t). \tag{18.21}$$

Defining $m(s) = \left(\frac{-s^2}{EI+sc_d I}\right)^{\frac{1}{4}}$, taking the Laplace transform of (18.21) with zero initial conditions, and then applying the boundary conditions, yields transfer function

$$G_{beam1}(s) = \frac{s\left[\cosh\left(Lm(s)\right)\sin\left(Lm(s)\right) - \sinh\left(Lm(s)\right)\cos\left(Lm(s)\right)\right]}{EIm^3(s)\left[1+\cosh\left(Lm(s)\right)\cos\left(Lm(s)\right)\right]}.$$

All poles are in the left half-plane, so G_{beam1} is analytic on the right half-plane. However it is not bounded in the right-hand plane [3] and so $G_{beam1} \notin \mathcal{H}_\infty$. This implies lack of external stability. Small changes in a control could lead to large changes in measurement. Also, the magnitude of the frequency response increases with frequency, which is unrealistic. The mistake lies in the omission of the effect of damping on the moment in the boundary conditions (18.20). The moment M of an undamped Euler-Bernoulli beam is $EI\frac{\partial w^2}{\partial x^2}$. Kelvin-Voigt damping affects the moment, which becomes

$$M(x,t) = EI\frac{\partial^2 w}{\partial x^2} + c_d I\frac{\partial^3 w}{\partial x^2 \partial t}.$$

The correct boundary conditions for a free end are not (18.20) but

$$M(L,t) = 0, \qquad \frac{\partial M}{\partial x}(L,t) = u(t) \qquad t \geq 0. \tag{18.22}$$

With the boundary conditions (18.22) at $x = L$ and the original ones (18.19) at $x = 0$, the transfer function is

$$G_{beam2}(s) = \frac{s\left[\cosh\left(Lm(s)\right)\sin\left(Lm(s)\right) - \sinh\left(Lm(s)\right)\cos\left(Lm(s)\right)\right]}{m^3(s)\left(EI+sc_d I\right)\left[1+\cosh\left(Lm(s)\right)\cos\left(Lm(s)\right)\right]}.$$

This transfer function has the same poles as the previous one G_{beam1} except for an extra pole at $-E/c_d$ and so it is analytic in the closed right half-plane. But now it is bounded in the right-hand plane so $G_{beam2}(s) \in \mathcal{H}_\infty$. This is the transfer function of an externally stable system.

Example 18.4 Sensors for vibrations [12] Consider the following general model for the dynamics of a second-order system

$$\ddot{w}(t) + A_o w(t) + D\dot{w}(t) = B_o u(t). \tag{18.23}$$

Here A_o is a positive definite, and D a positive semidefinite operator on a Hilbert space $\mathscr{W}$. The control operator B_o is assumed here to be a bounded operator from the control space $\mathscr{U}$ into $\mathscr{W}$. For example, in the beam Example 18.3, A_o and D are $\frac{\partial^4}{\partial x^4}$ (times a constant) and

$$\mathscr{W} = \{w;\, w, w', w'' \in \mathscr{L}^2(0, L),\ w(0) = 0, w'(0) = 0\}$$

with norm $\|w\| = \left(\int_0^L |w(x)|^2 dx + \int_0^L |w'(x)|^2 dx + \int_0^L |w''(x)|^2 dx\right)^{\frac{1}{2}}$. If the state is chosen to be $z = \begin{bmatrix} w & \dot{w} \end{bmatrix}$, (18.23) can be written in state-space form as

$$\begin{bmatrix} \dot{w}(t) \\ \ddot{w}(t) \end{bmatrix} = \underbrace{\begin{bmatrix} 0 & I \\ -A_o & -D \end{bmatrix}}_{A} \begin{bmatrix} w(t) \\ \dot{w}(t) \end{bmatrix} + \begin{bmatrix} 0 \\ B_o \end{bmatrix} u(t), \tag{18.24}$$

with, defining $\mathscr{Z} = \mathscr{W} \times \mathscr{L}^2(0, L)$, $D(A) = \{(w, v) \in \mathscr{Z};\ v \in \mathscr{W},\ A_o w + Dv \in \mathscr{L}^2(0, L)\}$. With natural assumptions on A_o and D, the model is well-posed on $\mathscr{Z}$. The quantity $\|z\|^2_{\mathscr{Z}}$ is proportional to the potential and kinetic energies.

One measurement type is the position at $0 < x_0 < L$: $y(t) = w(x_0, t)$. Defining $C_0 w = w(x_0)$, $C_p = \begin{bmatrix} C_0 & 0 \end{bmatrix}$,

$$y(t) = C_p z(t).$$

The observation operator C_p is bounded from the state space $\mathscr{Z}$.

Another common sensor is an accelerometer. Acceleration involves the second time derivative and cannot easily be written in terms of the states. One approach is to to write

$$y(t) = \begin{bmatrix} 0 & C_o \end{bmatrix} \left(\begin{bmatrix} 0 & I \\ -A_o & -D \end{bmatrix} \begin{bmatrix} w(t) \\ \dot{w}(t) \end{bmatrix} + \begin{bmatrix} 0 \\ B_o \end{bmatrix} u(t) \right).$$

However, unless very strong assumptions, not satisfied by most applications, are made on stiffness A_o and damping D, this does not lead to well-posed observation with respect to the energy space $\mathscr{Z}$ [1, 12]. Small changes in the state $(w, \dot{w})$ lead to large changes in the observation y. The model is ill-posed.

However, sensors used to measure acceleration have dynamics. A common type is a micro-electro-mechanical system (MEMS) where a mass is suspended between two capacitors and the measured voltage is proportional to the mass position. Letting $F(t)$ be the force applied by the structure to the accelerometer, and a the deflection of the accelerometer mass, and m, d and k accelerometer parameters,

$$m\ddot{a}(t) + ka(t) + d\dot{a}(t) = F(t).$$

Using Hamilton's principle to obtain a description for the dynamics of a structure coupled to an accelerometer leads to

$$m\ddot{a}(t)+k\big(a(t)-C_o w(t)\big)+d\big(\dot{a}(t)-C_o\dot{w}(t)\big)=0,$$
$$\rho\ddot{w}(t)+A_o w(t)+D\dot{w}(t)+kC_o^*\big(C_o w(t)-a(t)\big)+dC_o^*\big(C_o\dot{w}(t)-\dot{a}(t)\big)=B_o u(t)$$
$$y(t)=\alpha(C_o w(t)-a(t)),$$

where C_o indicates position measurement at a point and α is a parameter. The observation operator is now bounded on the natural (energy) state-space $\mathscr{X}\times\mathbb{R}^2$.

18.4 Actuator and Sensor Location

For DPS, the locations of the control hardware, the actuators and sensors, are design variables. Performance depends on these locations; see for example Fig. 18.2. The best locations are often different from those that would be predicted from immediate physical intuition, see [2, 7, 15, 16].

A common approach is to place actuators at locations that maximize controllability (and place sensors to maximize observability) in some sense; see for example [21, 24], the review articles [10, 14, 25] and the books [13, 23]. However, the objective of a controller design is generally some other objective such as minimizing response to disturbances. Furthermore, it is very rare that the system needs to reach all points in the state space. Minimizing the energy to reach a state that is never a target is generally less useful than for instance, disturbance rejection or reducing settling time. Points of maximum controllability are generally not optimal with respect to a control objective; see Fig. 18.3 and [27] for more detail. Also, there are numerical difficulties associated with the fact that the PDE model is at best approximately, but not exactly controllable [27].

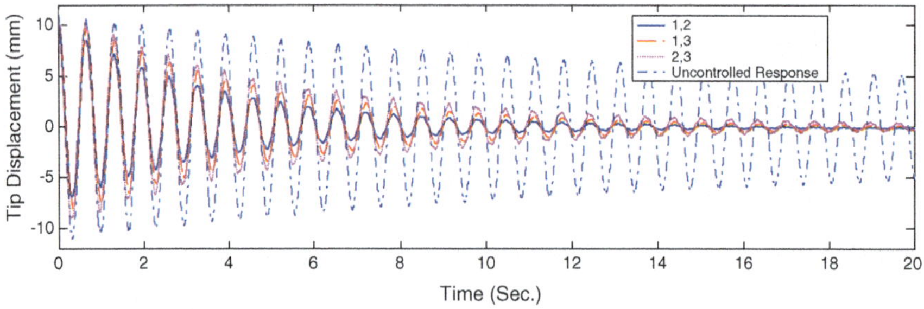

Fig. 18.2 Experimental results for use of 2 piezoelectric actuators in linear-quadratic control of a cantilevered beam. The same estimator was used in all cases; the controller is optimal and different for each pair of locations. Performance is strongly dependent on the actuator location. Reproduced with permission from ©2013 IOP Publishing Ltd [7]

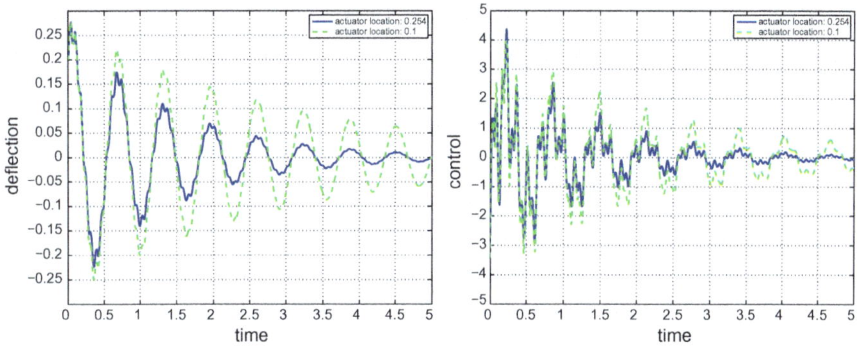

(a) beam deflection at centre (b) control signal

Fig. 18.3 LQ-optimal control of a simply supported beam ($Q = I$, $R = 1$). The location $x = 0.1$ optimizes controllability (green); $x = 0.254$ optimizes LQ-cost (blue). The response with the best LQ control for each location is shown. Both deflection and control signal are smaller for an actuator placed at the LQ-optimal actuator location than at the location of optimal controllability. Source: Figure used from the published article https://www.sciencedirect.com/science/article/pii/S0022460X15003892

An alternative approach is to place the actuators using the same criterion used to design the controller and to similarly place sensors using the same criterion used for estimator design. This can lead to considerably better performance, as illustrated in Fig. 18.3. Integration of actuator placement with controller design was first considered in [8, 9] with a linear-quadratic cost. Results have been obtained for linear quadratic, $\mathscr{H}_\infty$ and $\mathscr{H}_2$ cost functions for optimal actuator location [15–17] and for minimum variance optimal sensor location [26, 28].

18.5 Summary

Determining the correct boundary conditions is an aspect of modelling that arises in DPS and not in lumped parameter systems. Boundary conditions have a fundamental effect on the dynamics of the system; they affect the eigenvalues of the generator A and hence stability. This was illustrated here with the heat equation and also acoustic waves in a duct.

Modelling of sensing and actuation can determine whether or not the resulting model is well-posed. A well-posed model is one for which there is a unique solution for every initial condition and control signal (in a given class) and furthermore, this solution depends continuously on the initial condition and control. Well-posedness is not straightforward for DPS. Even the simplest model, such as the heat equation with Dirichlet control can fail to be well-posed. Acceleration measurement of vibrations is a more complex example of the same point.

The model of the actuators (and sensors) can also affect the nature of the control operator B, in particular whether or not it is bounded into the state-space. This affects not only well-posedness but also observability and stabilizability; see [6, 18, 22].

The location of control hardware is another issue that arises in control and estimation of DPS. Actuator location can have an effect on control system performance comparable to that of using control; sensor location has a similar effect on estimator performance. Furthermore, due to advances in materials such as piezo-ceramics and shape memory alloys the shape of the hardware can also be a design variable. This leads to complex mathematical and computational issues that are being explored.

References

1. Banks, H.T., Morris, K.A.: Input-output stability of accelerometer control systems. Control Theory Adv. Technol. **10**(1), 1–17 (1994)
2. Chen, K.K., Rowley, C.W.: H_2-optimal actuator and sensor placement in the linearised complex Ginzburg-Landau system. J. Fluid Mech. **681**(241–260) (2011)
3. Cheng, A., Morris, K.A.: Well-posedness of boundary control systems. SIAM J. Control Optim. **42**(4), 1244–1265 (2003)
4. Curtain, R.F., Morris, K.A.: Transfer functions of distributed parameter systems: a tutorial. Automatica **45**(5), 1101–1116 (2009)
5. Curtain, R.F., Weiss, G.: Well-posedness of triples of operators (in the sense of linear systems theory. In: Control and Estimation of Distributed Parameter Systems. International Series of Numerical Mathematics, vol. 91, pp. 41–59. Birkhäuser (1989)
6. Curtain, R.F., Zwart, H.: An Introduction to Infinite-Dimensional Linear Systems Theory. Springer, Berlin (1995)
7. Darivandi, N., Morris, K., Khajepour, A.: An algorithm for LQ-optimal actuator location. Smart Mater. Struct. **22**(3), 035001 (2013)
8. Demetriou, M.A.: Numerical investigation on optimal actuator/sensor location of parabolic pde's. In: Proceedings of the American Control Conference, vol. 3, pp. 1722–1726. IEEE, San Diego, CA, USA (1999) (Location-parametrization performance index)
9. Fahroo, F., Demetriou, M.A.: Optimal actuator/sensor location for active noise regulator and tracking control problems. J. Comp. Appl. Math. **114**(1), 137–158 (2000)
10. Frecker, M.I.: Recent advances in optimization of smart structures and actuators. J. Intell. Mater. Syst. Struct. **14**(4–5), 207–216 (2003)
11. Guenther, R.B., Lee, J.W.: Partial Differential Equations of Mathematical Physics and Integral Equations. Prentice-Hall (1988)
12. Jacob, B., Morris, K.A.: Second-order systems with acceleration measurements. IEEE Trans. Autom. Control **57**, 690–700 (2012)
13. El Jai, A., Pritchard, A.J.: Sensors and Controls in the Analysis of Distributed Systems. Halsted Press (1988)
14. Kubrusly, C.S., Malebranche, H.: Sensors and controllers location in distributed systems—a survey. Automatica **21**(2), 117–128 (1985)
15. Morris, K.A.: H_∞-optimal actuator location. IEEE Trans. Autom. Control **58**(10), 2522–2535 (2013)
16. Morris, K.A.: Linear quadratic optimal actuator location. IEEE Trans. Autom. Control **56**, 113–124 (2011)
17. Morris, K.A., Demetriou, M.A., Yang, S.D.: Using H_2-control performance metrics for infinite-dimensional systems. IEEE Trans. Autom. Control **60**(2), 450–462 (2015)
18. Morris, K.A., Ozer, A.O.: Modeling and stabilizability of voltage-actuated piezoelectric beams with magnetic effects. SIAM J. Control Optim. submitted (2013)

19. Pazy, A.: Semigroups of Linear Operators and Applications to Partial Differential Equations. Springer (1983)
20. Pierce, A.D.: Acoustics: An Introduction to Its Physical Principles and Applications. McGraw-Hill (1981)
21. Privat, Y., Trélat, E., Zuazua, E.: Optimal observation of the one-dimensional wave equation. J. Fourier Anal. Appl. **19**(3), 514–544 (2013)
22. Tucsnak, M., Weiss, G.: Observation and Control for Operator Semigroups. Birkhauser (2009)
23. Uciński, D.: Optimal Measurement Methods for Distributed Parameter System Identification. Systems and Control Series. CRC Press (2005)
24. Vaidya, U., Rajaram, R., Dasgupta, S.: Actuator and sensor placement in linear advection PDE with building system application. J. Math. Anal. Appl. **394**(1), 213–224 (2012)
25. van de Wal, M., de Jager, B.: A review of methods for input/output selection. Automatica **37**, 487–510 (2001)
26. Wu, X., Jacob, B., Elbern, H.: Optimal control and observation locations for time-varying systems on a finite-time horizon. SIAM J. Control Optim. **54**(1), 291–316 (2015)
27. Yang, S.D., Morris, K.A.: Comparision of actuator placement criteria for control of beam vibrations. J. Sound Vib. **353**, 1–18 (2015)
28. Zhang, M., Morris, K.A.: Sensor choice for minimum error variance estimation. under review
29. Zimmer, B.J., Lipshitz, S.P., Morris, K.A., Vanderkooy, J., Obasi, E.E.: An improved acoustic model for active noise control in a duct. ASME J Dyn. Syst. Meas. Contr. **125**(3), 382–395 (2003)

Chapter 19
Privacy in Networks of Interacting Agents

H. Vincent Poor

Abstract Many applications involve networks of interacting agents, each of whom is interested in making an individual inference or decision, the performance of which can be enhanced by the exchange of information with other agents. Though such agents can clearly benefit from exchanging information, they may also wish to maintain a degree of privacy in that information exchange. Such situations give rise to a notion of competitive privacy, which can be explored through a combination of information theory and game theory. In particular, information theory can be used to characterize the trade-off between privacy of data and the usefulness of that data for an individual agent, while game theory can be used to model the interactions between multiple agents each of whom is mindful of that trade-off. These ideas are explored in this chapter, first in a general setting, and then particularly in the context of data exchange for distributed state estimation, in which specific solutions can be obtained. Interesting open issues and other potential applications will also be discussed. Much of this abstract was originally used as the abstract of the author's academic keynote address to the Workshop on Advances in Network Localization and Navigation in London in 2015 (http://anln.spsc.tugraz.at/Program2015).

19.1 Introduction

Consider the situation depicted in Fig. 19.1, in which each of two agents measures a (generally noisy) function of two state sequences, $X_1^n = X_{1,1}, \ldots, X_{1,n}$ and $X_2^n = X_{2,1}, \ldots, X_{2,n}$. Agent 1 is interested in estimating X_1^n, and agent 2 is interested in estimating X_2^n. Since each agent has measurements relating to both state sequences, the agents can clearly benefit from sharing measurements. However, if these agents

H. V. Poor (✉)
Department of Electrical Engineering, Princeton University, Princeton, NJ, USA
e-mail: poor@princeton.edu

R. Tempo et al. (eds.), *Emerging Applications of Control and Systems Theory*, Lecture Notes in Control and Information Sciences - Proceedings,
https://doi.org/10.1007/978-3-319-67068-3_19

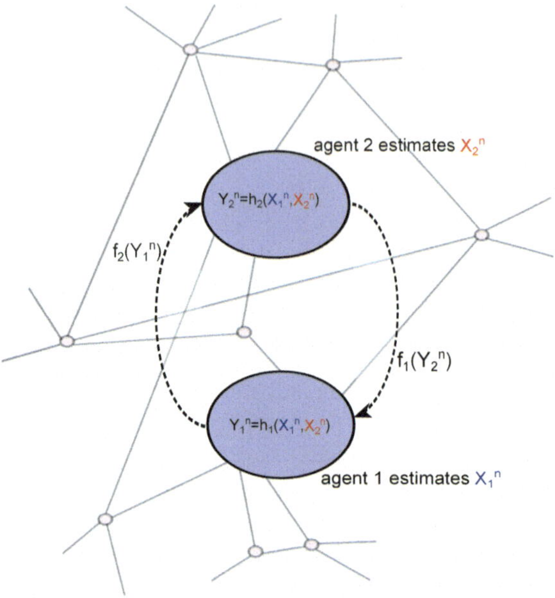

Fig. 19.1 Two interacting agents. (Used with permission from © IEEE [1].)

are competitors in some way, they may also wish to preserve the privacy of their own states to the extent possible.

Such situations can arise in a variety of applications. For example, the agents might be electric utilities, each of which would like to estimate the state of the electricity grid in its own business area. The grid couples these utilities, and so they can benefit from sharing data, while they would naturally like to maintain as much privacy as possible due to their competitive posture. Alternatively, the agents might be two companies exploring for mineral resources; each would like to get the most accurate picture possible of the resources they are seeking, without helping their competitors more than is absolutely necessary. Or, as a third example, the agents might be untrustworthy allies, each with sensors (e.g., radars) that can help pinpoint a common enemy. They can each benefit from information sharing, but would do so only with caution.

A natural question that arises in such situations is, how can these agents trade off these conflicting concerns, namely, maintaining the privacy of their own states versus the utility gained in state estimation by sharing measurements? In this chapter, we will address this question. We will begin, in Sect. 19.2, with a discussion of a general information theoretic formalism for characterizing the trade-off between privacy and utility in information systems. Then, in Sect. 19.3, we will consider specifically the problem of state estimation in competitive situations, showing in a basic linear- Gaussian model that the optimal mechanism for information exchange in such situations follows straightforwardly from classical information theoretic results. However, knowing this optimal mechanism does not specify the degree to which agents should be motivated to exchange information, a problem that we address using game theory in Sect. 19.4. Finally, in Sect. 19.5, we provide some concluding remarks, including a discussion of some other applications in which similar considerations arise.

19.2 Privacy-Utility Trade-offs

In this section, we describe briefly a general framework for examining the type of privacy-utility trade-off described in the preceding section. To motivate this discussion, consider a source of data, such as sensor measurements or an administrative database, collected for some practical purpose. Publishing the data openly would assure that it is maximally useful for that purpose. On the other hand, the data may contain information that should be kept private, and so publishing it openly would also leak this private information. One way of maintaining privacy would be to simply store the data somewhere in a vault and not let it be accessed by anyone. This, of course, would prohibit the data being used for its intended purpose. In reality, there should be some middle ground between these two extremes of openly publishing the data and hiding it away, in which a reasonable degree of privacy is sacrificed in the interest of doing something useful with the data. That is, there is a fundamental dichotomy here between the privacy and utility of data—the so-called *privacy-utility trade-off*. And, a basic question in designing and operating information systems is, how can we characterize this fundamental trade-off? One way of addressing this question is to examine this issue in an information theoretic setting.

To consider this problem, we can think of the data source as a random process consisting of two types of variables, public and private, which in general may be correlated or dependent. The public ones can be revealed without any privacy concerns, while the private ones should be kept hidden to the extent possible. These two types of variables might not necessarily be disjoint, i.e., there could be overlap between the two types. For example, a financial database might contain 16-digit credit card numbers, which should be kept private in full, while the last four digits might be revealed for various purposes without concern. But even if there is no overlap, revealing only the public data can leak information about the private data because of the dependence between them. So, there will be a trade-off due to this dependence, even without overlap between public and private variables.

Given the above statistical model, how can we characterize the trade- off between utility and privacy? We can first consider utility. Clearly, if we are going to protect aspects of the data, any publication of the data will not reveal the entire data source. This means that there will be distortion between the true data source and the data source as revealed to users. So, we can measure utility in terms of this distortion introduced in the public variables in whatever information is revealed to a user of the data. This will be an inverse measure of utility since low distortion means higher utility. Distortion could be measured, for example, in terms of the mean square error between the public variables and the best reconstructions of those variables from revealed data. Similarly, we can measure privacy in terms of the information leaked about the private variables in information revealed to a user.[1] For example, we might consider the entropy of the private variables conditioned on information revealed to a user, known as the *equivocation*. This is a direct measure of privacy:

[1]For other ways of characterizing privacy in data sources, see, e.g., [3, 6, 7, 9, 15].

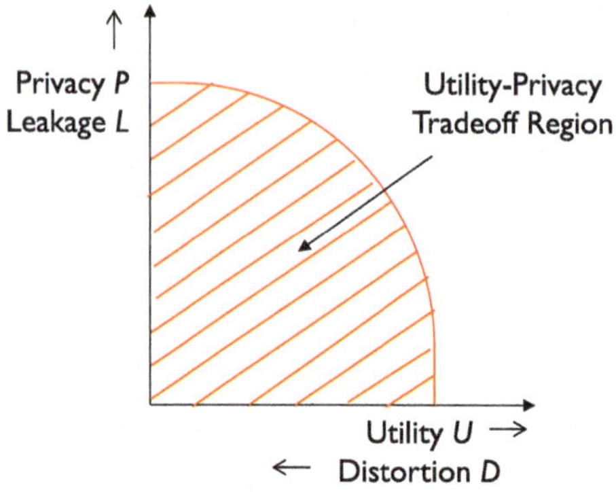

Fig. 19.2 Privacy-utility trade-off region

higher equivocation means greater privacy. And, if the equivocation were to equal the unconditioned entropy of the private variables, then there would be no privacy leakage at all. Other information theoretic measures of information leakage can be used as well.

Once we specify measures of distortion and information leakage and a statistical model for the data source, we thus have a clearly defined mathematical problem to solve. Namely, we would like to find all the possible distortion-leakage pairs for that particular statistical model, as illustrated in Fig. 19.2. Having determined that region, its boundary will give us the *efficient frontier* in the privacy-utility trade-off. This frontier specifies the minimal amount of privacy that must be sacrificed for a given level of utility, or conversely, the maximal utility that can be gained within a specific constraint on privacy leakage. As discussed in [14], this problem can be cast within the framework of a secure source coding problem of Yamamoto [19], which can facilitate its solution. Specific examples of the application of these ideas in various settings can be found in [13, 14, 16, 17, 20].

19.3 Competitive Privacy: A Basic Model

We now turn to a variation on the privacy-utility trade-off in which there are multiple interacting parties, each of which has its own trade-off while interactions among the parties affect these individual trade-offs.

Again, we consider the situation illustrated in Fig. 19.1, in which each of multiple agents is individually interested in estimating its own state, while keeping it as private as possible from competing agents. More reliable state estimates can be obtained if the agents exchange information with one another, but there may be economic competition or other reasons for keeping such measurements private. So this gives rise to a privacy-utility trade-off, but in a competitive setting. This problem is called *competitive privacy*.

To develop some insight into such situations, we can consider a simple model in which each agent's state is summarized in a single scalar, and the agents observe the states through a noisy linear model. That is, the mth of M agents has a state X_m, and each agent measures a noisy linear combination of those states, denoted by Y_j for the jth agent. So, we have a linear measurement model

$$Y_j = \sum_{m=1}^{M} H_{j,m} X_m + Z_j, \; j = 1, 2, \ldots, M, \tag{19.1}$$

where the $H_{j,m}$'s are the coefficients in the linear model and $\{Z_j\}$ represents measurement noise.

The exchange of measurements among agents will lead to inevitable leakage of state information, although all agents already have some knowledge of all states through the measurements (19.1). We can examine this problem in the privacy-utility trade-off framework by defining utility and privacy measures for each agent. A natural utility measure for the jth agent is mean square error occurred in estimating its own state, X_j, while privacy for the jth agent can be measured in terms of leakage of information about its own state to other agents.

It is illuminating to focus specifically on the case of two agents, and assume that each agent has n independent and identically distributed (i.i.d.) observations of the model (19.1), i.e.,

$$\begin{aligned} Y_{1,i} &= X_{1,i} + \alpha X_{2,i} + Z_{1,i}, i = 1, \ldots, n \\ Y_{2,i} &= \beta X_{1,i} + X_{2,i} + Z_{2,i}, i = 1, \ldots, n, \end{aligned} \tag{19.2}$$

where we assume that $\{X_{1,i}\}$, $\{X_{2,i}\}$, $\{Z_{1,i}\}$ and $\{Z_{2,i}\}$ are mutually independent, with the $X_{j,i}$'s i.i.d. $\mathcal{N}(0, 1)$ and the $Z_{j,i}$'s i.i.d. $\mathcal{N}(0, \sigma_j)$. Thus, in this model we have four parameters: the two "channel gains", α and β, and the two noise variances, σ_1^2 and σ_2^2.

Within this model, we can straightforwardly consider the trade-off between utility and privacy leakage described above. Each agent $j \in \{1, 2\}$ measures its utility as follows:

$$D_j = \frac{1}{n} \sum_{i=1}^{n} E\left[\left(X_{j,i} - \hat{X}_{j,i}\right)^2\right], \tag{19.3}$$

where $\hat{X}_{j,i}$ is agent j's estimate of $X_{j,i}$, its own state at time i. Furthermore, each agent j measure its privacy leakage as follows:

$$L_j = \frac{1}{n} I\left(X_j^n; f_{3-j}(Y_j^n), Y_{3-j}^n\right), \tag{19.4}$$

where $I(\cdot;\cdot)$ denotes mutual information, $X_j^n = X_{j,1}, \ldots, X_{j,n}$ (and similarly for Y_{3-j}^n) and $f_{3-j}(Y_j^n)$ denotes the information transferred from agent j to its counterpart agent $3 - j$, as illustrated in Fig. 19.1.

As a first step in understanding this problem, we can state the following result from [12] that characterizes optimal information exchange between the agents as follows:

Theorem *Wyner–Ziv coding maximizes privacy (i.e., minimizes L_1 and L_2) for a fixed level of utility at each agent (i.e., fixed D_1 and D_2).*

Recall that Wyner–Ziv coding [2, 18] refers to optimal distributed source coding of correlated sources, and thus that it minimizes the rate of information transfer (i.e., privacy leakage) for fixed levels of distortion in (19.2) should not come as a surprise.

So the optimal way to exchange information is through Wyner–Ziv coding. That is, this is the most efficient way of trading information in terms of providing the most reduction in mean square distortion for a given level of privacy leakage. Or, alternatively, it maximizes privacy for any desired fixed distortion levels. In other words, wherever the efficient frontier lies, Wyner–Ziv coding will be the way to implement information exchange.

This result does not solve the problem, however, because the leakage of one agent depends on the distortion of its counterpart, not on its own distortion. So it is not clear how the agents should behave; if an agent gives information to its counterpart it only helps the counterpart, unless there is a quid pro quo transfer of information from the counterpart. This suggests that this problem can be profitably viewed in the context of game theory, an approach explored in [1], and which is described in the following section.

19.4 A Game Theoretic Model

To impose a game theoretic framework on this problem we need to establish a set of actions and a payoff function for each agent. Of course, the actions of an agent involve the transfer of information to its counterpart. Such actions can be characterized straightforwardly using the property that, in the linear-Gaussian model of (19.1), the rate privacy leakage of one agent as a function of the corresponding distortion experienced by the counterpart agent is monotonically decreasing. So, we can in fact think about the action of a given agent in terms of how much distortion it inflicts on its counterpart agent; i.e., an action a_j of agent $j \in \{1, 2\}$ can be specified by D_{3-j}. We will assume that there is a maximal level of distortion for each agent, say $\bar{D}_j$ for agent j, and that each agent must release at least enough information to its counterpart so that the distortion achieved by the counterpart is at most this maximal level. (Such a requirement might be imposed by utility regulators, say, in the power grid example mentioned in Sect. 19.1.)

A reasonable payoff for agent j, which of course depends on both its own action and that of its counterpart, is the following:

$$u_j(a_j, a_{3-j}) = -L(a_j) + \frac{q_j}{2} \log\left(\frac{\bar{D}_j}{a_{3-j}}\right), \tag{19.5}$$

where $L(a_j)$ is agent j's privacy leakage (19.4) due to action a_j, which is penalized in the payoff; $\log\left(\frac{\bar{D}_j}{a_{3-j}}\right)$ is a logarithmic payoff that accrues to agent j when its distortion is lower than the maximum $\bar{D}_j$—it represents the information rate of the data received from the other agent; and $q_j > 0$ is a weighting factor that balances the importance of these two components. The properties of games with these payoff functions are examined in [1], as summarized in the following paragraphs.

It can be shown that the Nash game with the payoff (19.5) leads to a classical prisoner's dilemma, in which there is no incentive for either agent to share any information beyond what minimal amount is necessary to achieve the maximal allowed distortion at its counterpart. That is, the only Nash equilibrium of the game is achieved at $(a_1, a_2) = (\bar{D}_2, \bar{D}_1)$.

Clearly, other incentives are necessary if greater information exchange is desired. One way to provide such incentives is to encourage quid pro quo behavior by considering a multistage game, in which the game is repeated over multiple time periods. So, what an agent does for its counterpart today might effect what the counterpart does for tomorrow. We can examine this possibility by looking at a T -stage game with $T > 1$, in which the payoff is given by

$$\sum_{t=1}^{T} \rho^{t-1} u_j\left(a_j^{(t)}, a_{3-j}^{(t)}\right) \tag{19.6}$$

where $a_j^{(t)}$ and $a_{3-j}^{(t)}$ are the actions taken by the two parties at time t, and ρ is a discount or "forgetting" factor. Again, however, with finite T this problem is a prisoner's dilemma problem, in which the only Nash equilibrium[2] is $\left(a_1^{(t)}, a_2^{(t)}\right) = (\bar{D}_2, \bar{D}_1), \forall t$. On the other hand, if T is infinite, i.e., when the game will be played indefinitely, there are nontrivial Nash equilibria. In particular, for sufficiently large $\rho < 1$, any pair of strategies $\left(a_1^{(t)}, a_2^{(t)}\right) = (D_2^*, D_1^*), \forall t$, for which

$$u_j(D_j^*, D_{3-j}^*) > u_j(\bar{D}_j, \bar{D}_{3-j}), \;\; j = 1, 2, \tag{19.7}$$

is a (subgame perfect) Nash equilibrium. Figure 19.3 illustrates the range of ρ for which various action pairs are equilibria in a specific example.

Another type of incentive that might be applied is pricing. For example, the payoff of (19.5) could be replaced by

$$\tilde{u}_j(a_j, a_{3-j}) = u_j(a_j, a_{3-j}) + p_j \log\left(\frac{\bar{D}_{3-j}}{a_j}\right), \tag{19.8}$$

[2]In this multiparty game, the equilibrium of interest is a subgame perfect equilibrium, which essentially is an equilibrium for every subset of time periods [4].

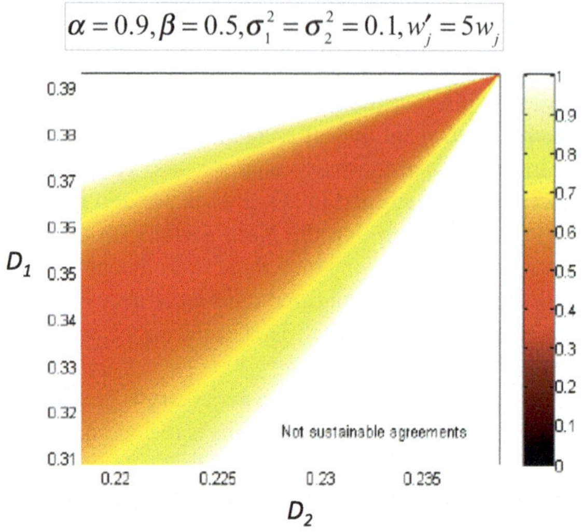

Fig. 19.3 Minimal discount factors for sustaining an equilibrium. (Used with permission from © IEEE [1].)

where p_j is a price paid to agent j for improving its counterpart's distortion. With this payoff, essentially any desired behavior can be incentivized by choosing appropriate values of p_1 and p_2, as one might expect. Other nontrivial cooperative behaviors can also be induced through a common payoff function (i.e., a *common goal* game), such as

$$u_{sys}(a_1, a_2) = -L(a_1) - L(a_2) + \frac{q}{2} \log \left(\frac{\bar{D}_1 + \bar{D}_2}{a_1 + a_2} \right), \tag{19.9}$$

where, again, $q > 0$ is weighting factor. This particular payoff gives rise to a so-called *potential game* [8], and various nontrivial equilibria can be achieved, depending on the value of q. (See [1] for details.) Depending on the application, such incentives might be imposed by regulation, or in the case of pricing by a market structure.

19.5 Conclusion

In this chapter, we have considered the trade-off between privacy and utility associated with data exchange among multiple interacting agents. We have seen that information theoretic and game theoretic principles can be combined to provide insights into this problem. In particular, using a simple two-agent model with Gaussian states and measurement noise, we have seen that Wyner–Ziv coding provides the optimal means of information exchange among the parties. This result tells us how to exchange information, but not how much information to exchange. To determine how much information to exchange, game theory is useful. In this context, we have examined several types of games—single-stage games (with and without pricing, common goal) and multistage games with finite and infinite horizons—seeing that the range of equilibria varies considerably among these types of games. In particular, the one shot and finite multi-stage games without pricing are basically prisoner's dilemma problems, while the common goal and infinite multistage games yield more interesting behavior.

This work suggests many open areas for future research. Extending the analysis to larger numbers of agents M is of particular interest, including examining large-M asymptotics and introducing the possibility of coalition formation [10]. Also, considering more general problems than state estimation and more general models than the simple noisy linear model treated here is of interest, as well as examining the human element through prospect theoretic analysis [5, 11]. Note that these ideas have a number of potential applications, some of which were noted in Sect. 19.1. Examples of other potential settings in which competitive privacy issues arise naturally include social networking, e-commerce, and online gaming.

Acknowledgements This work was prepared while the author was visiting Stanford University under the support of the Precourt Institute for Energy. The support of the U. S. National Science Foundation under Grants ECCS-1549881 and ECCS-1647198 is also gratefully acknowledged.

References

1. Belmega, E.V., Sankar, L., Poor, H.V.: Enabling data exchange in interactive state estimation under privacy constraints. IEEE J. Sel. Top. Signal Process. **9**, 1285–1297 (2015)
2. Berger, T.: Multiterminal source coding. In: Longo, G. (ed.) Information Theory Approach to Communications. Springer, New York (1978)
3. Dwork, C., Roth, A.: The algorithmic foundations of differential privacy. Found. Trends Comput. Sci. **9**, 211–407 (2014)
4. Fudenberg, D., Tirole, J.: Game Theory. MIT Press, Cambridge, MA (1991)
5. Kahneman, D., Tversky, A.: Prospect theory: an analysis of decision under risk. Econometrica **47**, 263–291 (1979)
6. Kargupta, H., Datta, S., Wang, Q., Sivakumar, K.: Random data perturbation techniques and privacy preserving datamining. J. Knowl. Inf. Syst. **7**, 387–414 (2005)
7. Machanavajjhala, A., Kifer, D., Gehrke, J., Venkitasubramaniam, M.: L-diversity: privacy beyond k-anonymity. ACM Trans. Knowl. Discov. Data **1**, 24 (2007)
8. Monderer, D., Shapley, L.S.: Potential games. Games Econ. Behav. **14**, 124–143 (1996)
9. Raghunathan, T.E., Reiter, J.P., Rubin, D.B.: Multiple imputation for statistical disclosure limitation. IEEE Trans. Inf. Theory **43**, 1877–1894 (1997)
10. Ray, D.: A Game-Theoretic Perspective on Coalition Formation. Oxford University Press, Oxford, UK (2007)
11. Saad, W., Glass, A., Mandayam, N., Poor, H.V.: Towards a consumer-centric grid: a behavioral perspective. Proc. IEEE **104**, 865–882 (2016)
12. Sankar, L., Kar, S.K., Tandon, R., Poor, H.V.: Competitive privacy in the smart grid: an information-theoretic approach. In: Proceedings IEEE International Conference Smart Grid Communication. IEEE, Piscataway, NJ (2011)
13. Sankar, L., Rajagopalan, S.R., Mohajer, S., Poor, H.V.: Smart meter privacy: a theoretical framework. IEEE Trans. Smart Grid **4**, 837–846 (2013)
14. Sankar, L., Rajagopalan, S.R., Poor, H.V.: Utility-privacy tradeoffs in databases: an information-theoretic approach. IEEE Trans. Inf. Forensics Secur. **8**, 838–852 (2013)
15. Sweeney, L.: k-anonymity: a model for protecting privacy. Int. J. Uncertain. Fuzziness Knowl. Syst. **10**, 557–570 (2002)

16. Tan, O., Gündüz, D., Poor, H.V.: Increasing smart meter privacy through energy harvesting and storage devices. IEEE J. Sel. Areas Commun. **31**, 1331–1341 (2013)
17. Tandon, R., Sankar, L., Poor, H.V.: Discriminatory lossy source coding: side information privacy. IEEE Trans. Inf. Theory **59**, 5665–5677 (2013)
18. Wyner, A.D., Ziv, J.: The rate-distortion function for source coding with side information at the decoder. IEEE Trans. Inf. Theory IT **22**, 1–10 (1976)
19. Yamamoto, H.: A source coding problem for sources with additional outputs to keep secret from the receiver or wiretappers. IEEE Trans. Inf. Theory **29**, 918–923 (1983)
20. Yang, L., Chen, X., Zhang, J., Poor, H.V.: Cost-effective and privacy-preserving energy management for smart meters. IEEE Trans. Smart Grid **6**, 486–495 (2015)

Chapter 20
Excitable Behaviors

Rodolphe Sepulchre, Guillaume Drion and Alessio Franci

Abstract This chapter revisits the concept of excitability, a basic system property of neurons. The focus is on excitable systems regarded as behaviors rather than dynamical systems. By this, we mean open systems modulated by specific interconnection properties rather than closed systems classified by their parameter ranges. Modeling, analysis, and synthesis questions can be formulated in the classical language of circuit theory. The input–output characterization of excitability is in terms of the local sensitivity of the current-voltage relationship. It suggests the formulation of novel questions for nonlinear system theory, inspired by questions from experimental neurophysiology.

20.1 Introduction

In his 1996 survey paper [1], George Zames credits Charles Desoer and Mathukumalli Vidyasagar for writing the *ultimate text* on input–output theory of nonlinear feedback systems. This textbook was largely inspired by the engineering question of analyzing and designing nonlinear electrical circuits, a popular topic at the time. In the last decades of the century, the dominant driving application of nonlinear control theory moved from electrical circuits to robotics. The present chapter is a tribute to one of the pioneers of the input–output theory of nonlinear feedback systems. It is entirely motivated by a nonlinear electrical circuit model published in 1952 to explain the biophysical foundation of nerve excitability. The landmark paper of

R. Sepulchre (✉)
Department of Engineering, University of Cambridge, Cambridge, UK
e-mail: rs771@cam.ac.uk

G. Drion
Institut Montefiore, Université de Liege, Liege, Belgium
e-mail: g.drion@ulg.ac.be

A. Franci
Department of Mathematics, Universidad Nacional Autónoma de México, Mexico City, México
e-mail: afranci@ciencias.unam.mx

R. Tempo et al. (eds.), *Emerging Applications of Control and Systems Theory*, Lecture Notes in Control and Information Sciences - Proceedings,
https://doi.org/10.1007/978-3-319-67068-3_20

Hodgkin and Huxley [2] defined circuit theory as the modeling language of neurophysiology. Most today's questions of experimental neurophysiologists are still very naturally formulated in the language of circuit theory. But the computational push for neurophysiology *in silico* has progressively favored the replacement of circuit models by their state-space representations, in the form of high-dimensional models of nonlinear differential equations. The growing ease at simulating those state-space models on a personal computer is only matched by the increasing difficulty to analyse them and to resolve their inherent fragility. The difficulty of translating robustness and sensitivity questions from input–output models to state-space models is familiar to control theorists. Bridging the two worlds has been at the core of the developments of linear system theory. But progresses in the nonlinear world have been slow and limited. This bottleneck is severely restricting the possibility to analyze neuronal circuits with the tools of nonlinear state-space theory. At a broader level, this bottleneck is contributing to the gap that separates experimental neurophysiology from computational neuroscience.

The discussion of excitability in this chapter is an attempt to revisit one of the most basic properties of biological systems in the classical language of nonlinear circuit theory. The discussion complements the presentation of excitability found in textbooks of neurodynamics. The experience of the authors in their recent work on neuromodulation [3–5] suggests that there is value in reopening the ultimate text of input–output nonlinear feedback systems to model excitability. We stress the importance of *localized ultrasensitivity*, a defining feature of excitability that singles out a highly specific property of excitable behaviors. This property is tractable because it is amenable to local analysis. It should be acknowledged in any system theory of excitability.

20.2 What Is Excitability?

Excitability is the property of a system to exhibit all-or-none response to pulse inputs. It is defined at a stable equilibrium, meaning that small inputs cause small outputs. But beyond a given threshold, the response is a large and stereotyped output. This property is primarily observed in neurons, muscle cells, and endocrine cells, where it refers to an *electrical* phenomenon: the input is current, and the output is voltage. The large stereotyped output observed in response to a current stimulus is called an action potential, or a spike. Excitability is central to physiological signaling. Ultimately, it is instrumental in converting sensory signals into motor actions. Not surprisingly, excitability is usually modeled in the language of circuit theory (Fig. 20.1).

We regard excitability as a *behavioral* property in the sense of Willems [6]. An excitable behavior is the set of trajectories $(I(t), V(t))$ of a one-port electrical circuit. Those trajectories are those that are observed by an electrophysiologist; trains of pulses for the current, and trains of spikes for the voltage. A behavioral theory of excitable systems is about modeling the relationship between them with the aim of addressing questions that are system theoretic in nature: control (what are the

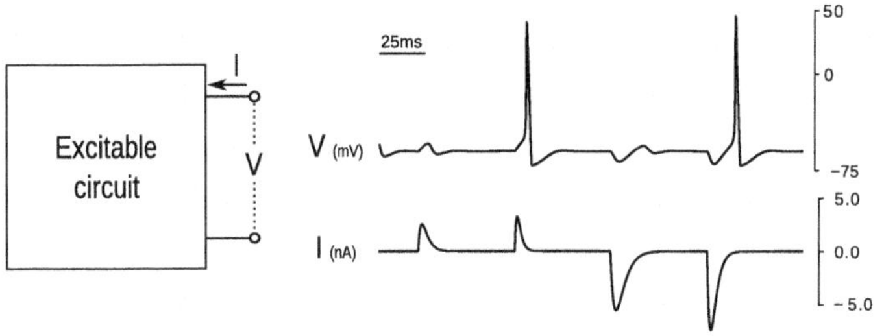

Fig. 20.1 An excitable behavior is a set of current pulses and voltage spikes defining the trajectories of a nonlinear one-port circuit

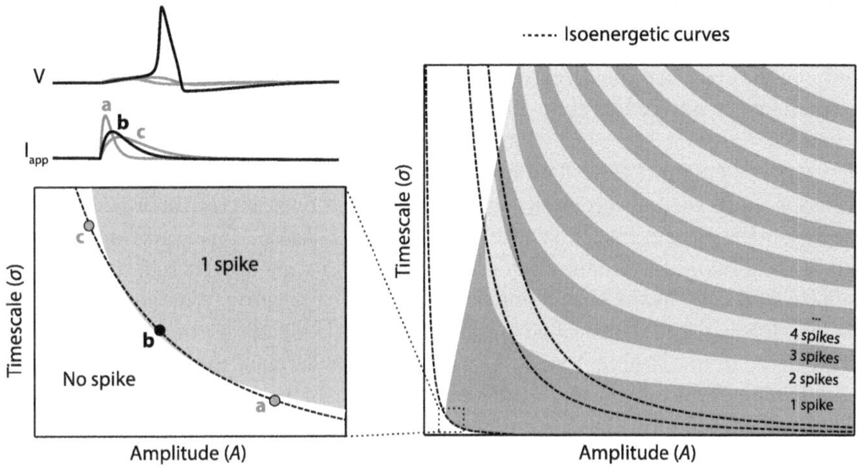

Fig. 20.2 The threshold property of an excitable circuit converts an analog pulse into a discrete number of spikes. It is localized both in amplitude and time

mechanisms that shape the behavior?), robustness (how robust is the behavior to uncertainty?), and interconnections (how to predict the behavior of the whole from the behavior of the parts?).

An excitable behavior is essentially nonlinear because of the all-or-none nature of the spike. Behavioral theory is a mature theory for linear time-invariant behaviors but a theory in its infancy for nonlinear behaviors. Figure 20.2 suggests a simple way of characterizing the excitability property of a one-port circuit, in analogy to the step response of a linear time-invariant behavior. Here we consider a pulse current trajectory parametrized by an amplitude A and a time duration σ. The figure indicates the number of spikes in the corresponding voltage trajectory.

This representation of excitability is general and model independent. It captures the fundamental quantification property of an excitable circuit. Spikes can be regarded as discrete quantities but their number continuously depends on analog properties of the current pulses. The threshold property of an excitable system is well captured by the figure. The energy threshold is the minimum amount of charge that is necessary to trigger a spike. It endows the circuit with a characteristic *scale* (A^*, σ^*), both in *amplitude* and in *time*. For a current pulse above the energy threshold, the family of pulses that can trigger a spike is *localized* both in *amplitude* and in *time*. For a fixed suprathreshold energy, pulses that are only localized in time or in amplitude do not trigger a spike. Energy levels are themselves quantified, meaning that an excitable circuit has a maximal spiking frequency.

The spike itself is a discrete quantity in Fig. 20.2 because its all-or-none nature makes it independent of the input. The input only triggers a transient excursion between an OFF state and an ON state of the circuit. The OFF state is a stable equilibrium or operating point of the circuit. The ON state is a stable limit cycle of fixed amplitude, or, less frequently, a distinct equilibrium at a significantly higher potential than the OFF state. This signature is easily identified experimentally by studying the stationary behavior of the circuit for current pulses of long duration.

There exists a large literature about the analysis of excitable models as nonlinear dynamical systems, see e.g. [7] and references therein. Assuming that the law of the excitable circuit is described by a nonlinear differential equation, an excitable behavior is regarded as a (closed) dynamical system by studying the trajectories of the dynamical system for a given (usually constant) current. Phase portrait analysis and bifurcation theory are the central analysis tools in this approach. Excitability is then characterized by the bifurcation that governs the transition from the OFF state to the ON state using the fixed value of the current as a bifurcation parameter. Different bifurcations define different types of excitability, associated to distinct phase portraits when modeled by second-order differential equations. While neurodynamics has been central to the development of mathematical physiology, it also suffers from limitations inherent to the dynamical systems approach. The mathematical classification is not always easy to reconcile with the neurophysiological (or behavioral) classification, and the complexity of the analysis rapidly grows with the dimension of the model. Questions pertaining to modulation, robustness, and interconnections are not easy to address in the framework of neurodynamics and call for complementary approaches.

20.3 The Inverse of an Excitable System

The key advance in modeling neuronal excitability came from the voltage clamp experiment, one of the first scientific applications of the feedback amplifier. The voltage clamp experiment assigns a step trajectory to the voltage of an excitable circuit by means of a high gain feedback amplifier. The required current provides the corresponding current trajectory. In the language of system theory, the current trajectory is nothing but the step response of the inverse system.

The step response in Fig. 20.3a is typical of a transfer function with a fast right-half plane zero and a slow left-half plane pole: the sign of the low frequency (or static) gain is opposite to the sign of the high frequency (or instantaneous) gain.

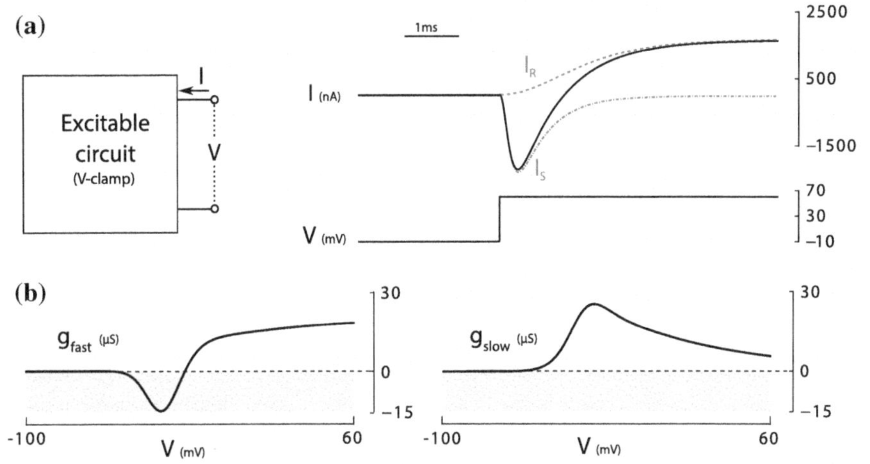

Fig. 20.3 The voltage experiment was key to modeling neuronal excitability. It provides the step response of the inverse system (**a**). Dynamic input conductances (DICs) extract key properties of the inverse system response as a function of voltage (**b**). The trajectory and DICs shown are computed from the Hodgkin Huxley model

This property led Hodgkin and Huxley to identify the distinct roles of a slow and of a fast currents (I_{early} and I_{late} in the terminology of [8]) as a key mechanism of excitability. The fast right-half plane zero of the voltage-driven circuit corresponds to a fast unstable pole of the current-driven excitable system, whereas the slow left-half plane pole corresponds to a slow left-half plane zero. Hodgkin and Huxley also observed that this essential feature of the step response is voltage-dependent. It holds for a step voltage around the resting potential, but it disappears if the step voltage is repeated around a higher potential. At higher values of the potential, the step response becomes the stable response of a slow first-order system (not shown here, but abundantly illustrated in [8]). In other words, the non-minimum phase nature of the step response shown in Fig. 20.3 only holds in a narrow voltage range.

To date, the voltage clamp experiment remains the fundamental experiment by which a neurophysiologist studies the effect of neuromodulators or the role of particular ion channels in a specific neuron. The recent paper [4] by the authors proposes that modulation and robustness properties can indeed be studied efficiently via the dynamic conductances of the neuron. Dynamic conductances extract from small step voltage clamp trajectories the *quasi-static* conductance $\frac{\Delta I}{\Delta V}$ in different timescales. Figure 20.3b shows the *fast* (g_{fast}) and *slow* (g_{slow}) dynamic conductances of Hodgkin and Huxley model. The non-minimum phase voltage step response translates into a voltage range where the fast (or instantaneous) conductance is negative, whereas the slow conductance is positive. The dynamic input conductance curves quantify the temporal and voltage dependence of the conductances. The fast dynamic conductance is negative close to the resting potential, whereas the slow dynamic conductance is positive everywhere. Those features are the essential signature of an excitable circuit. In particular, the zero crossing of the fast conductance is an excellent predictor of the threshold voltage and the fundamental signature of the *localized sensitivity* of the circuit. A small conductance means ultrasensitivity of the circuit with respect to current variations. The voltage clamp experiment identifies the temporal and amplitude window of ultrasensitivity of an excitable circuit.

20.4 A Circuit Representation and a Balance of Positive and Negative Conductances

An excitable circuit is made of three distinct elements: a passive circuit, a switch, and a regulator. Each element is itself a one port circuit and the three elements are interconnected according to Kirchoff law.

The *passive circuit* accounts for the passive properties of the excitable behavior. Its static behavior is strictly resistive, hence the monotone current-voltage (I-V) relationship. Its dynamic behavior is strictly passive, meaning that the circuit can only dissipate energy. In the language of dissipativity theory, the change in internal energy stored in the capacitor is upper-bounded by the externally supplied power [9].

The *switch* accounts for the large voltage transient of the spike. Its static behavior is characterized by a range of negative conductance. It is the destabilizing element of the excitable circuit and it requires an active source. The activation is however *localized*, meaning that the negative conductance of the switch can overcome the positive conductance of the passive circuit only within a local amplitude and temporal range.

The *regulator* accounts for the repolarization of the circuit following a spike. In particular, it ensures a refractory period following the spike, which contributes to the *all-or-none* nature of the spike and to its temporal scale: two consecutive spikes are always separated by a minimal time interval. The regulatory element is a distinct source of dissipation, that continuously modulates the balance between the positive conductance of the passive circuit and the negative conductance of the switch (Fig. 20.4).

The static model of an excitable circuit is a nonlinear resistor, characterized by its so-called $I - V$ curve. This curve can be either monotonically increasing, or hys-

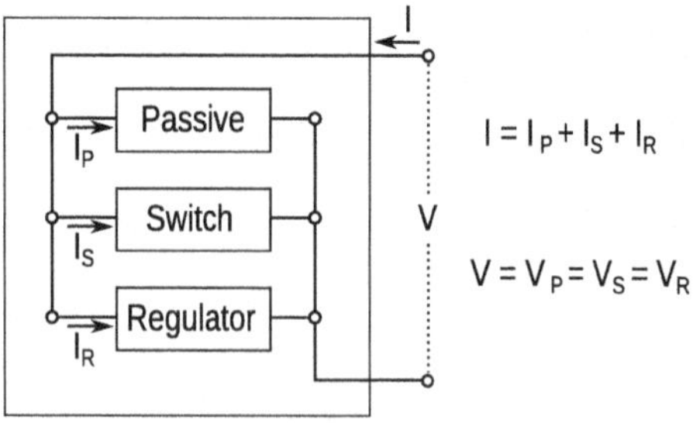

Fig. 20.4 The three circuit elements of an excitable one-port circuit

teretic if the negative resistance of the switch locally overcomes the positive resistance in static conditions. Hysteresis of the I-V curve is not necessary for excitability [5], a clear evidence that excitability is a *dynamical* phenomenon. In classical circuit theory, the dynamics of the circuit is captured by a small-signal analysis around operating points. The *admittance* of the circuit at a given operating point is the dynamic generalization of its conductance. It is the frequency response of the linearized behavior $\delta I = G(j\omega)\delta V$ around a given operating point (I, V). The admittance is a complex number that depends both on the amplitude V and of the frequency ω.

The admittance of a passive circuit is positive real, that is, its real part is nonnegative at all frequencies. The regulatory element preserves this property if it is itself passive. In contrast, the switch element creates an amplitude and frequency range where the real part of the admittance becomes negative. It is the only destabilizing element of the circuit. A clear local signature of excitability is therefore a localized amplitude and frequency range where the real part of the admittance becomes negative.

The characterization of an excitable circuit from its admittance properties is not limited to low-dimensional models amenable to phase portrait analysis. The dynamic input conductances discussed in Sect. 20.2 are snapshots of the admittance in different timescales.

20.5 A Mixed Feedback Motif and a Robust Balance Property

Due to its negative conductance, the switch of an excitable circuit acts as a source of positive feedback. Due to its positive conductance, the regulator of an excitable circuit acts a source of negative feedback. As a consequence, an excitable circuit always admits the representation of a passive system surrounded by two distinct feedback loops of opposite sign. This representation is important because it coincides with the excitatory-inhibitory (E-I) feedback motif found in many biological models. The excitatory feedback loop often models an autocatalytic process whereas the inhibitory feedback loop often corresponds to a regulatory process.

The balance between positive and negative feedback is key to the localized sensitivity of an excitable circuit. The static picture is the one of the mixed feedback amplifier illustrated in Fig. 20.5. When negative feedback dominates, the circuit is purely resistive and the behavior is analog. More negative feedback enlarges the linearity range of the circuit and decreases its input–output sensitivity. In contrast, when positive feedback dominates, the circuit is hysteretic and the behavior is quantized. More positive feedback enlarges the hysteretic range and decreases its input–output sensitivity in the OFF and ON mode. When positive and negative feedback balances each other, the circuit becomes characterized by a tiny range of ultrasensitivity.

An excitable circuit offers a versatile architecture to tune ultrasensitivity by balancing positive and negative feedback. The switch ensures a local range in time and

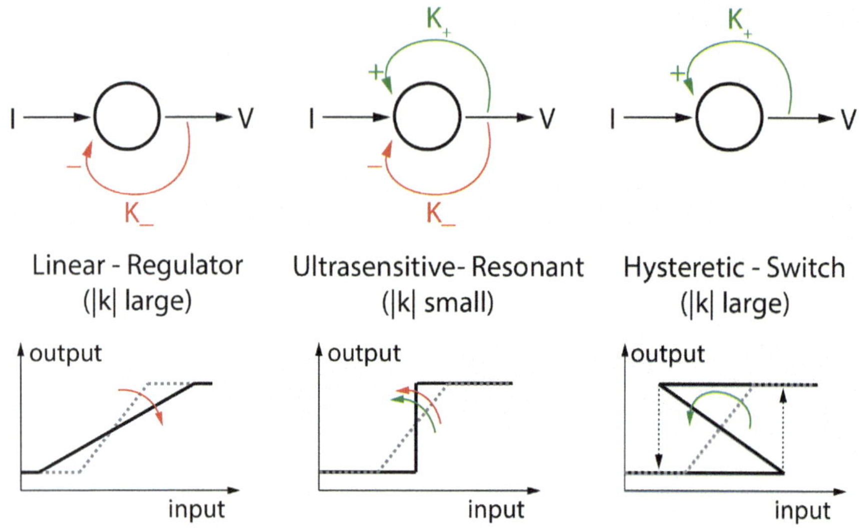

Fig. 20.5 Regulating the balance between positive and negative feedback can switch a system between linear, ultrasensitive and hysteretic states. Top, sketches of the systems composed of a negative feedback (left), a positive feedback (right), or both (center). Bottom, input/output relationships in the three cases. The dashed gray lines show the open-loop relationships, and the full black lines the closed-loop relationships

amplitude where the circuit behaves as a hysteretic switch. The regulator ensures that, away from a local range, the circuit behavior is resistive and linear. By continuity of the feedback gain, the circuit must be ultrasensitive along trajectories that connect the switch-like and the linear-like behaviors. Suprathreshold current pulses and the corresponding spikes are examples of such trajectories.

The feedback representation of an excitable circuit highlights the robustness of achieving ultrasensitivity by a balance of feedback. Ultrasensitivity must exist provided that there exists a local range in amplitude and time where positive feedback dominates negative feedback. For the admittance of the circuit along its I-V curve, this means that there must exist a voltage range and a frequency range where the negative real part of the switch admittance exceeds the positive real part of the passive admittance. For the supplied energy, this means that there must exist trajectories in a local amplitude and temporal range where the overall circuit is active, that is, the energy supplied by the switch exceeds the energy absorbed by the passive admittance.

Time-scale separation between a fast switch and a slow regulator enhances the robustness of excitability, creating a two time-scale circuit that behaves as a bistable switch in the fast timescale and as a linear resistive circuit in the slow timescale. In the spike of Fig. 20.1, the fast behavior is the upstroke, whereas the repolarization is the slow behavior. As the time-scale separation between the switch and the regulator decrease, the distinction between the "switch-like" and "linear-like" parts of an excitable behavior become progressively blurred. The localization of excitability requires a hierarchy between the positive and the negative feedback: the range where the positive feedback gain exceeds the negative must be narrow relative to the negative feedback range. The feedback motif of an excitable circuit is thus *fast and localized positive* feedback balanced by *slow and global negative* feedback.

20.6 Models of Excitability

20.6.1 FitzHugh–Nagumo Circuit

An elementary circuit fitting the requirements of Fig. 20.4 is the parallel connection of a capacitor (the passive element), a static diode with a cubic $I - V$ characteristic (the switch), and an RC branch (the regulator). This circuit was first studied by Nagumo et al. [10], following the proposal of FitzHugh [11] to study excitability through a minor modification of Van der Pol oscillator. The motivation in both papers was to extract a simple qualitative model of Hodgkin–Huxley model. FitzHugh–Nagumo model admits the state-space representation

$$\begin{aligned} C\dot{V} &= -i_s - i_r + I \\ L\dot{i}_r &= -Ri_r + V, \\ i_s &= \tfrac{V^3}{3} - kV \end{aligned} \tag{20.1}$$

Its phase portrait has been a key paradigm to explain excitability ever since. See for instance [12] and references therein.

In the configuration where the static conductance of the regulator exceeds the negative conductance of the diode, i.e., $k < \frac{1}{R}$, the static I-V curve is monotone. The circuit is excitable when the capacitance C is small relative to the time constant $\tau = \frac{L}{R}$ of the regulatory element. The circuit can then be analyzed as a fast-slow system. The fast subsystem is made of the capacitor and the switch element. Its static I-V curve is the cubic characteristic of the diode. This circuit is a simple bistable device. The slow subsystem is made of the regulatory inductive element, which is a linear first-order lag. The fast-slow behavior is ultrasensitive in the amplitude and voltage range where the real part of the admittance

$$G(j\omega; \bar{V}) = Cj\omega + (\bar{V}^2 - k) + \frac{1}{Lj\omega + R}$$

is close to zero. Sensitive trajectories include fast current pulses applied near the local extrema of the cubic characteristic of the switch.

20.6.2 Hodgkin–Huxley Circuit

The first figure from the landmark paper of Nobel prize winners Hodgkin and Huxley [2] is also a circuit fitting the requirements of Fig. 20.4. The circuit models the excitability of a neuron. It is made of the parallel connection of a capacitor with several distinct resistive branches, each modeling the flow of a specific ion through a specific ion channel type. The passive element is the RC circuit made of a capacitor modeling the cellular membrane and a "leak" current I_L. The switch element is the sodium current I_{Na}. The regulatory element is the potassium current I_K. Using the voltage clamp experiment, the authors identified the following model for the two ionic currents:

$$\begin{aligned} I_K &= \bar{g}_K n^4 (V - V_K) \\ \tau_n(V)\dot{n} &= -n + n_\infty(V) \end{aligned}$$

and

$$\begin{aligned} I_{Na} &= \bar{g}_{Na} m^3 h (V - V_{Na}) \\ \tau_m(V)\dot{m} &= -m + m_\infty(V) \\ \tau_h(V)\dot{h} &= -h + h_\infty(V) \end{aligned}$$

The state variables m, n, and h are *gating variables* in the range $[0, 1]$ introduced to model the amplitude and temporal dependence of the ionic conductances. The voltage-dependent time constants and gains of the gating variables were obtained by curve fitting, see Fig. 20.6. The admittance of the potassium current is

$$g_K(\bar{V}; j\omega) = \bar{g}_K(\bar{V} - V_K) 4n^3(\bar{V}) \frac{n'_\infty(\bar{V})}{1 + \tau_n(\bar{V}) j\omega}$$

It is positive real provided that $\bar{V} \geq V_K$, that is, whenever the potassium current is an outward current, which is always the case in physiological conditions. The admittance of the sodium current is

$$g_{Na}(\bar{V}; j\omega) = \bar{g}_{Na} m^2(\bar{V})(\bar{V} - V_{Na})(3h(\bar{V}) \frac{m'_\infty(\bar{V})}{1 + \tau_m(\bar{V}) j\omega} + m(\bar{V}) \frac{h'_\infty(\bar{V})}{1 + \tau_h(\bar{V}) j\omega})$$

At any voltage and any frequency, it is the sum of two terms of opposite real part. Whenever the sodium current is an inward current, i.e., $\bar{V} \leq V_{Na}$, which is always the case in physiological conditions, the first term is negative real whereas the second term is positive real. Looking at Fig. 20.6, it is obvious that the negative real term largely dominates the positive real term in a voltage range that includes the resting potential (around -70 mV and the high frequency range $\approx 1\ \mathrm{ms}^{-1}$). The sodium current thus acts as a negative resistance switch in the fast dynamic range of the

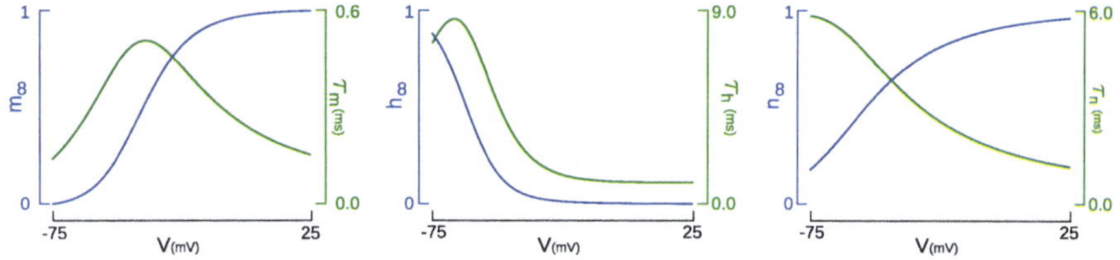

Fig. 20.6 Voltage dependence of the time constants and the static gains of the Hodgkin–Huxley model. The blue curves correspond to the sodium steady-state activation $m_\infty(V_m)$, the sodium steady-state inactivation $h_\infty(V_m)$ and the potassium steady-state activation $n_\infty(V_m)$. The green curves correspond to the sodium activation time constant $\tau_m(V_m)$, the sodium inactivation time constant $\tau_h(V_m)$ and the potassium activation time constant $\tau_n(V_m)$

activation variable m, whose time constant is about ten times smaller than the other gating time constants near the resting potential. In turn, the sodium *inactivation* variable h and the potassium *activation* variables both contribute to the negative feedback that regulates the refractory period of the spike.

It is important to observe that the balance of positive and negative feedback responsible for the ultrasensitivity of the circuit is robust to the details of the modeling. It rests entirely on the *localization* of the positive feedback in a specific voltage range (*near* the resting potential) and in a specific frequency range (about one decade above the cutoff frequency of the regulatory elements). This localization makes the excitability robust, by ensuring a range of ultrasensitivity (i.e., balance between positive and negative feedback) independent of the modeling details of the circuit.

FitzHugh–Nagumo circuit captures the excitability of Hodgkin–Huxley circuit by modeling the sodium activation as an instantaneous negative resistance diode and the sodium inactivation and potassium activation as a slow linear regulatory feedback. The reader will notice that this simplification introduces two artifacts. First, the timescale of the positive feedback must be fast relative to the timescale of negative feedback but should not be constrained to be instantaneous. In fact, this timescale is a critical feature of excitability as it sets the temporal localized window of excitability. Second, the time-scale separation between slow and fast gating variables in Hodgkin–Huxley circuit is only observed in a narrow voltage range around the resting potential. It vanishes at higher voltages, which means that the ultrasensitivity region is confined to the resting potential voltage range. There is no ultrasensitivity during the spike. In contrast, because the voltage dependence of time constants is ignored in FitzHugh-Nagumo circuit, the spikes have a range of ultrasensitivity both in the subthreshold and suprathreshold voltage ranges.

20.7 Conclusion

Excitability is a behavior at the core of biology. The presentation in this chapter emphasizes that the core property of an excitable circuit is a localized ultrasensitivity of the current–voltage relationship: small current variations are largely amplified in a specific temporal and amplitude range. This property can be quantified by the elementary concept of dynamic input conductance, which is nothing but the local gain of the inverse system computed at a given voltage and in a given timescale. Excitable circuits can be modulated by shaping their dynamic conductance, that is, by localizing the windows of low conductance (i.e., high sensitivity). Excitable circuits can be interconnected to create behaviors with localized and overlapping windows of ultrasensitivity. The recent study by the authors of modulation and robustness of bursting [3] is the first step in that direction.

Acknowledgements The research leading to these results has received funding from the European Research Council under the Advanced ERC Grant Agreement Switchlet n.670645. A. Franci was also supported by DGAPA-PAPIIT(UNAM) grant IA105816-RA105816.

References

1. Zames, G.: Input-output feedback stability and robustness, 1959–85. IEEE Control Syst. **16**(3), 61–66 (1996)
2. Hodgkin, A.L., Huxley, A.F.: A quantitative description of membrane current and its application to conduction and excitation in nerve. J. Physiol. **117**(4), 500–544 (1952)
3. Franci, A., Drion, G., Sepulchre, R.: Modeling the modulation of neuronal bursting: a singularity theory approach. SIAM J. Appl. Dyn. Syst. **13**(2), 798–829 (2014)
4. Drion, G., Franci, A., Dethier, J., Sepulchre, R.: Dynamic input conductances shape neuronal spiking. Eneuro **2**(1) (2015)
5. Drion, G., O'Leary, T., Marder, E.: Ion channel degeneracy enables robust and tunable neuronal firing rates. Proc. Natl. Acad. Sci. **112**(38), E5361–E5370 (2015)
6. Willems, J.C.: The behavioral approach to open and interconnected systems. IEEE Control Syst. **27**(6), 46–99 (2007)
7. Izhikevich, E.M.: Dynamical Systems in Neuroscience. MIT press (2007)
8. Hodgkin, A.L., Huxley, A.F.: Currents carried by sodium and potassium ions through the membrane of the giant axon of loligo. J. Physiol. **116**(4), 449 (1952)
9. Willems, J.C.: Dissipative dynamical systems part i: general theory. Arch. Ration. Mech. Anal. **45**(5), 321–351 (1972)
10. Nagumo, J., Arimoto, S., Yoshizawa, S.: An active pulse transmission line simulating nerve axon. Proc. IRE **50**(10), 2061–2070 (1962)
11. FitzHugh, R.: Impulses and physiological states in theoretical models of nerve membrane. Biophys. J. **1**(6), 445 (1961)
12. Izhikevich, E.M., FitzHugh, R.: Fitzhugh–nagumo model. Scholarpedia **1**(9), 1349 (2006)

Chapter 21
Electrical Network Synthesis: A Survey of Recent Work

Timothy H. Hughes, Alessandro Morelli and Malcolm C. Smith

Abstract The field of electrical network synthesis has a number of long-standing unanswered questions. The most perplexing concern minimality in the context of resistor, inductor and capacitor (RLC) network synthesis. In particular, the famous Bott–Duffin networks appear highly non-minimal from a system theoretic perspective. We survey some recent developments on this topic. These include results establishing the minimality of the Bott–Duffin networks and their simplifications for realising certain impedances; enumerative approaches to the analysis of RLC networks within given classes of restricted complexity; and new necessary and sufficient conditions for a (not necessarily controllable) system to be passive. Finally, some remaining open questions are discussed.

21.1 Introduction

The purpose of this paper is to survey some recent developments in electrical network synthesis. Despite the rich history of this field, there remain several significant open questions which have never been fully resolved. This is particularly true for RLC (resistor, inductor, capacitor) network synthesis. Notably, it is not known how to design an RLC network to realise a given driving-point behaviour while using the least possible number of elements; and classical methods of RLC network synthesis (such as the Bott–Duffin procedure [2]) appear highly non-minimal from a system

T. H. Hughes · A. Morelli · M. C. Smith (✉)
Department of Engineering, University of Cambridge, Cambridge CB2 1PZ, UK
e mail: mcs@eng.cam.ac.uk

T. H. Hughes
e-mail: thh22@cam.ac.uk

A. Morelli
e-mail: am2422@cam.ac.uk

R. Tempo et al. (eds.), *Emerging Applications of Control and Systems Theory*, Lecture Notes in Control and Information Sciences - Proceedings,
https://doi.org/10.1007/978-3-319-67068-3_21

theoretic perspective. Indeed, the need for a fresh examination of these issues, and an improved understanding of minimality in the context of passive systems, has recently been emphasised in [4, 5, 22, 39]. Additional practical motivation for studying these questions follows from the recent invention of the *inerter*, which completed the electrical–mechanical analogy [33]. Using this analogy, there is a one-to-one correspondence between RLC networks and mechanical networks containing springs, dampers and inerters, which allows the results of RLC network synthesis to be directly applied to mechanical control. Applications of this approach to the design of vibration absorption systems, such as vehicle suspension, train suspension, motorcycle steering compensators and building suspension, are described in [6, 7, 10, 21, 33, 36–38].

In this paper, we begin with a brief description of the objectives of electrical network synthesis (Sect. 21.2). We then discuss the classical methods of RLC network synthesis, notably the Bott–Duffin method [2] and its simplifications [8, 27, 30] (Sect. 21.3), and we present some newly discovered alternatives to Bott–Duffin [11]. In Sect. 21.4, we discuss some surprising results which establish the minimality of the Bott–Duffin networks and their simplifications for realising certain impedances [11, 14]. Sections 21.5 and 21.6 describe recent progress in the enumeration and analysis of RLC networks within a given class of restricted complexity [17–19, 22, 26]. Section 21.7 summarises results from [12] on passivity for systems which need not be controllable (as is the case for the Bott–Duffin networks). It is shown that the well-known positive-real condition is not sufficient for a (not necessarily controllable) system to be passive, and an alternative necessary and sufficient condition is presented (Theorem 21.2). Finally, some open questions are discussed in Sect. 21.8.

The notation is as follows. $\mathbb{R}$ and $\mathbb{C}$ denote the real and complex numbers. $\mathbb{C}_+$ and $\overline{\mathbb{C}}_+$ denote the open and closed right half plane. If $\lambda \in \mathbb{C}$, then $\Re(\lambda)$ and $\Im(\lambda)$ denote the real and imaginary parts of λ, and $\bar{\lambda}$ denotes its complex conjugate. $\mathbb{R}(s)$ and $\mathbb{R}[s]$ denote the rational functions and polynomials in the indeterminate s with real coefficients; and $\mathbb{R}^{m \times n}[s]$ denotes the $m \times n$ matrices with entries from $\mathbb{R}[s]$ (n is omitted if $n = 1$).

21.2 RLC Networks and Passivity

An RLC network comprises an interconnection of resistors, inductors and capacitors. The behaviours of these elements correspond to the set of solutions to the equations $v = iR$, $v = L\frac{di}{dt}$, and $i = C\frac{dv}{dt}$, respectively, where v denotes the voltage across and i the current through an element, and $R, L, C > 0$. Interconnection results in additional constraints corresponding to Kirchhoff's current law (the sum of currents entering any vertex is zero) and Kirchhoff's voltage law (the sum of voltages around any closed circuit is zero). In addition, energy can be exchanged with the environment at one or more ports, which correspond to pairs of terminals across which a voltage is applied and through which a current flows. Denoting the vector of port currents and corresponding voltages by **i** and **v**, respectively, then the driving-point behaviour of the network is the set of solutions to a differential equation of the form:

$$P(\tfrac{d}{dt})\mathbf{i} = Q(\tfrac{d}{dt})\mathbf{v}, \text{ with } \mathbf{i}, \mathbf{v} \text{ locally integrable, and } P, Q \in \mathbb{R}^{n \times n}[s]. \tag{21.1}$$

The driving-point behaviour is *passive*, in accordance with the definition below.

Definition 21.1 (*Passive system* [12]) The system in (21.1) is called *passive* if, for any given $(\mathbf{i}, \mathbf{v})$ satisfying (21.1) and $t_0 \in \mathbb{R}$, there exists a $K \in \mathbb{R}$ (dependent on $(\mathbf{i}, \mathbf{v})$ and t_0) such that, if $(\hat{\mathbf{i}}, \hat{\mathbf{v}})$ satisfies (21.1) and $(\hat{\mathbf{i}}(t), \hat{\mathbf{v}}(t)) = (\mathbf{i}(t), \mathbf{v}(t))$ for all $t < t_0$, then $-\int_{t_0}^{t_1} \hat{\mathbf{i}}^T(t)\hat{\mathbf{v}}(t)dt < K$ for all $t_1 \geq t_0$.

This definition formalises an important property of RLC networks: it is not possible to extract unlimited energy from the network from the present time (t_0) onward.

In the case of one-port networks (with the exception of an open circuit), the driving-point behaviour takes the form $p(\frac{d}{dt})i = q(\frac{d}{dt})v$ with $p, q \in \mathbb{R}[s]$ and $q \neq 0$ [15]. The impedance $Z := p/q$ of the network is then positive-real, in accordance with the following definition:

Definition 21.2 (*Positive-real*) $Z \in \mathbb{R}(s)$ is called *positive-real* if (i) Z is analytic in $\mathbb{C}_+$; and (ii) $\Re(Z(\lambda)) \geq 0$ for all $\lambda \in \mathbb{C}_+$.

RLC network synthesis is concerned with variations on the general problem: *given a positive-real Z, find an RLC network whose impedance is equal to Z*. As many different RLC networks can realise the same impedance, then we also seek network realisations which are *minimal*, i.e. which use the least possible numbers of the various types of element.

21.3 Bott–Duffin and Its Simplifications

One of the most striking results of electric circuit theory is that a given $Z \in \mathbb{R}(s)$ can be realised as the impedance of an RLC network if and only if Z is positive-real. The necessity of this condition was first proved by Otto Brune in [3], where the positive-real concept was first defined. Sufficiency was later proved in a famous one page paper by Bott and Duffin [2]. That paper gave an algorithm which, for any given positive-real function Z, provides an RLC network whose impedance is Z.

The starting point for the Bott–Duffin procedure is the *Foster preamble*. This provides a sequence of simplifications to the given positive-real function, each corresponding to the series or parallel extraction of a single resistor, inductor or capacitor, or an inductor–capacitor pair. This sequence of simplifications terminates with a so-called *minimum function*:

Definition 21.3 (*Minimum function*) $Z \in \mathbb{R}(s)$ is called a *minimum function* if (i) Z is positive-real; (ii) Z has no poles or zeros on the extended imaginary axis; and (iii) there exists $\omega_0 > 0$ such that $\Re(Z(j\omega_0)) = 0$.

The contribution of Bott and Duffin was to show that any given minimum function Z can be realised by an RLC network containing precisely six reactive elements (three inductors and three capacitors) and two further networks whose impedances are positive-real functions with McMillan degrees at least two fewer than that of Z. Thus, by the inductive application of this procedure (together with the Foster preamble), an RLC network is obtained to realise any given positive-real function.

The Bott–Duffin construction is based on the following theorem of Richards [31]:

Theorem 21.1 *If $Z \in \mathbb{R}(s)$ is positive-real and $\mu > 0$, then*

$$F(s) := \frac{\mu Z(s) - sZ(\mu)}{\mu Z(\mu) - sZ(s)}$$

is positive-real, and the McMillan degree of F does not exceed that of Z.

By inverting the relationship in Theorem 21.1, we obtain

$$Z(s) = \left(\frac{F(s)}{Z(\mu)} + \frac{\mu}{Z(\mu)s}\right)^{-1} + \left(\frac{1}{Z(\mu)F(s)} + \frac{s}{Z(\mu)\mu}\right)^{-1}, \tag{21.2}$$

whereupon a network realisation for Z is immediately obtained from network realisations for $F(s)/Z(\mu)$ and $1/(Z(\mu)F(s))$.

If Z is a minimum function, then there exists $\omega_0 > 0$ such that $Z(j\omega_0) = j\omega_0 X$, where $0 \neq X \in \mathbb{R}$. Bott and Duffin then considered the two cases: (i) $X > 0$; and (ii) $X < 0$. In case (i), it is easily shown that there exists a $\mu > 0$ such that $X = Z(\mu)/\mu$. For this value of μ, then F in Theorem 21.1 has the property that $1/F$ is positive-real and has poles at $\pm j\omega_0$. By a well-known property of positive-real functions, it follows that there exists a positive-real G and an $\alpha > 0$ such that

$$\frac{1}{F(s)} = \frac{1}{G(s)} + \frac{2\alpha s}{s^2 + \omega_0^2}, \tag{21.3}$$

and the McMillan degree of G is two fewer than that of F. In this manner, network realisations for $F(s)/Z(\mu)$ and $1/(Z(\mu)F(s))$ are immediately obtained from network realisations for $G(s)/Z(\mu)$ and $1/(Z(\mu)G(s))$. Thus, from (21.2)–(21.3), if $X > 0$, then Z is realised by the network on the left of Fig. 21.1. A similar argument for case (ii) concludes that, if $X < 0$, then Z is realised by the network on the right of Fig. 21.1.

One notable feature of the Bott–Duffin networks is the apparently excessive number of reactive elements. For example, if Z is biquadratic (i.e. the McMillan degree of Z is two), then the Bott–Duffin procedure uses six reactive elements. Yet, as will be discussed in Sects. 21.5–21.6, many positive-real functions of McMillan degree

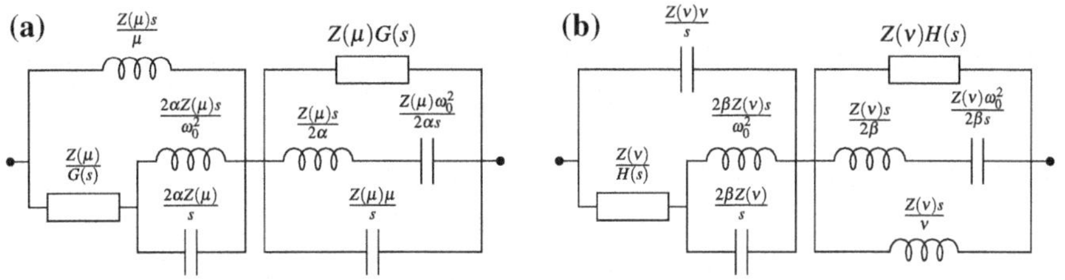

Fig. 21.1 Bott–Duffin networks for realising a minimum function Z with $Z(j\omega_0) = Xj$: **a** $X > 0$, and **b** $X < 0$. In case (**a**), μ, α and G are as defined in Sect. 21.3. In case (**b**), then $\nu > 0$ satisfies $-\omega_0^2 X = \nu Z(\nu)$, and $\beta > 0$ and H satisfy $(\nu Z(s) - sZ(\nu))/(\nu Z(\nu) - sZ(s)) = 1/H(s) + 2\beta s/(s^2 + \omega_0^2)$ with H positive-real and the McMillan degree of H at least two fewer than that of Z

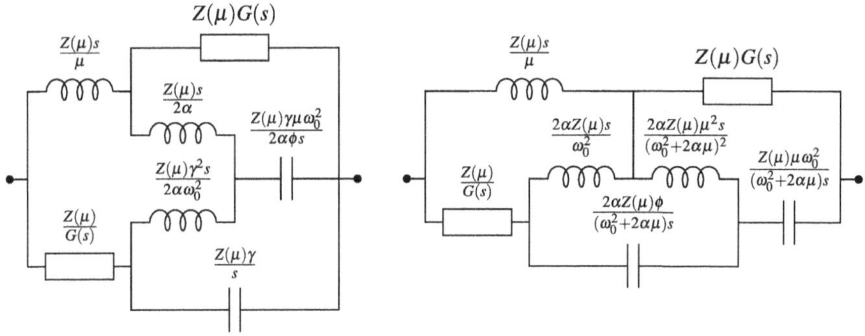

Fig. 21.2 Realisations of a minimum function Z with $Z(j\omega_0) = Xj$, $X > 0$. Here, μ, α and G are as defined in Sect. 21.3, $\gamma = \mu + 2\alpha$, and $\phi = \omega_0^2 + 2\alpha\mu + \mu^2$

two can be realised with only two reactive elements. This motivates the search for simpler alternatives to the Bott–Duffin networks.

Somewhat surprisingly, the best known improvements to the Bott–Duffin networks (for realising a minimum function) only save a single reactive element in each inductive step. The first such improvement was discovered simultaneously by several authors in the 1950s [8, 27, 30]. More recently, a second class of networks was discovered [11], which contain the same number of reactive elements and resistors as the networks in [8, 27, 30]. For a single step in the realisation of a general minimum function Z with $\Im(Z(j\omega_0)) > 0$, these networks are shown in Fig. 21.2.

21.4 Electric Networks and Minimality

From a systems and control perspective, it is particularly interesting to ask: *what is the least number of reactive elements required to realise a given biquadratic positive-real function?* The behaviour of an electric circuit is determined by a first-order linear differential algebraic equation whose degree is equal to the number of reactive elements in the circuit. More specifically, if $\mathbf{x}$ comprises the currents and voltages in the circuit's inductors, capacitors and resistors, and i and v are the port current and voltage, then there exist real matrices A, B, C, D and E such that $E\frac{d\mathbf{x}}{dt} = A\mathbf{x} + Bi$ and $v = C\mathbf{x} + Di$, and the rank of E is equal to the number of reactive elements. It might, therefore, be assumed that only two reactive elements are required to realise a biquadratic positive-real function. But it has been proven that this is not the case. In fact, the following two surprising results have recently been shown:

1. The Bott–Duffin networks contain the least possible number of reactive elements for realising a biquadratic minimum function with *series–parallel* networks [14];
2. The networks in [8, 11, 27, 30] contain the least possible number of reactive elements for realising almost all biquadratic minimum functions with arbitrary RLC networks [11].

These results were inspired by [32], which aimed to show that the networks in [8, 27, 30] contain the least possible number of elements for realising certain biquadratic minimum functions. However, it is shown in [9, 11] that [32] overlooked certain networks, and did not determine the minimality of these networks in the number of reactive elements used.

In [14], the proof of 1 exploits the special form of the impedance of a series–parallel network and the properties of positive-real functions. A key observation is that the impedance of a series–parallel network is a composition of sums and inverses of positive-real functions. More specifically, any series–parallel network N is either a series or a parallel connection of two series–parallel networks N_1 and N_2. In the first case, the impedance Z satisfies $Z = Z_1 + Z_2$, where Z_1 and Z_2 are the impedances of N_1 and N_2, respectively. Since Z_1 and Z_2 are positive-real, then: (a) since $\Re(Z(j\omega_0)) = 0$, then $\Re(Z_1(j\omega_0)) = 0$ and $\Re(Z_2(j\omega_0)) = 0$; (b) since Z has no imaginary axis poles, then neither Z_1 nor Z_2 has any imaginary axis poles; and (c) since $\Im(Z(j\omega_0)) \neq 0$, then either $\Im(Z_1(j\omega_0)) \neq 0$ or $\Im(Z_2(j\omega_0)) \neq 0$. It can then be shown that N must contain at least five reactive elements. Also, in [14], all of the series–parallel networks containing exactly five reactive elements which satisfy conditions (a)–(c) are identified, and it is shown that none of these can realise a biquadratic minimum function. A similar argument holds in the case that N is a parallel connection of two series–parallel networks, and proves 1.

An entirely different argument is required to prove 2. The proof in [11] considers a general RLC network N whose impedance Z is a minimum function, and a sinusoidal trajectory of N at the minimum frequency. For such a trajectory, it is shown that the net amount of energy dissipated by N over a single period must be zero, so only the reactive elements can transmit current. Further constraints are placed on N by virtue of the fact that Z has no imaginary axis poles or zeros. These observations were used in [11] to identify all of the RLC networks which realise a minimum function and contain fewer than five reactive elements. It was then shown that these RLC networks can only realise a small subset of the biquadratic minimum functions.

21.5 Enumeration Approach

As discussed in Sect. 21.3, the realisation method introduced by Bott and Duffin in 1949 can prove to be apparently extravagant in the number of elements used. Following this result, many questions in passive network synthesis were still unanswered, and no better theory to solve the problem of minimal realisation of positive-real functions was found. An entirely different research strategy for this problem consists in enumerating and analysing all RLC networks within a given restricted class. Following the renewed interest in minimal realisations for passive mechanical control [33], Jiang and Smith adopted this approach to study the class of all five-element networks and six-element series–parallel networks [17, 18]. In [17], the concept of regularity was introduced, which greatly facilitates the classification of impedances, as follows:

Definition 21.4 (*Regular*) A positive-real function $Z(s)$ is defined to be *regular* if the smallest value of $\Re(Z(j\omega))$ or $\Re(Z^{-1}(j\omega))$ occurs at $\omega = 0$ or $\omega = \infty$.

The *Ladenheim Catalogue*, i.e. the set of all essentially distinct RLC networks with at most two reactive elements and at most three resistors, was studied in [17]. There are 108 essentially distinct such networks, which all realise biquadratic (or possibly bilinear or constant) functions of the form

$$Z(s) = \frac{As^2 + Bs + C}{Ds^2 + Es + F}, \tag{21.4}$$

where $A, B, C, D, E, F \geq 0$. These networks were first enumerated and studied by Edward Ladenheim, a student of Foster, in [23]. For each network in the class, the value of the impedance is first computed in [23], then the inverse process is performed, i.e. given the impedance, an expression for each element of the network is found. There is, however, no derivation in [23] of conditions which ensure positivity of the values computed in the last step.

It was shown in [17] that all but two of the Ladenheim networks can only realise regular impedances, and that the remaining two cover some, but not all, of the non-regular positive-real biquadratics. It was also proven that just six networks from the catalogue are needed to realise any regular biquadratic. By considering the class of all five-element networks instead, with an arbitrary number of reactive elements, it was finally shown that only another six networks, with three reactive elements, can realise some, but again not all, non-regular biquadratics.

A well-known result in circuit theory is that the class of impedances that can be realised using only two reactive elements is not increased by using more than three resistors [19]. This fundamental result is known as Reichert's theorem, and its first proof, which uses a complicated topological argument, can be found in Reichert's original German language publication [29]. This proof was later reworked in [19], and some parts of the argument were expanded. Recently, an alternative proof based on a result in [5] was provided in [40].

The class of six-element series–parallel networks was studied in [18], following earlier work [34, 35]. Based on Reichert's theorem, it is not useful to consider two-reactive six-element networks, as they do not expand the set of impedances realised by the Ladenheim class. A set of networks which can realise all the non-regular biquadratics that may be realised by the class was found in [18], consisting of networks with either three or four reactive elements.

Based on the results in [18], a four-reactive seven-element network was proposed in [20] which is very effective in realising most, but not all, non-regular biquadratics.

21.6 The Ladenheim Catalogue Revisited

The analysis carried out in [17] provided a complete understanding of the realisation power of the Ladenheim class of networks, through the concept of regularity. Such an analysis, however, offers no insight as to how structured the set is and whether further results or patterns might emerge from a more complete study of the realisability conditions for each network in the class.

The need for a fresh study of the catalogue was independently advocated in 2010 by Kalman, who considered the catalogue 'valuable experimental data, to be checked for accuracy and explained by theory' [22]. In his latest work, Kalman defined the concept of generic (or minimal) networks and approached the problem of electrical network synthesis using algebraic invariant theory [24].

The open questions regarding the structure of the Ladenheim class motivated a fresh study of the catalogue, which was carried out recently [25, 26]. A rigorous and complete classification of the catalogue was performed, by defining a group action and equivalence relation on the set. Equivalence relations include certain well-known network transformations (namely Y-Δ and *Zobel* transformations). More generally two networks are defined to be equivalent if they realise the same set of impedance functions. Group actions are based on the following transformations on the impedance $Z(s)$: frequency inversion $s \to s^{-1}$ and impedance inversion $Z \to Z^{-1}$. These notions allow the catalogue to be partitioned into 24 *subfamilies* of networks, each comprising a certain number of *equivalence classes* and *orbits*. The set of realisable impedances for one network in each subfamily is derived in explicit form as a semialgebraic set, in terms of necessary and sufficient conditions. This leads to a systematic derivation of the realisability conditions for every network of the Ladenheim class. By way of illustration, Fig. 21.3 shows one of the five-element subfamilies of networks, with the structure that emerged from the classification of the catalogue.

The new study of the Ladenheim catalogue in [26] demonstrates that there are no new equivalence relations which were not known classically. Yet, the question remains open whether further patterns might still be revealed which might shed some light on the question of realisability for higher order networks.

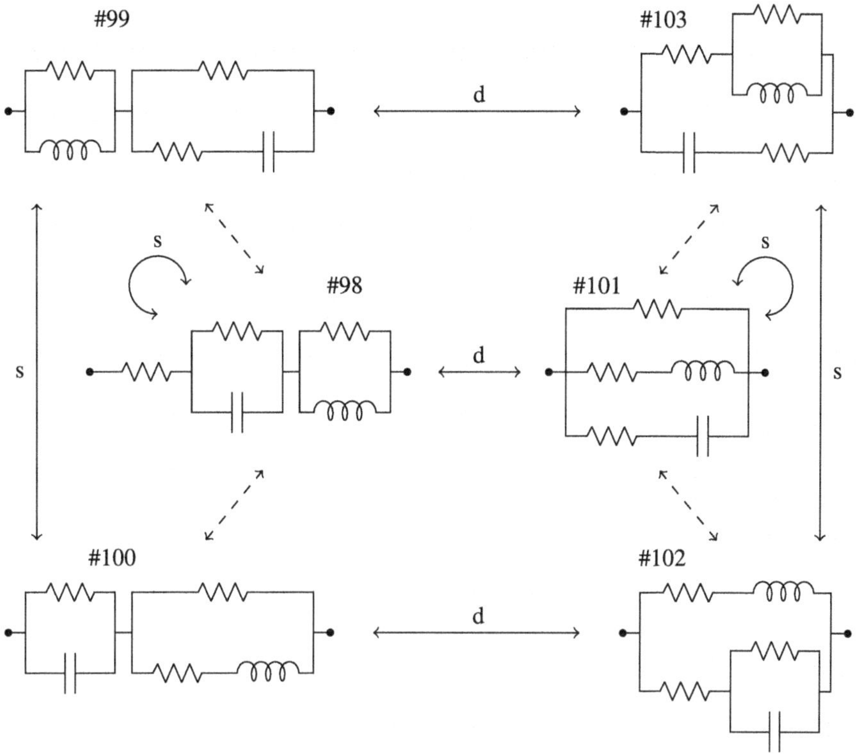

Fig. 21.3 Networks of subfamily V_F in the catalogue arranged according to the subfamily structure. *Zobel* transformations are represented by dashed arrows and define two equivalence classes. Duality and frequency inversion relations (d and s, respectively) are indicated through arrows, while the composition of the two and the identity (which complete the Klein four-group) are not shown. Here, the group action induces two *orbits* which comprise four and two networks respectively

21.7 Realisation of Behaviours

The necessity of the apparently extravagant numbers of reactive elements in the Bott–Duffin and related networks motivates a fresh examination of passivity for systems which are nonminimal in a system theoretic sense. A natural framework for examining these issues is the behavioural approach [28], and the associated concept of (behavioural) controllability [28, Definition 5.2.2]. Indeed, the driving-point behaviours of the Bott–Duffin networks and their simplifications are examples of uncontrollable behaviours. For example, from [15, Sect. 7], the driving-point behaviour of the Bott–Duffin network on the left of Fig. 21.1 for realising the biquadratic minimum function $Z(s) = (s^2 + s + 1)/(s^2 + s + 4)$ is the set of solutions to the equation:

$$\left(\tfrac{d}{dt} + 1\right)\left(\tfrac{d^2}{dt^2} + \tfrac{d}{dt} + 4\right)v = \left(\tfrac{d}{dt} + 1\right)\left(\tfrac{d^2}{dt^2} + \tfrac{d}{dt} + 1\right)i. \qquad (21.5)$$

That this behaviour is uncontrollable is a result of the common differential operator $(\frac{d}{dt} + 1)$ acting on both sides of the above equation [28, Theorem 5.2.10]. Note that the set of solutions to (21.5) contains the set of solutions to

$$\left(\frac{d^2}{dt^2} + \frac{d}{dt} + 4\right) v = \left(\frac{d^2}{dt^2} + \frac{d}{dt} + 1\right) i. \tag{21.6}$$

However, there are solutions to (21.5) which are not solutions to (21.6). It follows that the passivity of (21.5) guarantees the passivity of (21.6), but the converse implication does not hold. In particular, for a behaviour $p(\frac{d}{dt})i = q(\frac{d}{dt})v$ $(q \not\equiv 0)$ to be passive, it is necessary *but not sufficient* for the transfer function p/q to be positive-real.

By focusing on the driving-point behaviour of an electric network, as opposed to its impedance, we obtain another class of unanswered network synthesis problems (see [4] and [39, Sect. 12]). For example, it is not known *what are the necessary and sufficient conditions (akin to the positive-real condition) for a (not necessarily controllable) behaviour to be realised as the driving-point behaviour of an RLC network*. A more fundamental question is *what are the necessary and sufficient conditions for a (not necessarily controllable) behaviour to be passive*. This question was recently answered in [12], in the context of multi-port networks, as follows:

Theorem 21.2 ([12] Theorem 11) *The system in (21.1) is passive if and only if the following three conditions hold:*

1. $P(\lambda)Q(\bar{\lambda})^T + Q(\lambda)P(\bar{\lambda})^T \geq 0$ *for all* $\lambda \in \overline{\mathbb{C}}_+$.
2. *rank*$([P \;\; -Q](\lambda)) = n$ *for all* $\lambda \in \overline{\mathbb{C}}_+$.
3. *If* $\mathbf{p} \in \mathbb{R}^n[s]$ *and* $\lambda \in \mathbb{C}$ *satisfy* $\mathbf{p}(s)^T(P(s)Q(-s)^T + Q(s)P(-s)^T) = 0$ *and* $\mathbf{p}(\lambda)^T[P \;\; -Q](\lambda) = 0$, *then* $\mathbf{p}(\lambda) = 0$.

It was shown in [12] that if any one of the conditions in Theorem 21.2 does not hold then an arbitrarily large amount of energy can be extracted from the system. Thus, these conditions are necessary for a system to be passive. That these conditions are sufficient to guarantee passivity was shown using an algebraic argument to construct a *storage function* for the system, which provided an upper bound to the energy that can be extracted. Moreover, this storage function can be used in the *reactance extraction* approach to network analysis and synthesis [1] to prove that the system in (21.1) is passive if and only if it can be realised as the driving-point behaviour of a network containing resistors, inductors, capacitors, transformers and gyrators [13].

21.8 Open Questions

The discussion in the preceding sections motivates the following three questions (of which, 1 and 2 were first posed in [16]):

1. Among RLC realisations, what is the minimum number of reactive elements required to realise all positive-real functions of McMillan degree n?

2. In order to synthesise the entire class of impedances realisable by RLC networks containing n reactive elements, what is the least number of resistors required?
3. What is the class of impedances realisable by RLC networks containing at most n reactive elements and $n + 1$ resistors?

All of these questions have been answered in the case $n = 2$. Five reactive elements are required to realise all positive-real functions of McMillan degree 2 (see Sects. 21.3–21.4). Three resistors are required to synthesise the entire class of impedances realisable by RLC networks containing 2 reactive elements (see Sect. 21.5). The class of impedances realisable by RLC networks containing at most 2 reactive elements and 3 resistors (namely the class corresponding to the Ladenheim catalogue) is stated explicitly in [17, 26] (see Sect. 21.6). However, these three questions remain unanswered for $n > 2$.

We also restate the following open problem from Sect. 21.7 on the realisation of behaviours, which is a slight variation on [4, Problems 1 and 2]:

4. What are the necessary and sufficient conditions for the behaviour $p(\frac{d}{dt})i = q(\frac{d}{dt})v$ to be realised as the driving-point behaviour of an RLC network?

21.9 Conclusions

It is now known that minimality in the context of RLC network synthesis differs from minimality in a system theoretic sense. In the absence of a general theory for the minimal realisation of RLC network behaviours, the enumeration approach provides a valuable research strategy for investigating networks of restricted complexity. To date, this has proved productive for investigating RLC networks whose impedances have McMillan degree less than or equal to two. A remaining challenge is to extend these results to higher order networks.

Acknowledgements T. H. Hughes is grateful for the support of the Cambridge Philosophical Society in funding his Henslow Research Fellowship at the University of Cambridge. A. Morelli is grateful to acknowledge the support of the MathWorks for their funding of the MathWorks studentship in Engineering at the University of Cambridge.

References

1. Anderson, B.D.O., Vongpanitlerd, S.: Network Analysis and Synthesis. Prentice-Hall, NJ (1973)
2. Bott, R., Duffin, R.J.: Impedance synthesis without use of transformers. J. Appl. Phys. **20**, 816 (1949)
3. Brune, O.: Synthesis of a finite two-terminal network whose driving-point impedance is a prescribed function of frequency. J. Math. Phys. **10**, 191–236 (1931)
4. Çamlibel, M.K., Willems, J.C., Belur, M.N.: On the dissipativity of uncontrollable systems. In: Proceedings IEEE 42nd Conference Decision Control, pp. 1645–1650 (2003)

5. Chen, M.Z.Q., Smith, M.C.: Electrical and mechanical passive network synthesis In: Blondel, V.D., et al. (eds.) Recent Advances in Learning and Control, pp. 35–50 (2008)
6. Evangelou, S., Limebeer, D.J.N., Sharp, R.S., Smith, M.C.: Control of motorcycle steering instabilities–passive mechanical compensators incorporating inerters. IEEE Control Syst. Mag. **26**(5), 78–88 (2006)
7. Evangelou, S., Limebeer, D.J.N., Sharp, R.S., Smith, M.C.: Mechanical steering compensation for high-performance motorcycles. J. Appl. Mech. Trans. ASME **74**(2), 332–345 (2007)
8. Fialkow, A., Gerst, I.: Impedance synthesis without mutual coupling. Q. Appl. Math. **12**, 420–422 (1955)
9. Foster, R.M.: Minimum biquadratic impedances. IEEE Trans. Circuit Theory **10**(4), 527 (1963)
10. Hu, Y., Chen, M.Z.Q., Shu, Z.: Passive vehicle suspensions employing inerters with multiple performance requirements. J. Sound Vib. **333**(8), 2212–2225 (2014)
11. Hughes, T.H.: Why RLC realizations of certain impedances need many more energy storage elements than expected. IEEE Trans. Autom. Control **62**(9), 4333–4346 (2017)
12. Hughes, T.H.: A theory of passive linear systems with no assumptions. Automatica **86**, 87–97 (2017)
13. Hughes, T.H.: Passivity and electric circuits: a behavioral approach. In: Proceedings 20th IFAC World Congress (2017)
14. Hughes, T.H.: Passivity and electric circuits: a behavioral approach. In: Proceedings 20th IFAC World Congress, IFAC-Papers Online. **50**(1), 15500–15505 (2017)
15. Hughes, T.H., Smith, M.C.: Controllability of linear passive network behaviors. Syst. Control Lett. **101**, 58–66 (2017)
16. Hughes, T.H., Jiang, J.Z., Smith, M.C.: Two problems on minimality in RLC circuit synthesis. In: Workshop on Dynamics and Control in Networks, Lund University. http://www.lccc.lth.se/media/2014/malcolm.pdf (2014). Cited 29 Mar 2017
17. Jiang, J.Z., Smith, M.C.: Regular positive-real functions and five-element network synthesis for electrical and mechanical networks. IEEE Trans. Autom. Control **56**(6), 1275–1290 (2011)
18. Jiang, J.Z., Smith, M.C.: Series-parallel six-element synthesis of biquadratic impedances. IEEE Trans. Circuits Syst. **59**(11), 2543–2554 (2012)
19. Jiang, J.Z., Smith, M.C.: On the theorem of Reichert. Syst. Control Lett. **61**(12), 1124–1131 (2012)
20. Jiang, J.Z., Zhang, S.Y.: Synthesis of biquadratic impedances with a specific seven-element network. In: 2014 UKACC International Conference on Control, Loughborough, pp. 139–144 (2014)
21. Jiang, J.Z., Matamoros-Sanchez, A.Z., Goodall, R.M., Smith, M.C.: Passive suspensions incorporating inerters for railway vehicles. Veh. Syst. Dyn. **50**, 263–276 (2012)
22. Kalman, R.E.: Old and new directions of research in system theory. In: Perspectives in Mathematical System Theory, Control, and Signal Processing, vol. 398, pp. 3–13 (2010)
23. Ladenheim, E.L.: A synthesis of biquadratic impedances. Master's thesis, Polytechnic Institute of Brooklyn, New York (1948)
24. Lin, S., Oeding, L., Sturmfels B.: Electric Network Synthesis, UC Berkeley. http://www.acritch.com/media/bass/electric-network-synthesis-v2.pdf (2011). Cited 29 Mar 2017
25. Morelli, A., Smith, M.C.: Ladenheim's catalogue: group action, equivalence and realisation. In: 22nd International Symposium on Mathematical Theory of Networks and Systems (MTNS), Minneapolis (MN), pp. 729–732 (2016)
26. Morelli, A., Smith, M.C.: Passive Network Synthesis: The Ladenheim Catalogue, in preparation
27. Pantell, R.H.: A new method of driving point impedance synthesis. Proc. IRE (Correspondence) **42**(5), 861 (1954)
28. Polderman, J.W., Willems, J.C.: Introduction to Mathematical Systems Theory: A Behavioral Approach. Springer, New York (1998)
29. Reichert, M.: Die kanonisch und übertragerfrei realizierbaren Zweipolfunktionen zweiten Grades (Transformerless and canonic realisation of biquadratic immittance functions). Arch. Elek. übertragung **23**, 201–208 (1969)
30. Reza, F.M.: Synthesis without ideal transformers. J. Appl. Phys. **25**, 807–808 (1954)

31. Richards, P.I.: A special class of functions with positive real part in a half-plane. Duke J. Math. **21**, 777–786 (1942)
32. Seshu, S.: Minimal realizations of the biquadratic minimum function. IRE Trans. Circuit Theory **6**(4), 345–350 (1959)
33. Smith, M.C.: Synthesis of mechanical networks: the inerter. IEEE Trans. Autom. Control **47**(10), 1648–1662 (2002)
34. Vasiliu, C.G.: Series-parallel six-element synthesis of the biquadratic impedances. IEEE Trans. Circuit Theory **17**(1), 115–121 (1970)
35. Vasiliu, C.G.: Correction to 'Series-parallel six-element synthesis of the biquadratic impedances'. IEEE Trans. Circuit Theory **18**(1), 207–207 (1971)
36. Wang, F.C., Hong, M.F., Chen, C.W.: Performance analysis of building suspension control with inerters. In: Proceedings IEEE 46th Conference Decision Control, pp. 3786–3791 (2007)
37. Wang, F.C., Liao, M.K.: The lateral stability of train suspension systems employing inerters. Veh. Syst. Dyn. **48**(5), 619–643 (2010)
38. Wang, F.C., Hsieh, M.R., Chen, H.J.: Stability and performance analysis of a full-train system with inerters. Veh. Syst. Dyn. **50**(4), 545–571 (2012)
39. Willems, J.C.: Dissipative dynamical systems. Eur. J. Control **13**, 134–151 (2007)
40. Zhang, S.Y., Jiang, J.Z., Smith, M.C.: A new proof of Reichert's theorem. In: Proceedings IEEE 55th Conference Decision Control, pp. 2615–2619 (2016)

Chapter 22
Examples of Computation of Exact Moment Dynamics for Chemical Reaction Networks

Eduardo D. Sontag

Abstract The study of stochastic biomolecular networks is a key part of systems biology, as such networks play a central role in engineered synthetic biology constructs as well as in naturally occurring cells. This expository paper reviews in a unified way a pair of recent approaches to the finite computation of statistics for chemical reaction networks.

22.1 Introduction

The study of biochemical networks is of great interest not only for the understanding of natural biological systems, but also in the engineering design of biological control systems, and specifically in the field of synthetic biology. Chemical systems are inherently stochastic, as reactions depend on thermally induced random effects. For large systems, deterministic mean-field models are appropriate, but such models cannot account for random fluctuations, and stochastic models, and specifically the Chemical Master Equation (CME), a discrete-space continuous-time Markov process that describes stochastic chemical kinetics, are required for a more accurate description. Tools from dynamical systems and from control theory play key roles in the analysis of the CME. The CME is typically an infinite-dimensional linear differential equation, and even its steady-state solutions are very difficult to compute in closed form. Various techniques, typically moment closure tools based on the"mass fluctuation kinetics" and "fluctuation-dissipation" ideas are used to approximate solutions or moments [5, 10, 11, 14]. In this expository paper, we first introduce the setup, and then review in a unified way results for two types of stochastic chemical reaction systems for which moments can be effectively computed: *feedforward networks (FFN)*, treated in [12], and *complex balanced networks (CBN)*, treated in [13], and provide several worked examples.

E. D. Sontag (✉)
Northeastern University, Boston, MA, USA
e-mail: eduardo.sontag@gmail.com

R. Tempo et al. (eds.), *Emerging Applications of Control and Systems Theory*, Lecture Notes in Control and Information Sciences - Proceedings,
https://doi.org/10.1007/978-3-319-67068-3_22

22.2 Preliminaries

We start by reviewing standard concepts regarding master equations for biochemical networks, see for instance [11].

Chemical Reaction Networks. Chemical reaction networks involve interactions among a finite set of *species* $\mathscr{S} = \{S_i,\ i = 1, 2, \ldots n\}$ where one thinks of the S_i's as counting the numbers of molecules of a certain type (or individuals in an ecological model, or cells in a cell population model): $S_i(t) = k_i =$ number of units of species i at time t. In stochastic models, one thinks of these as random variables, which interact with each other. The complete vector $S = (S_1, \ldots, S_n)'$ is called the *state* of the system at time t, and it is probabilistically described as a Markov stochastic process which is indexed by time $t \geq 0$ and takes values in $\mathbb{Z}^n_{\geq 0}$. Thus, $S(t)$ is a $\mathbb{Z}^n_{\geq 0}$-valued random variable, for each $t \geq 0$. (Abusing notation, we also write $S(t)$ to represent an outcome of this random variable on a realization of the process.) We will denote $p_k(t) = \mathbb{P}[S(t) = k]$ for each $k \in \mathbb{Z}^n_{\geq 0}$. Then $p(t) = (p_k)_{k \in \mathbb{Z}^n_{\geq 0}}$ is the discrete probability density (also called the "probability mass function") of $S(t)$. To describe the Markov process, one needs to formally introduce chemical reaction networks.

A *chemical reaction network* is a finite set $\mathscr{R} = \{R_j,\ j = 1, 2, \ldots, m\}$ of formal transformations or *reactions*

$$R_j: \quad \sum_{i=1}^{n} a_{ij} S_i \longrightarrow \sum_{i=1}^{n} b_{ij} S_i\,, \quad j \in \{1, 2, \ldots, m\} \tag{22.1}$$

among species, together with a set of m functions $\rho_j : \mathbb{Z}^n_{\geq 0} \to \mathbb{R}_{\geq 0}$, $j = 1, \ldots, m$, with $\rho_j(0) = 0$, the *propensity functions* for the respective reactions R_j. The coefficients a_{ij} and b_{ij} are nonnegative integers, the *stoichiometry coefficients*, and the sums are understood informally, indicating combinations of elements. The intuitive interpretation is that $\rho_j(S_1, \ldots, S_n)dt$ is the probability that reaction R_j takes place, in a short interval of length dt, provided that the complete state was $S = (S_1, \ldots, S_n)$ at the beginning of the interval. In principle, the propensities can be quite arbitrary functions, but we will focus on mass-action kinetics, for which the functions ρ_j are polynomials whose degree is the sum of the a_{ij}'s in the respective reaction. Before discussing propensities, we introduce some more notations and terminology.

The linear combinations $\sum_{i=1}^n a_{ij} S_i$ and $\sum_{i=1}^n b_{ij} S_i$ appearing in the m reactions are the *complexes* involved in the reactions. For each reaction R_j, we collect the coefficients appearing on its left-hand side and on its right-hand side into two vectors, respectively: $\mathbb{S}(R_j) = a_j := (a_{1j}, \ldots, a_{nj})'$ and $\mathbb{T}(R_j) = b_j := (b_{1j}, \ldots, b_{nj})'$ (prime indicates transpose). We call $\mathbb{S}, \mathbb{T} : \mathscr{R} \to \mathscr{C}$ the *source* and *target* functions, where $\mathscr{C} \subseteq \mathbb{Z}^n_{\geq 0}$ is the set of all vectors $\{a_j, b_j,\ j = 1 \ldots m\}$. We identify complexes with elements of $\mathscr{C}$. The *reactants* S_i of the reaction R_j are those species appearing with a nonzero coefficient, $a_{ij} \neq 0$ in its left-hand side and the *products* S_i of reaction R_j are those species appearing with a nonzero coefficient $b_{ij} \neq 0$ in its right-hand side.

For every vector of nonnegative integers $v = (v_1, \ldots, v_n) \in \mathbb{Z}^n_{\geq 0}$, let us write the sum of its entries as $\oplus v := v_1 + \cdots + v_n$. In particular, for each $j \in \{1, \ldots, m\}$, we define the *order* of the reaction R_j as $\oplus a_j := \sum_{i=1}^n a_{ij}$, which is the total number of units of all species participating in the reaction R_j.

The $n \times m$ *stoichiometry matrix* $\Gamma = \{\gamma_{ij}\}$ is defined as the matrix whose entries are defined as follows: $\gamma_{ij} := b_{ij} - a_{ij},\ i = 1, \ldots, n,\ j = 1, \ldots, m$. The integer γ_{ij} counts the net change (positive or negative) in the number of units of species S_i each time that the reaction R_j takes place. We will denote by γ_j the jth column of Γ. With these notations, $\gamma_j = b_j - a_j,\ j = 1, \ldots, m$. We will assume that $\gamma_j \neq 0$ for all j (each reaction changes at least some species).

For example, suppose that $n = 4$, $m = 2$, and the reactions are $R_1 : S_1 + S_2 \to S_3 + S_4$, $R_2 : 2S_1 + S_3 \to S_2$ which have orders $1 + 1 = 2$ and $2 + 1 = 3$, respectively. The set $\mathscr{C}$ has four elements, which list the coefficients of the species participating in the reactions: $\mathscr{C} = \{(1, 1, 0, 0)', (0, 0, 1, 1)', (2, 0, 1, 0)', (0, 1, 0, 0)'\}$ with $\mathbb{S}(R_1) = a_1 = (1, 1, 0, 0)'$, $\mathbb{S}(R_2) = a_2 = (2, 0, 1, 0)'$, $\mathbb{T}(R_1) = b_1 = (0, 0, 1, 1)'$, $\mathbb{T}(R_1) = b_2 = (0, 1, 0, 0)'$ and $\gamma_1 = (-1, -1, 1, 1)'$, $\gamma_2 = (-2, 1, -1, 0)'$. The reactants of R_1 are S_1 and S_2, the reactants of R_2 are S_1 and S_3, the products of R_1 are S_3 and S_4, the only product of R_2 is S_2, and the stoichiometry matrix is (using MATLAB-like notation, listing row by row): $\Gamma = [-1, -2; -1, 1; 1, -1; 1, 0]$.

It is sometimes convenient to write $\sum_{i=1}^n a_{ij} S_i \xrightarrow{\rho_j(S)} \sum_{i=1}^n b_{ij} S_i$ to show that the propensity ρ_j is associated to the reaction j, and to combine two reactions R_j and R_k that are the reverse of each other (complexes are transposed): $\mathbb{S}(R_j) = \mathbb{T}(R_k)$ and $\mathbb{S}(R_k) = \mathbb{T}(R_j)$, using double arrows: $\sum_{i=1}^n a_{ij} S_i \underset{\rho_k(S)}{\overset{\rho_j(S)}{\rightleftharpoons}} \sum_{i=1}^n b_{ij} S_i$. When propensities are given by mass-action kinetics, as discussed below, one simply writes on the arrows the kinetic constants instead of the full form of the kinetics.

Chemical Master Equation. A *Chemical Master Equation (CME)*, which is the differential form of the Chapman–Kolmogorov forward equation, is a system of linear differential equations that describes the time evolution of the joint probability distribution of the $S_i(t)$'s:

$$\frac{dp_k}{dt} = \sum_{j=1}^{m} \rho_j(k - \gamma_j)\, p_{k-\gamma_j} - \sum_{j=1}^{m} \rho_j(k)\, p_k\,, \quad k \in \mathbb{Z}^n_{\geq 0} \tag{22.2}$$

where, for notational simplicity, we omitted the time argument "t" from p, and the function ρ_j has the property that $\rho_j(k - \gamma_j) = 0$ unless $k \geq \gamma_j$ (coordinatewise inequality). There is one equation for each $k \in \mathbb{Z}^n_{\geq 0}$, so this is an infinite system of linked equations. When discussing the CME, we will assume that an initial probability vector $p(0)$ has been specified, and that there is a unique solution of (22.2) defined for all $t \geq 0$. (See [9] for existence and uniqueness results.) A different CME results for each choice of propensity functions, a choice that is dictated by physical chemistry considerations. The most commonly used propensity functions, and the ones best-justified from elementary physical principles, are *ideal mass-action kinetics* propensities, defined as follows (see [4]), proportional to the number of ways in which species can combine to form the jth source complex:

$$\rho_j(k) = \kappa_j \prod_{i=1}^{n} \binom{k_i}{a_{ij}} \mathscr{H}(k - a_j) \quad j = 1, \ldots, m. \tag{22.3}$$

where, for any scalar or vector, we denote $\mathscr{H}(u) = 1$ if $u \geq 0$ (coordinatewise) and $\mathscr{H}(u) = 0$ otherwise. In other words, the expression can only be nonzero provided that $k_i \geq a_{ij}$ for all $i = 1, \ldots, n$ (and thus the combinatorial coefficients are well-defined). Observe that the expression in the right-hand side makes sense even if $k \not\geq 0$, in the following sense. In that case, $k_i < 0$ for some index i, so the factorial is not well-defined, but on the other hand, $k_i - a_{ij} \leq k_i < 0$ implies that $\mathscr{H}(k - a_j) = 0$. So $\rho_j(k)$ can be thought of as defined by this formula for all $k \in \mathbb{Z}^n$, even if some entries of k are negative, but is zero unless $k \geq 0$, and the combinatorial coefficients can be arbitrarily defined for $k \not\geq 0$. (In particular, $\rho_j(k - \gamma_j) = 0$ unless $k \geq \gamma_j$ in (22.2).) The m nonnegative "kinetic constants" are arbitrary, and they represent quantities related to the volume, shapes of the reactants, chemical, and physical information, and temperature. The model described here assumes that temperature and volume are constant, and that the system is well-mixed (no spatial heterogeneity).

Derivatives of Moments Expressed as Linear Combinations of Moments. Notice that $\rho_j(k)$ can be expanded into a polynomial in which each variable k_i has an exponent less or equal to a_{ij}. In other words, $\rho_j(k) = \sum_{c_j \leq a_j} \kappa_{c_j} k^{c_j}$ ("$\leq$" is understood coordinatewise, and by definition $k^{c_j} = k_1^{c_{1j}} \ldots k_n^{c_{nj}}$ and $r^0 = 1$ for all integers), for suitably redefined coefficients κ_{c_j}'s. Suppose given a function $M : \mathbb{Z}^n_{\geq 0} \to \mathbb{R}$ (to be taken as a monomial when computing moments). The expectation of the random variable $M(S)$ is by definition $\mathbb{E}[M(S(t))] = \sum_{k \in \mathbb{Z}^n_{\geq 0}} p_k(t)\, M(k)$, since $p_k(t) = \mathbb{P}[S(t) = k]$. Let us define, for any $\gamma \in \mathbb{Z}^n$, the new function $\Delta_\gamma M$ given by $(\Delta_\gamma M)(k) := M(k + \gamma) - M(k)$. With these notations,

$$\frac{d}{dt}\mathbb{E}[M(S(t))] = \sum_{j=1}^{m} \mathbb{E}\left[\rho_j(S(t))\, \Delta_{\gamma_j} M(S(t))\right] \tag{22.4}$$

(see [11] for more details). We next specialize to a monomial function: $M(k) = k^u = k_1^{u_1} k_2^{u_2} \ldots k_n^{u_n}$ where $u \in \mathbb{Z}^n_{\geq 0}$. There results $(\Delta_{\gamma_j} M)(k) = \sum_{\nu \in \mathscr{I}(u,j)} d_\nu k^\nu$ for appropriate coefficients d_ν, where

$$\mathscr{I}(u, j) := \left\{ \nu \in \mathbb{Z}^n_{\geq 0} \;\middle|\; \begin{array}{c} \nu = u - \mu, \ u \geq \mu \neq 0 \\ \mu_i = 0 \text{ for each } i \text{ such that } \gamma_{ij} = 0 \end{array} \right\}$$

(inequalities "$\geq$" in $\mathbb{Z}^n_{\geq 0}$ are understood coordinatewise). Thus, for (22.3):

$$\frac{d}{dt}\mathbb{E}\left[S(t)^u\right] = \sum_{j=1}^{m} \sum_{c_j \leq a_j} \sum_{\nu \in \mathcal{I}(u,j)} d_\nu \kappa_{c_j} \mathbb{E}\left[S(t)^{\nu + c_j}\right]. \tag{22.5}$$

In other words, we can recursively express the derivative of the moment of order u as a linear combination of other moments. This results in an infinite set of coupled linear ordinary differential equations, so it is natural to ask whether, for given a particular moment or order u of interest, there is a finite set of moments, including the desired one, that satisfies a finite set of differential equations. This question can be reformulated combinatorially, as follows. For each multi-index $u \in \mathbb{Z}^n_{\geq 0}$, let us define $\mathcal{R}^0(u) = \{u\}, \mathcal{R}^1(u) := \left\{\nu + c_j\,,\ 1 \leq j \leq m, c_j \leq a_j,\ \nu \in \mathcal{I}(u, j)\right\}$, and, more generally, for any $\ell \geq 1$, $\mathcal{R}^{\ell+1}(u) := \mathcal{R}^1(\mathcal{R}^\ell(u))$ where, for any set U, $\mathcal{R}^\ell(U) := \bigcup_{u \in U} \mathcal{R}^\ell(u)$. Finally, we set $\mathcal{R}(u) := \bigcup_{i=0}^{\infty} \mathcal{R}^i(u)$. Each set $\mathcal{R}^\ell(u)$ is finite, but the cardinality $\#(\mathcal{R}(u))$ may be infinite. It is finite if and only if there is some $L \geq 0$ such that $\mathcal{R}(u) = \bigcup_{i=0}^{L} \mathcal{R}^i(u)$, or equivalently $\mathcal{R}^{L+1}(u) \subseteq \bigcup_{i=0}^{L} \mathcal{R}^i(u)$.

Equation (22.5) says that the derivative of the u-th moment can be expressed as a linear combination of the moments in the set $\mathcal{R}^1(u)$. The derivatives of these moments, in turn, can be expressed in terms of the moments in the set $\mathcal{R}^1(u')$, for each $u' \in \mathcal{R}^1(u)$, i.e. in terms of moments in the set $\mathcal{R}^2(u)$. Iterating, we have the following: "Finite reachability implies linear moment closure" observation:

Lemma. Suppose $N := \#(\mathcal{R}(u)) < \infty$, and $\mathcal{R}(u) = \{u = u_1, \ldots, u_N\}$. Then, with $x(t) := (\mathbb{E}[S^{u_1}(t)], \ldots, \mathbb{E}[S^{u_N}(t)])'$, there is an $A \in \mathbb{R}^{N \times N}$ so that $\dot{x}(t) = Ax(t), t \geq 0$.

A classical case is when all reactions have order 0 or 1, i.e., $\oplus a_j \in \{0, 1\}$. Since $\mu \neq 0$ in the definition of $\mathcal{I}(u, j)$, it follows that $\oplus a_j \leq \oplus\mu$ for every index j. Therefore, $\oplus(\nu + a_j) = \oplus u + \oplus a_j - \oplus\mu \leq \oplus u$ for all u, and the same holds for $\nu + c_j$ if $c_j \leq a_j$. So all elements in $\mathcal{R}(u)$ have degree $\leq \oplus u$, and thus $\#(\mathcal{R}(u)) < \infty$. A more general case is as follows.

22.3 Feedforward Networks

A chemical network is of *feedforward type (FFN)* if one can partition its n species $S_i, i \in \{1, 2, \ldots, n\}$ into p layers $\mathbf{S}_1, \ldots, \mathbf{S}_p$ and there are a total of $m' = m + d$ reactions, where d of the reactions are "pure degradation" (or "dilution") reactions D_j : $S_{i_j} \to 0, j \in \{1, \ldots, d\}$ and the additional m reactions $R_j, j \in \{1, 2, \ldots, m\}$ can be partitioned into $p \geq 1$ layers $\mathbf{R}_1, \ldots, \mathbf{R}_p$ in such a manner that, in the each reaction layer R_π there may be any number of order-zero or order-one reactions involving species in layer π, but every higher order reaction at a layer $\pi > 1$ must have the form: $a_{i_1 j} S_{i_1} + \cdots a_{i_q j} S_{i_q} \to a_{i_1 j} S_{i_1} + \cdots a_{i_q j} S_{i_q} + b_{i_{q+1} j} S_{i_{q+1}} + \cdots b_{i_{q+q'} j} S_{i_{q+q'}}$, where all the species $S_{i_1}, \ldots, S_{i_q}$ belong to layers having indices $< \pi$, and the species $S_{i_{q+1}}, \ldots, S_{i_{q+q'}}$ are in layer π. In other words, multimers of species in "previous" layers can "catalyze" the production of species in the given layer, but are not affected by these reactions. This can be summarized by saying that for reactions at any given layer π, the only species that appear as reactants in nonlinear reactions are those in layers $< \pi$ and the only ones that can change are those in layer π.

A more formal way to state the requirements is as follows. The reactions R_j that belong to the first layer $\mathbf{R}_1$ are all of order-zero or one, i.e. they have $\oplus a_j \in \{0, 1\}$ (this first layer might model several independent separate chemical subnetworks; we collect them all as one larger network), and

$$\text{if } R_j \in \mathbf{R}_\pi : \begin{cases} a_{ij} \neq 0 \text{ and } \oplus a_j > 1 \Rightarrow S_i \in \bigcup_{1 \leq s < \pi} \mathbf{S}_\pi \\ \gamma_{ij} \neq 0 \qquad\qquad\qquad \Rightarrow S_i \in \mathbf{S}_\pi \,. \end{cases} \tag{22.6}$$

FFN's have the finite reachability property ([12]): given any desired moment u, there is a linear differential equation $\dot{x}(t) = Ax(t)$ for a suitable set of N moments $x(t) := (\mathbb{E}[S^{u_1}(t)], \ldots, \mathbb{E}[S^{u_N}(t)])'$, which contains the moment u of interest. Notice that steady-state moments can then be computed by solving $Ax = 0$. The proof uses a Lyapunov-like construction. In practice, we simply compute (22.5) starting from the desired moment, then recursively apply the same rule to the moments appearing on the right-hand side, and so forth until no new moments appear. The integer N at which the system closes might be very large, but the procedure is guaranteed to stop. The last section of the paper [12] explains how certain non-feedforward networks also lead to moment closure, provided that conservation laws ensure that variables appearing in nonlinear reactions take only a finite set of possible values.

Steady States of CME. Often, the interest is in long-time behavior, after a transient, that is to say in the probabilistic *steady state* of the system: the joint distribution of the random variables $S_i = S_i(\infty)$ that result in the limit as $t \to \infty$ (provided that such a limit exists in an appropriate technical sense). This joint distribution is a solution of the steady-state CME (ssCME), the infinite set of linear equations obtained by setting the right-hand side of the CME to zero, that is:

$$\sum_{j=1}^{m} \rho_j(k - \gamma_j)\, p_{k-\gamma_j} = \sum_{j=1}^{m} \rho_j(k)\, p_k \,, \quad k \in \mathbb{Z}^n_{\geq 0} \tag{22.7}$$

with the convention that $\rho_j(k - \gamma_j) = 0$ unless $k \geq \gamma_j$. When substituting mass-action propensities $\rho_j(k) = \kappa_j \prod_{i=1}^{n} \binom{k_i}{a_{ij}} \mathscr{H}(k - a_j)$ the steady-state equation (22.7) becomes:

$$\sum_{j=1}^{m} \kappa_j \prod_{i=1}^{n} \binom{k_i - \gamma_{ij}}{a_{ij}} \mathscr{H}(k - b_j)\, p_{k-\gamma_j} = \sum_{j=1}^{m} \kappa_j \prod_{i=1}^{n} \binom{k_i}{a_{ij}} \mathscr{H}(k - a_j)\, p_k \tag{22.8}$$

for all $k \in \mathbb{Z}^n_{\geq 0}$. Equivalently, for all $k \in \mathbb{Z}^n_{\geq 0}$:

$$\sum_{j=1}^{m} \widetilde{\kappa}_j \prod_{i=1}^{n} \frac{(k_i - \gamma_{ij})!}{(k_i - b_{ij})!} \mathscr{H}(k - b_j)\, p_{k-\gamma_j} = \sum_{j=1}^{m} \widetilde{\kappa}_j \prod_{i=1}^{n} \frac{k_i!}{(k_i - a_{ij})!} \mathscr{H}(k - a_j)\, p_k \tag{22.9}$$

when introducing new constants $\widetilde{\kappa}_j := \kappa_j / \prod_{i=1}^n (a_{ij}!)$. Writing $\lambda^k := \lambda_1^{k_1} \dots \lambda_n^{k_n}$ and $k! := k_1! \dots k_n!$ for each $k = (k_1, \dots, k_n) \in \mathbb{Z}^n_{\geq 0}$ and $\lambda = (\lambda_1, \dots, \lambda_n) \in \mathbb{R}^n_{>0}$, (22.9) is:

$$\sum_{j=1}^{m} \widetilde{\kappa}_j \frac{(k - \gamma_j)!}{(k - b_j)!} \mathscr{H}(k - b_j)\, p_{k-\gamma_j} = \sum_{j=1}^{m} \widetilde{\kappa}_j \frac{k!}{(k - a_j)!} \mathscr{H}(k - a_j)\, p_k\,, \quad k \in \mathbb{Z}^n_{\geq 0} \tag{22.10}$$

Since (22.10) is a linear equation on the $\{p_k, k \in \mathbb{Z}^n_{\geq 0}\}$, any rescaling p_k's will satisfy the same equation; for probability densities, one normalizes to a unit sum.

If there are conservation laws satisfied by the system then steady-state solutions will not be unique, and the equation $Ax = 0$ must be supplemented by a set of linear constraints that uniquely specify the solution. For example, consider a reversible reaction $S_1 \underset{\kappa_2}{\overset{\kappa_1}{\rightleftharpoons}} S_2$ (propensities are mass-action, $\rho_i(S_1, S_2) = \kappa_i S_i$). The first moments (means) satisfy $\dot{x}_1 = \kappa_2 x_2 - \kappa_1 x_1$ and $\dot{x}_2 = \kappa_1 x_1 - \kappa_2 x_2$. Any vector $(\bar{\xi}_1, \bar{\xi}_2)$ with $\kappa_1 \bar{\xi}_1 = \kappa_2 \bar{\xi}_2$ is a steady state of these equations. However, the sum of the numbers of molecules S_1 and S_2 is conserved in the reactions. Given a particular total number, β, the differential equations can be reduced to just one equation, say for x_1: $\dot{x}_1 = \kappa_2(\beta - x_1) - \kappa_1 x_1 = -(\kappa_1 + \kappa_2)x_1 + \kappa_2 \beta$, which has the affine form $\dot{x} = Ax + b$. At steady state, we have the unique solution $\bar{\xi}_1 = \beta\kappa_2/(\kappa_1 + \kappa_2)$, $\bar{\xi}_2 = \beta\kappa_1/(\kappa_1 + \kappa_2)$ obtained by imposing the constraint $\bar{\xi}_1 + \bar{\xi}_2 = \beta$. It can easily be proved (see e.g. [13]) that at steady state, S_1 is a binomial random variable $B(\beta, p)$ with $p = \frac{1}{1+\mu}$, where $\mu = \kappa_1/\kappa_2$. We later discuss further conservation laws.

A Worked Example. For networks with only zero and first-order reactions, which are feedforward, it is well known that one may compute all moments in closed form. For example, start with a reversible reaction $S_1 \underset{\delta}{\overset{\kappa}{\rightleftharpoons}} S_2$ with mass-action propensities, thinking of S_1 as the active form of a certain gene and S_2 as the inactive form of this gene. Transcription and translation are summarized, for simplicity, as one reaction $S_1 \overset{\rho}{\rightarrow} S_1 + S_3$ and degradation or dilution of the gene product S_3 is a linear reaction $S_3 \overset{\eta}{\rightarrow} \emptyset$. The stoichiometry matrix is $\Gamma = [-1, 1, 0, 0; 1, -1, 0, 0; 0, 0, 1, -1]$. Suppose, we are interested in the mean and variance of S_3 subject to the conservation law $S_1 + S_2 = \beta$, for some fixed positive integer β. A linear differential equation for these second-order moments: $\mathscr{M} = (E[S_1], E[S_1^2], E[S_1 S_3], E[S_3], E[S_3^2])'$ is $\dot{\mathscr{M}} = A\mathscr{M} + b$, where $A = [-\delta - \kappa, 0, 0, 0, 0; \kappa - \delta + 2\,\delta\,\beta, -2\,\delta - 2\,\kappa, 0, 0, 0; 0, \rho, -\delta - \eta - \kappa, \delta\,\beta, 0; \rho, 0, 0, -\eta, 0; \rho, 0, 2\,\rho, \eta, -2\,\eta]$ and $b = [\delta\,\beta; \delta\,\beta; 0; 0; 0]$. One can then solve $A\overline{\mathscr{M}} + b = 0$ to obtain steady state moments.

A Simple Nonlinear Example. We consider a feedforward system with three species; S_1 catalyzes production S_2, and S_1 and S_2 are both needed to produce S_3: $0 \xrightarrow{\kappa_1} S_1 \xrightarrow{\delta_1} 0$, $S_1 \xrightarrow{\kappa_2} S_1 + S_2$, $S_2 \xrightarrow{\delta_2} 0$, $S_1 + S_2 \xrightarrow{\kappa_3} S_1 + S_2 + S_3$, $S_3 \xrightarrow{\delta_3} 0$. Computing $E[S_3]$, the mean of S_3, requires a minimal differential equation of order 5, for the moments $\mathscr{M} = (E[S_3], E[S_1S_2], E[S_2], E[S_1^2], E[S_1])'$ and has form $\dot{\mathscr{M}} = A\mathscr{M} + b$, where $A = [-\delta_2, \kappa_3, 0, 0, 0; 0, -\delta_1 - \delta_2, \kappa_1, \kappa_2, 0; 0, 0, -\delta_2, 0, \kappa_2; 0, 0, 0, -2\,\delta_1, 2\,\kappa_1 + \delta_1; 0, 0, 0, 0, -\delta_1]$ and $b = [0; 0; 0; \kappa_1; \kappa_1]$,

22.4 Poisson-Like Solutions and Complex Balanced Networks

We observe that for any given positive vector $\bar{\lambda} \in \mathbb{R}^n_{>0}$, the set of numbers

$$\Pi = \left\{p_k = \bar{\lambda}^k/k!\,, \quad k \in \mathbb{Z}^n_{\geq 0}\right\} \tag{22.11}$$

satisfies the ssCME equations (22.10) if and only if

$$\sum_{j=1}^{m} \widetilde{\kappa}_j \frac{\bar{\lambda}^{k-\gamma_j}}{\left(k - b_j\right)!} \mathscr{H}(k - b_j) = \sum_{j=1}^{m} \widetilde{\kappa}_j \frac{\bar{\lambda}^{k}}{\left(k - a_j\right)!} \mathscr{H}(k - a_j)\,, \quad k \in \mathbb{Z}^n_{\geq 0}\,, \tag{22.12}$$

Rewriting this as:

$$\sum_{c\in\mathscr{C}} \sum_{\{j|b_j=c\}} \widetilde{\kappa}_j \frac{\bar{\lambda}^{k-\gamma_j}}{\left(k - b_j\right)!} \mathscr{H}(k - b_j) = \sum_{c\in\mathscr{C}} \sum_{\{j|a_j=c\}} \widetilde{\kappa}_j \frac{\bar{\lambda}^{k}}{(k - c)!} \mathscr{H}(k - a_j), k \in \mathbb{Z}^n_{\geq 0}, \tag{22.13}$$

a sufficient condition for (22.11) to be a solution is that

$$\sum_{\{j|b_j=c\}} \widetilde{\kappa}_j \frac{\bar{\lambda}^{k-\gamma_j}}{(k - c)!} \mathscr{H}(k - b_j) = \sum_{\{j|a_j=c\}} \widetilde{\kappa}_j \frac{\bar{\lambda}^{k}}{(k - c)!} \mathscr{H}(k - a_j)\,, \quad k \in \mathbb{Z}^n_{\geq 0}$$

for each individual complex $c \in \mathscr{C}$, or, equivalently,

$$\frac{\mathscr{H}(k-c)}{(k-c)!} \sum_{\{j|b_j=c\}} \widetilde{\kappa}_j \bar{\lambda}^{k-\gamma_j} = \frac{\mathscr{H}(k-c)}{(k-c)!} \sum_{\{j|a_j=c\}} \widetilde{\kappa}_j \bar{\lambda}^{k}\,, \quad k \in \mathbb{Z}^n_{\geq 0}\,.$$

A sufficient condition for this to hold is that, for all complexes:

$$\sum_{\{j|b_j=c\}} \widetilde{\kappa}_j \bar{\lambda}^{a_j} = \sum_{\{j|a_j=c\}} \widetilde{\kappa}_j \bar{\lambda}^{a_j}\,, \quad k \in \mathbb{Z}^n_{\geq 0} \tag{22.14}$$

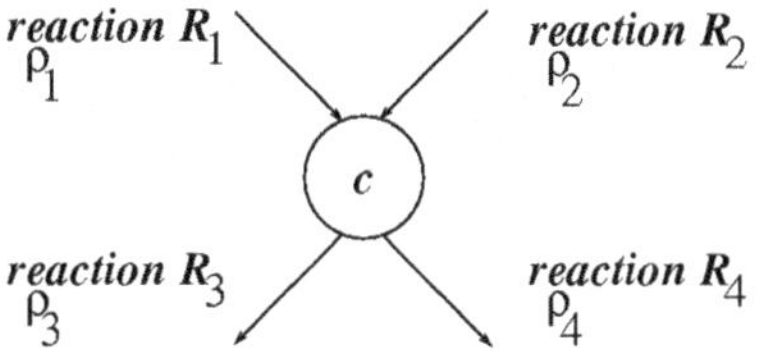

Fig. 22.1 Complex balancing: outflows and inflows must balance at each complex c. The left-hand side of (22.14) is $\widetilde{\kappa}_3\bar{\lambda}^{a_3} + \widetilde{\kappa}_4\bar{\lambda}^{a_4}$ and the right-hand side is $\widetilde{\kappa}_1\bar{\lambda}^{a_1} + \widetilde{\kappa}_2\bar{\lambda}^{a_2}$

(conversely, this last condition is necessary for all complexes for which $k \geq c$). One can write "$\bar{\lambda}^c$" and bring this term outside of the sum, in the right-hand side.

When property (22.14) holds for every complex, one says that $\bar{\lambda}$ is a *complex balanced steady state* of the associated *deterministic* chemical reaction network. (That is, the system of differential equations $\dot{x} = \Gamma Q(x)$, where $Q(x)$ is a column vector of size m whose jth entry is $\rho_j(x)$ and $x(t) \in \mathbb{R}^n_{\geq 0}$ for all t.) Complex balancing means that, for each complex, outflows and inflows balance out. This is a Kirschoff current law (in-flux = out-flux, at each node). See Fig. 22.1.

Foundational results in deterministic chemical network theory were obtained by Horn, Jackson, and Feinberg ([2, 3]). One of the key theorems is that a sufficient condition for the existence of a complex balanced steady state is that the network be *weakly reversible* and have *deficiency zero*. The deficiency is computed as $n_c - \ell - r$, where n_c is the number of complexes, r is the rank of the matrix Γ, and ℓ is the number of "linkage classes" (connected components of the reaction graph). Weak reversibility means that each connected component of the reaction graph must be strongly connected. One of the most interesting features of this theorem is that no assumptions need to be made about the kinetic constants. (Of course, the choice of the vector $\bar{\lambda}$ will depend on the kinetic constants.) We refer the reader to the citations for details on deficiency theory, as well as, of interest in the present context, several examples discussed in [13]. The theorems for weakly reversible deficiency zero networks are actually far stronger, and they show that every possible steady state of the corresponding deterministic network is complex balanced, and that they are asymptotically stable relative to stoichiometry classes. The connection with ssCME solutions was a beautiful observation made in [1], but can be traced to the "nonlinear traffic equations" from queuing theory, described in Kelly's textbook [7], Chap. 8 (see also [8] for a discussion),

The elements of Π given by formula (22.11) add up to:

$$\sum_{k \in \mathbb{Z}^n_{\geq 0}} p_k = \sum_{k_1=0}^{\infty} \cdots \sum_{k_n=0}^{\infty} \frac{\bar{\lambda}_1^k}{k_1!} \cdots \frac{\bar{\lambda}_n^k}{k_n!} = Z := e^{\bar{\lambda}_1} \cdots e^{\bar{\lambda}_n}$$

Thus, normalizing by the total, $\{p_k/Z, k \in \mathbb{Z}^n_{\geq 0}\}$ is a probability distribution. However, because of stoichiometric constraints, solutions are typically not unique, and general solutions appear as convex combinations of solutions corresponding to invariant subsets of states. A solution with only a finite number of nonzero p_k's will then have a different normalization factor Z.

Conservation Laws, Complex Balanced Case. When steady states do not form an irreducible Markov chain, the solutions of the form (22.11) are not the only solutions in the complex balanced case. Restrictions to each component of the Markov chain are also solutions, as are convex combinations of such restrictions. To formalize this idea, suppose that there is some subset $\mathscr{Z}_0 \subseteq \mathbb{Z}^n$ with the following stoichiometric invariance property: $k \in \mathscr{Z}_0 \Rightarrow k \pm \gamma_j \in \mathscr{Z}_0$ for all $j = 1, \ldots, m$. (The same property is then true for the complement of $\mathscr{Z}_0$.) Consider, the set $\mathscr{Z} := \mathscr{Z}_0 \bigcap \mathbb{Z}^n_{\geq 0}$. For each $k \in \mathscr{Z}$, the left-hand side term in Eq. (22.12) either involves an index $k - \gamma_j > 0$, and hence, in $\mathscr{Z}$, or it is zero (because $k - b_j \geq 0$ implies $k - \gamma_j \geq 0$) and so it does not matter that $k - \gamma_j \notin \mathscr{Z}$. Thus,

$$p_k = \frac{\bar{\lambda}^k}{k!} \text{ if } k \in \mathscr{Z}, \quad = 0 \text{ if } k \in \mathbb{Z}^n_{\geq 0} \setminus \mathscr{Z} \tag{22.15}$$

is also a solution, in the complex balanced case (observe that, for indices in $\mathbb{Z}^n_{\geq 0} \setminus \mathscr{Z}$, Eq. (22.12) is trivially satisfied, since both sides vanish). So we need to divide by the sum Z of the elements in (22.15) in order to normalize to a probability distribution. The restriction to $\mathscr{Z}$ will the unique steady-state distribution provided that the restricted Markov chain has appropriate irreducibility properties.

In particular, suppose that the nullspace of $\mathscr{A} = (\alpha_{ij}) \in \mathbb{R}^{m \times n}$ includes $\mathscr{C}$ (for example, $\mathscr{A}$ could be the orthogonal complement of the "stoichiometric subspace" spanned by $\mathscr{C}$), and pick any vector $\beta = (\beta_1, \ldots, \beta_q)' \in \mathbb{R}^q$. Then $\mathscr{Z}_0 = \{k | \mathscr{A} k = \beta\}$ has the invariance property, and the sum of the elements in (22.15) is:

$$Z(\beta_1, \ldots, \beta_q) = \sum_{\substack{k_1, \ldots, k_n \geq 0 \\ \mathscr{A} k = \beta}} \frac{\lambda_1^{k_1}}{k_1!} \frac{\lambda_2^{k_2}}{k_2!} \cdots \frac{\lambda_n^{k_n}}{k_n!}$$

(zero if sum empty). The normalized form of (22.15) has $p_k = 0$ for $k \in \mathbb{Z}^n_{\geq 0} \setminus \mathscr{Z}$, and

$$p_k = \frac{1}{Z(\beta_1, \ldots, \beta_q)} \frac{\lambda_1^{k_1}}{k_1!} \frac{\lambda_2^{k_2}}{k_2!} \cdots \frac{\lambda_n^{k_n}}{k_n!} \tag{22.16}$$

for $k \in \mathscr{Z}$. A probabilistic interpretation is as follows. Suppose given n independent Poisson random variables, S_i, $i = 1, \ldots, n$, with parameters λ_i respectively, so

$$\mathbb{P}[S_1 = k_1, S_2 = k_2, \ldots, S_n = k_n] = e^{-(\lambda_1 + \ldots + \lambda_n)} \frac{\lambda_1^{k_1}}{k_1!} \frac{\lambda_2^{k_2}}{k_2!} \cdots \frac{\lambda_n^{k_n}}{k_n!} \tag{22.17}$$

for $k \geq 0$ (and zero otherwise). Let us introduce the following new random variables: $Y_j := \sum_{i=1}^n \alpha_{ji} S_i,\ j = 1, \ldots, q$. Observe that $\mathbb{P}\left[Y_1 = \beta_1, \ldots, Y_m = \beta_q\right]$ equals

$$\sum_{\substack{k_1, \ldots, k_n \geq 0 \\ \alpha_{11}k_1 + \ldots + \alpha_{1n}k_n = \beta_1, \ \ldots, \ \alpha_{q1}k_1 + \ldots + \alpha_{qn}k_n = \beta_q}} \mathbb{P}[S_1 = k_1, S_2 = k_2, \ldots, S_n = k_n]$$

which is $e^{-(\lambda_1+\ldots+\lambda_n)} Z(\beta_1, \ldots, \beta_q)$. Therefore, for each $k \in \mathscr{X}$, p_k in (22.16) equals the conditional probability $\frac{\mathbb{P}[S_1=k_1, S_2=k_2, \ldots, S_n=k_n]}{\mathbb{P}[Y_1=\beta_1, \ldots, Y_q=\beta_q]}$, which is the same as

$$\mathbb{P}\left[S_1 = k_1, S_2 = k_2, \ldots, S_n = k_n \,\middle|\, Y_1 = \beta_1, \ldots, Y_q = \beta_q\right].$$

If our interest is in computing this conditional probability, the main effort goes into computing $Z(\beta_1, \ldots, \beta_q)$. The main contribution of the paper [13] was to provide effective algorithms for the computation of $Z(\beta_1, \ldots, \beta_q)$ recursively on the β_i's. A package for that purpose, called MVPoisson, was included with that paper.

Conditional moments $E[S_j^r \mid Y_1 = \beta_1, \ldots, Y_m = \beta_q]$, $r \geq 1$, including the conditional expectation (when $r = 1$), as well as centered moments such as the conditional variance, can be computed once that these conditional probabilities are known. It is convenient for that purpose to first compute the factorial moments. Recall that, the rth factorial moment $E[W^{(r)}]$ of a random variable W is defined as the expectation of $W!/(W - r)!$. For example, when $r = 1$, $E[W^{(r)}] = E[W]$, and for $r = 2$, $E[W^{(r)}] = E[W^2] - E[W]$, and thus, the mean and variance of W can be obtained from these. We denote the conditional factorial moment of S_i given $Y = \beta$, as $E[S_j^{(r)} \mid Y]$. It is not difficult to see (Theorem 2 in [13]) that:

$$E[S_j^{(r)} \mid Y] \;=\; \lambda_j^r \cdot \frac{Z(\beta_1 - r\alpha_{1j}, \beta_2 - r\alpha_{2j}, \ldots, \beta_q - r\alpha_{qj})}{Z(\beta_1, \ldots, \beta_q)}$$

when all $\beta_i - r\alpha_{ij} \geq 0$ and zero otherwise. The paper [13] discusses mixed moments such as covariances too. For example, for $r = 1$ we have the conditional mean:

$$E[S_j \mid Y] \;=\; \lambda_j \cdot \frac{Z(\beta_1 - \alpha_{1j}, \beta_2 - \alpha_{2j}, \ldots, \beta_q - \alpha_{qj})}{Z(\beta_1, \ldots, \beta_q)} \tag{22.18}$$

when all $\beta_i \geq \alpha_{ij}$, and zero otherwise, and for $r = 2$ the conditional second moment:

$$E[S_j^2 \mid Y] \;=\; \lambda_j^2 \cdot \frac{Z(\beta_1 - 2\alpha_{1j}, \beta_2 - 2\alpha_{2j}, \ldots, \beta_q - 2\alpha_{qj})}{Z(\beta_1, \ldots, \beta_q)} \;+\; E[S_j \mid Y]$$

when all $\beta_i \geq 2\alpha_{ij}$, and zero otherwise. We next work out a concrete example.

Worked Example: Simple Binding. Suppose that two molecules of species S_1 and S_2 can reversibly combine through a bimolecular reaction to produce a molecule of species S_3: $S_1 + S_2 \underset{\kappa_2}{\overset{\kappa_1}{\rightleftharpoons}} S_3$. Since the deficiency of this network is $n_c - \ell - r = 2 - 1 - 1 = 0$ and it is reversible and hence weakly reversible as well, we know that there is a complex balanced equilibrium (and every equilibrium is complex balanced). We may pick, for example, $\bar{\lambda} = (1, 1, K)$, where $K := \kappa_1/\kappa_2$. The count of S_1 molecules goes down by one every time that a reaction takes place, at which time the count of S_3 molecules goes up by one. Thus, the sum of the number of S_1 molecules plus the number of S_3 molecules remains constant in time, equal to their starting value, which we denote as p. Similarly, the sum of the number of S_2 molecules plus the number of S_3 molecules remains constant, equal to some number n. (In the general notations, we have $a_{11} = a_{13} = 1$, $a_{22} = a_{23} = 1$, $a_{12} = a_{21} = 0$, $\beta_1 = p, \beta_2 = n$.) In the steady-state limit as $t \to \infty$, these constraints persist. In other words, all p_k should vanish except those corresponding to vectors $k = (k_1, k_2, k_3)$ such that $k_1 + k_3 = p$ and $k_2 + k_3 = n$. The set consisting of all such vectors is invariant, so

$$p_k = \begin{cases} \dfrac{\bar\lambda_1^{k_1}}{k_1!} \dfrac{\bar\lambda_2^{k_2}}{k_2!} \dfrac{\bar\lambda_3^{k_3}}{k_3!} & \text{if } k_1 + k_3 = p \text{ and } k_2 + k_3 = n \\ 0 & \text{otherwise} \end{cases}$$

is a solution of the ssCME. In order to obtain a probability density, we must normalize by the sum $Z(p, n)$ of these p_k's. Because of the two constraints, the sum can be expressed in terms of just one of the indices, let us say k_1. Observe that, since $k + k_3 = p$ and $k_3 \geq 0$, necessarily $k \leq p$. Since $k_2 = n - k_3 = n + k - p$ must be nonnegative, we also have the constraint $k \geq \max\{0, p - n\}$. So the only nonzero terms are for $k \in \{\max\{0, p - n\}, \ldots, p\}$. With $k_3 = p - k, k_2 = n - k_3 = n + k - p$, we have:

$$Z(p, n) = \sum_{\ell=\max\{0,p-n\}}^{p} \frac{K^{p-\ell}}{\ell!\,(n + \ell - p)!\,(p - \ell)!} = \sum_{\ell=0}^{\min\{p,n\}} \frac{K^{\ell}}{(p - \ell)!\,(n - \ell)!\,\ell!} \tag{22.19}$$

The second form if the summation makes it obvious that $Z(p, n) = Z(n, p)$.

When $n \geq p$, we can also write

$$Z(p, n) = \frac{1}{n!p!} \sum_{\ell=0}^{p} \frac{n!}{(n - p + \ell)!} \binom{p}{\ell} K^{p-\ell} \tag{22.20}$$

which shows the expression as a rational function in which the numerator is a polynomial of degree p on n. This was derived assuming that $n \geq p$, and the factorials in the denominator do not make sense otherwise. However, let us think of each term $\frac{n!}{(n-p+\ell)!}$ as the product $n(n - 1) \ldots (n - p + \ell + 1)$, which may include zero as well as negative numbers. With this understanding, the formula in (22.20) makes sense even when $n < p$. Observe that such a term vanishes for any index $\ell < p - n$. Thus, for $n < p$, (22.20) reduces to: $\frac{1}{p!} \sum_{\ell=p-n}^{p} \frac{1}{(n-p+\ell)!} \binom{p}{\ell} K^{p-\ell}$ or equivalently, with a change of indices $\ell = p - \ell$ and then using $\binom{p}{p-\ell} = \binom{p}{\ell}$:

$$\frac{1}{p!} \sum_{\ell=0}^{n} \frac{1}{(n - \ell)!} \binom{p}{p - \ell} K^{\ell} = \frac{1}{p!} \sum_{\ell=0}^{n} \frac{1}{(n - \ell)!} \binom{p}{\ell} K^{\ell} = \sum_{\ell=0}^{n} \frac{K^{\ell}}{(n - \ell)!(p - \ell)!\ell!}.$$

In this last form, we have the same expression as the last one in (22.19). In conclusion, provided that we interpret the quotient of combinatorial numbers in (22.20) as a product that may be zero, formula (22.20) is valid for all n and p, not just for $n \geq p$. In particular, we have; $Z(0,n) = \frac{1}{n!}$, $Z(1,n) = \frac{1}{n!}(Kn+1)$, $Z(2,n) = \frac{1}{2n!}\left(K^2n^2 + \left(-K^2+2\,K\right)n+1\right)$, etc. In terms of the Gauss's hypergeometric function ${}_2\mathrm{F}_0$, we can also write: $Z(p,n) = \frac{1}{p!\,n!}\,{}_2\mathrm{F}_0(-n,-p;\ ;\ K)$. The recursion on n obtained by using the package MVPoisson from [13] is as follows (by symmetry, a recursion on p can be found by exchanging n and p):

$$Z(p,n+2) \;=\; \frac{K}{n+2}\,Z(p,n) + \frac{-Kn+Kp-K+1}{n+2}\,Z(p,n+1)\,.$$

Now (22.18) gives the conditional mean of the first species, S_1 ($j=1$ for this index, $r=1$ for the first moment, and $\lambda_1^1 = 1^1 = 1$) as zero if $p<1$ or $n<0$ and otherwise

$$\varphi(p,n) \;:=\; E[S_1 \mid S_1+S_3=p,\, S_2+S_3=n] \;=\; \frac{Z(p-1,n)}{Z(p,n)}\,.$$

For example, $\varphi(1,n) = \frac{1}{Kn+1}$, $\varphi(2,n) = \frac{2(Kn+1)}{K^2n^2+(-K^2+2\,K)n+1}$.

Worked Example: Synthesis and Degradation, and Binding. Suppose molecules of species S_1 can be randomly created and degraded, and they can also reversibly combine with molecules of S_2 through a bimolecular reaction to produce molecules of species S_3: $\emptyset \xrightarrow{\kappa_1} S_1 \xrightarrow{\kappa_2} \emptyset$, $S_1+S_2 \underset{\kappa_4}{\overset{\kappa_3}{\rightleftharpoons}} S_3$. There are $n_c = 4$ complexes: $\emptyset, S_1, S_1+S_2$, and S_3, and $\ell = 2$ linkage classes. The stoichiometry matrix $\Gamma = [1,-1,-1,1;\ 0,0,-1,1;\ 0,0,1,-1]$ has rank $r=2$, so the deficiency of this weakly reversible network is $n_c - \ell - r = 4-2-2 = 0$. Thus, there is a complex balanced equilibrium (and every equilibrium is complex balanced). We may pick, for example, $\bar{\lambda} = (\lambda, 1, \mu)$, where $\lambda := \frac{\kappa_1}{\kappa_2}$ and $\mu := \frac{\kappa_1\kappa_3}{\kappa_2\kappa_4}$. Notice that, there is only one nontrivial conserved quantity, $S_2+S_3=n$, since S_1 is not conserved. We have:

$$Z(n) \;=\; \sum_{\substack{k_1,k_2,k_3\geq 0\\ k_2+k_3=n}} \frac{\lambda_1^{k_1}\,\lambda_2^{k_2}\,\lambda_3^{k_3}}{k_1!\,k_2!\,k_3!} \;=\; \sum_{k_1=0}^{\infty}\frac{\lambda^{k_1}}{k_1!}\sum_{k_2=0}^{n}\frac{\mu^{n-k_2}}{k_2!(n-k_2)!} \;=\; \frac{e^{\lambda}}{n!}(1+\mu)^n\,.$$

The normalized probability (22.15), for $k=(k_1,k_2,k_3)\geq 0$ with $k_2+k_3=n$, is: $p_k = \frac{1}{Z(n)}\frac{\lambda^{k_1}}{k_1!}\frac{1}{k_2!}\frac{\mu^{k_3}}{k_3!} = \frac{n!}{e^{\lambda}(1+\mu)^n}\frac{\lambda^{k_1}\mu^{k_3}}{k_1!k_2!k_3!}$ and as discussed earlier, this is the conditional probability $\mathbb{P}\left[S_1=k_1, S_2=k_2, S_3=k_3 \mid S_2+S_3=n\right]$. Using this expression, we may compute, for example, the conditional marginal distribution of S_2:

$$\mathbb{P}\left[S_2=r \mid S_2+S_3=n\right] = \sum_{k_1=0}^{\infty}\frac{n!}{e^{\lambda}(1+\mu)^n}\,\frac{\lambda^{k_1}\mu^{(n-r)}}{k_1!r!(n-r)!} \;=\; \binom{n}{r}\,p^r(1-p)^{(n-r)}$$

(where we use $p := 1/(1+\mu)$, so $\mu = \frac{1-p}{p}$), which shows this conditional marginal distribution is a binomial random variable with parameters n and $p = \frac{\kappa_2\kappa_4}{\kappa_2\kappa_4+\kappa_1\kappa_3}$.

Worked Example: Competitive Binding. We consider the following example (using now $A, B, \ldots$ for species to simplify notations):

$$A + B \underset{\kappa_{B,-}}{\overset{\kappa_{B,+}}{\rightleftharpoons}} D \qquad A + C \underset{\kappa_{C,-}}{\overset{\kappa_{C,+}}{\rightleftharpoons}} E$$

so for the associated deterministic system, the steady states satisfy $\kappa_{B,+}AB = \kappa_{B,-}D$ and $\kappa_{C,+}AC = \kappa_{C,-}E$, so one such equilibrium is $(1, 1, 1, \lambda, \mu)$ where $\lambda := \frac{\kappa_{B,+}}{\kappa_{B,-}}$, $\mu := \frac{\kappa_{C,+}}{\kappa_{C,-}}$. The following quantities are conserved: $A + D + E = n_A$, $B + D = n_B$, $C + E = n_C$ and subject to these constraints, one may pick the partition function:

$$Z(n_A, n_B, n_C) = \sum_{(k_A,k_B,k_C,k_D,k_E)\in\mathscr{S}} \frac{1}{k_A!}\frac{1}{k_B!}\frac{1}{k_C!}\frac{\lambda^{k_D}}{k_D!}\frac{\mu^{k_E}}{k_E!}$$

$$\mathscr{S} = \{(k_A, k_B, k_C, k_D, k_E) \geq 0 \mid k_A + k_D + k_E = n_A,\ k_B + k_D = n_B,\ k_C + k_E = n_C\}.$$

In order to rewrite this function as a double sum, we first show that $\mathscr{S}$ is equal to the following set:

$$\mathscr{S}' = \{(k_A, k_B, k_C, k_D, k_E) \mid 0 \leq k_D \leq n_B,\ 0 \leq k_E \leq \min\{n_A - k_D, n_C\},\ k_A = n_A - (k_D + k_E),\ k_B = n_B - k_D,\ k_C = n_C - k_E\}.$$

Indeed, suppose that $(k_A, k_B, k_C, k_D, k_E) \in \mathscr{S}$. Then $k_D \geq 0$ and from $k_B + k_D = n_B$ we have that $k_D = n_B - k_B \leq n_B$. Also, $k_E \geq 0$, and from $k_C + k_E = n_C$ we have that $k_E = n_C - k_C \leq n_C$ and from $k_A + k_D + k_E = n_A$ we have that $k_D + k_E = n_A - k_A \leq n_A$ and hence, $k_E \leq n_A - k_D$, so $k_E \leq \min\{n_A - k_D, n_C\}$. Thus, $(k_A, k_B, k_C, k_D, k_E) \in \mathscr{S}'$.

Conversely, suppose that $(k_A, k_B, k_C, k_D, k_E) \in \mathscr{S}'$. We have that k_D and k_E are nonnegative. From $k_E \leq n_A - k_D$, it follows that $k_A = n_A - (k_D + k_E) \geq 0$, from $k_D \leq n_B$, it follows $k_B = n_B - k_D \geq 0$, and from $k_E \leq n_C$, we have $k_C = n_C - k_E > 0$.

Therefore, we may rewrite the partition function as follows (using (i, j) instead of (k_D, k_E) as indices):

$$\begin{aligned} Z(n_A, n_B, n_C) &= \sum_{i=0}^{n_B} \frac{\lambda^i}{(n_B - i)!\,i!} \sum_{j=0}^{\min\{n_A-i,n_C\}} \frac{\mu^j}{((n_A - i) - j)!\,(n_C - j)!\,j!} \\ &= \sum_{i=0}^{n_B} \frac{\lambda^i}{(n_B - i)!\,i!} Q(n_A - i, n_C) = \frac{1}{n_A!} \sum_{i=0}^{n_B} \binom{n_A}{i} \frac{\lambda^i}{(n_B - i)!} \widetilde{Q}(n_A - i, n_C). \end{aligned}$$

where

$$Q(p,n) := \sum_{\ell=0}^{\min\{p,n\}} \frac{\mu^\ell}{(p-\ell)!(n-\ell)!\ell!}, \quad \widetilde{Q}(p,n) := p!Q(p,n) = \sum_{\ell=0}^{\min\{p,n\}} \binom{p}{\ell} \frac{\mu^\ell}{(n-\ell)!}.$$

The sum in $\widetilde{Q}$ is numerically better behaved than that in Q when p is large and n is small. We find that Q is itself the partition function $Z(p,n)$ given by formula (22.19) for the simpler binding example $S_1 + S_2 \rightleftharpoons S_3$ and can also be written as $\frac{1}{p!n!}\,{}_2F_0(-p, -n; \; ; \mu)$, in terms of ${}_2F_0$, Gauss's hypergeometric function.

For example, when $n_B = 0$ or 1, the formula specializes to: $Z(n_A, 0, n_C) = Q(n_A, n_C)$, $Z(n_A, 1, n_C) = Q(n_A, n_C) + \lambda\, Q(n_A - 1, n_C)$ (the first of these is not surprising, as when $n_B = 0$ the species B can only be zero, so the system reduces to the previous example, with $S_1 = A$, $S_2 = C$, and $S_3 = E$), and the mean of species D given the constraints $(n_A, 1, n_C)$ is by Eq. (22.18):

$$E[D \mid n_A, 1, n_C] = \lambda \frac{Z(n_A - 1, 0, n_C)}{Z(n_A, 1, n_C)} = \lambda \frac{Q(n_A - 1, n_C)}{Q(n_A, n_C) + \lambda\, Q(n_A - 1, n_C)}.$$

Using $\widetilde{Q}$, we may write, alternatively, $Z(n_A, 0, n_C) = \frac{1}{n_A!}\widetilde{Q}(n_A, n_C)$, $Z(n_A, 1, n_C) = \frac{1}{n_A!}\left(\widetilde{Q}(n_A, n_C) + \lambda\, n_A\, \widetilde{Q}(n_A - 1, n_C)\right)$ and thus, cancelling the $n_A!$ terms, and using that $Z(n_A - 1, 0, n_C) = \frac{n_A}{n_A!}\widetilde{Q}(n_A - 1, n_C)$, $E[D \mid n_A, 1, n_C] = \lambda \frac{n_A\, \widetilde{Q}(n_A-1, n_C)}{\widetilde{Q}(n_A, n_C) + \lambda\, n_A\, \widetilde{Q}(n_A-1, n_C)}$, which is far better behaved numerically when n_A is large.

We also remark that there is a third-order recursion for Z, obtained by the algorithm MVPoisson from [13].

In order to conveniently display the recurrences, let us use the following notations. We will write Z instead of $Z(b_1, b_2, b_3)$, and a notation like $Z_i^{+\cdots+}$ means a shift of the ith argument by the indicated number of plus signs. For example, Z_3^{++} means $Z(b_1, b_2, b_3 + 2)$. There are three recurrences of order three, as follows, for each of the three arguments: $(3 + b_1)Z_1^{+++} = \lambda\mu Z - (\lambda\mu b_1 - \lambda\mu b_2 - \lambda\mu b_3 + \lambda\mu - \lambda - \mu)Z_1^{+} - (\lambda b_1 - \lambda b_2 + \mu b_1 - \mu b_3 + 2\lambda + 2\mu - 1)Z_1^{++}$, $M(3 + b_3)(b_2 + 2)Z_2^{+++} = (\lambda^2 - \lambda\mu)Z + (\lambda^2 b_1 - \lambda^2 b_2 - \lambda\mu b_1 + 2\lambda\mu b_2 + \lambda\mu b_3 - \lambda^2 + 3\lambda\mu + \lambda - \mu)$ $Z_2^{+} + (\lambda\mu b_1 b_2 - \lambda\mu b_2{}^2 - \lambda\mu b_2 b_3 + 2\lambda\mu b_1 - 4\lambda\mu b_2 - 2\lambda\mu b_3 - 4\lambda\mu - \lambda b_2 + 2\mu b_2 - 2\lambda + 4\mu)Z_2^{++}$, $L(3 + b_3)(b_3 + 2)Z_3^{+++} = (-\lambda\mu + \mu^2)Z + (-\lambda\mu b_1 + \lambda\mu b_2 + 2\lambda\mu b_3 + \mu^2 b_1 - \mu^2 b_3 + 3\lambda\mu - \mu^2 - \lambda + \mu)Z_3^{+} + (\lambda\mu b_1 b_3 - \lambda\mu b_2 b_3 - \lambda\mu b_3{}^2 + 2\lambda\mu b_1 - 2\lambda\mu b_2 - 4\lambda\mu b_3 - 4\lambda\mu + 2\lambda b_3 - \mu b_3 + 4\lambda - 2\mu)Z_3^{++}$.
The algorithm provides 27 initial conditions, the values of Z for the triples $(1, 1, 1)$, $(1, 1, 2)$, $(1, 1, 3)$, ...$(3, 3, 3)$ listed in that order. We display them as three matrices, respectively shown below. The first matrix lists the elements of the form $(1, \star, \star)$, the next one $(2, \star, \star)$, and the last one $(3, \star, \star)$. In each matrix, elements are listed in the usual matrix order: $(\star, i, j)$ is the (i, j)th entry of the matrix.

$$\begin{bmatrix} \lambda+\mu+1 & \frac{\lambda}{2}+\mu+\frac{1}{2} & \frac{\lambda}{6}+\frac{\mu}{2}+\frac{1}{6} \\ \lambda+\frac{\mu}{2}+\frac{1}{2} & \frac{\lambda}{2}+\frac{\mu}{2}+\frac{1}{4} & \frac{\lambda}{6}+\frac{\mu}{4}+\frac{1}{12} \\ \frac{\lambda}{2}+\frac{\mu}{6}+\frac{1}{6} & \frac{\lambda}{4}+\frac{\mu}{6}+\frac{1}{12} & \frac{\lambda}{12}+\frac{\mu}{12}+\frac{1}{36} \end{bmatrix}$$

$$\begin{bmatrix} (\mu+1)\lambda+\mu+\frac{1}{2} & (\mu+\frac{1}{2})\lambda+\frac{1}{2}\mu^2+\mu+\frac{1}{4} & \kappa_1 \\ \frac{1}{2}\lambda^2+(\mu+1)\lambda+\frac{\mu}{2}+\frac{1}{4} & \frac{1}{4}\lambda^2+(\mu+\frac{1}{2})\lambda+\frac{1}{4}\mu^2+\frac{\mu}{2}+\frac{1}{8} & \kappa_2 \\ \frac{1}{2}\lambda^2+\frac{1}{2}(\mu+1)\lambda+\frac{\mu}{6}+\frac{1}{12} & \frac{1}{4}\lambda^2+\frac{1}{2}(\mu+\frac{1}{2})\lambda+\frac{1}{12}\mu^2+\frac{\mu}{6}+\frac{1}{24} & \kappa_3 \end{bmatrix}$$

$$\begin{bmatrix} \frac{1}{2}(2\mu+1)\lambda+\frac{\mu}{2}+\frac{1}{6} & \gamma_1 & \gamma_2 \\ \frac{1}{2}(\mu+1)\lambda^2+\frac{1}{2}(2\mu+1)\lambda+\frac{\mu}{4}+\frac{1}{12} & \beta_1 & \beta_2 \\ \frac{1}{6}\lambda^3+\frac{1}{2}(\mu+1)\lambda^2+\frac{1}{4}(2\mu+1)\lambda+\frac{\mu}{12}+\frac{1}{36} & \alpha_1 & \alpha_2 \end{bmatrix}$$

where we are using these notations:

$$\kappa_1 = (\frac{\mu}{2}+\frac{1}{6})\lambda+\frac{1}{2}\mu^2+\frac{\mu}{2}+\frac{1}{12}$$

$$\kappa_2 = \frac{1}{12}\lambda^2+(\frac{\mu}{2}+\frac{1}{6})\lambda+\frac{1}{4}\mu^2+\frac{\mu}{4}+\frac{1}{24}$$

$$\kappa_3 = \frac{1}{12}\lambda^2+\frac{1}{2}(\frac{\mu}{2}+\frac{1}{6})\lambda+\frac{1}{12}\mu^2+\frac{\mu}{12}+\frac{1}{72}$$

$$\gamma_1 = \frac{1}{2}(\mu^2+2\mu+\frac{1}{2})\lambda+\frac{1}{2}\mu^2+\frac{\mu}{2}+\frac{1}{12}$$

$$\gamma_2 = \frac{1}{2}(\mu^2+\mu+\frac{1}{6})\lambda+\frac{1}{6}\mu^3+\frac{1}{2}\mu^2+\frac{\mu}{4}+\frac{1}{36}$$

$$\beta_1 = \frac{1}{2}(\mu+\frac{1}{2})\lambda^2+\frac{1}{2}(\mu^2+2\mu+\frac{1}{2})\lambda+\frac{1}{4}\mu^2+\frac{\mu}{4}+\frac{1}{24}$$

$$\beta_2 = \frac{1}{2}(\frac{\mu}{2}+\frac{1}{6})\lambda^2+\frac{1}{2}(\mu^2+\mu+\frac{1}{6})\lambda+\frac{1}{12}\mu^3+\frac{1}{4}\mu^2+\frac{\mu}{8}+\frac{1}{72}$$

$$\alpha_1 = \frac{1}{12}\lambda^3+\frac{1}{2}(\mu+\frac{1}{2})\lambda^2+\frac{1}{4}(\mu^2+2\mu+\frac{1}{2})\lambda+\frac{1}{12}\mu^2+\frac{\mu}{12}+\frac{1}{72}$$

$$\alpha_2 = \frac{1}{36}\lambda^3+\frac{1}{2}(\frac{\mu}{2}+\frac{1}{6})\lambda^2+\frac{1}{4}(\mu^2+\mu+\frac{1}{6})\lambda+\frac{1}{36}\mu^3+\frac{1}{12}\mu^2+\frac{\mu}{24}+\frac{1}{216}$$

so, reading-out entries from the matrices above we have, for example:

$$Z(1,1,1)=\lambda+\mu+1\,,\quad Z(2,2,2)=\lambda^2/4+(\mu+1/2)\lambda+\mu^2/4+\mu/2+1/8\,,\quad Z(3,2,3)=\beta_2\,.$$

We remark that the reduced indices for the sums defining the partition function can be obtained in a more systematic form, through the use of Smith canonical forms. Suppose that P is a matrix in $\mathbb{Z}^{q\times n}$ that represents q conservation laws on n species. For instance, $P = [1, 0, 0, 1, 1; 0, 1, 0, 1, 0; 0, 0, 1, 0, 1]$ in the competitive binding example. We assume, as in this and other examples, that $q \leq n$ and that the matrix P has full row rank q. Under this assumption, the integer matrix P can be represented in Smith canonical form (see, for example, [6]), meaning that there exist two unimodular (that is to say, invertible over the ring of integers) matrices $U \in \mathbb{Z}^{q\times q}$ and $V \in \mathbb{Z}^{n\times n}$ so that $UPV = [\Delta\ 0]$, where $\Delta = \text{diag}\,(\delta_1, \ldots, \delta_q)$, 0 is a $q \times (n-q)$ matrix of zeroes, and the δ_i's are the *elementary divisors* of the matrix P. The elementary divisors are unique up to sign change, there are formulas that express then in terms of the minors of P (see [6] for details). For example, for the above example, we have $U = I$ (3×3 identity matrix), $V = [1, 0, 0, -1, -1; 0, 1, 0, -1, 0; 0, 0, 1, 0, -1; 0, 0, 0, 1, 0; 0, 0, 0, 0, 1]$ and $\delta_1 = \delta_2 = \delta = 3 = 1$, so $UPV = [I\ 0]$. In general, if we wish to find nonnegative integer solutions of $Ak = b$, for a given (nonnegative) integer vector b, we use that $UPVV^{-1}k = Ub$, so, using the indices $\ell = V^{-1}k$, $[\Delta\ 0]\ell = Ub$, which means that the last $n - q$ indices ℓ are free, and the constraint $V\ell \geq 0$ is imposed to insure nonnegativity of k. For instance, in the competitive binding example, and recalling that $U = I$ and $\Delta = I$, the equation $[\Delta\,0]\ell = Ub$ gives that $\ell_1 = b_1$, $\ell_2 = b_2$, $\ell_3 = b_3$, and $\ell_4 = i$, $\ell_5 = j$ are arbitrary. Thus, we can express the sum as a sum over the two indices $k_4 = i$ and $k_5 = j$, with $k_1 = b_1 - (i + j)$, $k_2 = b_2 - i$, and $k_3 = b_3 - j$. The nonnegativity condition $V\ell \geq 0$, applied with the above matrix V, says that these expressions must be nonnegative: which means that the sum can be reexpressed as a sum over $i \geq 0$, $j \geq 0$, subject to $i \leq b_2$, $j \leq b_3$, and $i + j \leq b_1$. This is exactly the same as the set $\mathscr{S}'$ computed by hand.

References

1. Anderson, D.F., Craciun, G., Kurtz, T.G.: Product-form stationary distributions for deficiency zero chemical reaction networks. Bull. Math. Biol. **72**(8), 1947–1970 (2010)
2. Feinberg, M.: Chemical reaction network structure and the stability of complex isothermal reactors—I. The deficiency zero and deficiency one theorems. Chem. Eng. Sci. **42**, 2229–2268 (1987)
3. Feinberg, M.: The existence and uniqueness of steady states for a class of chemical reaction networks. Arch. Ration. Mech. Anal. **132**, 311–370 (1995)
4. Gillespie, D.T.: The chemical Langevin equation. J. Chem. Phys. **113**(1), 297–306 (2000)
5. Gomez-Uribe, C.A., Verghese, G.C.: Mass fluctuation kinetics: capturing stochastic effects in systems of chemical reactions through coupled mean-variance computations. J. Chem. Phys. **126**, 024109 (2007)
6. Jacobson, N.: Basic Algebra I, 2 edn. Freeman and Company (1995)
7. Kelly, F.: Reversibility and Stochastic Networks. Wiley, New York (1979)
8. Mairesse, J., Nguyen, H.-T.: Deficiency zero Petri nets and product form. In: Franceschinis, G., Wolf, K. (eds.) Applications and Theory of Petri Nets, pp. 103–122. Springer (2009)
9. Meyn, S.P., Tweedie, R.L.: Stability of Markovian Processes III: Foster-Lyapunov criteria for continuous-time processes. Adv. Appl. Prob. **25**, 518548 (1993)

10. Singh, A., Hespanha, J.P.: A derivative matching approach to moment closure for the stochastic logistic model. Bull. Math. Biol. **69**, 1909–1925 (2007)
11. Sontag, E.D.: Lecture notes on mathematical systems biology, Rutgers University, 2002–2015. http://www.math.rutgers.edu/sontag/FTP_DIR/systems_biology_notes.pdf
12. Sontag, E.D., Singh, A.: Exact moment dynamics for feedforward nonlinear chemical reaction networks. IEEE Life Sci. Lett. **1**, 26–29 (2015)
13. Sontag, E.D., Zeilberger, D.: A symbolic computation approach to a problem involving multivariate Poisson distributions. Adv. Appl. Math. **44**, 359–377 (2010)
14. Van Kampen, N.G.: Stochastic Processes in Physics and Chemistry. Elsevier Science (2001)

Chapter 23
Design Theory of Distributed Controllers via Gradient-Flow Approach

Kazunori Sakurama, Sun-ichi Azuma and Toshiharu Sugie

Abstract This paper describes a unified design methodology of distributed controllers of multi-agent systems for general tasks based on the authors' recent work. First, a complete characterization is given to the distributed controllers via the gradient-flow approach. It is stressed that not edges but cliques (i.e., complete subgraphs) of network topologies are the crucial components of this characterization. Next, an optimal distributed controller is introduced, which achieves a given task as long as the network satisfies a certain condition. Even if the network does not satisfy the condition, the best approximate result to the task is achieved. Then, it is shown that the connection structure between the cliques plays an important role in achieving the task. While the conventional distributed controller design can handle specific tasks based on the edges of networks, the introduced approach provides us a systematic design methodology applicable to general tasks by using cliques.

23.1 Introduction

In recent years, control of large-scale systems has been eagerly investigated due to increasing the scale of social infrastructures, e.g., power grids and traffic networks; engineering systems, e.g., sensor networks and swarm robotics [2, 3, 5, 7]. Since large-scale systems consist of many components, it is difficult to observe and con-

K. Sakurama
Graduate School of Engineering, Tottori University, 4-101 Koyama-Minami, Tottori 680-8552, Japan
e-mail: sakurama@mech.tottori-u.ac.jp

S.-i. Azuma
Graduate School of Engineering, Nagoya University, Furu-cho, Chikusa-ku, Nagoya 464-8603, Japan
e-mail: shunichi.azuma@mae.nagoya-u.ac.jp

T. Sugie (✉)
Graduate School of Informatics, Kyoto University, Yoshida-honmachi, Sakyo-ku, Kyoto 606-8501, Japan
e-mail: sugie@i.kyoto-u.ac.jp

R. Tempo et al. (eds.), *Emerging Applications of Control and Systems Theory*, Lecture Notes in Control and Information Sciences - Proceedings,
https://doi.org/10.1007/978-3-319-67068-3_23

trol these systems in a centralized manner. Therefore, *distributed control*, a different approach from conventional control theories, is inevitable [14, 15, 21]. This control strategy depends only on the local information obtained from mutual communications and sensing between components. Distributed control has the advantage of the scalability in the sense that the computation burden on each component is independent of the scale of the systems.

On the other hand, each of large-scale systems has a task to achieve for the overall profit. For example, the supply and consumption balance has to be maintained in power grids; a specific formation pattern should be formed by swarm robots. However, it might be impossible to achieve a given task due to the limited information under distributed control. Therefore, the feasibility of tasks should be clarified, which closely relates to the network topology describing the connections of communication and sensing between components. Actually, the consensus problem is solvable if and only if the network topology is connected [17]. Now, assume that a network topology is given. Then, can we say whether a given task is achievable by distributed control or not? If the task is unachievable over the network topology, what can we do close to the task?

To answer these questions, this paper introduces a design methodology of distributed controllers developed by the authors [19]. First, we give a complete characterization of the class of distributed controllers over a given network topology derived by the gradient-flow approach. It is shown that the key of the characterization is not edges but cliques (i.e., complete subgraphs) in the network topology. Next, we design an optimal distributed controller in terms of a certain performance index. By the designed controller, if a given task is achievable over the network topology, the task is really achieved; otherwise, the best approximate result to the task is achieved. Then, a required network topology to achieve a certain task is discussed based on the authors' other paper [20], which declares that the connection structure between the cliques play an important role in the success of the task. Finally, the effectiveness of the proposed method is demonstrated through a numerical example.

Various tasks of multi-agent systems have been investigated including consensus (rendezvous) [17], attitude synchronizing [10, 18], enclosing [9, 11], coverage [6], formation with free scale [4, 13], and distance-based formation [1, 12], which are surveyed by [14, 16]. While the conventional distributed controllers are designed for specific tasks based on the edges in network topologies, this paper's approach provides us a unified design methodology in a systematic manner by using cliques. In this sense, this approach is expected to innovate in the technology of distributed control.

Notations: Let $\mathbb{R}$ be the set of all real numbers, $\mathbb{R}_+$ be the set of all nonnegative real number. The cardinal number and the power set of a set are denoted as $|\cdot|$ and $\mathrm{pow}(\cdot)$, respectively. The Euclidean norm of a vector is denoted as $\|\cdot\|$. For n vector-valued functions $x_i(t) \in \mathbb{R}^d$ $(i = 1, 2, \ldots, n)$ and a subset of natural numbers $\mathcal{S} \subset$

$\{1, 2, \dots, n\}$, the notation $[x_i(t)]_{i\in\mathcal{S}} = [x_{s_1}^\top(t)\ x_{s_2}^\top(t)\ \cdots\ x_{s_{|\mathcal{S}|}}^\top(t)]^\top$ is used, where the sequence $s_1, s_2, \dots, s_{|\mathcal{S}|} \in \mathcal{S}$ is strictly monotonically increasing. Similarly, for n matrices $M_i \in \mathbb{R}^{d\times m}$ $(i = 1, 2, \dots, n)$, $[M_i]_{i\in\mathcal{S}} = [M_{s_1}^\top\ M_{s_2}^\top\ \cdots\ M_{s_{|\mathcal{S}|}}^\top]^\top$ is defined. The orthogonal projection of a set $\mathcal{D} \subset \mathbb{R}^n$ to the $[x_i]_{i\in\mathcal{S}}$-space is defined as

$$P_{\mathcal{S}}(\mathcal{D}) = \{y \in \mathbb{R}^{|\mathcal{S}|d} : \exists x \in \mathcal{D} \text{ s.t. } y = [x_i]_{i\in\mathcal{S}}\}.$$

The zero set of a function $f : \mathbb{R}^n \to \mathbb{R}^m$ is denoted as $f^{-1}(0) = \{x \in \mathbb{R}^n : f(x) = 0\}$. The distance from a vector $x \in \mathbb{R}^n$ to a set $\mathcal{D} \subset \mathbb{R}^n$ and the Hausdorff distance from a set $\mathcal{D}_1 \subset \mathbb{R}^n$ to another set $\mathcal{D}_2 \subset \mathbb{R}^n$ are, respectively, defined as follows:

$$\operatorname{dist}(x, \mathcal{D}) = \inf_{y\in\mathcal{D}} \|x - y\|, \quad \operatorname{dist}_{\mathrm{H}}(\mathcal{D}_1, \mathcal{D}_2) = \sup_{x\in\mathcal{D}_1} \operatorname{dist}(x, \mathcal{D}_2).$$

23.2 Problem Formulation

23.2.1 Target System and Distributed Controllers

Consider, a multi-agent system, which is an abstract model of various large-scale systems. The multi-agent system consists of a large number of components, called agents. Let n be the number of the agents, numbered from 1 to n. The set of the agents is denoted by $\mathcal{V} = \{1, 2, \dots, n\}$. The dynamics of agent $i \in \mathcal{V}$ is governed by the single integrator

$$\dot{x}_i(t) = u_i(t), \tag{23.1}$$

where $x_i(t), u_i(t) \in \mathbb{R}^d$ are the state and the input of agent i in the d-dimensional space, respectively.

A part of the agent pairs exchange information over a network of communication, sensing, and/or so forth. The set of such pairs is described by $\mathcal{E} \subset \operatorname{pow}(\mathcal{V})$; say if $\{i, j\} \in \mathcal{E}$, agents i and j can bilaterally exchange their information. Then, the network topology is described by a graph $G = (\mathcal{V}, \mathcal{E})$ with node set $\mathcal{V}$ and edge set $\mathcal{E}$. We assume that G is simple and undirected. Let $\mathcal{N}_i \subset \mathcal{V}$ be the adjacent set of agent i, defining the set of agents who can exchange information with agent i as

$$\mathcal{N}_i = \{j \in \mathcal{V} : \{i, j\} \in \mathcal{E}\}.$$

Each agent can obtain information through the network. For agent $i \in \mathcal{V}$, the own state $x_i(t)$ and the states $[x_j(t)]_{j\in\mathcal{N}_i}$ of the agents on the adjacent set are available. Then, the control input $u_i(t)$ has to be of the form

$$u_i(t) = f_i(x_i(t), [x_j(t)]_{j\in\mathcal{N}_i}) \tag{23.2}$$

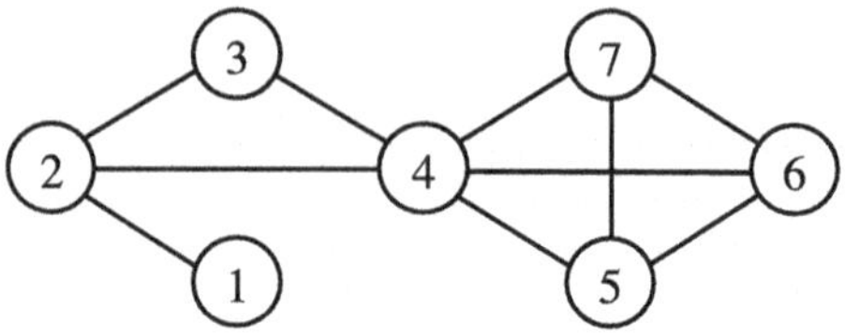

Fig. 23.1 Example of a graph

with some function $f_i : \mathbb{R}^d \times \mathbb{R}^{|\mathcal{N}_i|d} \to \mathbb{R}^d$, which is called a *distributed controller* over graph G.

Example 23.1 Consider, the graph in Fig. 23.1, whose node set $\mathcal{V}$, edge set $\mathcal{E}$, and adjacent sets $\mathcal{N}_i$ $(i = 1, 2, 3, 4)$ are given as follows:

$$\begin{aligned}\mathcal{V} &= \{1, 2, 3, 4, 5, 6, 7\},\\ \mathcal{E} &= \{\{1, 2\}, \{2, 3\}, \{2, 4\}, \{3, 4\}, \{4, 5\}, \{4, 6\}, \{4, 7\}, \{5, 6\}, \{5, 7\}, \{6, 7\}\},\\ \mathcal{N}_1 &= \{2\},\ \mathcal{N}_2 = \{1, 3, 4\},\ \mathcal{N}_3 = \{2, 4\},\ \mathcal{N}_4 = \{2, 3, 5, 6, 7\}.\end{aligned}$$

Distributed controllers of agents 1, 2, 3, and 4 are of the following forms:

$$\begin{aligned}u_1(t) &= f_1(x_1(t), x_2(t)),\ u_2(t) = f_2(x_2(t), x_1(t), x_3(t), x_4(t)),\\ u_3(t) &= f_3(x_3(t), x_2(t), x_4(t)),\ u_4(t) = f_4(x_4(t), x_2(t), x_3(t), x_5(t), x_6(t), x_7(t)).\end{aligned}$$

□

23.2.2 General Description of Tasks

The multi-agent system ought to achieve a given task. Assume that the task is described by a target set $\mathcal{D} \subset \mathbb{R}^{nd}$ of $x(t)$ as[1]

$$\lim_{t \to \infty} \mathrm{dist}(x(t), \mathcal{D}) = 0, \tag{23.3}$$

where $x(t) = [x_1^\top(t)\ x_2^\top(t)\ \cdots\ x_n^\top(t)]^\top \in \mathbb{R}^{nd}$ is the collective states of all agents. Therefore, $x(t)$ can converge to any points on $\mathcal{D}$ in (23.3), which correspond to the freedom of the task determined via agent communication. For three agents in a one-dimensional space $(n = 3, d = 1)$, a geometric explanation of this task is described as Fig. 23.2 (I).

This formulation provides a general description of tasks including various conventional tasks as follows.

[1] Additionally, any points on $\mathcal{D}$ have to be (periodically) approached by some $x(t)$; say, for any $x_d \in \mathcal{D}$, there exist $x(t)$ satisfying $\lim_{t\to\infty} x(t) = x_d$ or $\lim_{k\to\infty} x(t_k) = x_d$ for some $t_1, t_2, \ldots$ $(0 < t_1 < t_2 < \cdots)$. Without this condition, (23.3) would be a trivial problem by predetermining a convergent point of $x(t)$ on $\mathcal{D}$.

Example 23.2 Consider, the consensus problem [17], which requires that the states $x_i(t)$ of all agents $i \in \mathcal{V}$ agree, namely,

$$\lim_{t\to\infty}(x_i(t) - x_j(t)) = 0$$

for any $i, j \in \mathcal{V}$. This is described by (23.3) with the target set

$$\mathcal{D} = \{x \in \mathbb{R}^{nd} : x_1 = x_2 = \cdots = x_n\},$$

where $x = [x_1^\top\ x_2^\top\ \cdots\ x_n^\top]^\top$. In the case of $d = 1$, the set $\mathcal{D}$ is the line in $\mathbb{R}^n$ with the slope 1 passing through the origin as depicted in Fig. 23.2 (II) for $n = 3$. Note that $x(t)$ can converge to any point on the line, which corresponds to the freedom of the consensus value. □

Example 23.3 Consider the formation problem with free scale [4, 13]. As shown in Fig. 23.3, agents are expected to achieve the desired formation pattern with any scale. For $d = 1$, this problem is described as

$$\lim_{t\to\infty}\left(\frac{x_i(t) - x_j(t)}{x_{ij}^*} - \frac{x_k(t) - x_\ell(t)}{x_{k\ell}^*}\right) = 0$$

for any $i, j, k, \ell \in \mathcal{V}\,(i \neq j, k \neq \ell)$, where $x_{ij}^* \in \mathbb{R}\backslash\{0\}$ is the desired relative position between agents i and j. This is formulated as (23.3) with the target set

$$\mathcal{D} = \left\{x \in \mathbb{R}^n : \frac{x_i - x_j}{x_{ij}^*} = \frac{x_k - x_\ell}{x_{k\ell}^*}\ \forall i, j, k, \ell \in \mathcal{V}\ (i \neq j, k \neq \ell)\right\}.$$

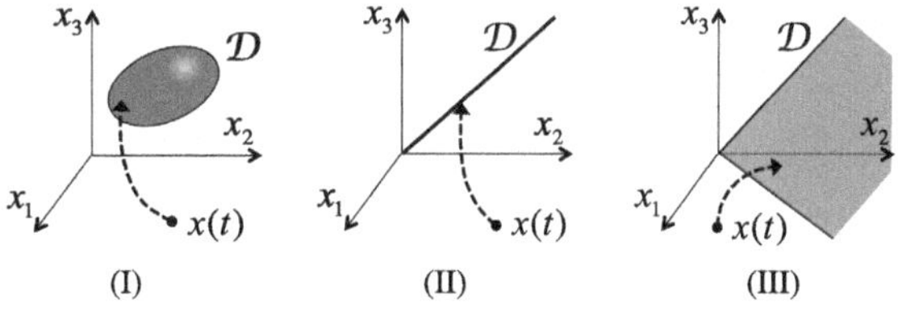

Fig. 23.2 Geometric explanation of task achievement

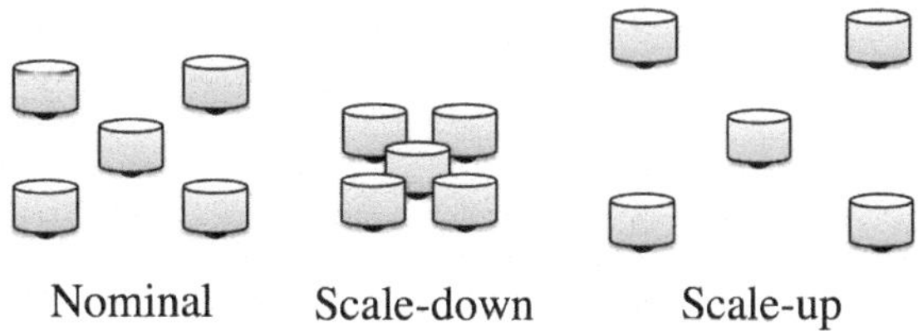

Fig. 23.3 Formation with free scale

This problem has the two-degrees of freedom: the position and the scale of the formation, which correspond to the dimension two of the set $\mathcal{D}$. Actually, $\mathcal{D}$ is given by a plane as Fig. 23.2 (III) for $n = 3$. □

Example 23.4 Consider the distance-based formation [1, 12], which requires the relative distance between $x_i(t)$ and $x_j(t)$ to converge to a desired distance $d_{ij}^* > 0$, namely,

$$\lim_{t\to\infty} \|x_i(t) - x_j(t)\| = d_{ij}^*$$

for all $i, j \in \mathcal{V}$. This is formulated by (23.3) with the target space

$$\mathcal{D} = \{x \in \mathbb{R}^{nd} : \|x_i - x_j\| = d_{ij}^* \ \forall i, j \in \mathcal{V}\ (i \neq j)\}.$$

The freedom allowed to this task is the position, rotation, and flip of the formation. □

The question here is whether the task described by the general form (23.3) is achievable or not by distributed controllers. If the task is achievable, we will design a distributed controller which achieves the task. Otherwise, we can just design a distributed controller with which the task is most similarly achieved.

23.3 Main Result

23.3.1 *Characterization of the Class of Distributed Controllers*

In this section, we employ the gradient-flow approach, which provides the gradient-type controller

$$u_i(t) = -\left(\frac{\partial V}{\partial x_i}(x(t))\right)^\top \tag{23.4}$$

with a differentiable function $V : \mathbb{R}^{nd} \to \mathbb{R}_+$. The function $V(x)$, called an *objective function*, evaluates the achievement of a task. We just have to design $V(x)$ that takes the minimum, zero, when the task is achieved. The time-derivative of $V(x(t))$ is reduced to

$$\dot{V}(x(t)) = \sum_{i=1}^{n} \frac{\partial V}{\partial x_i}(x(t))\dot{x}_i(t) = -\sum_{i=1}^{n} \left\|\frac{\partial V}{\partial x_i}(x(t))\right\|^2 \tag{23.5}$$

for the system (23.1) with the control input (23.4). Then, $V(x(t))$ is monotonically decreasing with respect to t and is locally minimized to zero, where the task is achieved.

It should be stressed that the gradient-type controller (23.4) is not necessarily distributed. To design a distributed controller, an objective function having a *distributed gradient* is important. We say a function $V : \mathbb{R}^{nd} \to \mathbb{R}_+$ has a distributed gradient over graph G if it is differentiable and there exist functions $f_i : \mathbb{R}^d \times \mathbb{R}^{|\mathcal{N}_i|d} \to \mathbb{R}^d$ satisfying

$$\left(\frac{\partial V}{\partial x_i}(x)\right)^\top = -f_i(x_i, [x_j]_{j\in\mathcal{N}_i}) \tag{23.6}$$

for all $i \in \mathcal{V}$. If a strict class of objective functions having distributed gradients is determined, we can restrict objective functions to this class when designing controllers. We consider the following problem to specify such a class of objective functions.

Problem 23.1 *Derive a necessary and sufficient condition under which a function $V(x)$ has a distributed gradient over graph G.* □

The key to solve this problem is a *clique* [8]. A node subset $\mathcal{I} \subset \mathcal{V}$ is called a clique of graph G if it induces a complete subgraph over G, namely $\{i, j\} \in \mathcal{E}$ holds for any $i, j \in \mathcal{I}$ $(i \neq j)$. We say a function $V : \mathbb{R}^{nd} \to \mathbb{R}_+$ is *clique-wise decomposable* over graph G if there exists a set of cliques $\mathcal{C} \subset \mathrm{pow}(\mathcal{V})$ of G and functions $W_\mathcal{I} : \mathbb{R}^{|\mathcal{I}|} \to \mathbb{R}_+$ $(\mathcal{I} \in \mathcal{C})$ such as

$$V(x) = \sum_{\mathcal{I}\in\mathcal{C}} W_\mathcal{I}([x_i]_{i\in\mathcal{I}}). \tag{23.7}$$

Example 23.5 The graph in Fig. 23.1 contains cliques $\{1, 2\}$, $\{2, 3, 4\}$, and $\{4, 5, 6, 7\}$. Note that edges are cliques of order two, and that clique $\{4, 5, 6, 7\}$ contains the cliques of smaller order: $\{4, 5, 6\}$, $\{4, 5, 7\}$, $\{4, 6, 7\}$, and $\{5, 6, 7\}$. The following function is clique-wise decomposable:

$$V(x) = W_{12}(x_1, x_2) + W_{234}(x_2, x_3, x_4) + W_{4567}(x_4, x_5, x_6, x_7). \tag{23.8}$$

□

The clique-wise decomposable function (23.7) is partially differentiated with respect to x_i as

$$\frac{\partial V}{\partial x_i}(x) = \sum_{\mathcal{I}\in\mathcal{C}_i} \frac{\partial W_\mathcal{I}}{\partial x_i}([x_j]_{j\in\mathcal{I}}), \tag{23.9}$$

where $\mathcal{C}_i \subset \mathcal{C}$ is the set of cliques containing $i \in \mathcal{V}$, namely $\mathcal{C}_i = \{\mathcal{I} \in \mathcal{C} : i \in \mathcal{I}\}$. From the definition of the clique, the right-hand side of (23.9) depends only on x_i and $[x_j]_{j\in\mathcal{N}_i}$. Thus, $V(x)$ of the form (23.7) satisfies (23.6) for any $i \in \mathcal{V}$ and has a distributed gradient; say, if $V(x)$ is clique-wise decomposable, it has a distributed gradient. Importantly, the converse relation holds, namely, only if $V(x)$ is clique-wise decomposable, it has a distributed gradient. This relation is not trivial and leads to the following theorem.

Theorem 23.1 ([19]) *A continuously differentiable function $V : \mathbb{R}^{nd} \to \mathbb{R}_+$ has a distributed gradient over graph G if and only if $V(x)$ is clique-wise decomposable over G.* □

Theorem 23.1 guarantees that all objective functions having distributed gradients are of the form (23.7). By regarding $W_{\mathcal{I}}([x_i]_{i\in\mathcal{I}})$ as designable parameters, we can see (23.7) as a parametrization of all objective functions to design distributed controllers. We do not have to consider other objective functions because they do not generate distributed controllers. Conventional studies mainly deal with *edge-wise decomposable* objective functions $V(x) = \sum_{\{i,j\}\in\mathcal{E}} W_{ij}(x_i, x_j)$ to design distributed controllers. Theorem 23.1 shows that the class of available objective functions is wider than this and (23.7) gives the strict class. The function (23.8) in Example 23.5 is in this wider class due to the terms $W_{234}(x_2, x_3, x_4)$ and $W_{4567}(x_4, x_5, x_6, x_7)$.

23.3.2 Design of Optimal Distributed Controller

We consider the task (23.3) and design a distributed controller to achieve it. Whether the task is achievable or not depends on graph G. For example, the consensus problem is solvable if and only if the graph is connected. Thus, a certain index is introduced which can evaluate control performance even if the task is unachievable over G. Then, we design a distributed controller optimal in terms of the performance index.

Consider, the following function as such a performance index:

$$J(V) = \mathrm{dist}_{\mathrm{H}}(V^{-1}(0), \mathcal{D}), \tag{23.10}$$

which evaluates the achievement of (23.3) by $V(x)$ via the gradient-type controller (23.4). Since $V^{-1}(0) \subset \mathbb{R}^{nd}$ is the set where $V(x)$ takes the minimum, the state $x(t)$ locally converges to $V^{-1}(0)$. If $V(x)$ satisfies $J(V) = 0$, $V^{-1}(0) = \mathcal{D}$ holds from (23.10), and (23.3) is locally achieved. On the other hand, if $J(V) \neq 0$, the undesired zero set $V^{-1}(0)\backslash\mathcal{D}$ is not empty. In this case, even if $V(x(t)) \to 0$ is achieved as t grows, $x(t)$ possibly converges to a point out of $\mathcal{D}$. Then, the performance index $J(V)$ evaluates the width of the undesired zero set of $V(x)$. Now, we consider the following problem.

Problem 23.2 *Find a function $V(x)$ which minimizes $J(V)$ of all functions having distributed gradients.* □

If $V(x) = (\mathrm{dist}(x, \mathcal{D}))^2$ was chosen, $J(V) = 0$ would be obtained from (23.10). This function, however, does not have a distributed gradient in general. The core of this problem is to design an objective function which appropriately evaluates the discrepancy between x and $\mathcal{D}$ with keeping a distributed gradient. From Theorem 23.1, we just have to consider the objective function (23.7), where only $W_{\mathcal{I}}([x_i]_{i\in\mathcal{I}})$ for cliques $\mathcal{I}$ are designable. However, $W_{\mathcal{I}}([x_i]_{i\in\mathcal{I}})$ cannot directly evaluate the discrepancy between x and $\mathcal{D}$. The idea to overcome this issue is to evaluate it after projecting x and $\mathcal{D}$ to the $[x_i]_{i\in\mathcal{I}}$-space, which yields

$$W_{\mathcal{I}}([x_i]_{i\in\mathcal{I}}) = (\mathrm{dist}([x_i]_{i\in\mathcal{I}}, P_{\mathcal{I}}(\mathcal{D})))^2. \tag{23.11}$$

Actually, a distributed controller optimal in terms of $J(V)$ is designed as follows.

Theorem 23.2 ([19]) *The function $V(x)$ in (23.7) with assigning $W_{\mathcal{I}}([x_i]_{i\in\mathcal{I}})$ as (23.11) and $\mathcal{C}$ as the set of all cliques (or all maximal cliques) minimizes the performance index $J(V)$ in (23.10) of all functions having distributed gradients.* □

23.4 Graph Topology Analysis for High-Dimensional Target Subspace

Consider the τ-dimensional subspace ($\tau < n$) as the target set $\mathcal{D}$ for $d = 1$, namely,

$$\mathcal{D} = \{x \in \mathbb{R}^n : \exists\theta \in \mathbb{R}^\tau \text{ s.t. } x = T\theta\}, \tag{23.12}$$

where the matrix $T \in \mathbb{R}^{n\times\tau}$ specifies the coordination pattern of the agents. We assume that for the rows $T_i \in \mathbb{R}^{1\times\tau}$ $(i \in \mathcal{V})$ of T, $[T_i]_{i\in\mathcal{I}} \in \mathbb{R}^{|\mathcal{I}|\times\tau}$ is full-rank for any $\mathcal{I} \subset \mathcal{V}$. This target set describes various tasks including the consensus and the formation with free scale in Examples 23.2 and 23.3. We consider a graph topology required to achieve the task (23.3) with (23.12).

For the target set (23.12), from Theorem 23.2, the optimal distributed controller is derived as (23.4) with (23.7), (23.11), and (23.12). This is reduced to the linear controller

$$u_i(t) = -k_{ii}x_i(t) + \sum_{j\in\mathcal{N}_i} k_{ij}x_j(t) \tag{23.13}$$

with the gains

$$k_{ii} = \sum_{\mathcal{I}\in\mathcal{C}_{\{i\}}} (1 - T_i([T_\ell]_{\ell\in\mathcal{I}}^\top[T_\ell]_{\ell\in\mathcal{I}})^{-1}T_i^\top), \quad k_{ij} = \sum_{\mathcal{I}\in\mathcal{C}_{\{i,j\}}} T_i([T_\ell]_{\ell\in\mathcal{I}}^\top[T_\ell]_{\ell\in\mathcal{I}})^{-1}T_j^\top,$$

where $\mathcal{C}_{\mathcal{J}} = \{\mathcal{I} \in \mathcal{C} : \mathcal{J} \subset \mathcal{I}\}$ represents the set of all cliques containing the node subset $\mathcal{J} \subset \mathcal{V}$. Then, from (23.5), the state $x(t)$ governed by (23.1) with the controller (23.13) globally converges to $V^{-1}(0)$, where $V(x)$ is given by (23.7) with (23.11). The global convergence to $V^{-1}(0)$ is guaranteed from the convexity of the set (23.12).

Now, to globally guarantee (23.3), we just have to investigate the graph topology such that this $V^{-1}(0)$ is equivalent to $\mathcal{D}$ in (23.12). From (23.7), (23.11), and (23.12),

$$V^{-1}(0) = \bigcap_{\mathcal{I}\in\mathcal{C}} \{x \in \mathbb{R}^n : [x_i]_{i\in\mathcal{I}} \in P_{\mathcal{I}}(\mathcal{D})\} = \bigcap_{\mathcal{I}\in\mathcal{C}} \{x \in \mathbb{R}^n : \exists\theta_{\mathcal{I}} \in \mathbb{R}^\tau \text{ s.t. } [x_i]_{i\in\mathcal{I}} = [T_i]_{i\in\mathcal{I}}\theta_{\mathcal{I}}\}$$

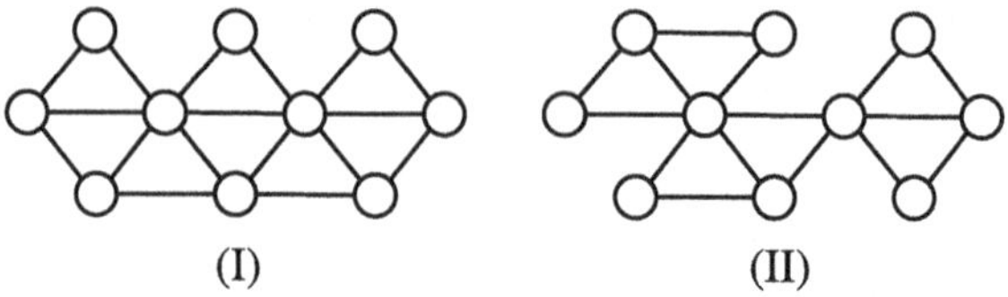

Fig. 23.4 Examples of graphs (I) satisfying and (II) not satisfying the condition

is derived. Thus, for any $x \in V^{-1}(0)$ and two cliques $\mathcal{I}, \mathcal{J} \in \mathcal{C}$ ($\mathcal{I} \neq \mathcal{J}$), $[x_i]_{i\in\mathcal{I}} = [T_i]_{i\in\mathcal{I}}\theta_{\mathcal{I}}$ and $[x_i]_{i\in\mathcal{J}} = [T_i]_{i\in\mathcal{J}}\theta_{\mathcal{J}}$ hold, which yield $[x_i]_{i\in(\mathcal{I}\cap\mathcal{J})} = [T_i]_{i\in(\mathcal{I}\cap\mathcal{J})}\theta_{\mathcal{I}} = [T_i]_{i\in(\mathcal{I}\cap\mathcal{J})}\theta_{\mathcal{J}}$. Then, from the assumption of T, $\theta_{\mathcal{I}} = \theta_{\mathcal{J}}$ holds when two cliques $\mathcal{I}$ and $\mathcal{J}$ are connected via at least τ nodes, namely $|\mathcal{I} \cap \mathcal{J}| \geq \tau$. If (i) all cliques in $\mathcal{C}$ are connected in this way and (ii) all nodes belong to any of the cliques in $\mathcal{C}$, then all $\theta_{\mathcal{I}}$ ($\mathcal{I} \in \mathcal{C}$) agree and $V^{-1}(0) = \mathcal{D}$ holds. Then, (23.3) is globally achieved for (23.12). The graph in Fig. 23.4 (I) satisfies the conditions for $\tau = 2$, consisting of cliques of order three connected via two nodes, while the graph in Fig. 23.4 (II) does not. See [20] for more details including a necessary and sufficient condition of G.

23.5 Numerical Example

Consider $n = 30$ agents in the three-dimensional space ($d = 3$). The dynamics of agent $i \in \mathcal{V}$ is governed by (23.1). The network topology of the agents is depicted by the lines in Fig. 23.5. The task is to achieve the desired coordination described by the dots in Fig. 23.5, consisting of two horizontal ellipses whose major and minor axises agree with either X- or Y-axis and the centers are common. This task allows

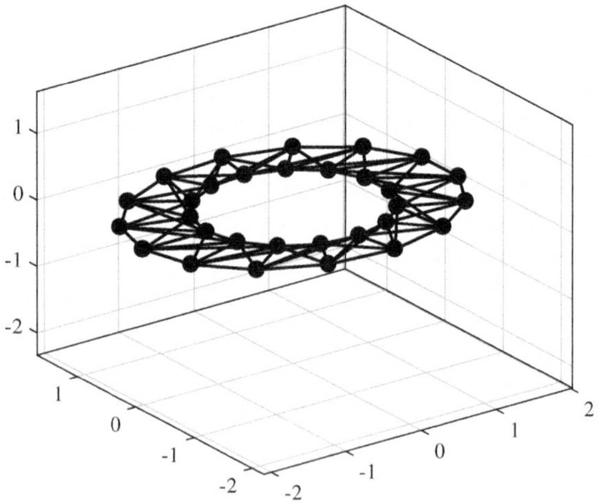

Fig. 23.5 Desired coordination (dots) and the network connections (lines) for the numerical example

any lengths of the major and minor axises and any position of the common centers of the ellipses. Let $\mathcal{V}_o$ and $\mathcal{V}_e$ be the sets of the odd and even numbers, respectively, satisfying $\mathcal{V} = \mathcal{V}_o \cup \mathcal{V}_e$. We expect that agents $i \in \mathcal{V}_o$ form one ellipse, and agents $i \in \mathcal{V}_e$ form the other one.

This task is described by (23.3) with the target set

$$\mathcal{D} = \big\{x \in \mathbb{R}^{90} : \exists \theta_c \in \mathbb{R}^3, \theta_o, \theta_e \in \mathbb{R}^2 \\ \text{s.t. } x_i = \theta_c + C_i\theta_o \ \forall i \in \mathcal{V}_o,\ \ x_i = \theta_c + C_i\theta_e \ \forall i \in \mathcal{V}_e\} \quad (23.14)$$

for the matrix

$$C_i = \begin{bmatrix} \cos(2\pi\lceil i/2\rceil/15 + 0.1) & 0 \\ 0 & \sin(2\pi\lceil i/2\rceil/15 + 0.1) \\ 0 & 0 \end{bmatrix},$$

where $\lceil\cdot\rceil$ is the ceiling function. Note that in (23.14), the parameter θ_c corresponds to the center of the ellipses, θ_o corresponds to the lengths of the axises of one ellipse, and θ_e corresponds to those of the other ellipse.

The set (23.14) is reduced to the axis-wise description

$$\mathcal{D} = \{x \in \mathbb{R}^{90} : [x_{i1}]_{i\in\mathcal{V}} \in \mathcal{D}_1,\ [x_{i2}]_{i\in\mathcal{V}} \in \mathcal{D}_2,\ [x_{i3}]_{i\in\mathcal{V}} \in \mathcal{D}_3\}$$

with $x_i = [x_{i1}\ x_{i2}\ x_{i3}]^\top \in \mathbb{R}^3$ and $\mathcal{D}_k = \{y \in \mathbb{R}^{30} : \exists\theta_k \in \mathbb{R}^{\tau_k},\ y = T_k\theta_k\}$, where $\tau_1 = \tau_2 = 3$, $\tau_3 = 1$, $T_3 \in \mathbb{R}^{30}$ is the vector whose all components are 1, and

$$T_1 = \begin{bmatrix} 1 & 1 & 1 & 1 & 1 & \cdots & 1 & 1 \\ c_1 & 0 & c_2 & 0 & c_3 & \cdots & c_{15} & 0 \\ 0 & c_1 & 0 & c_2 & 0 & \cdots & 0 & c_{15} \end{bmatrix}^\top, \quad T_2 = \begin{bmatrix} 1 & 1 & 1 & 1 & 1 & \cdots & 1 & 1 \\ s_1 & 0 & s_2 & 0 & s_3 & \cdots & s_{15} & 0 \\ 0 & s_1 & 0 & s_2 & 0 & \cdots & 0 & s_{15} \end{bmatrix}^\top$$

for $c_\ell = \cos(2\pi\ell/15 + 0.1)$ and $s_\ell = \sin(2\pi\ell/15 + 0.1)$ ($\ell \in \{1, 2, \ldots, 15\}$). Then, this task is decomposed into $\lim_{t\to\infty} \operatorname{dist}([x_{ik}(t)]_{i\in\mathcal{V}}, \mathcal{D}_k) = 0$ for $k = 1, 2, 3$, corresponding to X-, Y-, and Z-axises, each of which is equivalent to the problem discussed in Sect. 23.4. Thus, the distributed controller of the form (23.13) and the derived condition on the graph topology are available with $\tau = 3 (= \tau_1 = \tau_2)$. The network topology depicted in Fig. 23.5 satisfies this condition.

Figures 23.6 (I) and (II) show the transitions of the agent positions with the dynamics (23.1) and the control input (23.13) from different initial positions. It is observed that the agents achieve the desired coordination, consisting of two ellipses with the common centers, in each of (I) and (II), whereas the lengths of the major and minor axises and the positions of the centers of the ellipses are different between (I) and (II).

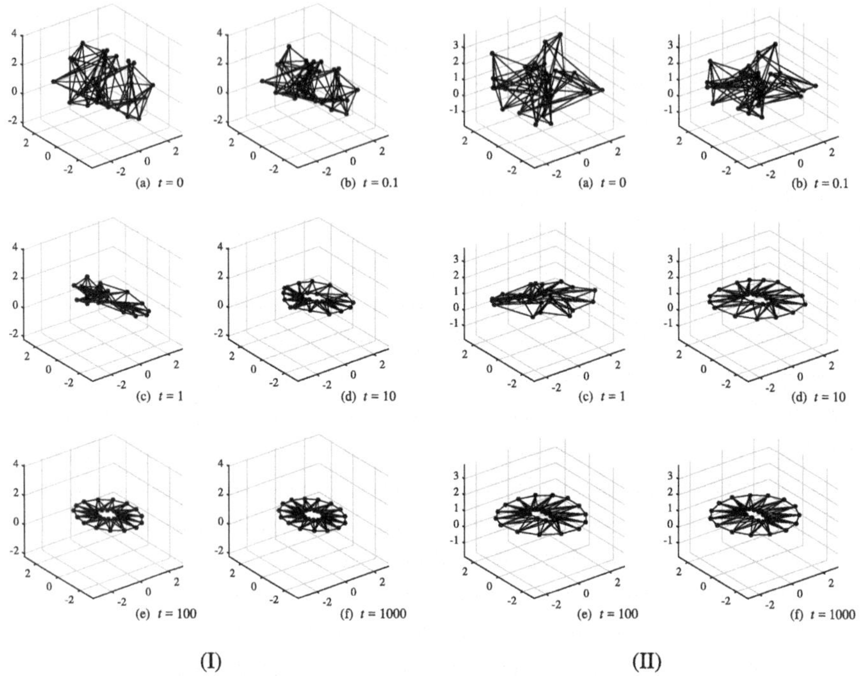

Fig. 23.6 Transitions of the agent positions in the simulation results from different initial positions

23.6 Conclusion

This paper introduced a unified design methodology of distributed controllers of multi-agent systems for general tasks based on the authors' recent work. First, a complete characterization was given for distributed controllers via the gradient-flow approach, where not edges but cliques play a crucial role. This result indicates an important relation between the cliques in the graph theory and the distributed controllers in the control engineering. Next, an optimal distributed controller was introduced in terms of the performance index evaluating the achievement of a given task. Since the task is described by the general formulation through the target set, this approach is applicable to various practical tasks. In this sense, this approach is expected to be a key technology to distributed control of large-scale systems.

Acknowledgements A part of this work was supported by JSPS KAKENHI Grant Number 15K06143 and JST CREST.

References

1. Anderson, B., Yu, C., Fidan, B., Hendrickx, J.: Rigid graph control architectures for autonomous formations. IEEE Control Syst. Mag. **28**(6), 48–63 (2008)
2. Barca, J.C., Sekercioglu, Y.A.: Swarm robotics reviewed. Robotica **31**(3), 345–359 (2013)
3. Campos-Nañez, E., Garcia, A., Li, C.: A game theoretic approach to efficient power management in sensor networks. Operat. Res. **56**(3), 552–561 (2008)
4. Coogan, S., Arcak, M.: Scaling the size of a formation using relative position feedback. Automatica **48**(10), 2677–2685 (2012)
5. Coogan, S., Arcak, M.: A compartmental model for traffic networks and its dynamical behavior. IEEE Trans. Autom. Control **60**(10), 2698–2703 (2015)
6. Cortés, J., Martinez, S., Karatas, T., Bullo, F.: Coverage control for mobile sensing networks. IEEE Trans. Robot. Autom. **20**(2), 243–255 (2004)
7. Gkatzikis, L., Koutsopoulos, I., Salonidis, T.: The role of aggregators in smart grid demand response markets. IEEE J. Sel. Areas Commun. **31**(7), 1247–1257 (2013)
8. Godsil, C., Royle, G.F.: Algebraic Graph Theory. Springer (2001)
9. Guo, J., Yan, G., Lin, Z.: Local control strategy for moving-target-enclosing under dynamically changing network topology. Syst. Control Lett. **59**, 654–661 (2010)
10. Igarashi, Y., Hatanaka, T., Fujita, M., Spong, M.W.: Passivity-based attitude synchronization in $SE(3)$. IEEE Trans. Control Syst. Technol. **17**(5), 1119–1134 (2009)
11. Kim, T.H., Hara, S., Hori, Y.: Cooperative control of multi-agent dynamical systems in target-enclosing operations using cyclic pursuit strategy. Int. J. Control **83**(10), 2040–2052 (2010)
12. Krick, L., Broucke, M.E., Francis, B.A.: Stabilisation of infinitesimally rigid formations of multi-robot networks. Int. J. Control **82**(3), 423–439 (2009)
13. Lin, Z., Wang, L., Chen, Z., Fu, M., Han, Z.: Necessary and sufficient graphical conditions for affine formation control. IEEE Trans. Autom. Control **61**(10), 2877–2891 (2016)
14. Martínez, S., Cortés, J., Bullo, F.: Motion coordination with distributed information. IEEE Control Syst. Mag. **27**(4), 75–88 (2007)
15. Mesbahi, M., Egerstedt, M.: Graph Theoretic Methods in Multiagent Networks. Princeton University Press (2010)
16. Oh, K.K., Park, M.C., Ahn, H.S.: A survey of multi-agent formation control. Automatica **53**(3), 424–440 (2015)
17. Olfati-Saber, R., Fax, J.A., Murray, R.M.: Consensus and cooperation in networked multi-agent systems. Proc. IEEE **95**(1), 215–233 (2007)
18. Ren, W.: Distributed cooperative attitude synchronization and tracking for multiple rigid bodies. IEEE Trans. Control Syst. Technol. **18**(2), 383–392 (2010)
19. Sakurama, K., Azuma, S., Sugie, T.: Distributed controllers for multi-agent coordination via gradient-flow approach. IEEE Trans. Autom. Control **60**(6), 1471–1485 (2015)
20. Sakurama, K., Azuma, S., Sugie, T.: Multi-agent coordination to high-dimensional target subspaces. IEEE Trans. Control Netw, Syst (2017)
21. Shamma, J. (ed.): Cooperative control of distributed multi-agent systems. Wiley-Interscience (2008)

Chapter 24
Machine Learning for Joint Classification and Segmentation

Jeremy Lerner, Romeil Sandhu, Yongxin Chen and Allen Tannenbaum

Abstract In this note, we consider the use of 3D models for visual tracking in controlled active vision. The models are used for a joint 2D segmentation/3D pose estimation procedure in which we automatically couple the two processes under one energy functional. Further, employing principal component analysis or locally linear embedding from statistical learning, one can train our tracker on a catalog of 3D shapes, giving a priori shape information. The segmentation itself is information based, which allows us to track in uncertain adversarial environments. Our methodology is demonstrated on realistic scenes, which illustrate its robustness on challenging scenarios.

24.1 Introduction

2D image segmentation and 3D pose estimation have been studied separately in computer vision literature, and we will argue that it is advantageous to study them simultaneously. The 3D model gives shape information of the 2D projections for segmentation, and the segmentation gives key information about the pose and classification of the 3D object. Thus, we combine both approaches in a single variational framework. Specifically, if one has a 3D model of a target, the model can be used

---

J. Lerner
Department of Applied Mathematics & Statistics, Stony Brook University, New York, NY, USA
e-mail: jeremy.lerner@stonybrook.edu

R. Sandhu
Department of Biomedical Informatics, Stony Brook University, New York, NY, USA
e-mail: romeil.sandhu@stonybrook.edu

Y. Chen
Department of Medical Physics, Memorial Sloan Kettering Cancer Center, New York, NY, USA
e-mail: chen2468@umn.edu

A. Tannenbaum (✉)
Department of Computer Science at Stony Brook, New York, NY, USA
e-mail: allen.tannenbaum@stonybrook.edu

R. Tempo et al. (eds.), *Emerging Applications of Control and Systems Theory*, Lecture Notes in Control and Information Sciences - Proceedings,
https://doi.org/10.1007/978-3-319-67068-3_24

to guide the segmentation of the object as well as estimate its location and pose in the world. The ability to measure the location and pose of a target can drastically improve the usefulness of tracking results over those obtained without knowledge of the target's 3D characteristics.

The ability to acquire 3D models has become very accessible. Models can be obtained off-line using methods such as high accuracy laser scanning and multi-view stereo reconstruction. Additionally, models can be learned online by registering multiple views of an object acquired by 3D imaging modalities such as light detection and ranging (LADAR) and stereoscopic vision [17].

In this work, using statistical learning techniques, we show how to train a tracker based on a catalog of 3D shapes, which enables one to employ a shape prior in 3D tracking tasks. Moreover, this allows us to detect and track deformable objects even in cluttered and noisy environments. The image segmentation aspect of this method is crucial, and so we briefly outline here what is involved.

Segmentation, in the two-class case, is the task of separating an object from the background in an image, and Geometric Active Contours, GAC, have been used successfully to solve this problem, see [12, 19]. Note that segmentation can be extended to scenarios with multiple objects, see [11]. Using GAC originally involved evolving a 3D surface, i.e., a level set function, based on local image information (such as edges) near the zero-level set. However, local image information is highly susceptible to corruption with even limited noise or missing information, which can result in a poor segmentation. Region-based segmentation methods are much more robust and resistant to noise. For example, using only mean intensities, many classes of images can be successfully segmented [2], and for more complicated images, higher statistical moments can be used as well. To utilize these higher moments, we base our model on statistical information theory, in particular, the Bhattacharyya coefficient [11]. Although this improves segmentation results, one may still encounter problems with tracking in cluttered adversarial scenarios. Thus, we will use a shape prior to restrict the evolution of the active contour using principal component analysis (PCA) [9] and, in future work, locally linear embedding (LLE) [16]. To this end, we derive a novel 3D shape prior from a dictionary of 3D shapes to do the 2D image segmentation, rather than to derive a 2D shape prior from a collection of 2D images. As a result, we are able to reduce computational complexity in statistical shape learning approaches through a compact shape representation.

24.2 Tracking

We first describe how our framework applies to 3D tracking. We begin by assuming that we have a smooth 3D surface, $S \subset \mathbb{R}^3$. $\mathbf{X} = [X, Y, Z]^T$ define the spatial coordinates that are measured with respect to the imaging camera's referential. The (outward) unit normal to S at each point $\mathbf{X} \in S$ is $\mathbf{N} = [N_1, N_2, N_3]^T$. Moreover, we assume a pinhole camera realization $\pi : \mathbb{R}^3 \mapsto \Omega; \mathbf{X} \mapsto \mathbf{x}$, where $\mathbf{x} = [x, y]^T = [X/Z, \ Y/Z]^T$, and $\Omega \subset \mathbb{R}^2$ denotes the domain of the image I with the corresponding area element $d\Omega$. From this, we define $R = \pi(S)$ to be the region onto which S is projected. Similarly, we can form the complementary region (the area outside the projection) and boundary, or "silhouette" curve, as $R^c = \Omega \setminus R$ and $\hat{c} = \partial R$, respectively. Alternatively stated, if we define the "occluding" curve C to be the intersection of the visible and non-visible regions of S, then the silhouette curve can interpreted as $\hat{c} = \pi(C)$.

Next, let $\mathbf{X}_0$ and $S_0 \in \mathbb{R}^3$ be the coordinates and surface that correspond to the 3D world respectively, [10]. S_0 is given by the zero-level surface of the PCA functional: $\hat{\varphi}(\mathbf{X}_0, w) = \overline{\varphi}(\mathbf{X}_0) + \sum_0^k w_i \psi_i(\mathbf{X}_0)$, where $\varphi_1, ..., \varphi_n$ are the signed distance functions representing the 3D models, $\overline{\varphi} = \left(\frac{1}{n}\right) \sum_{i=1}^n \varphi_i$ and ψ_i are the orthogonal modes of shape variation. We find ψ_i by defining $\tilde{\varphi}_i = \varphi_i - \overline{\varphi}$, letting $M = \{\tilde{\varphi}_1 | \tilde{\varphi}_2 | ... | \tilde{\varphi}_n\}$ and then finding the Singular Value Decomposition of $\frac{1}{n} M M^T = U \Sigma U^T$, where the columns of $U = \{\psi_1 | \psi_2 | ... | \psi_n\}$, [18]. Then we have, $S_0 = \{\mathbf{X}_0 \in \mathbb{R}^3 : \hat{\varphi}(\mathbf{X}_0, w) = 0\}$. Finally, one can locate S in the camera referential via the transformation $g \in SE(3)$, such that $S = g(S_0)$. Writing this point-wise yields $\mathbf{X} = g(\mathbf{X}_0) = \mathbf{R}\mathbf{X}_0 + \mathbf{T}$, where $\mathbf{R} \in SO(3)$ and $\mathbf{T} \in \mathbb{R}^3$.

24.2.1 Gradient Flow for 3D Tracking

We follow the setup of [5], to which we refer the reader for all of the details. First, assume that if the correct 3D pose and shape are given, then the projection of the occluding curve, i.e., $\hat{c} = \pi(C)$, delineates the boundary that optimally separates, or segments, the 2D object from the background. Further, we assume that the image statistics between the 2D projection and the background are distinct and generally separable. Therefore, the energy we wish to minimize may be written in the following general manner:

$$E = \int_R r_o(I(\mathbf{x}), \hat{c}) d\mathbf{x} + \int_{R^c} r_b(I(\mathbf{x}), \hat{c}) d\mathbf{x}, \tag{24.1}$$

where $r_o\left(\chi, \hat{c}\right) : \mathbf{x} \mapsto \mathbb{R}$ and $r_b\left(\chi, \hat{c}\right) : \mathbf{x} \mapsto \mathbb{R}$ measure the similarity of the image pixels with a statistical model over the regions R and R^c, respectively, and χ corresponds to the photometric variable of interest (for example, gray scale intensity, color, or texture vector). E measures the discrepancy between the pixels in R and the pixels in R^c, and we seek a global minimum for E in order to maximize that discrepancy, [4]. In Sect. 24.3.1, we will give details on how to measure this discrepancy based on the Bhattacharyya distance.

We use gradient descent to optimize (24.1) with respect to a finite parameter set denoted as $\xi = \{\xi_1, \xi_2, ..., \xi_m\}$, with m being the number of elements in the respective set, [7]:

$$\frac{\partial E}{\partial \xi_i} = \int_{\hat{c}} \Big(r_o(I(\mathbf{x}), \hat{c}) - r_b(I(\mathbf{x}), \hat{c}) \Big) \Big\langle \frac{\partial \hat{c}}{\partial \xi_i}, \hat{\mathbf{n}} \Big\rangle d\hat{s}, \tag{24.2}$$

where the silhouette curve is parameterized by the arc length $\hat{s}$ with the corresponding outward normal $\hat{\mathbf{n}}$.

Assuming that parameter ξ_i acts on 3D coordinates, the above line integral may be difficult to compute since $\hat{c}$ and $\hat{\mathbf{n}}$ lie in the 2D image plane. Hence, it is more convenient to express the above line integral around the occluding curve C, which is parameterized by s. We refer the reader to [4, 5] for the details; though an an outline of how this can be done is as follows, write

$$\Big\langle \frac{\partial \hat{c}}{\partial \xi}, \hat{\mathbf{n}} \Big\rangle d\hat{s} = \Big\langle \frac{\partial \pi(C)}{\partial \xi}, J \frac{\partial \pi(C)}{\partial s} \Big\rangle ds, \tag{24.3}$$

where $J = \begin{bmatrix} 0 & 1 \\ -1 & 0 \end{bmatrix}$, and this yields the following expression:

$$\Big\langle \frac{\partial \hat{c}}{\partial \xi}, \hat{\mathbf{n}} \Big\rangle d\hat{s} = \frac{\|\mathbf{X}\|}{Z^3} \sqrt{\frac{\kappa_{\mathbf{X}} \kappa_{\mathbf{t}}}{K}} \Big\langle \frac{\partial \mathbf{X}}{\partial \xi_i}, \mathbf{N} \Big\rangle ds, \tag{24.4}$$

where K denotes the Gaussian curvature, and $\kappa_{\mathbf{X}}$ and $\kappa_{\mathbf{t}}$ denote the normal curvatures in the directions $\mathbf{X}$ and $\mathbf{t}$ respectively, with $\mathbf{t}$ being the vector tangent to the curve C at the point $\mathbf{X}$, i.e., $\mathbf{t} = \frac{\partial \mathbf{X}}{\partial s}$. If we now plug the result of (24.4) into (24.2), we arrive at the following flow:

$$\frac{\partial E}{\partial \xi_i} = \int_C \Big(r_o\big(I(\pi(\mathbf{X})), \hat{c}\big) - r_b\big(I(\pi(\mathbf{X})), \hat{c}\big) \Big) \cdot \frac{\|\mathbf{X}\|}{Z^3} \sqrt{\frac{\kappa_X \kappa_t}{K}} \Big\langle \frac{\partial \mathbf{X}}{\partial \xi_i}, \mathbf{N} \Big\rangle ds. \tag{24.5}$$

As we will show, in order to minimize the Bhattacharyya coefficient, we use $r_o\left(I\left(\pi\left(\mathbf{X}\right)\right), \hat{c}\right) - r_b\left(I\left(\pi\left(\mathbf{X}\right)\right), \hat{c}\right) = V(\mathbf{x})$, where $\mathbf{x} = \pi\left(\mathbf{X}\right) \in \Omega$, $I(\mathbf{x})$ has N channels and $V(\mathbf{x})$ is defined as given by either (24.13) or (24.14). Note, that in the above derivation we made no assumption about the type of finite set. That is, we show that the overall framework is essentially "blind" to whether we optimize over the shape weights or pose parameters.

24.3 Information-Theoretic Approach to Segmentation

The next ingredient in our statistical learning tracking scheme is an information-theoretic approach to segmentation. We follow the approach in [11] to which we refer the interested reader for all of the details. To segment an image, we treat the pixels inside and outside the current zero-level set of a level-set function as two probability distributions, and we derive an energy functional based on maximizing the Bhattacharyya distance between them. Because of the region-based nature of the flow, it is very useful for target tracking and can easily be combined with statistical learning as described above. In particular, given the values of a photometric variable, which is to be used for classifying the image pixels, the active contours are designed to converge to the shape that results in *maximal* discrepancy, measured by the Bhattacharyya distance, between the distributions of the photometric variable inside and outside of the contours.

24.3.1 Bhattacharyya Flow

For the segmentation side of our problem, we are simply partitioning the domain $\Omega \subset \mathbb{R}^2$ of an image $I(\mathbf{x})$ (with $\mathbf{x} \in \Omega$) into two mutually exclusive and complementary subsets Ω_- and Ω_+. Given a level set function $\Psi(\mathbf{x})$, its zero-level set, $Z_L(\Psi) = \{\mathbf{x} \mid \Psi(\mathbf{x}) = 0, \mathbf{x} \in \Omega\}$, is used to implicitly represent a curve, as in [12, 19]. In this case, the objective of active contour-based image segmentation is, given an initialization $\Psi_0(\mathbf{x})$, to construct a *convergent* sequence of level set functions $\{\Psi_t(\mathbf{x})\}_{t>0}$ (with $\Psi_t(\mathbf{x})|_{t=0} = \Psi_0(\mathbf{x})$) such that the zero-level set of $\Psi_T(\mathbf{x})$ coincides with the boundary of the object of interest for some $T > 0$.

We define $\Psi(\mathbf{x})$ to be negative inside the 2D projection of the 3D model and positive outside. We denote the subset Ω_- as the domain containing the object of interest, while Ω_+ is the background. To solve the 3D pose estimation in conjunction with the 2D segmentation, we will solve the joint energy functional (24.5), such that the final 2D projection of the 3D object exactly separates between the object of interest and the background.

We construct our joint energy functional via a gradient flow that maximizes the Bhattacharyya distance between the distributions inside and outside the 2D projection of the 3D model. First, the following two *kernel-based estimates* of the probability density functions are computed:

$$P_-(z \mid \Psi) = \frac{\int_\Omega K_-(z - I(\mathbf{x}))\, H(-\Psi(\mathbf{x}))\, d\mathbf{x}}{\int_\Omega H(-\Psi(\mathbf{x}))\, d\mathbf{x}}, \qquad (24.6)$$

and

$$P_+(z \mid \Psi) = \frac{\int_\Omega K_+(z - I(\mathbf{x}))\, H(\Psi(\mathbf{x}))\, d\mathbf{x}}{\int_\Omega H(\Psi(\mathbf{x}))\, d\mathbf{x}}, \qquad (24.7)$$

where $z \in \mathbb{R}^N$, $K_-(z)$ and $K_+(z)$ are two scalar-valued *kernels* having compact or effectively compact supports and normalized to have unit integrals, and $H(\tau)$ is the standard Heaviside function.

The key idea underpinning the segmentation approach of [11] is that for a properly selected subset of image features, the "overlap" between the informational contents of the object and of the background is minimal. In other words, if one thinks of the silhouette curve as the *discriminator* that separates the image pixels into two subsets, then the optimal contour should minimize the *mutual information* between these subsets. Note that for the case at hand, minimizing the mutual information is equivalent to maximizing the Kullback–Leibler divergence between the pdfs associated with the "inside" and "outside" subsets of pixels. However, because of computational efficiency, instead of the divergence, we instead maximize the Bhattacharyya distance between the pdfs, which is defined to be $-\log(\tilde{B})$, where $\tilde{B}$ defines the *Bhattacharyya coefficient* and is given in (24.8). Specifically, the optimal 2D projection (i.e., segmentation) is defined as $\Psi^\star = \arg\min_\Psi\{\tilde{B}(\Psi)\}$, where

$$\tilde{B}(\Psi) = \int_{z\in\mathbb{R}^N} \sqrt{P_-(z\mid\Psi)\,P_+(z\mid\Psi)}\,dz, \tag{24.8}$$

where $P_-(z\mid\Psi)$ and $P_+(z\mid\Psi)$ are given by (24.6) and (24.7), respectively.

Gradient Flow

In order to derive a scheme for minimizing (24.8), we need to compute its first variation (with respect to Ψ), which is given by

$$\begin{aligned}\frac{\delta\tilde{B}(\Psi)}{\delta\Psi}(\mathbf{x}) = \frac{1}{2}\int_{z\in\mathbb{R}^N} & \frac{\partial P_-(z\mid\Psi)}{\partial\Psi}(\mathbf{x})\sqrt{\frac{P_+(z\mid\Psi)}{P_-(z\mid\Psi)}} \\ & + \frac{\partial P_+(z|\Psi)}{\partial\Psi}(\mathbf{x})\sqrt{\frac{P_-(z\mid\Psi)}{P_+(z\mid\Psi)}}\,dz.\end{aligned} \tag{24.9}$$

Differentiating (24.6) and (24.7) with respect to $\Psi(\mathbf{x})$, one obtains

$$\frac{\partial P_-(z\mid\Psi)}{\partial\Psi}(\mathbf{x}) = \delta(\Psi(\mathbf{x}))\left(\frac{P_-(z\mid\Psi) - K_-(z - I(\mathbf{x}))}{A_-}\right), \tag{24.10}$$

and

$$\frac{\partial P_+(z\mid\Psi)}{\partial\Psi}(\mathbf{x}) = \delta(\Psi(\mathbf{x}))\left(\frac{K_+(z - I(\mathbf{x})) - P_+(z\mid\Psi)}{A_+}\right), \tag{24.11}$$

where $\delta(\cdot)$ is the delta function, and A_- and A_+ are the areas of Ω_- and Ω_+, respectively.

By substituting (24.10) and (24.11) into (24.9) and combining the corresponding terms, one can arrive at

$$\frac{\delta\tilde{B}(\Psi)}{\delta\Psi}(\mathbf{x}) = \delta(\Psi(\mathbf{x}))\,V(\mathbf{x}), \tag{24.12}$$

where

$$V(\mathbf{x}) = \frac{1}{2}\tilde{B}(\Psi)(A_-^{-1} - A_+^{-1}) +$$
$$\frac{1}{2}\int_{z\in\mathbb{R}^N} K_+(z - I(\mathbf{x}))\,\frac{1}{A_+}\sqrt{\frac{P_-(z\,|\Psi)}{P_+(z\,|\,\Psi)}}\,dz -$$
$$\frac{1}{2}\int_{z\in\mathbb{R}^N} K_-(z - I(\mathbf{x}))\,\frac{1}{A_-}\sqrt{\frac{P_+(z\,|\,\Psi)}{P_-(z\,|\,\Psi)}}\,dz. \tag{24.13}$$

Assuming the same kernel $K(z)$ is used for computing the last two terms in (24.13), i.e., $K(z) = K_-(z) = K_+(z)$, the latter can be further simplified to the following form:

$$V(\mathbf{x}) = \frac{1}{2}\tilde{B}(\Psi)(A_-^{-1} - A_+^{-1}) + \frac{1}{2}\int_{z\in\mathbb{R}^N} K(z - I(\mathbf{x}))\, L(z\,|\,\Psi)\, dz, \tag{24.14}$$

where

$$L(z\,|\,\Psi) = \frac{1}{A_+}\sqrt{\frac{P_-(z\,|\,\Psi)}{P_+(z\,|\,\Psi)}} - \frac{1}{A_-}\sqrt{\frac{P_+(z\,|\,\Psi)}{P_-(z\,|\,\Psi)}}. \tag{24.15}$$

Thus, to utilize the Bhattacharyya coefficient in the 2D-3D pose estimation problem, we define, for $\hat{c} = Z_L(\Psi)$ and

$$r_o(I(\mathbf{x}), \hat{c}) = \frac{A_-^{-1}}{2}\left(\tilde{B}(\Psi) - \int_{\mathbb{R}^N} K_-(z - I(\mathbf{x}))\sqrt{\frac{P_+(z|\,\Psi)}{P_-(z|\,\Psi)}}dz\right) \tag{24.16}$$

$$r_b(I(\mathbf{x}), \hat{c}) = \frac{A_+^{-1}}{2}\left(\tilde{B}(\Psi) - \int_{\mathbb{R}^N} K_+(z - I(\mathbf{x}))\sqrt{\frac{P_-(z|\,\Psi)}{P_+(z|\,\Psi)}}dz\right) \tag{24.17}$$

resulting in $r_o\left(I(\pi(\mathbf{X})), \hat{c}\right) - r_b\left(I(\pi(\mathbf{X})), \hat{c}\right) = V(\mathbf{x})$, where $V(\mathbf{x})$ is defined as given by either (24.13) or (24.14). We do not require the $\delta(\Psi(\mathbf{x}))$ term because in (24.5), we integrate only over the occluding curve, where $\Psi(\mathbf{x})$ is equal for all points.

24.4 Results

In this section, we present experimental tracking results that demonstrate the algorithm's ability to segment realistic objects where the color is not easily distinguishable from the background, as in Figs. 24.1 and 24.2. Moreover, we have calibrated the camera that is responsible for acquiring the images. That is, the focal length is 671 with principle point to be roughly the center of the image. However, for these examples, we present only the rigid case. That is, we only evolve translation and rotation parameters, and we have not used the full power of the algorithm. Using the LLE algorithm and exploring other methods of manifold learning for classification of multiple types of ships will be the subject of future work.

As in the work of [21], the task of tracking can be decoupled into two fundamental parts: **deformotion**, which is a finite group acting on the target, and **deformation**, which is the small, but infinite dimensional, perturbations that occur. However, because we approach tracking with only a finite set of parameters, namely the Euclidean group of $g \in SE(3)$, we do not need to directly identify the types of motion. Very importantly, our methodology enables us to return the 3D pose of the object, which is a drawback to the method proposed in typical 2D tracking algorithms [13].

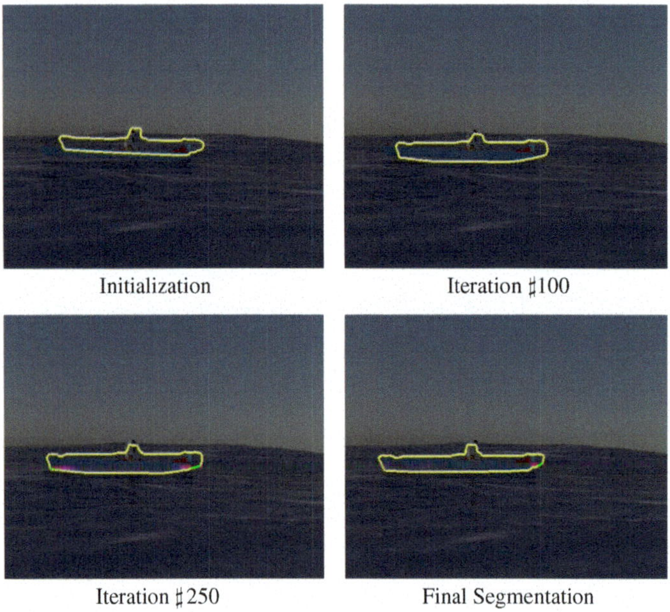

Fig. 24.1 Ocean Scene 1

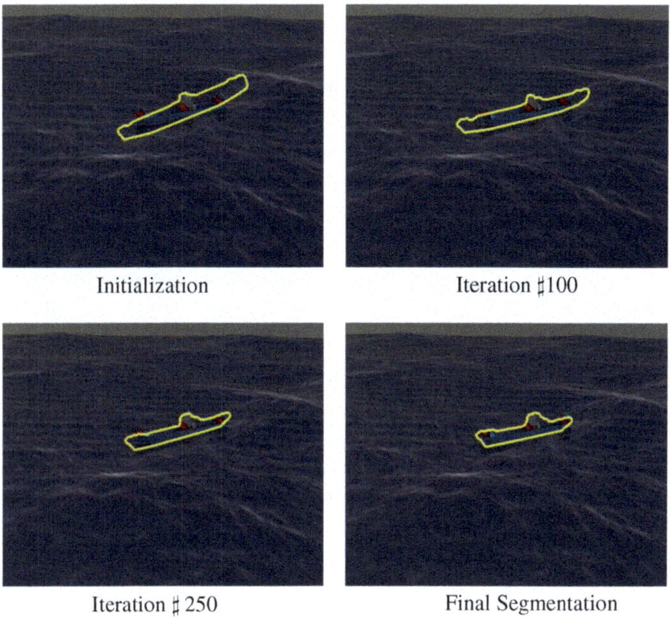

Fig. 24.2 Ocean Scene 2

24.5 Future Work

24.5.1 Manifold Learning

Given a number of observations describing different states of a complex system, we seek the effective number of independent parameters of the system (i.e., its intrinsic dimensionality) and the way these are coupled. The goal of manifold learning, a set of new geometric methods that have been developed in the machine learning community over the past decade, is to solve this problem. These methods are based upon the assumption that the observed data, i.e., a point cloud in some n-dimensional space, lie on or are close to a submanifold $M \subset \mathbf{R}^d$.

Naturally, many different methods are necessary to meet the variety of challenging scenarios. Nevertheless, many of the most popular approaches are essentially spectral methods. Note that the term spectral method is ambiguous and used differently within different communities; for example, in numerical partial differential equations it often involves the use of the fast Fourier transform. Specifically, for manifold learning, a spectral method involves deriving a symmetric matrix from point cloud data, whose eigenvectors are used to obtain the solution to a given optimization problem. The geometric optimization problems that lead to a spectral technique are mostly of a least squares nature and include the following standard problems:

(1) Find the k-dimensional subspace that best approximates the point cloud, in a least squares sense.
(2) Find the embedding of the point cloud in k-dimensional space that best preserves the distances between the points, in a least squares sense.

The first problem is better known as principal component analysis (PCA), as it asks for the principal directions (components) of the data. It is essentially a data quantization technique: every data point is replaced by its projection onto the best approximating k-dimensional subspace. The loss incurred by the quantization is the variance of the data in the directions orthogonal to this best approximating k-dimensional subspace. As long as this variance is small, PCA can also be seen as denoising the original data. Many machine learning techniques, including clustering, classification, semi-supervised learning, and near neighbor indexing and search can benefit from such denoising, [8].

The second problem is called multi-dimensional scaling, MDS. An important application of MDS is the visualization of point cloud data: the data points are embedded into two- or three-dimensional space, where they can be directly visualized. The main purpose of visualization is to use the human visual system to acquire insights into the structure of the point cloud data, e.g., the existence of clusters or, for data points labeled with discrete attributes, relations between these attributes. MDS visualization remains a popular tool for point cloud data analysis; however, if the intrinsic dimension of the data set is larger than three, a great deal of information will be lost (and, in general, cannot be restored by the human visual system).

Recently, the focus in point cloud data analysis has shifted: more emphasis has been put on detecting nonlinear features in the data, although processing the data for visual inspection is still important. What drives this shift in focus is the insight that most features are based on local correlations of the data points, but PCA and MDS both have only a global view of the point cloud data. The shift toward local correlations was pioneered by two techniques called ISOMAP [20] and locally linear embedding (LLE) [16]. Fortunately, with these and other local correlation methods, the global picture is not lost. For example, the global intrinsic dimension of the data can be estimated from local information; whereas, when the data are embedded nonlinearly, it is often not possible to derive this information from a purely global analysis. ISOMAP, LLE and their successors (some of which we will also discuss in future work) can all be used for the traditional purposes: data quantization and data visualization. In general, they preserve more information of the data (than PCA and MDS) while achieving a similar quantization error or when targeting the same embedding dimension for data visualization, respectively. The LLE methodology has proven quite useful for shape analysis in several contexts [13], so we will now consider LLE in some detail.

24.5.2 Locally Linear Embedding

In [16], the authors proposed an unsupervised locally linear embedding (LLE) algorithm that computes low dimensional, neighborhood preserving embeddings of high-dimensional data. LLE attempts to discover nonlinear structure in high-dimensional data by exploiting the local symmetries of linear combinations. It has been used in many pattern recognition problems for classification.

The LLE algorithm is based on certain simple geometric principles. If we assume that the data, n vectors φ_i of dimension $N^2 \times 1$, are sampled from some underlying smooth manifold, provided there is sufficient data, we expect each data point and its neighbors to lie on or close to a locally linear patch of the manifold. We can characterize the local geometry of these patches by a set of coefficients, or weights, that reconstruct each data point only from its nearest neighbors. In the simplest formulation of LLE, one identifies the k nearest neighbors for each data point. Reconstruction error is then measured by the cost function: $E(W) = \left(\varphi - \sum_{j=1}^{n} \alpha_j \varphi_j\right)^2$, which is minimized subject to the constraint that the weights α_j that lie outside the appropriate neighborhood are zero and $\sum_j \alpha_j = 1$. With these constraints, the weights for points in the neighborhood of φ can be obtained as in [15]:

$$E(W) = \sum_{j=1}^{k} \left(\varphi - \alpha_j \varphi_j\right)^2 = \sum_{j=1}^{k} \sum_{m=1}^{k} \alpha_j \alpha_m Q_{jm}$$

$$\Rightarrow \alpha_j = \frac{\sum_{m=1}^{k} R_{jm}}{\sum_{p=1}^{k} \sum_{q=1}^{k} R_{pq}}, \tag{24.18}$$

$$\text{where } Q_{jm} = (\varphi - \varphi_j)^T (\varphi - \varphi_m) \text{ and } R = (Q)^{-1}. \tag{24.19}$$

For our application, we choose to represent the 3D models as signed distance functions, $\varphi_1, ..., \varphi_n$, where the zero-level set, $Z_L(\varphi_i)$, represents the surface S_0. In order to calculate the k nearest neighbors, we define the distance between surfaces as in [14]:

$$d^2\left(\varphi_i, \varphi_j\right) = \sum_{p \in Z_L(\varphi_i)} \left|\varphi_j(p)\right| + \sum_{p \in Z_L(\varphi_j)} \left|\varphi_i(p)\right| \tag{24.20}$$

To apply this LLE algorithm to the joint classification and segmentation problem at hand, we first find the k nearest neighbors and the ideal corresponding weights, α_j, $j = 1...n$, of some initialization, and now define the surface S_0 as the zero-level surface of $\hat{\varphi}(\mathbf{X}_0, \alpha) = \sum_{j=1}^{n} \alpha_j \varphi_i$, with $\alpha = [\alpha_1, ..., \alpha_n]^T$. Then we evolve the weights according to (24.5), where

$$\frac{\partial}{\partial \alpha_i}\hat{\varphi}(\mathbf{X}_0, \alpha) = \left(\nabla_{\mathbf{X}_0}\hat{\varphi}\right) \cdot \left(\frac{\partial \mathbf{X}_0}{\partial \alpha_i}\right) + \frac{\partial \hat{\varphi}}{\partial \alpha_i}$$
$$= \left(\left\|\nabla_{\mathbf{X}_0}\hat{\varphi}\right\| \mathbf{N}_0\right) \cdot \left(\frac{\partial \mathbf{X}_0}{\partial \alpha_i}\right) + \varphi_i(\mathbf{X}_0) = 0. \quad (24.21)$$

Thus, we have

$$\left\langle \frac{\partial \mathbf{X}}{\partial \alpha_i}, \mathbf{N} \right\rangle = -\frac{\varphi_i(\mathbf{X}_0)}{\left\|\nabla_{\mathbf{X}_0}\hat{\varphi}\right\|}. \quad (24.22)$$

Additionally, to introduce the nonlinearity to this methodology, every so often we check which k signed distance functions are the current nearest neighbors of $\hat{\varphi}(\mathbf{X}_0, \alpha)$. For the neighbors that change, the appropriate weights are changed to correspond to new signed distance functions.

References

1. Belkin, M., Niyogi, P.: Laplacian eigenmaps for dimensionality reduction and data representation. Neural Comput. **15**, 1373–139615 (2003)
2. Chan, T., Sandberg, B., Vese, L.: Active contours without edges for vector-valued images. J. Visual Commun. Image Represent. **11**(2), 130–141 (2000)
3. Cipolla, R., Blake, A.: Surface shape from the deformation of apparent contours. Int. J. Comp. Vis. **9**(2), 83–112 (1992)
4. Dambreville, S., Sandhu, R., Yezzi, A., Tannenbaum, A.: Robust 3D pose estimation and efficient 2D region-based segmentation from a 3D shape prior. In: Proceedings of ECCV, pp. 169–182 (2008)
5. Dambreville, S., Sandhu, R., Yezzi, A., Tannenbaum, A.: A geometric approach to joint 2D region-based segmentation and 3D pose estimation using a 3D shape prior. SIAM Imaging Sci. **3**, 110–132 (2010)
6. Donoho, D.L., Grimes, C.: "Hessian eigenmaps: new locally linear embedding techniques for high-dimensional data," Technical Report TR-2003-08. Stanford University, Department of Statistics (2003)
7. Flanders, H.: Differentiation under the integral sign. Am. Math. Monthly **80**(6), 615–627 (1973)
8. Hein, M., Maier, M.: Manifold denoising. NIPS:561–568 (2006)
9. Leventon, M., Grimson, E., Faugeras, O.: Statistical shape influence in geodesic active contours. In: IEEE CVPR, pp. 1316–1324 (2000)
10. Ma, Y., Stefano, S., Kosecka, J., Sastry, S.: An invitation to 3-d vision: from images to geometric models. In: Springer Science and Business Media, p. 26 (2012)
11. Michailovich, O., Rathi, Y., Tannenbaum, A.: Image segmentation using active contours driven by the Bhattacharyya gradient flow. IEEE Trans. Image Process. **16**, 2787–2801 (2007)
12. Osher, S., Fedkiw, R.: Level Set Methods and Dynamic Implicit Surfaces. Springer (2003)
13. Rathi, Y., Tannenbaum, A.: A generic framework for tracking using particle filter with dynamic shape prior. IEEE Trans. Image Process. **16**, 1370–1382 (2007)
14. Rathi, Y., Vaswani, N., Tannenbaum, A.: A generic framework for tracking using particle filter with dynamic shape prior. IEEE Trans. Image Process. **16**(5), 1370–1382 (2007)
15. Ridder, D., Duin, R.: Locally linear embedding for classification. Technical report. PH-2002-01, Pattern Recognition Group, Delft University of Technology (2002)
16. Roweis, S., Saul, L.: Nonlinear dimensionality reduction by locally linear embedding. Science **290**(5500), 2323–2326 (2000)

17. Sandhu, R., Dambreville, S., Tannenbaum, A.: Particle filtering for registration of 2D and 3D point sets with stochastic dynamics. In: IEEE CVPR (2008)
18. Sandhu, R., Dambreville, S., Yezzi, A., Tannenbaum, A.: Non-rigid 2D-3D pose estimation and 2D image segmentation. In: IEEE CVPR (2009)
19. Sethian, J.A.: Level Set Methods and Fast Marching Methods, 2nd edn. Cambridge University Press (1999)
20. Tenenbaum, J.B., de Silva, V., Langford, J.C.: A global geometric framework for nonlinear dimensionality reduction. Science **290**, 2319–2323 (2000)
21. Yezzi, A., Soatto, S.: Deformotion: Deforming motion, shape average and the joint registration and approximation of structures in images. Int. J. Comput. Vis. **53**, 153–167 (2003)

Chapter 25
Networked Parallel Algorithms for Robust Convex Optimization via the Scenario Approach

Keyou You and Roberto Tempo

Abstract This chapter proposes a parallel computing framework to distributedly solve robust convex optimization (RCO) when the constraints are affected by nonlinear uncertainty. To this end, we adopt a scenario approach by randomly sampling the uncertainty set. However, the number of samples to attain high levels of probabilistic guarantee of robustness may be large, which results in a large number of constraints in the scenario problem (SP). Instead of using a single processor, we resort to many processors that are distributed among different nodes of time-varying unbalanced digraphs. Then, we propose a random projected algorithm to distributedly solve the SP, which is given in an explicitly recursive form with simple iteration. We show that if the sequence of digraphs are *uniformly jointly strongly connected* (U-JSC), each node asymptotically converges to a common optimal solution to the SP. That is, the RCO is effectively solved in a distributed parallel way.

25.1 Introduction

This chapter is concerned with the robust convex optimization (RCO) which has an infinite number of constraints parameterized by uncertainties. RCO is widely used in control analysis and synthesis of complex systems. To solve it, we have to address the nonlinear and complicated dependence of the constraints on the uncertainties. Thus,

---

Portions of this chapter were previously presented at the 2016 ACC and the material is used with permission of the AACC. (Keyou You, Roberto Tempo, "Parallel algorithms for robust convex programs over networks", in Proceedings of the 2016 American Control Conference, https://doi.org/10.1109/ACC.2016.7525215)
R. Tempo—Author deceased.

K. You (✉)
Department of Automation and TNList, Tsinghua University, Beijing 100084, China
e-mail: youky@tsinghua.edu.cn

R. Tempo
CNR-IEIIT, Politecnico di Torino, 10129 Torino, Italy

R. Tempo et al. (eds.), *Emerging Applications of Control and Systems Theory*, Lecture Notes in Control and Information Sciences - Proceedings,
https://doi.org/10.1007/978-3-319-67068-3_25

RCO generally not easily solvable, and attracts significant interest in the literature, see, e.g., [4, 6, 24] and references therein.

Here we adopt a *scenario approach* [9, 25] to solve RCO by randomly sampling the uncertainty set, which results in a standard convex optimization called the scenario problem (SP). For an optimal solution of the SP, the good news is that the level of confidence for a small probability of violating the original constraints can be reduced by increasing the sample size, and the sample complexity can be computed a priori [1]. While for a single processor, it may be beyond its computational capability to solve such a large convex SP.

To this end, we consider to adopt a parallel computing framework with *many interconnected processors* which are distributed among different nodes of time-varying graphs. In this chapter, our key approach is to appropriately assign the computation task among nodes so that each node is able to collaboratively compute some common optimal solution of the SP with a low computational cost. Then, we break a large number of constraints in the SP into many small sets of *local* constraints that are easily handled in each node. Under local interactions, the SP is then jointly solved in every node via three key steps.

First, every node randomly samples the uncertainty set of RCO, with the sample size inversely proportional to the total number of nodes or being determined by its computation capacity. Although this step has been originally adopted in [12] to solve the SP, our approach is substantially different. Each node in [12] requires to completely solve a local SP at each iteration and exchanges the set of active constraints with its neighbors. The process continues until a consensus on the set of active constraints is reached among nodes, and every node solves its local SP under *all* active constraints. Clearly, the number of active constraints in every local SP is increasing with the number of iterations. In some extreme cases, each constraint in the SP can be active, and every node eventually engages to solve such a local SP that has the same number of constraints as the SP. In this view, the computation cost in each node cannot be reduced, and will certainly waste computation resources. Since an active constraint may become inactive in next iteration, identifying active constraints cannot be recursively completed and is very sensitive to numerical errors. In this work, each node only needs to handle a fixed number of local constraints, and recursively run an explicit algorithm with a simple structure.

Second, the SP is reformulated as a distributed optimization problem with many decoupled small sets of local constraints and a coupled constraint, which is specially designed in conform with the graph structure. This is the key idea for assigning computation task among nodes. If the number of nodes is large, each node only needs to deal with a very small number of local constraints. While the information circulation across the network is to address the coupled constraint, it should be well designed so that it can be locally addressed. A similar technique has been used to solve the distributed optimization problem, see, e.g., [7, 19]. However, they do not consider any robustness issue. Without considering the distributed computation, robust optimization has also attracted significant attention in many research areas [3, 15]. In this work, we are able to address both distributed and robust optimization problems simultaneously.

Third, each node keeps updating a local copy of an optimal solution by individually handling its local constraints and interacting with its neighbors to overcome the coupled constraint. To this end, we address the coupled constraint by adopting a consensus algorithm, and design a novel two-stage recursive algorithm. At the first stage, we are solving an unconstrained optimization problem which removes the decoupled local constraints in the reformulated distributed optimization, and obtain an intermedia state vector in each node. It is worth mentioning that we do not need a balanced digraph here, unlike [13, 14, 16, 18]. At the second stage, each node individually addresses its decoupled local constraints by adopting a generalized Polyak's random algorithm [20], which moves its intermedia state vector toward a randomly selected local constraint set. Combining these two stages and under some mild conditions, both consensus and feasibility of the iteration in each node are eventually achieved almost surely.

The rest of this chapter is organized as follows. In Sect. 25.2, we formulate RCO, after which the probabilistic approach to RCO is provided. In Sect. 25.3, we describe a parallel computing framework for distributedly solving the SP. In Sect. 25.4, we design a distributed random projected algorithm over time-varying digraphs to distributedly solve RCO and prove its convergence. Some brief concluding remarks are drawn in Sect. 25.5.

Notation: The sub-gradient of a vector function (a vector whose components are convex functions) $y = [y_1, \ldots, y_n]' \in \mathbb{R}^n$ with respect to an input vector $x \in \mathbb{R}^m$ is $\partial y = [\partial y_1, \ldots, \partial y_n]' \subseteq \mathbb{R}^{n\times m}$. For two vectors $a = [a_1, \ldots, a_n]'$ and $b = [b_1, \ldots, b_n]'$, the notation $a \succeq b$ means that a_i is greater than b_i for any $i \in \{1, \ldots, n\}$. Similar notations will be made for $\succ$, $\preceq$ and $\prec$. $\mathbf{1}$ denotes a vector with a compatible dimension and each element being one. $\otimes$ Kronecker product. In addition, $f(\theta)_+ = \max\{0, f(\theta)\}$ takes the positive part of f.

25.2 Robust Convex Optimization and Scenario Approach

25.2.1 Robust Convex Optimization

Consider a robust convex optimization (RCO) of the form

$$\min_{\theta\in\Theta} c'\theta \ \text{ subject to } f(\theta, q) \leq 0, \forall q \in \mathcal{Q}, \tag{25.1}$$

where $\Theta \subseteq \mathbb{R}^n$ is a convex and closed set with non-empty interior, and the scalar-valued function $f(\theta, q) : \mathbb{R}^n \times \mathcal{Q} \to \mathbb{R}$ is convex in the design vector θ for any $q \in \mathcal{Q} \subseteq \mathbb{R}^\ell$. The uncertainty q enters into the constraint function $f(\theta, q)$ without assuming any structure, except for the Borel measurability [2] of $f(\theta, \cdot)$ for any fixed θ. In particular, $f(\theta, \cdot)$ may be affected by nonlinear parametric uncertainty. For simplicity, the objective function $c'\theta \in \mathbb{R}$ is set to be linear in θ, which does not loose generality. For example, consider a convex objective function $f_0(\theta)$ and introduce an auxiliary variable t. Then, the optimization in (25.1) is equivalent to

$$\min_{\theta \in \Theta, t \in \mathbb{R}} t \text{ subject to } f_0(\theta) - t \leq 0 \text{ and } f(\theta, q) \leq 0, \forall q \in \mathcal{Q}.$$

Obviously, the objective function becomes linear in the augmented decision variable (θ, t) and is of the same form as (25.1).

25.2.2 Scenario Approach for RCO

The design constraint $f(\theta, q) \leq 0$ for all possible $q \in \mathcal{Q}$ is crucial in the study of robustness of complex systems, e.g., $\mathcal{H}_\infty$ performance of a system affected by the parametric uncertainty and the design of uncertain model predictive control. However, obtaining worst-case solutions has been proved to be computationally difficult, even NP-hard as the uncertainty q may enter into $f(\theta, q)$ in a complicated manner.

In fact, it is generally very difficult to explicitly characterize the constraint set with uncertainty, i.e.,

$$\{\theta \,|\, f(\theta, q) \leq 0, \forall q \in \mathcal{Q}\}, \tag{25.2}$$

which renders it impossible to directly solve RCO. There are only few cases that the uncertainty set is tractable [15]. Moreover, this approach also introduces undesirable *conservatism*. For these reasons, we adopt the widely accepted scenario approach [25] to solve RCO. Instead of satisfying the hard constraint in (25.2), the idea of the scenario approach is to derive an *approximation* by means of a finite number of random constraints, i.e.,

$$\{\theta \,|\, f(\theta, q^{(i)}) \leq 0, i = 1, \ldots, N_{bin}\} \tag{25.3}$$

where N_{bin} is a positive integer representing the constraint size, and $\{q^{(i)}\} \subseteq \mathcal{Q}$ are independent identically distributed (i.i.d.) samples extracted according to an arbitrary absolutely continuous (w.r.t. the Lebesgue measure) distribution $\mathbb{P}_q(\cdot)$ over $\mathcal{Q}$.

Under the constraint in (25.3), we only guarantee that most, albeit not all, possible uncertainty constraints in RCO are not violated. By the randomness of $\{q^{(i)}\}$, (25.3) may be very close to its counterpart (25.2) in the sense of obtaining a low *violation probability*, which is now formally defined below.

Definition 25.1 *(Violation probability)* Given a design vector $\theta \in \mathbb{R}^n$, the violation probability $V(\theta)$ is defined as

$$V(\theta) := \mathbb{P}_q\{q \in \mathcal{Q} | f(\theta, q) > 0\}.$$

The multi-sample $q^{1:N_{bin}} := \{q^{(1)}, \ldots, q^{(N_{bin})}\}$ is called a *scenario* and the resulting optimization problem under the constraint (25.3) is referred to as a *scenario problem* (SP)

$$\min_{\theta \in \Theta} c'\theta \ \text{ subject to}$$
$$f(\theta, q^{(i)}) \leq 0, i = 1, \ldots, N_{bin}. \tag{25.4}$$

In the sequel, let Θ^* be the set of optimal solutions to the SP and Θ_0 be the set of feasible solutions, i.e.,

$$\Theta_0 = \{\theta \in \Theta | f(\theta, q^{(i)}) \leq 0, i = 1, \ldots, N_{bin}\}. \tag{25.5}$$

For the SP, we make the following assumption to describe its relationship with RCO in (25.1) from a probabilistic point of view.

Assumption 25.1 *(Optimal solutions and interior point)* The SP in (25.4) is feasible for any multisample extraction and has a non-empty set of optimal solutions. In addition, there exists a vector $\theta_0 \in \Theta$ such that

$$f(\theta_0, q^{(i)}) < 0, \forall i = 1, \ldots, N_{bin}. \tag{25.6}$$

The interior condition (often called Slater's constraint qualification) in (25.6) implies that there is no duality gap between the primal and dual problems of (25.4) and the dual problem contains at least an optimal solution [5]. We remark that in robust control it is common to study strict inequalities [21], e.g., when dealing with the robust asymptotic stability of a system and therefore this is not a serious restriction. In fact, the set of the feasible solutions to (25.1) is a subset of that of the SP in (25.4), and the feasibility assumption can also be relaxed in the analysis of the SP by using the approach introduced in [10]. The main result of the scenario approach for RCO is stated below.

Lemma 25.1 ([11]) Assume that there exists a unique solution to (25.4). Let ε, $\delta \in (0, 1)$, and N_{bin} satisfy the following inequality:

$$\sum_{i=0}^{n-1} \binom{N_{bin}}{i} \varepsilon^i (1-\varepsilon)^{N_{bin}-i} \leq \delta. \tag{25.7}$$

With probability at least $1 - \delta$, the solution θ_{sc} of the SP in (25.4) satisfies $V(\theta_{sc}) \leq \varepsilon$, i.e.,

$$\mathbb{P}_q\{V(\theta_{sc}) \leq \varepsilon\} \geq 1 - \delta.$$

The uniqueness condition can be also relaxed in most cases by introducing a tie-breaking rule, see Sect. 4.1 of [8]. If the sample complexity N_{bin} satisfies (25.7), the solution θ_{sc} to (25.4) approximately solve the RCO in (25.1) with certain probabilistic guarantee. A subsequent problem is to compute the smallest sample complexity, which dictates the number of constraints required in the SP in (25.4). Fortunately, this has been well addressed in [1] via the following improved bound

$$N_{bin} \geq \frac{e}{\varepsilon(e-1)}\left(\ln\frac{1}{\delta}+n-1\right) \tag{25.8}$$

where e is the Euler constant. Thus, RCO in (25.1) can be well approximately solved via the SP in (25.4) with a sufficiently large N_{bin}.

The remaining problem is to effectively solve the SP in (25.4).

25.3 Networked Parallel Scheme for SPs

25.3.1 Networked Computing Nodes

Although RCO in (25.1) can be attacked via the SP, clearly N_{bin} may be large to achieve a high confidence level with small violation probability. For example, in a problem with $n = 32$ variables, setting probability levels $\varepsilon = 0.001$ and $\delta = 10^{-6}$, it follows from (25.8) that the number of constraints in the SP is $N_{bin} \geq 70898$. For such a large sample complexity N_{bin}, the computation cost for solving the SP (25.4) becomes very high, which may be far from the computation capacity of a cheap processor.

In this section, we exploit the idea of *solving large problems with many small solvers*. Specifically, we propose to use m computing units (nodes) which cooperatively solve the SP in (25.4) in a parallel fashion. Then, the number of design constraints for node j is reduced to n_j, provided that $\sum_{j=1}^m n_j \geq N_{bin}$.

A heuristic approach is to assign the number of constraints in (25.4) among nodes proportional to their computing power. In practice, each node can declare the total number of constraints that can be handled. If the number of nodes is comparable to the scenario size N_{bin}, the number of constraints for every node j is significantly reduced, e.g., $n_j \ll N_{bin}$, and n_j can be even as small as one.

The problem is how to parallelize the computational task across multiple nodes to cooperatively solve the SP. To this end, we introduce a digraph $\mathscr{G} = (\mathscr{V}, \mathscr{E})$ to model interactions between the computing nodes where $\mathscr{V} := \{1, \ldots, m\}$ denotes the set of nodes, and the set of interaction links is represented by $\mathscr{E}$. A directed edge $(i, j) \in \mathscr{E}$ exists if node i can directly receive information from node j. The digraph $\mathscr{G}$ can also be described a *row-stochastic* weighting matrix $A = \{a_{ij}\} \in \mathbb{R}^{m\times m}$, e.g., $a_{ij} > 0$ if $(i, j) \in \mathscr{E}$ and 0, otherwise, and $a_{jj} = 1 - \sum_{i=1}^m a_{ji}$ for all $j \in \mathscr{V}$.

In this work, we are more interested in time-varying digraphs $\{\mathscr{G}^k = (\mathscr{V}, \mathscr{E}^k)\}$ to model node interactions, and the objective of this paper is to solve the SP in (25.4) over time-varying digraphs under the following setup:

(a) Every node j of limited computation capability is able to independently generate n_j i.i.d. samples with an absolutely continuous distribution $\mathbb{P}_q$, and is not allowed to share their samples with other nodes.
(b) Every node is able to broadcast a finite dimensional data via directed edges.
(c) The vector c in the objective function, the constraint function $f(\theta, q)$ and the set Θ are visible to every node.

In contrast with [12], the dimension of the data for communication is independent of both the scenario size N_{bin} and the number of nodes m. In addition, every node j only needs to address a fixed number n_j of constraints, while in [12] each node requires to completely solve a *local* SP with an increasing number of constraints.

25.3.2 Reformulation of the SP

Let $q^{(j1)}, \ldots, q^{(jn_j)}$ be the samples that are independently generated in node j according to the distribution $\mathbb{P}_q$. For brevity, the local constraint functions are collectively rewritten in a vector form

$$f_j(\theta) := \begin{bmatrix} f(\theta, q^{(j1)}) \\ \vdots \\ f(\theta, q^{(jn_j)}) \end{bmatrix} \in \mathbb{R}^{n_j}.$$

Then, the SP in (25.4) can be equivalently solved via the following constrained minimization problem:

$$\min_{\theta_1,\ldots,\theta_m \in \Theta} \sum_{j=1}^{m} c'\theta_j \text{ subject to } \theta_1 = \ldots = \theta_m, \; f_j(\theta) \preceq 0, \forall j \in \mathscr{V}, \tag{25.9}$$

where $f_j(\theta)$ is only known to node j. Hence, we are sufficient to distributedly solve the above optimization over time-varying graphs.

25.4 Networked Random Projected Algorithms over Time-Varying Digraphs

25.4.1 Distributed Random Projected Algorithm

Consider that each node has very limited computation capability, the computation for each node should be easy to implement with a low computation cost. We utilize the distributed sub-gradient descend to achieve consensus and adopt the idea of Polyak's random algorithm [22] to ensure feasibility of the iterate in each node. Note that their algorithms are centralized and do not touch the distributed implementation, which are resolved in this work by adapting to the network structure. Specifically, we propose the following two-stage distributed random projected algorithm

$$v_j^k = \sum_{i=1}^{m} a_{ji}^k \theta_i^k - \zeta^k \cdot c, \tag{25.10}$$

$$\theta_j^{k+1} = \Pi_{\Theta}(v_j^k - \beta \cdot \frac{f(v_j^k, q^{(jw_j^k)})_+}{\|d_j^k\|^2} d_j^k), \tag{25.11}$$

where $\zeta^k > 0$ is a deterministic stepsize and satisfies the condition

$$\zeta^k > 0, \ \sum_{k=0}^{\infty} \zeta^k = \infty, \ \sum_{k=0}^{\infty} (\zeta^k)^2 < \infty, \tag{25.12}$$

and $\beta \in (0, 2)$ is a deterministic parameter. $w_j^k \in \{1, \ldots, n_j\}$ is a random variable. The vector $d_j^k \in \partial f(v_j^k, q^{(jw_j^k)})_+$ if $f(v_j^k, q^{(jw_j^k)})_+ > 0$ and $d_j^k = d_j$ for some $d_j \neq 0$ if $f(v_j^k, q^{(jw_j^k)})_+ = 0$.

The intuition of the above algorithm is illustrated as follows. (25.10) is to distributedly solve an unconstrained optimization, i.e., the optimization by removing the constraints in (25.9), see [18] for details. Note that their work assumes the double stochasticity of A, which is unnecessary as shown in this work. (25.11) is to drive the intermediate state v_j^k toward a randomly selected local constraint set $\Theta \cap \Theta_j^{w_j^k}$, where $\Theta_j^{w_j^k} := \{\theta | f(\theta, q^{(jw_j^k)}) \leq 0\}$. If β is sufficiently small, it follows from [5, Proposition 6.3.1] that $d(\theta_j^{k+1}, \Theta \cap \Theta_j^{w_j^k}) \leq d(v_j^k, \Theta \cap \Theta_j^{w_j^k})$. That is, θ_j^{k+1} is closer to the local constraint set $\Theta \cap \Theta_j^{w_j^k}$ than v_j^k. If w_j^k is uniformly selected at random from $\{1, \ldots, n_j\}$, we conclude that θ_j^{k+1} is closer to the local constraint set $\Theta \cap \Theta_j$ than v_j^k in the average sense. Once the consensus is achieved among nodes, the state vector θ_j^k in each node asymptotically converges to a point in the feasible set Θ_0.

It is stressed that most existing works on distributed optimization require the underlying graph to be *balanced* of the form that A^k is doubly stochastic, see, e.g., [13, 14, 16, 18]. Clearly, a balanced graph is much more restrictive and complicates the topology design. Furthermore, we see no reason at all to enforce such a strong condition. This issue has been recently resolved either by combining the gradient descent and the push-sum consensus [17] or augmenting an additional variable for each agent to record the state updates [26]. In comparison, the algorithm in [17] only focuses on the *unconstrained* optimization, involves nonlinear iterations and requires the updates of four vectors. The algorithm in [26] requires an additional "surplus" vector to record the state update, which increases the computation and

Algorithm 1: Networked random projection algorithm for the SP with directed graphs

1: **Initialization:** Each node $j \in \mathcal{V}$ sets $\theta_j = 0$.
2: **repeat**
3: **Local information exchange:** Every node $j \in \mathcal{V}$ broadcasts θ_j to its out-neighboring nodes.
4: **Local variables update**: Every node $j \in \mathcal{V}$ receives the state vector θ_i from its in-neighbor $i \in \mathcal{N}_j^{in}$ and updates as follows:

- $v_j = \sum_{i \in \mathcal{N}_j^{in}} a_{ji}^k \theta_i - \zeta c$ where the stepsize ζ is given in (25.12).
- Draw $w_j \in \{1, \ldots, n_j\}$ uniformly at random.
- $\theta_j \leftarrow \Pi_\Theta(v_j - \beta \cdot \frac{f(v_j, q^{(jw_j)})_+}{\|d_j\|^2} d_j)$ where d_j is defined in (25.11).

5: **Set** $k = k + 1$.
6: **until** a predefined stopping rule is satisfied.

communication cost. From this point of view, the proposed algorithm has a simpler structure and it is easier to implement, see also Algorithm 1 for details.

25.4.2 *Convergence of Algorithm 1*

To prove the convergence, we impose the following assumptions on time-varying graphs and the boundedness of sub-gradients.

Assumption 25.2 (a) (UJSC) Time-varying graphs $\{\mathcal{G}^k\}$ are uniformly jointly strongly connected, i.e., there exists an positive integer B such that the joint graph $\mathcal{G}^k \cup \ldots \mathcal{G}^{k+B-1}$ is strongly connected for any $k \geq 0$. (b) There exists a scalar $\gamma > 0$ such that $a_{ij}^k \geq \gamma$ if $a_{ij}^k > 0$ for any $i, j \in \mathcal{V}$ and $k \geq 0$.

Assumption 25.3 (a) $\{w_j^k\}$ is an i.i.d. sequence that is uniformly distributed over the set $\{1, \ldots, n_j\}$ for any $j \in \mathcal{V}$, and is also independent over the index j. (b) The sub-gradients d_j^k are uniformly bounded, i.e., there exists a scalar $\bar{d}$ such that

$$\|d_j^k\| \leq \bar{d}, \forall j \in \mathcal{V}$$

Clearly, Assumption 25.2 is common in the literature on distributed algorithms over time-varying graphs. The designer is free to choose any distribution for drawing samples w_j^k. Thus, Assumption 25.3(a) is easy to satisfy. By the property of the sub-gradient and (25.11), a sufficient condition for Assumption 25.3(b) is that Θ is bounded.

Theorem 25.1 (Almost sure convergence) *Under Assumptions 25.1–25.3, the sequence $\{\theta_j^k\}$ of Algorithm 1 converges almost surely to some common point in the set Θ^* of optimal solutions to (25.4).*

The proof is roughly divided into three parts. The first part establishes a sufficient condition to ensure asymptotic *consensus*, under which the sequence $\{\theta_j^k\}$ converges to the same value for all $j \in \mathscr{V}$. The second part demonstrates the asymptotic *feasibility* of the state vector θ_j^k. Finally, the last part illustrates the *optimality* by showing that the distance of θ_j^k to any optimal point θ^* is "stochastically" decreasing. Combing these three parts, we show that $\{\theta_j^k\}$ converges to some common point in Θ^* almost surely.

Lemma 25.2 ([27]). Under Assumption 25.2, consider the following sequence:

$$\theta_j^{k+1} = \sum_{i=1}^{n} a_{ji}^k \theta_i^k + \varepsilon_j^k, \quad \forall j \in \mathscr{V}. \tag{25.13}$$

If $\lim_{k\to\infty} \|\varepsilon_j^k\| = 0$ for any $j \in \mathscr{V}$, it follows that

$$\lim_{k\to\infty} \|\theta_j^k - \bar{\theta}^k\| = 0, \quad \forall j \in \mathscr{V}. \tag{25.14}$$

The second result essentially ensures the local feasibility.

Lemma 25.3 ([27]) Let $\{\theta_j^k\}$ be generated Algorithm 1. Define λ_j^k and μ_j^k as follows

$$\lambda_j^k = \sum_{i=1}^{n} a_{ji}^k \theta_i^k, \text{ and } \mu_j^k = \Pi_{\Theta_0}(\lambda_j^k), \tag{25.15}$$

where Θ_0 is defined in (25.5). If $\lim_{k\to\infty} \|\lambda_j^k - \mu_j^k\| = 0$, then $\lim_{k\to\infty} \|\mu_j^k - \theta_j^{k+1}\| = 0$.

Finally, the last part is a stochastically "decreasing" result.

Lemma 25.4 ([28]) Let $\mathscr{F}^k$ be the sigma-field generated by the random variables $\{w_j^t, j \in \mathscr{V}\}$ up to time k, i.e.,

$$\mathscr{F}^k = \sigma\{w^0, \ldots, w^k\}. \tag{25.16}$$

Under Assumptions 25.1 and 25.3, it holds almost surely that for all $j \in \mathscr{V}$ and $k \geq \tilde{k}$, which is a sufficiently large number,

$$\begin{aligned}\mathbb{E}[\|\theta_j^{k+1} - \theta^*\|^2 | \mathscr{F}_k] \leq \left(1 + O(\zeta^k)^2\right) \|\lambda_j^k - \theta^*\|^2 \\ -2\zeta^k c'(\mu_j^k - \theta^*) - O(\|\lambda_j^k - \mu_j^k\|^2) + O(\zeta^k)^2,\end{aligned} \tag{25.17}$$

where $\theta^* \in \Theta^*$, λ_j^k and μ_j^k are given in (25.15), and $O(a^k)$ means that there exists a positive constant M such that $O(a_k) \leq M a^k$ for all $k \geq 0$.

The proof also relies crucially on the super-martingale convergence Theorem [23].

Theorem 25.2 (Super-martingale Convergence). *Let $\{v_k\}$, $\{u_k\}$, $\{a_k\}$ and $\{b_k\}$ be sequences of nonnegative random variables such that*

$$\mathbb{E}\left[v_{k+1}|\mathscr{F}_k\right] \leq (1+a_k)v_k - u_k + b_k \tag{25.18}$$

where $\mathscr{F}_k$ denotes the collection $v_0, \ldots, v_k$, $u_0, \ldots, u_k$, $a_0, \ldots, a_k$, $b_0, \ldots, b_k$. Let $\sum_{k=0}^{\infty} a_k < \infty$ and $\sum_{k=0}^{\infty} b_k < \infty$ almost surely. Then, we have $\lim_{k\to\infty} v_k = v$ for a random variable $v \geq 0$ and $\sum_{k=0}^{\infty} u_k < \infty$ almost surely.

Proposition 25.1 ([27]) Under Assumptions 25.1–25.3, and λ_j^k and μ_j^k are given in (25.15). Let $\bar{\lambda}^k = \sum_{j=1}^n \pi_j^{k+1}\mu_j^k$, $\bar{\mu}^k = \sum_{j=1}^n \pi_j^{k+1}\mu_j^k$, and $\bar{\theta}^k = \sum_{j=1}^n \pi_j^k \theta_j^k$. For any $\theta^* \in \Theta^*$ and $j \in \mathscr{V}$, the following statements hold in the almost sure sense:

(a) $\{\sum_{j=1}^n \pi_j^k \|\theta_j^k - \theta^*\|^2\}$ converges as $k \to \infty$.
(b) $\liminf_{k\to\infty} c^T\bar{\mu}^k = c^T\theta^*$.
(c) $\lim_{k\to\infty} \|\mu_j^k - \lambda_j^k\| = 0$.
(d) $\lim_{k\to\infty} \|\mu_j^k - \theta_j^{k+1}\| = \lim_{k\to\infty} \|\lambda_j^k - \theta_j^{k+1}\| = 0$.
(e) $\lim_{k\to\infty} \|\bar{\mu}^k - \bar{\theta}^{k+1}\| = \lim_{k\to\infty} \|\bar{\lambda}^k - \bar{\theta}^{k+1}\| = 0$.

Proof of Theorem 25.1. Notice that $\lambda_j^k = \sum_{i=1}^n a_{ji}^k \theta_j^k$, it follows from Proposition 25.1(d) that $\lim_{k\to\infty} \|\theta_j^{k+1} - \sum_{i=1}^n a_{ji}^k \theta_j^k\| = 0$. Then it holds from Lemma 25.2 almost surely that $\lim_{k\to\infty} \|\theta_j^k - \bar{\theta}^k\| = 0$. Together with Proposition 25.1(a) and the row-stochasticity of A^k, we obtain that $\{\|\bar{\theta}^k - \theta^*\|\}$ converges. We know from Proposition 25.1(e) that $\bar{\mu}^k \to \bar{\theta}^{k+1}$ as $k \to \infty$, then $\{\|\bar{\mu}^k - \theta^*\|\}$ converges as well. By Proposition 25.1(b), it implies that there exists a subsequence $\{\bar{\mu}^k | k \in \mathscr{K}\}$ that converges almost surely to some point in the optimal set Θ^*, which is denoted as θ_0^*, and it holds clearly that

$$\lim_{k\in\mathscr{K}, k\to\infty} \|\bar{\mu}^k - \theta_0^*\| = 0.$$

Since $\{\|\bar{\mu}^k - \theta_0^*\|\}$ converges, it follows that $\lim_{k\to\infty} \|\bar{\mu}^k - \theta_0^*\| = 0$. Finally, we note that $\|\theta_j^{k+1} - \theta_0^*\| \leq \|\theta_j^{k+1} - \bar{\theta}^{k+1}\| + \|\bar{\theta}^{k+1} - \bar{\mu}^k\| + \|\bar{\mu}^k - \theta_0^*\|$, which converges almost surely to zero as $k \to \infty$. Therefore, there exists $\theta_0^* \in \Theta^*$ such that $\lim_{k\to\infty} \theta_j^k = \theta_0^*$ for all $j \in \mathscr{V}$ with probability one. Thus, Theorem 25.1 is proved. ∎

25.5 Conclusion

In this work, we developed a parallel computing framework to collaboratively solve RCO via the SP with a large number of constraints over time-varying graphs. A distributed parallel algorithm with simple structure was provided for time-varying directed graphs. Compared with the existing results, the complexity per iteration in our algorithms is significantly reduced. In future work, we shall study how the network topology affects the convergence speed of the proposed algorithms.

Acknowledgements This work was supported by the National Natural Science Foundation of China (41576101), Tsinghua University Initiative Scientific Research Program, and CNR International Joint Lab COOPS.

References

1. Alamo, T., Tempo, R., Luque, A., Ramirez, D.R.: Randomized methods for design of uncertain systems: sample complexity and sequential algorithms. Automatica **52**, 160–172 (2015)
2. Ash, R., Doléans-Dade, C.: Probability and Measure Theory. Academic Press, Cambridge (2000)
3. Ben-Tal, A., El Ghaoui, L., Nemirovski, A.: Robust Optimization. Princeton University Press, New Jersey (2009)
4. Ben-Tal, A., Nemirovski, A.: Robust convex optimization. Math. Oper. Res. **23**(4), 769–805 (1998)
5. Bertsekas, D.P: Nonlinear Programming. Athena Scientific, Nashville (1999)
6. Bertsimas, D., Brown, D.B., Caramanis, C.: Theory and applications of robust optimization. SIAM Rev. **53**(3), 464–501 (2011)
7. Boyd, S., Parikh, N., Chu, E., Peleato, B., Eckstein, J.: Distributed optimization and statistical learning via the alternating direction method of multipliers. Found. Trends® Mach. Learn. **3**(1), 1–122 (2011)
8. Calafiore, G.C., Campi, M.C.: Uncertain convex programs: randomized solutions and confidence levels. Math. Program. **102**, 25–46 (2004)
9. Calafiore, G.C., Dabbene, F., Tempo, R.: Research on probabilistic methods for control system design. Automatica **47**(7), 1279–1293 (2011)
10. Calafiore, G.C.: Random convex programs. SIAM J. Optim. **20**, 3427–3464 (2010)
11. Campi, M.C., Garatti, S.: The exact feasibility of randomized solutions of uncertain convex programs. SIAM J. Optim. **19**(3), 1211–1230 (2008)
12. Carlone, L., Srivastava, V., Bullo, F., Calafiore, G.C: Distributed random convex programming via constraints consensus. SIAM J. Control Optim. **52**(1), 629–662, 2014
13. Duchi, J.C., Agarwal, A., Wainwright, M.J.: Dual averaging for distributed optimization: convergence analysis and network scaling. IEEE Trans. Autom. Control **57**(3), 592–606 (2012)
14. Gharesifard, B., Cortés, J.: Distributed continuous-time convex optimization on weight-balanced digraphs. IEEE Trans. Autom. Control **59**(3), 781–786 (2014)
15. Gorissen, B.L., Yanıkoğlu, İ., den Hertog, D.: A practical guide to robust optimization. Omega **53**, 124–137 (2015)
16. Lee, S., Nedich, A.: Asynchronous gossip-based random projection algorithms over networks. IEEE Trans. Autom. Control **61**(4), 953–968 (2016)
17. Nedich, A., Olshevsky, A.: Distributed optimization over time-varying directed graphs. IEEE Trans. Autom. Control **60**(3), 601–615 (2015)
18. Nedich, A., Ozdaglar, A.: Distributed subgradient methods for multi-agent optimization. IEEE Trans. Autom. Control **54**(1), 48–61 (2009)
19. Nedich, A.: Convergence rate of distributed averaging dynamics and optimization in networks. Found. Trends® Syst. Control **2**(1), 1–100 (2015)
20. Nedich, A.: Random algorithms for convex minimization problems. Math. Program. **129**(2), 225–253 (2011)
21. Petersen, I.R., Tempo, R.: Robust control of uncertain system: classical results and recent developments. Automatica **50**, 1315–1335 (2014)
22. Polyak, B.T.: Random algorithms for solving convex inequalities. Stud. Comput. Math. **8**, 409–422 (2001)
23. Robbins, H., Siegmund, D.: A convergence theorem for non negative almost supermartingales and some applications. In: Herbert Robbins Selected Papers, pp. 111–135. Springer, Berlin (1985)

24. Scherer, C.W.: Relaxations for robust linear matrix inequality problems with verifications for exactness. SIAM J. Matrix Anal. Appl. **27**(2), 365–395 (2005)
25. Tempo, R., Calafiore, G., Dabbene, F.: Randomized Algorithms for Analysis and Control of Uncertain Systems: With Applications. Springer, Berlin (2012)
26. Xi, C., Khan, U.A.: Directed-distributed gradient descent. In: 53rd Annual Allerton Conference on Communication, Control, and Computing, Monticello, IL, USA. (2015)
27. Xie, P., You, K., Tempo, R., Song, S., Wu, C.: Distributed convex optimization with inequality constraints over time-varying unbalanced digraphs (2017). arXiv preprint arXiv:1612.09029
28. You, K., Tempo, R.: Networked parallel algorithms for robust convex optimization via the scenario approach (2017). arXiv preprint arXiv:1607.05507

Chapter 26
On the Bipartite Consensus of Higher-Order Multi-agent Systems with Antagonistic Interactions and Switching Topologies

Maria Elena Valcher and Pradeep Misra

Abstract In this paper we, investigate the bipartite consensus of higher-order multi-agent systems, by assuming that the interactions among agents are either cooperative or antagonistic and that the communication graph switches among a finite number of possible configurations. We first show that the "lifting approach", proposed in [3] to model opinion dynamics in case of antagonistic interactions and agents modeled as integrators, can be extended to the case of higher order multi-agent systems with cooperative/antagonistic interactions and switching topologies. Subsequently, we are able to translate the bipartite consensus problem into a consensus problem among cooperative agents with switching topologies. This allows us to make use of the results obtained in [13] and hence to solve the bipartite consensus problem.

26.1 Introduction

Multi-agents systems and consensus problems have been the subject of an impressive number of papers in the last 10–15 years. A common assumption in the majority of these papers is that agents aim at achieving consensus through collaboration, namely by exchanging information with a common goal in mind. However, there is a number of situations where two agents regard each other as opponents rather than collaborators, and hence even when one has access to information about the decisions of the other, it does not use it to align its decision to the competitor's one but, on the contrary, to move to the opposite direction. This is what typically happens in the context of markets and social networks [2], similar conditions are also encountered

M. E. Valcher (✉)
Dipartimento di Ingegneria dell'Informazione, Università di Padova,
via Gradenigo 6/B, 35131 Padova, Italy
e-mail: meme@dei.unipd.it

P. Misra
Department of Electrical Engineering, Wright State University,
Russ Engineering Center 424, 3640 Colonel Glenn Hwy., Dayton, OH, USA
e-mail: pradeep.misra@wright.edu

R. Tempo et al. (eds.), *Emerging Applications of Control and Systems Theory*, Lecture Notes in Control and Information Sciences - Proceedings,
https://doi.org/10.1007/978-3-319-67068-3_26

when modeling the behavior of two competing teams in sport disciplines or robot competitions. Each individual or robot, respectively, collects information regarding both the teammates and the antagonists, and processes this data in order to take decisions (position, speed, elevation, etc.) that are in agreement with those of their teammates.

In [1] Altafini first addressed the problem of investigating consensus and bipartite consensus among agents whose mutual interactions are either cooperative or antagonistic. By considering the classical example of homogeneous agents modeled as simple integrators, he introduced the concept of structural balance and showed that if the signed, weighted and connected communication graph describing the agents' interactions is structurally balanced, then the agents reach bipartite consensus. Also, he proposed a result about bipartite consensus of agents modeled as integrators, under the assumption that the communication graph switches among a finite number of possible configurations. More recently, Hendrickx [3] proposed a "lifting approach" to model opinion dynamics in case of antagonistic interactions. This interesting approach reduces a multi-agent system with N agents communicating in a cooperative/antagonistic way to a system with $2N$ agents that cooperate with each other. In this way, in [3] and [16] consensus and bipartite consensus among agents with cooperative/antagonistic interactions and switching communication topologies have been fully explored, by assuming again that agents are modeled as integrators. Some recent results about this problem, by assuming a discrete time version of the agents' model, can be found in [8].

In a number of practical situations, the agents' status is represented by a vector rather than a scalar variable (e.g., position and velocity, price and production levels, etc.), and the agents' dynamics is described by a linear state-space model. Bipartite consensus among agents described by a higher order model, under the assumption that the communication graph is fixed, was first investigated in [14].

The aim of this paper is to show that the lifting approach proposed in [3] can be extended to the case of higher order multi-agent systems with cooperative/antagonistic interactions and switching topologies. This allows one to make use of the results about consensus among cooperative agents and switching topologies derived in [13] to solve the bipartite consensus problem.

Notation. $\mathbb{R}_+$ is the semiring of nonnegative real numbers. For any $k, n \in \mathbb{Z}$, with $k \leq n$, $[k, n]$ is the set of integers $\{k, k+1, \ldots, n\}$. The (i, j)th entry of a matrix A is denoted by $[A]_{ij}$. A matrix (in particular, a vector) A with entries in $\mathbb{R}_+$ is called *nonnegative*, and denoted by $A \geq 0$. The *spectrum* of a square matrix A is denoted by $\sigma(A)$. Given $A \in \mathbb{R}^{n \times n}$, the symbol $|A|$ denotes the nonnegative $n \times n$ matrix whose (i, j)th entry is the absolute value of the (i, j)th entry of A.

26.2 Communications Graphs and Enlarged Communication Graphs

In this section, we review some existing results and present several new propositions that will be the foundation for main results proposed in subsequent sections.

26.2.1 Undirected, Signed and Weighted Graphs

An *undirected, signed and weighted graph* is a triple [9] $\mathcal{G} = (\mathcal{V}, \mathcal{E}, \mathcal{A})$, where $\mathcal{V} = \{v_1, \ldots, v_N\}$ is the set of vertices, $\mathcal{E} \subseteq \mathcal{V} \times \mathcal{V}$ the set of arcs, and $\mathcal{A} \in \mathbb{R}^{N \times N}$ the *adjacency matrix* of the weighted graph $\mathcal{G}$. An arc $(v_j, v_i) \in \mathcal{E}$ if and only if $[\mathcal{A}]_{ij} \neq 0$. As the graph is undirected, (v_i, v_j) belongs to $\mathcal{E}$ if and only if $(v_j, v_i) \in \mathcal{E}$, or, equivalently, $\mathcal{A}$ is a symmetric matrix. We assume that the graph $\mathcal{G}$ has no self-loops, i.e., $[\mathcal{A}]_{ii} = 0$ for every $i \in [1, N]$, and arcs in $\mathcal{E}$ have either positive or negative weights, namely the off-diagonal entries of $\mathcal{A}$ are either positive or negative.

A sequence $v_{j_1} \leftrightarrow v_{j_2} \leftrightarrow v_{j_3} \leftrightarrow \cdots \leftrightarrow v_{j_k} \leftrightarrow v_{j_{k+1}}$ is a *path* of length k connecting v_{j_1} and $v_{j_{k+1}}$ provided that $(v_{j_1}, v_{j_2}), (v_{j_2}, v_{j_3}), \ldots, (v_{j_k}, v_{j_{k+1}}) \in \mathcal{E}$. A graph is said to be *connected* if for every pair of distinct indices $i, j \in [1, N]$ there is a path connecting v_j and v_i. This is equivalent to the fact that $\mathcal{A}$ is an *irreducible* matrix, namely no permutation matrix Π can be found such that

$$\Pi^\top \mathcal{A} \Pi = \begin{bmatrix} \mathcal{A}_{11} & \mathcal{A}_{12} \\ 0 & \mathcal{A}_{22} \end{bmatrix},$$

where $\mathcal{A}_{11}$ and $\mathcal{A}_{22}$ are square (non-vacuous) matrices.

The undirected, signed and weighted graph $\mathcal{G} = (\mathcal{V}, \mathcal{E}, \mathcal{A})$ is *structurally balanced* [1, 4] if the set of vertices $\mathcal{V}$ can be partitioned into two disjoint subsets $\mathcal{V}_1$ and $\mathcal{V}_2$ such that for every $v_i, v_j \in \mathcal{V}_p$, $p \in [1, 2]$, $[\mathcal{A}]_{ij} \geq 0$, while for every $v_i \in \mathcal{V}_p, v_j \in \mathcal{V}_q, p, q \in [1, 2], p \neq q, [\mathcal{A}]_{ij} \leq 0$. Note that if $\mathcal{A}$ is a positive matrix, then $\mathcal{G}$ is trivially structurally balanced since the previous definition holds for $\mathcal{V}_1 = \mathcal{V}$ and $\mathcal{V}_2 = \emptyset$. A graph which is not structurally balanced is called *structurally unbalanced.*

We define the *(signed) Laplacian matrix* $\mathcal{L} \in \mathbb{R}^{N \times N}$ associated with the graph $\mathcal{G}$ as [1, 4] $\mathcal{L} := \mathcal{C} - \mathcal{A}$, where $\mathcal{C} \in \mathbb{R}_+^{N \times N}$ is a diagonal matrix with

$$[\mathcal{C}]_{ii} := \sum_{j=1}^{N} |[\mathcal{A}]_{ij}|, \qquad \forall\, i \in [1, N]. \tag{26.1}$$

Accordingly, the Laplacian matrix $\mathscr{L} = \mathscr{L}^\top$ takes the following form:

$$\mathscr{L} = \begin{bmatrix} \sum_{j=1}^N |[\mathscr{A}]_{1j}| & -[\mathscr{A}]_{12} & \dots & -[\mathscr{A}]_{1N} \\ -[\mathscr{A}]_{12} & \sum_{j=1}^N |[\mathscr{A}]_{2j}| & \dots & -[\mathscr{A}]_{2N} \\ \vdots & \vdots & \ddots & \vdots \\ -[\mathscr{A}]_{1N} & -[\mathscr{A}]_{2N} & \dots & \sum_{j=1}^N |[\mathscr{A}]_{Nj}| \end{bmatrix}. \tag{26.2}$$

Note that $\mathscr{L}$ is irreducible if and only if $\mathscr{A}$ is irreducible.

The Laplacian matrix $\mathscr{L}$ is a symmetric and positive semidefinite matrix, whose nonnegative real eigenvalues can always be sorted so that $0 \le \lambda_1(\mathscr{L}) \le \lambda_2(\mathscr{L}) \le \dots \le \lambda_N(\mathscr{L})$. Moreover, when $\mathscr{G}$ is connected, $\lambda_1(\mathscr{L}) = 0$ if and only if $\mathscr{G}$ is structurally balanced, and if this is the case then 0 is a simple eigenvalue of $\mathscr{L}$ (namely $\lambda_2(\mathscr{L}) > 0$). In particular, if $\mathscr{G}$ is connected and the adjacency matrix is positive (and irreducible), then 0 is necessarily a simple eigenvalue of $\mathscr{L}$.

26.2.2 Enlarged Graph of the Signed Graph $\mathscr{G}$

Following [3] and [16], we introduce the concept of enlarged graph as follows.

Definition 26.1 Given an undirected, signed and weighted graph $\mathscr{G} = (\mathscr{V}, \mathscr{E}, \mathscr{A})$, with $\mathscr{V} = \{v_1, \dots, v_N\}$, $\mathscr{E} \subseteq \mathscr{V} \times \mathscr{V}$ and $\mathscr{A} = \mathscr{A}^\top \in \mathbb{R}^{N \times N}$ (and $[\mathscr{A}]_{ij} < 0$ for at least one index pair (i, j)), we define the *enlarged graph associated with* $\mathscr{G}$ as $\bar{\mathscr{G}} = (\bar{\mathscr{V}}, \bar{\mathscr{E}}, \bar{\mathscr{A}})$, where $\bar{\mathscr{V}} = \{v_1^+, v_2^+, \dots, v_N^+, v_1^-, v_2^-, \dots, v_N^-\}$, and $\bar{\mathscr{E}} \subseteq \bar{\mathscr{V}} \times \bar{\mathscr{V}}$ is defined as follows:

- If there is an arc $(v_j, v_i) \in \mathscr{E}$ with positive weight (i.e., $[\mathscr{A}]_{ij} > 0$), then the arcs (v_j^+, v_i^+) and (v_j^-, v_i^-) both belong to $\bar{\mathscr{E}}$.
- If there is an arc $(v_j, v_i) \in \mathscr{E}$ with negative weight (i.e., $[\mathscr{A}]_{ij} < 0$), then the arcs (v_j^+, v_i^-) and (v_j^-, v_i^+) both belong to $\bar{\mathscr{E}}$.

Finally, if we define the (symmetric) matrices $\mathscr{A}^+$ and $\mathscr{A}^- \in \mathbb{R}_+^{N \times N}$ as follows

$$[\mathscr{A}^+]_{ij} := \max\{[\mathscr{A}]_{ij}, 0\}, \qquad [\mathscr{A}^-]_{ij} := \max\{-[\mathscr{A}]_{ij}, 0\}, \qquad \forall\, i, j \in [1, N],$$

so that $\mathscr{A} = \mathscr{A}^+ - \mathscr{A}^-$, then

$$\bar{\mathscr{A}} := \begin{bmatrix} \mathscr{A}^+ & \mathscr{A}^- \\ \mathscr{A}^- & \mathscr{A}^+ \end{bmatrix} = \bar{\mathscr{A}}^\top \in \mathbb{R}_+^{2N \times 2N}.$$

It is easy to see that the enlarged graph has all positive weights and hence it is an undirected, unsigned and weighted graph. Further, the Laplacians of $\mathcal{G}$ and $\bar{\mathcal{G}}$ are related by the following identity

$$\bar{\mathcal{L}} = \begin{bmatrix} \mathcal{C} & 0 \\ 0 & \mathcal{C} \end{bmatrix} - \begin{bmatrix} \mathcal{A}^+ & \mathcal{A}^- \\ \mathcal{A}^- & \mathcal{A}^+ \end{bmatrix},$$

where $\mathcal{C}$ is defined as in (26.1).

Lemma 26.1 *Let $\mathcal{G}$ be a connected, undirected, signed and weighted graph and let $\bar{\mathcal{G}}$ be the corresponding enlarged graph. The spectra of Laplacians $\mathcal{L}$ and $\bar{\mathcal{L}}$ of $\mathcal{G}$ and $\bar{\mathcal{G}}$, respectively, are mutually related by the following relationship:*

$$\sigma(\bar{\mathcal{L}}) = \sigma(\mathcal{L}) \cup \sigma(\mathcal{L}_{unsigned}), \tag{26.3}$$

where $\mathcal{L}_{unsigned} := \mathcal{C} - |\mathcal{A}|$. Consequently,

- *$\mathcal{G}$ is structurally balanced if and only if 0 is an eigenvalue of $\bar{\mathcal{L}}$ of algebraic multiplicity 2, and*
- *$\mathcal{G}$ is structurally unbalanced if and only if 0 is a simple eigenvalue of $\bar{\mathcal{L}}$.*

Proof Identity (26.3) follows immediately from the fact that $|\mathcal{A}| = \mathcal{A}^+ + \mathcal{A}^-$ and

$$\begin{bmatrix} I_N & -I_N \\ 0 & I_N \end{bmatrix} \begin{bmatrix} \mathcal{C} - \mathcal{A}^+ & -\mathcal{A}^- \\ -\mathcal{A}^- & \mathcal{C} - \mathcal{A}^+ \end{bmatrix} \begin{bmatrix} I_N & I_N \\ 0 & I_N \end{bmatrix} = \begin{bmatrix} \mathcal{C} - (\mathcal{A}^+ - \mathcal{A}^-) & 0 \\ -\mathcal{A}^- & \mathcal{C} - (\mathcal{A}^+ + \mathcal{A}^-) \end{bmatrix}$$
$$= \begin{bmatrix} \mathcal{C} - \mathcal{A} & 0 \\ -\mathcal{A}^- & \mathcal{C} - |\mathcal{A}| \end{bmatrix} = \begin{bmatrix} \mathcal{L} & 0 \\ -\mathcal{A}^- & \mathcal{L}_{unsigned} \end{bmatrix}.$$

Since $\mathcal{A}$ is irreducible, so is $|\mathcal{A}| \in \mathbb{R}_+^{N \times N}$ and this ensures that 0 is a simple eigenvalue of the Laplacian $\mathcal{L}_{unsigned}$. On the other hand, the connected signed graph $\mathcal{G}$ is structurally balanced if and only if 0 is an eigenvalue of $\mathcal{L}$, and if so it is a simple eigenvalue. Therefore two cases may occur: if $\mathcal{G}$ is structurally balanced then 0 is a simple eigenvalue of $\mathcal{L}$ and hence 0 is an eigenvalue of $\bar{\mathcal{L}}$ with algebraic multiplicity 2; if $\mathcal{G}$ is structurally unbalanced then 0 is not an eigenvalue of $\mathcal{L}$ and hence 0 is a simple eigenvalue of $\bar{\mathcal{L}}$.

The following fundamental relationships between the graphs $\mathcal{G}$ and $\bar{\mathcal{G}}$ hold.

Proposition 26.1 [16] Let $\mathcal{G}$ be an undirected, connected, signed and weighted graph and let $\bar{\mathcal{G}}$ be the corresponding enlarged graph. Then

(i) $\mathcal{G}$ is structurally balanced if and only if $\bar{\mathcal{G}}$ consists of two disjoint and connected components;
(ii) $\mathcal{G}$ is structurally unbalanced if and only if $\bar{\mathcal{G}}$ is connected.

Proposition 26.1, above, has been derived in [16] using purely graph-theoretic arguments. We show next that the same result can be derived using algebraic reasonings based on the adjacency matrix of the enlarged graph. The advantage of this solution is that one gains insight into the structure of the adjacency matrix and hence of the Laplacian associated with the enlarged graph that will be useful later.

Proposition 26.2 *Let $\mathscr{G} = (\mathscr{V}, \mathscr{E}, \mathscr{A})$ be a connected, undirected, signed and weighted graph and let $\bar{\mathscr{G}} = (\bar{\mathscr{V}}, \bar{\mathscr{E}}, \bar{\mathscr{A}})$ be the corresponding enlarged graph. Then*

(i) $\mathscr{G}$ is structurally balanced if and only if there exists a permutation matrix $\Pi \in \mathbb{R}_+^{2N\times 2N}$ such that

$$\Pi^\top \bar{\mathscr{A}} \Pi = \begin{bmatrix} |\mathscr{A}| & 0 \\ 0 & |\mathscr{A}| \end{bmatrix};$$

(ii) $\mathscr{G}$ is structurally unbalanced if and only if $\bar{\mathscr{A}}$ is irreducible.

Proof (i) Suppose first that $\mathscr{G}$ is structurally balanced. We can always reorder the N agents so that $\mathscr{V}_1 = \{v_i : i \in [1, r]\}$ and $\mathscr{V}_2 = \{v_i : i \in [r+1, N]\}$, for some $r \in [1, N-1]$, and the symmetric adjacency matrix $\mathscr{A}$ can be expressed [14] as

$$\mathscr{A} = \begin{bmatrix} \mathscr{A}_{11} & \mathscr{A}_{12} \\ \mathscr{A}_{12}^\top & \mathscr{A}_{22} \end{bmatrix}, \tag{26.4}$$

where $\mathscr{A}_{11} = \mathscr{A}_{11}^\top \in \mathbb{R}_+^{r\times r}$, $\mathscr{A}_{22} = \mathscr{A}_{22}^\top \in \mathbb{R}_+^{(N-r)\times(N-r)}$, while $-\mathscr{A}_{12} \in \mathbb{R}_+^{r\times(N-r)}$. Accordingly,

$$\mathscr{A}^+ = \begin{bmatrix} \mathscr{A}_{11} & 0 \\ 0 & \mathscr{A}_{22} \end{bmatrix}, \quad \mathscr{A}^- = \begin{bmatrix} 0 & -\mathscr{A}_{12} \\ -\mathscr{A}_{12}^\top & 0 \end{bmatrix}, \quad \bar{\mathscr{A}} = \begin{bmatrix} \mathscr{A}_{11} & 0 & 0 & -\mathscr{A}_{12} \\ 0 & \mathscr{A}_{22} & -\mathscr{A}_{12}^\top & 0 \\ 0 & -\mathscr{A}_{12} & \mathscr{A}_{11} & 0 \\ -\mathscr{A}_{12}^\top & 0 & 0 & \mathscr{A}_{22} \end{bmatrix}.$$

This clearly shows that if we permute the order of the $2N$ agents in graph $\bar{\mathscr{G}}$, so that the ordered vertex set is $\bar{\mathscr{V}}_{ord} = \{v_1^+, \dots, v_r^+, v_{r+1}^-, \dots, v_N^-, v_1^-, \dots, v_r^-, v_{r+1}^+, \dots, v_N^+\}$, namely we use the permutation matrix

$$\Pi = \begin{bmatrix} I_r & 0 & 0 & 0 \\ 0 & 0 & 0 & I_{N-r} \\ 0 & 0 & I_r & 0 \\ 0 & I_{N-r} & 0 & 0 \end{bmatrix} = \Pi^\top \in \mathbb{R}_+^{2N\times 2N},$$

then the adjacency matrix becomes

$$\bar{\mathscr{A}}_{ord} = \Pi^\top \bar{\mathscr{A}} \Pi = \begin{bmatrix} \mathscr{A}_{11} & -\mathscr{A}_{12} & 0 & 0 \\ -\mathscr{A}_{12}^\top & \mathscr{A}_{22} & 0 & 0 \\ 0 & 0 & \mathscr{A}_{11} & -\mathscr{A}_{12} \\ 0 & 0 & -\mathscr{A}_{12}^\top & \mathscr{A}_{22} \end{bmatrix} = \begin{bmatrix} |\mathscr{A}| & 0 \\ 0 & |\mathscr{A}| \end{bmatrix}.$$

Conversely, if $\bar{\mathscr{A}}$ takes the previous block diagonal structure after a suitable permutation Π, then

$$\Pi^\top \bar{\mathscr{L}} \Pi = \begin{bmatrix} \mathscr{C} - |\mathscr{A}| & 0 \\ 0 & \mathscr{C} - |\mathscr{A}| \end{bmatrix} = \begin{bmatrix} \mathscr{L}_{unsigned} & 0 \\ 0 & \mathscr{L}_{unsigned} \end{bmatrix}, \quad (26.5)$$

and the irreducibility of $|\mathscr{A}|$ ensures that 0 is an eigenvalue of $\bar{\mathscr{L}}$ of multiplicity 2 and consequently, by Lemma 26.1, $\mathscr{G}$ is structurally balanced.

(ii) By Lemma 26.1, $\mathscr{G}$ is structurally unbalanced if and only if 0 is a simple eigenvalue of $\bar{\mathscr{L}}$. But as proved in Corollary 2 of [11] for directed, unsigned graphs, 0 is a simple eigenvalue of $\bar{\mathscr{L}}$ if and only if $\bar{\mathscr{G}}$ has a spanning tree, that for an undirected graph is equivalent to saying that it is connected, i.e., $\bar{\mathscr{A}}$ is irreducible.

26.2.3 Dynamic Graphs

Define $\mathscr{P} := [1, p]$ and consider a finite family of undirected, signed and weighted graphs sharing the same vertex set $\mathscr{V}$, i.e., $\mathscr{G}_k := (\mathscr{V}, \mathscr{E}_k, \mathscr{A}_k)$, $k \in \mathscr{P}$. We can define the Laplacians $\mathscr{L}_k = \mathscr{C}_k - \mathscr{A}_k = \mathscr{C}_k - (\mathscr{A}_k^+ - \mathscr{A}_k^-)$ and the enlarged graphs $\bar{\mathscr{G}}_k = (\bar{\mathscr{V}}, \bar{\mathscr{E}}_k, \bar{\mathscr{A}}_k)$, for each $k \in \mathscr{P}$, and all the previous definitions and results apply.

If $\sigma : \mathbb{R}_+ \to \mathscr{P}$ is a piecewise constant, right-continuous switching signal, we define [13] the *dynamic graph* $\mathscr{G}_{\sigma(t)} := (\mathscr{V}, \mathscr{E}_{\sigma(t)}, \mathscr{A}_{\sigma(t)})$ as follows: for every $t \in \mathbb{R}_+$ it coincides with $\mathscr{G}_k = (\mathscr{V}, \mathscr{E}_k, \mathscr{A}_k)$, where k is the value taken by the switching signal σ at time t.

In the following we will assume that the graphs $\mathscr{G}_k$, $k \in \mathscr{P}$, and hence the dynamic graph $\mathscr{G}_{\sigma(t)}$ are *sign consistent*, i.e., [12] for every pair (i, j) the family of weights $\{[\mathscr{A}_k]_{ij} : k \in \mathscr{P}\}$ (and therefore the family of weights $\{[\mathscr{A}_{\sigma(t)}]_{ij} : t \in \mathbb{R}_+\}$) consists either of nonnegative or nonpositive elements, but it cannot include elements of opposite signs.

Let σ be a switching signal with switching times $0 = t_0 < t_1 < t_2 < \dots$, and consider a subsequence of the switching times $t_{i_0} < t_{i_1} < t_{i_2} < \dots$, i.e., $\{t_{i_k}\}_{k \in \mathbb{Z}_+} \subseteq \{t_i\}_{i \in \mathbb{Z}_+}$. We define the *union graph* [13, 16] $\mathscr{G}([t_{i_k}, t_{i_{k+1}}))$ as the undirected, unsigned and unweighted graph having $\mathscr{V}$ as vertex set and $\cup_{q=i_k}^{i_{k+1}-1} \mathscr{E}_{\sigma(t_q)}$ as set of arcs. The dynamic graph $\mathscr{G}_{\sigma(t)} = (\mathscr{V}, \mathscr{E}_{\sigma(t)}, \mathscr{A}_{\sigma(t)})$, with $\sigma : \mathbb{R}_+ \to \mathscr{P}$ a switching signal with switching times $0 = t_0 < t_1 < t_2 < \dots$, is said to be *uniformly connected over* $[0, +\infty)$ [7, 13] if there exist $T > 0$ and a subsequence of the switching times $t_{i_0} < t_{i_1} < t_{i_2} < \dots$, with $t_{i_{k+1}} - t_{i_k} \leq T$, such that each union graph $\mathscr{G}([t_{i_k}, t_{i_{k+1}}))$ is connected. A dynamic graph satisfying this property is also known in the literature [5, 10] as *jointly connected over the intervals* $[t_{i_k}, t_{i_{k+1}})$ (see also [13]). Clearly, a necessary condition for this property to hold true is that the graph $\mathscr{G} = (\mathscr{V}, \cup_{k \in \mathscr{P}} \mathscr{E}_k)$ is connected.

In the following, given a family of undirected, signed and weighted graphs $\mathscr{G}_k = (\mathscr{V}, \mathscr{E}_k, \mathscr{A}_k)$, $k \in \mathscr{P}$, that are signed consistent and such that the graph

$\mathcal{G} = (\mathcal{V}, \cup_{k\in\mathcal{P}}\mathcal{E}_k)$ is connected, we will denote by $\mathcal{S}_{dwell,uc}$ the set of switching signals $\sigma : \mathbb{R}_+ \to \mathcal{P}$ that have the following properties:

- σ has some dwell time $\tau = \tau(\sigma) > 0$ [6, 13, 15], meaning that if $0 = t_0 < t_1 < t_2 < \ldots$ is the sequence of switching instants, then $t_{i+1} - t_i \geq \tau$ for every $i \in [0, +\infty)$;
- the associated dynamic graph $\mathcal{G}_{\sigma(t)} = (\mathcal{V}, \mathcal{E}_{\sigma(t)}, \mathcal{A}_{\sigma(t)})$ is uniformly connected.

26.3 Bipartite Consensus: Problem Statement

We consider a multi-agent system consisting of N agents, each of them described by the same continuous-time state-space model. Specifically, $\mathbf{x}_i(t)$, the ith agent's state, $i \in [1, N]$, evolves according to the first-order differential equation

$$\dot{\mathbf{x}}_i(t) = A\mathbf{x}_i(t) + B\mathbf{u}_i(t), \tag{26.6}$$

where $\mathbf{x}_i(t) \in \mathbb{R}^n, \mathbf{u}_i(t) \in \mathbb{R}^m$, $A \in \mathbb{R}^{n\times n}$, and $B \in \mathbb{R}^{n\times m}$. The communication among the N agents is described by an *undirected, signed and weighted dynamic communication graph* [13] $\mathcal{G}_{\sigma(t)} = (\mathcal{V}, \mathcal{E}_{\sigma(t)}, \mathcal{A}_{\sigma(t)})$, where $\mathcal{V} = \{v_1, v_2, \ldots, v_N\}$ is the set of vertices, $\mathcal{E}_{\sigma(t)} \subseteq \mathcal{V} \times \mathcal{V}$ and $\mathcal{A}_{\sigma(t)} = \mathcal{A}_{\sigma(t)}^\top$ are the set of arcs and the adjacency matrix at time $t \in \mathbb{R}_+$, respectively, and σ is a right-continuous switching signal taking values in the finite set $\mathcal{P} = [1, p]$. The (i, j)th entry of $\mathcal{A}_{\sigma(t)}$, $[\mathcal{A}_{\sigma(t)}]_{ij}$, with $i \neq j$, is nonzero if and only if the information about the status of the jth agent is available to the ith agent at time t (and conversely). At time t, the interaction between the ith and the jth agents is cooperative if $[\mathcal{A}_{\sigma(t)}]_{ij} > 0$ and antagonistic if $[\mathcal{A}_{\sigma(t)}]_{ij} < 0$. In the following, to simplify the notation and when no confusion may arise, we will denote the (i, j)th entry of $\mathcal{A}_{\sigma(t)}$ either by $[\mathcal{A}_\sigma]_{ij}$ or by $[\mathcal{A}_k]_{ij}$ if the value k of σ at time t is specified.

We assume that the family of (undirected, signed and weighted) graphs $\mathcal{G}_k = (\mathcal{V}, \mathcal{E}_k, \mathcal{A}_k)$, $k \in \mathcal{P} = [1, p]$, is sign consistent. The sign consistency property ensures that whenever two agents i and j interact, they interact either always in a cooperative way or always in an antagonistic way, but they do not change over time the reciprocal attitude. All switching signals $\sigma : \mathbb{R}_+ \to \mathcal{P}$ considered in the sequel belong to $\mathcal{S}_{dwell,uc}$. Finally, the pair (A, B) is controllable and the matrix A is *simply stable* (equivalently, *marginally stable*), namely every eigenvalue of A has nonnegative real part, and the eigenvalues with zero real part are associated with Jordan blocks of size 1. We adopt a DeGroot type state feedback law for each agent [1, 13, 16] as follows:

$$\mathbf{u}_i(t) = -K \sum_{j:(j,i)\in\mathcal{E}_\sigma} |[\mathcal{A}_\sigma]_{ij}| \cdot [\mathbf{x}_i(t) - \text{sign}([\mathcal{A}_\sigma]_{ij})\mathbf{x}_j(t)], \qquad i \in [1, N], \tag{26.7}$$

where $K \in \mathbb{R}^{m\times n}$ is a feedback matrix to be designed, and sign$(\cdot)$ is the sign function. If we introduce the state and input vectors

$$\mathbf{x}(t) := \left[\mathbf{x}_1^\top(t)\ \mathbf{x}_2^\top(t)\ \ldots\ \mathbf{x}_N^\top(t)\right]^\top, \qquad \mathbf{u}(t) := \left[\mathbf{u}_1(t)\ \mathbf{u}_2(t)\ \ldots\ \mathbf{u}_N(t)\right]^\top,$$

the overall dynamics of the multi-agent system is described by the equations

$$\dot{\mathbf{x}}(t) = (I_N \otimes A)\mathbf{x}(t) + (I_N \otimes B)\mathbf{u}(t), \tag{26.8}$$
$$\mathbf{u}(t) = -(\mathscr{L}_\sigma \otimes K)\mathbf{x}(t), \tag{26.9}$$

or in compact form, by

$$\dot{\mathbf{x}}(t) = [(I_N \otimes A)\mathbf{x}(t) - (I_N \otimes B)(\mathscr{L}_\sigma \otimes K)]\,\mathbf{x}(t) = [(I_N \otimes A)\mathbf{x}(t) - \mathscr{L}_\sigma \otimes (BK)]\,\mathbf{x}(t), \tag{26.10}$$

where $\mathscr{L}_\sigma$ is the Laplacian of $\mathscr{A}_\sigma$. The aim of this paper is to investigate the following problem: given some switching signal $\sigma \in \mathscr{S}_{dwell,uc}$, under what conditions on the dynamic communication graph $\mathscr{G}_\sigma$ can a constant state feedback matrix K be found such that *bipartite consensus* (or, equivalently, *polarization* see [16]) is achieved for (almost[1]) every choice of the initial conditions, equivalently the agents' states satisfy the following conditions:

$$\lim_{t\to+\infty} |\mathbf{x}_i(t)| = \lim_{t\to+\infty} |\mathbf{x}_j(t)|, \qquad \forall\, i, j \in [1, N],$$
$$\lim_{t\to+\infty} \mathbf{x}_i(t) = -\lim_{t\to+\infty} \mathbf{x}_j(t), \qquad \exists\, i, j \in [1, N], i \neq j.$$

26.4 Lifting Approach

In order to provide a solution to the proposed problem, we use the lifting approach, first introduced in [3] and later adopted in [16], to address the simpler case when the agents' model is a scalar integrator $\dot{x}_i(t) = u_i(t), i \in [1, N]$, (this corresponds to assuming $A = 0$ and $B = 1$ in (26.6)) and hence the overall multi-agent system is described by the differential equation $\dot{\mathbf{x}}(t) = -\mathscr{L}(t)\mathbf{x}(t)$, where $\mathscr{L}(t)$ is the Laplacian of the communication graph at time t. We will show that the lifting approach can be extended to the case when the agents' dynamics are described by a general state-space model (26.6). This allows us to apply to the case of cooperative and antagonistic interactions some results available for consensus of cooperative higher order agents with switching communication topologies.

[1]This means that there may be a zero-measure set of initial conditions for which all agents' states converge to zero.

We observe that the ith agent's state, under the effect of the feedback control algorithm (26.7), evolves according to the following differential equation:

$$\dot{\mathbf{x}}_i(t) = A\mathbf{x}_i(t) - BK \sum_{j:(j,i)\in\mathscr{E}_\sigma} |[\mathscr{A}_\sigma]_{ij}| \cdot [\mathbf{x}_i(t) - \text{sign}([\mathscr{A}_\sigma]_{ij})\mathbf{x}_j(t)]. \quad (26.11)$$

If we refer to the (uniquely determined) positive matrices $\mathscr{A}_\sigma^+$ and $\mathscr{A}_\sigma^-$, with disjoint nonzero patterns (meaning that $\{(i,j) : [\mathscr{A}_\sigma^+]_{ij} > 0\} \cap \{(i,j) : [\mathscr{A}_\sigma^-]_{ij} > 0\} = \emptyset$), satisfying $\mathscr{A}_\sigma = \mathscr{A}_\sigma^+ - \mathscr{A}_\sigma^-$, the previous differential equation becomes

$$\begin{aligned}\dot{\mathbf{x}}_i(t) = A\mathbf{x}_i(t) - BK &\sum_{j:[\mathscr{A}_\sigma]_{ij}>0} [\mathscr{A}_\sigma^+]_{ij} \cdot [\mathbf{x}_i(t) - \mathbf{x}_j(t)] \quad (26.12)\\ - BK &\sum_{j:[\mathscr{A}_\sigma]_{ij}<0} [\mathscr{A}_\sigma^-]_{ij} \cdot [\mathbf{x}_i(t) - (-\mathbf{x}_j(t))].\end{aligned}$$

On the other hand, from (26.11) (equivalently, from (26.12)), we can easily deduce that the states $-\mathbf{x}_i(t)$, $i \in [1, N]$, evolve according to the following equation:

$$\begin{aligned}-\dot{\mathbf{x}}_i(t) = A(-\mathbf{x}_i(t)) - BK &\sum_{j:[\mathscr{A}_\sigma]_{ij}>0} [\mathscr{A}_\sigma^+]_{ij} \cdot [-\mathbf{x}_i(t) - (-\mathbf{x}_j(t))] \quad (26.13)\\ - BK &\sum_{j:[\mathscr{A}_\sigma]_{ij}<0} [\mathscr{A}_\sigma^-]_{ij} \cdot [-\mathbf{x}_i(t) - \mathbf{x}_j(t)].\end{aligned}$$

So, if we introduce the $2N$ variables, $\mathbf{z}_i(t)$, $i \in [1, 2N]$, where

$$\mathbf{z}_i(t) = \mathbf{x}_i(t), \qquad \mathbf{z}_{N+i}(t) = -\mathbf{x}_i(t), \qquad i \in [1, N],$$

it is easy to see that (26.12) and (26.13) can be rewritten as

$$\begin{aligned}\dot{\mathbf{z}}_i(t) = A\mathbf{z}_i(t) - BK &\sum_{j:[\mathscr{A}_\sigma]_{ij}>0} [\mathscr{A}_\sigma^+]_{ij} \cdot [\mathbf{z}_i(t) - \mathbf{z}_j(t)]\\ - BK &\sum_{j:[\mathscr{A}_\sigma]_{ij}<0} [\mathscr{A}_\sigma^-]_{ij} \cdot [\mathbf{z}_i(t) - \mathbf{z}_{N+j}(t))],\\ \dot{\mathbf{z}}_{N+i}(t) = A\mathbf{z}_{N+i}(t)) - BK &\sum_{j:[\mathscr{A}_\sigma]_{ij}>0} [\mathscr{A}_\sigma^+]_{ij} \cdot [\mathbf{z}_{N+i}(t) - \mathbf{z}_{N+j}(t))]\\ - BK &\sum_{j:[\mathscr{A}_\sigma]_{ij}<0} [\mathscr{A}_\sigma^-]_{ij} \cdot [\mathbf{z}_{N+i}(t) - \mathbf{z}_j(t)].\end{aligned}$$

It is a matter of simple calculations to verify that

$$\begin{bmatrix} \dot{\mathbf{z}}_1(t) \\ \vdots \\ \dot{\mathbf{z}}_N(t) \\ \hline \dot{\mathbf{z}}_{N+1}(t) \\ \vdots \\ \dot{\mathbf{z}}_{2N}(t) \end{bmatrix} = \left[\begin{array}{c|c} (I_N \otimes A) - (\mathscr{C}_\sigma - \mathscr{A}_\sigma^+) \otimes (BK) & \mathscr{A}_\sigma^- \otimes (BK) \\ \hline \mathscr{A}_\sigma^- \otimes (BK) & (I_N \otimes A) - (\mathscr{C}_\sigma - \mathscr{A}_\sigma^+) \otimes (BK) \end{array}\right] \begin{bmatrix} \mathbf{z}_1(t) \\ \vdots \\ \mathbf{z}_N(t) \\ \hline \mathbf{z}_{N+1}(t) \\ \vdots \\ \mathbf{z}_{2N}(t) \end{bmatrix}$$

that can be expressed in compact form as

$$\dot{\mathbf{z}}(t) = \left[(I_{2N} \otimes A) - \bar{\mathscr{L}}_\sigma \otimes (BK)\right] \mathbf{z}(t), \qquad (26.14)$$

with $\bar{\mathscr{L}}_\sigma$ the Laplacian associated with the enlarged graph $\bar{\mathscr{G}}_\sigma = (\bar{\mathscr{V}}, \bar{\mathscr{E}}_\sigma, \bar{\mathscr{A}}_\sigma)$ as follows:

$$\bar{\mathscr{L}}_\sigma = \begin{bmatrix} \mathscr{C}_\sigma & 0 \\ 0 & \mathscr{C}_\sigma \end{bmatrix} - \begin{bmatrix} \mathscr{A}_\sigma^+ & \mathscr{A}_\sigma^- \\ \mathscr{A}_\sigma^- & \mathscr{A}_\sigma^+ \end{bmatrix},$$

and $\mathscr{C}_\sigma = \text{diag}\{\sum_{j=1}^N |[\mathscr{A}_\sigma]_{1j}|, \dots, \sum_{j=1}^N |[\mathscr{A}_\sigma]_{Nj}|\}$. The following result extends Proposition 1 in [3] to higher order multi-agent systems.

Proposition 26.3 *For any switching signal $\sigma : \mathbb{R}_+ \to \mathscr{P}$, the state trajectory $\mathbf{x}(t)$, $t \in \mathbb{R}_+$ is a solution (in the sense of Caratheodory) of (26.10) corresponding to the initial condition $\mathbf{x}(0) \in \mathbb{R}^{Nn}$ if and only if $\mathbf{z}(t) = [\mathbf{x}(t)^\top \ -\mathbf{x}(t)^\top]^\top$ is a solution of (26.14) corresponding to the initial condition $\mathbf{z}(0) = [\mathbf{x}(0)^\top \ -\mathbf{x}(0)^\top]^\top$.*

By making use of the previous enlarged description of the agents' dynamics, in the next section, we will provide conditions for the problem solvability based on the results obtained in [13] for cooperative agents with switching topologies.

26.5 Bipartite Consensus: Problem Solution

In this section, we investigate the bipartite consensus problem for noncooperating multi-agent systems described in (26.8), under the assumptions that

A1. the pair (A, B) is stabilizable, and the matrix A is simply stable;
A2. the state feedback control algorithm is described in (26.7);
A3. the dynamic communication graph $\mathscr{G}_\sigma$ is undirected, signed, weighted and sign consistent;
A4. the switching signal σ is arbitrary in $\mathscr{S}_{dwell,uc}$.

Theorem 26.1 *Consider the multi-agent system described in (26.10), under the assumptions A1-A4. If there exists a partition of the vertex set $\mathscr{V} = \{v_1, v_2, \dots, v_N\}$ into two disjoint and non-empty subsets $\mathscr{V}_1$ and $\mathscr{V}_2$ such that at each time $t \in \mathbb{R}_+$ the* *graph $\mathcal{G}_\sigma$ is structurally balanced with respect to this partition, then there exists a constant feedback matrix K that solves the bipartite consensus problem.*

Proof Suppose that the partition of $\mathcal{V}$ exists. We may assume without loss of generality (see the proof of Proposition 26.2), $\mathcal{V}_1 = \{v_i : i \in [1, r]\}$ and $\mathcal{V}_2 = \{v_i : i \in [r+1, N]\}$. Using the same permutation matrix we used to prove Proposition 26.2 i.e.

$$\Pi = \begin{bmatrix} I_r & 0 & 0 & 0 \\ 0 & 0 & 0 & I_{N-r} \\ 0 & 0 & I_r & 0 \\ 0 & I_{N-r} & 0 & 0 \end{bmatrix} = \Pi^\top \in \mathbb{R}_+^{2N \times 2N},$$

it follows that

$$\Pi^\top \mathbf{z}(t) = \big[\mathbf{x}_1(t)^\top \dots \mathbf{x}_r(t)^\top \; -\mathbf{x}_{r+1}(t)^\top \dots -\mathbf{x}_N(t)^\top \mid \\ -\mathbf{x}_1(t)^\top \dots -\mathbf{x}_r(t)^\top \; \mathbf{x}_{r+1}(t)^\top \dots \mathbf{x}_N(t)^\top\big]^\top,$$

and if we apply the permutation matrix Π to the Laplacian of the enlarged adjacency matrix $\bar{\mathcal{A}}_\sigma$ we get (see (26.5))

$$\Pi^\top \bar{\mathcal{L}}_\sigma \Pi = \begin{bmatrix} \mathcal{C}_\sigma - |\mathcal{A}_\sigma| & 0 \\ 0 & \mathcal{C}_\sigma - |\mathcal{A}_\sigma| \end{bmatrix} = \begin{bmatrix} \mathcal{L}_{\sigma,unsigned} & 0 \\ 0 & \mathcal{L}_{\sigma,unsigned} \end{bmatrix}.$$

Therefore the dynamics of the variable $\Pi^\top \mathbf{z}(t)$ satisfies

$$\frac{d}{dt}\Big(\Pi^\top \mathbf{z}(t)\Big) = \left[\begin{array}{c|c} (I_N \otimes A) - (\mathcal{L}_{\sigma,unsigned}) \otimes (BK) & 0 \\ \hline 0 & (I_N \otimes A) - (\mathcal{L}_{\sigma,unsigned}) \otimes (BK) \end{array}\right]\Big(\Pi^\top \mathbf{z}(t)\Big).$$

Note that $\mathcal{L}_{\sigma,unsigned}$ is the Laplacian matrix of an undirected, weighted dynamic graph, with only nonnegative weights, which is uniformly connected over $[0, +\infty)$ (because, by assumption, $\mathcal{G}_\sigma$ is uniformly connected over $[0, +\infty)$). Moreover, assumptions A1-A4 hold. This ensures, by Theorem 2 in [13] that there exists a choice of K such that the first N entries of vector $\Pi^\top \mathbf{z}(t)$ converge to the same limit as t goes to $+\infty$. However, this means that

$$\lim_{t\to+\infty} |\mathbf{x}_i(t)| = \lim_{t\to+\infty} |\mathbf{x}_j(t)|, \qquad \forall\, i, j \in [1, N],$$
$$\lim_{t\to+\infty} \mathbf{x}_i(t) = -\lim_{t\to+\infty} \mathbf{x}_j(t), \qquad \forall i \in [1, r], \; j \in [r+1, N].$$

So, the bipartite consensus problem is solvable.

Remark 26.1 As shown in [13], an explicit solution of the bipartite consensus problem can always be found by proceeding in the following way:

1. Let T be a nonsingular matrix such that $TAT^{-1} = \begin{bmatrix} A_u & 0 \\ 0 & A_s \end{bmatrix}$, where A_u is an anti-symmetric matrix (with spectrum on the imaginary axis) and A_s is a Hurwitz matrix (i.e., all its eigenvalues have negative real part). Partition the matrix TB according to the partition of TAT^{-1}, namely as $TB = \begin{bmatrix} B_u \\ B_s \end{bmatrix}$.
2. Let P_u be a symmetric positive definite matrix satisfying $A_u^\top P_u + P_u A_u = 0$.
3. Then a possible solution K takes the form $K = \begin{bmatrix} B_u^\top P_u & 0 \end{bmatrix} T$.

References

1. Altafini, C.: Consensus problems on networks with antagonistic interactions. IEEE Trans. Aut. Contr. **58**, 935–946 (2013)
2. Easley, D., Kleinberg, J.: Networks, Crowds, and Markets: Reasoning About a Highly Connected World. Cambridge University Press, Cambridge (2010)
3. Hendrickx, J.M.: A lifting approach to models of opinion dynamics with antagonisms. In: Proceedings of the 53th IEEE Conference on Decision and Control, Los Angeles, USA, pp. 2118–2123. (2014)
4. Hou, Y., Li, J., Pan, Y.: On the Laplacian eigenvalues of signed graphs. Linear and Multilinear Algebra **51**, 21–30 (2003)
5. Jadbabaie, A., Lin, J., Morse, A.S.: Coordination of groups of mobile autonomous agents using nearest neighbor rules. IEEE Trans. Aut. Contr. **48**, 988–1001 (2003)
6. Liberzon, D.: Switching in Systems and Control. Volume in series Systems and Control: Foundations and Applications, Birkhauser, Boston (MA) (2003)
7. Lin, Z.: Coupled dynamic systems: from structure towards stability and stabilizability. Ph. D. thesis, University of Toronto, Toronto, Canada (2005)
8. Meng, Z., Shi, G., Johansson, K.H., Cao, M., Hong, Y.: Behaviors of networks with antagonistic interactions and switching topologies. arXiv:1402.2766v2 (2016)
9. Mohar, B.: The Laplacian spectrum of graphs. Graph Theory Comb. Appl. **2**, 871–898 (1991)
10. Ren, W., Beard, R.W.: Consensus seeking in multiagent systems under dynamically changing interaction topologies. IEEE Trans. Aut. Contr. **50**, 655–661 (2005)
11. Ren, W., Beard, R.W., McLain, T.W.: Coordination variables and consensus building in multiple vehicle systems. In: Leonard, N.E., Kumar, V., Morse A.S. (eds.) Cooperative Control. Lecture Notes in Control and Information Sciences, vol. 309, pp. 171–188. Springer (2004)
12. Shi, G., Proutiere, A., Johansson, M., Baras, J.S., Johansson, K.H.: Emergent behaviors over signed random dynamical networks: state-flipping model. IEEE Trans. Contr. Netw. Syst. **2**, 142–153 (2015)
13. Su, Y., Huang, J.: Stability of a class of linear switching systems with applications to two consensus problems. IEEE Trans. Aut. Contr. **57**, 1420–1430 (2012)
14. Valcher, M.E., Misra, P.: On the consensus and bipartite consensus in high-order multi-agent dynamical systems with antagonistic interactions. Syst. Control Lett. **66**, 94–103 (2014)
15. Wen, G., Duan, Z., Chen, G., Yu, W.: Consensus tracking of multi-agent systems with Lipschitz-type node dynamics and switching topologies. IEEE Trans. Circ. Syst. I: Regul. Pap. **61**, 499–511 (2014)
16. Xia, W., Cao, M., Johansson, K.H.: Structural balance and opinion separation in trust-mistrust social networks. IEEE Trans. Contr. Netw. Sys. **3**, 46–56 (2016)

Chapter 27
Hypertracking Beyond the Nyquist Frequency

Kaoru Yamamoto, Yutaka Yamamoto and Masaaki Nagahara

Abstract We study the problem of tracking or disturbance rejection for signals beyond the Nyquist frequency in the sampled-data context. Given the well-established sampling theorem, one is inclined to infer that it is possible to track or reject only those signals below the Nyquist frequency. However, such a high-frequency signal may be detected as an aliased signal, and this opens a new possibility in tracking and regulation. This article shows that a multirate processing with a suitable choice of a weighting function enables tracking or rejection of high-frequency signals beyond the Nyquist frequency. An example is given to illustrate the result.

Some portion of this chapter has been reproduced/derived with permission of the IEEE from the source Y. Yamamoto, K. Yamamoto, M. Nagahara, Tracking of signals beyond the Nyquist frequency, 2016 IEEE 55th Conference on Decision and Control (CDC), https://doi.org/10.1109/CDC.2016.7798875.
Kaoru Yamamoto is a member of the LCCC Linnaeus Center and the ELLIIT Excellence Center at Lund University.
Yutaka Yamamoto was supported in part by the Japan Society for the Promotion of Science under Grants-in-Aid for Scientific Research No. 15H04021 and 24360163. He wishes to thank DIGITEO, ICODE, and Laboratoire des Signaux et Systemes (L2S, UMR CNRS), CNRS-CentraleSupelec-University Paris-Sud and Inria Saclay for their financial support while part of this research was conducted.

K. Yamamoto
Automatic Control LTH, Lund University, Box 118, SE 221 00 Lund, Sweden
e-mail: k.yamamoto@ieee.org

Y. Yamamoto (✉)
Kyoto University, Kyoto 606-8501, Japan
e-mail: yy@i.kyoto-u.ac.jp

Y. Yamamoto
Labolatoire des Signaux et Systèmes L2S, CentraleSupélec, Plateau de Moulon, 3 Juliot-Curie, 91192 Gif Sur Yvette, France
e-mail: yy@i.kyoto-u.ac.jp

M. Nagahara
Institute of Environmental Science and Technology, The University of Kitakyushu, Fukuoka 808-0135, Japan
e-mail: nagahara@ieee.org

R. Tempo et al. (eds.), *Emerging Applications of Control and Systems Theory*, Lecture Notes in Control and Information Sciences - Proceedings,
https://doi.org/10.1007/978-3-319-67068-3_27

27.1 Preface

It is our great pleasure to be able to contribute this short note to celebrate the 70th birthday of our friend Mathukumalli Vidyasagar. We are pleased to dedicate herewith this chapter to him.

Sagar has been a constant source of inspiration for us through his amazingly versatile contributions to many aspects of control and system theory throughout his career. Among them, we were very much influenced by his work on robust control and tracking. For example, his factorization approach over the ring of stable rational matrices, graph topolog,y and its consequences on necessary and sufficient conditions for robustness margins (or for allowable perturbations).

An early investigation of robust tracking exhibits the fact that one should incorporate an internal model of the tracking signal generator into the loop to achieve robust tracking under plant perturbations. This has induced many directions of development, e.g., voltage control of electrical power supply, repetitive control, for periodic signals, etc.

In this note, we consider the tracking or disturbance rejection problem in the sampled-data control context. To track continuous-time reference signals, one should still incorporate an internal model, but its construction can be more intricate. We refer the reader to [12] where a rather thorough study is conducted.

However, when a precise continuous-time internal model cannot be implemented, we must resort to an approximate internal model constructed by a digital controller and a hold device. Furthermore, if we resort to such an approximation, there is also a limitation due to the limited resolution in frequency due to sampling. This limitation, due to the well-known sampling theorem, may appear as a severe restriction in designing digital control systems. However, this paper shows that this limitation is superficial, partly due to a wrong perception in the understanding of the sampling theorem.

27.2 Introduction

Let us recall the sampling theorem [9, 18]:

Theorem 27.1 *Suppose that the continuous-time signal $f(t)$ is band-limited to $(-\pi/h, \pi/h)$, that is, its Fourier transform $\hat{f}$ satisfies $\hat{f}(\omega) = \mathscr{F}[f](\omega) \equiv 0$ for $|\omega| \geq \pi/h$. Then $f(t)$ can be uniquely recovered by the formula*

$$f(t) = \sum_{n=-\infty}^{\infty} f(nh)\mathrm{sinc}(t - nh), \tag{27.1}$$

where

$$\mathrm{sinc}\,t := \frac{\sin \pi t/h}{\pi t/h}. \tag{27.2}$$

Shannon [9] placed this at the center of his signal recovery paradigm. From here on, we refer to this paradigm as the Shannon paradigm. The paper [9] has been extremely influential, and the paradigm has remained to be the guiding principle in most of sound/image processing applications. It also induced a rather universal (yet often unfortunate) understanding that under no circumstances it is possible to recover high-frequency components beyond the Nyquist frequency π/h. A close examination of the sampling theorem or Shannon's paper [9] would immediately tell us that this is a naive misunderstanding. What the sampling theorem tells us is that if there is no model for the frequency contents of the signals to be processed, and *if* we assume that there are no frequency components beyond the Nyquist frequency, then the signal can be uniquely recovered by formula (27.1). Clearly, Theorem 27.1 says nothing if there is a model for a class of signals we should process. While there have been various attempts to generalize the sampling theorem to various nonclassical contexts (see, e.g., [5]), it is interesting to observe that the Shannon paradigm still dominates most of the commercial applications of sounds and images in one way or another.

However, if we replace the band-limiting hypothesis by another one, then one can indeed obtain a different signal reconstruction result. For example, if we assume that the frequency components of $\hat{f}$ are limited only to the range $(-2\pi/h, -\pi/h) \cup (\pi/h, 2\pi/h)$, one can easily obtain the following reconstruction formula:

$$f(t) = \sum_{n=-\infty}^{\infty} f(nh)(2\mathrm{sinc}(2(t - nh)) - \mathrm{sinc}(t - nh)). \tag{27.3}$$

The proof is via direct calculation. This will also suggest the possibility of recovering high frequency signals if a different signal model is employed.

The present authors have proposed another signal processing direction in [16] and some commercial success has been made based on the patents [14, 15]. The novel idea there is that by assuming an analog signal generator that has a nonideal frequency decay curve, one can invoke H^∞ sampled-data control theory to optimally recover the intersampling behavior including high-frequency components beyond the Nyquist frequency.

The objective of the present paper is to consider tracking or rejection of signals having frequency components *beyond the Nyquist frequency*.

As we discussed above, this may appear impossible, because the sampling theorem demands that the signal recovery be limited only below the Nyquist frequency. However, this is also based on the band-limiting hypothesis. As we noted above, if we can assume a suitable analog model, there is a possibility of accomplishing such an objective.

Indeed, such a problem is not at all an artificial one. We here list two examples. One of them is the case of electric power supply, where an inverter is often used, and we need to regulate the frequency to some standard, e.g., 60 [Hz]. The sampling rate need not be taken high enough to cover this below the Nyquist frequency. Another example is the position control system for hard disk drives. The sampling rate there is also limited by the physical limitation, and it is not high enough to reject some disturbances; see, e.g., [2, 19] for details.

We will show that the above objectives are indeed attainable with a low sampling rate. By employing a proper analog model both for the plant and the tracking signal and using a multirate technique, we can show that an (sub)optimal tracking/rejection is possible. This is a new paradigm and has a great potential in many digital control systems.

Recall that modern sampled-data control theory, along with the introduction of lifting, e.g., [4, 12], has enabled us to optimize the intersample behavior with a discrete-time controller [3, 6, 13]. While these results are mainly established in the single-rate context, multirate systems have been effectively utilized in the signal processing literature [10]. While multirate control systems are also studied in the literature [1, 7, 8], these studies often make full use of upsampled output values in the fast rate. A crucial difference in this paper is that we do *not* execute extra fast sampling of the plant output, and upsampling is used only for computing the intersample control signals. See also [16] for the success in the signal reconstruction problem via the combination of multirate processing and H^∞ control theory.

27.3 High-Frequency Tracking Problem

We consider the sampled-data system given by Fig. 27.1, consisting of the linear, time-invariant, continuous-time plant $P(s)$ and the discrete-time, linear, time-invariant controller $K(z)$.

After sampled with period h, the error e is further upsampled by factor M. The upsampler $\uparrow M$ inserts $M - 1$ zeros in each sampling interval, and is defined by $(\uparrow M)(e)[kh + \ell] = e[kh]$ for $\ell = 0$, and $(\uparrow M)(e)[kh + \ell] = 0$ elsewhere, i.e., $\ell = h/M, \dots (M - 1)h/M$. $\mathcal{H}_{h/M}$ is the zero-order hold that holds the output as constant for the period of h/M.

Now consider the following *hypertracking* problem:

Problem *In the block diagram Fig.* 27.1, *let* $\sin \omega t$, $\omega > \pi/h$ *be a reference input. Design a discrete-time controller* $K(z)$ *such that the output* $y(t)$ *approximately tracks the delayed reference* $r(t - L) = \sin \omega(t - L)$ *for some* $L \geq 0$.

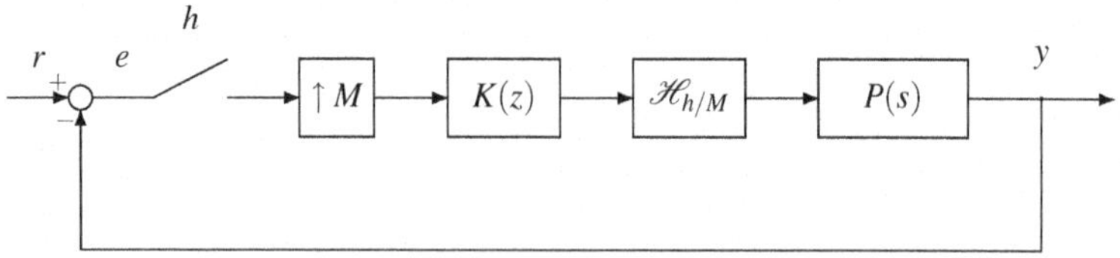

Fig. 27.1 Sampled feedback system. Reproduced with permission from © IEEE, Yamamoto et al. [17]

The objective is to track a signal beyond the Nyquist frequency, exceeding the limit of an ordinary tracking, hence called *hypertracking*.

To derive a solution, we first lift the system Fig. 27.1 to obtain a discrete-time model [3, 6, 12], characterize its invariant zeros, and then compute transmission zero directions. These directions determine the intersample tracking performance. This process will make it possible to formulate an H^∞ sampled-data control problem, and it will give us a desired solution.

27.4 Lifted Multirate System

By lifting the system in Fig. 27.1, we can obtain a single-rate time-invariant discrete-time system with sampling period h. We employ both continuous-time lifting and the discrete-time one (so-called *blocking* in the signal processing literature).

Let $P(s)$ and $K(z)$ be the following systems:

$$P(s): \begin{cases} \frac{d}{dt}x_c(t) &= A_c x_c(t) + B_c u(t) \\ y(t) &= C_c x_c(t) \end{cases} \tag{27.4}$$

$$K(z): \begin{cases} x_d[k+1] &= A_d x_d[k] + B_d w_d[k] \\ y_d[k] &= C_d x_d[k] + D_d w_d[k]. \end{cases} \tag{27.5}$$

In the sequel, we denote by $f(t)$ a continuous-time signal with parentheses, by $g[k]$ a discrete-time signal with brackets, where t and k denote a continuous-time variable and an integer variable, respectively.

Recall the continuous-time lifting [3, 4, 6, 12] as follows:

$$\begin{aligned} \mathscr{L}:\ & L^2_{loc}[0,\infty) \to \ell^2(L^2[0,h)) : x(\cdot) \mapsto \{x[k](\cdot)\}_{k=0}^{\infty}, \\ & x[k](\theta) := x(kh+\theta). \end{aligned} \tag{27.6}$$

The continuous-time plant $P(s)$ is then described by

$$\tilde{\Sigma}_P: \begin{cases} x_c[k+1] &= e^{A_c h} x_c[k] + \int_0^h e^{A_c(h-\tau)} B_c u[k](\tau) d\tau \\ y_c[k](\theta) &= C_c e^{A_c\theta} x_c[k] + \int_0^\theta C_c e^{A_c(\theta-\tau)} B_c u[k](\tau) d\tau. \end{cases} \tag{27.7}$$

Performing the discrete-time lifting (i.e., blocking) with period h, we obtain the following description for the discrete-time controller.

Proposition 27.4.1 *When lifted with period h, the discrete-time controller $K(z)$ is expressible as*

$$\tilde{\Sigma}_K : x_d[k+1] := x_d(kh+h) = A_d^M x_d[k] + A_d^{M-1} B_d e[k](0)$$
$$=: \overline{A_d} x_d[k] + \overline{B_d} e[k](0)$$

$$y_d[k] := \begin{bmatrix} y_d(kh) \\ y_d(kh+h/M) \\ \vdots \\ y_d(kh+(M-1)h/M) \end{bmatrix}$$
$$= \begin{bmatrix} C_d \\ C_d A_d \\ \vdots \\ C_d A_d^{M-1} \end{bmatrix} x_d[k] + \begin{bmatrix} D_d \\ C_d B_d \\ \vdots \\ C_d A_d^{M-2} B_d \end{bmatrix} e[k](0)$$
$$=: \overline{C_d} x_d[k] + \overline{D_d} e[k](0).$$

Introduce the generalized hold function $H(\theta)$ *as*

$$H(\theta) := [\chi_{[0,h/M)}(\theta), \chi_{[h/M,2h/M)}(\theta), \dots, \chi_{[(M-1)h/M,h)}(\theta)], \tag{27.8}$$

where $\chi_{[ih/M,(i+1)h/M)}(\theta)$, $i = 0, \dots, M-1$ *is the characteristic function of the interval* $[ih/M, (i+1)h/M)$. *The lifted input* $u[k](\theta)$ *for* P *then takes the simple form*

$$u[k](\theta) = H(\theta) y_d[k]. \tag{27.9}$$

Proof is omitted. See [17].

Define

$$B(\theta) := \int_0^\theta e^{A_c(\theta-\tau)} B_c H(\tau)\, d\tau. \tag{27.10}$$

Then the lifted $\tilde{\Sigma}_K$ and $\tilde{\Sigma}_P$ can be written as

$$\tilde{\Sigma}_K : x_d[k+1] =: \overline{A_d} x_d[k] + \overline{B_d} e[k](0) \tag{27.11}$$
$$y_d[k] =: \overline{C_d} x_d[k] + \overline{D_d} e[k](0)$$
$$u[k](\theta) =: H(\theta) y_d[k]$$
$$\tilde{\Sigma}_P : x_c[k+1] = e^{A_c h} x_c[k] + B(h) y_d[k] \tag{27.12}$$
$$y[k](\theta) = C_c e^{A_c\theta} x_c[k] + C_c B(\theta) y_d[k].$$

Observe that they are time-invariant discrete-time systems, and readily yield the following description (27.13) for the closed-loop system Fig. 27.1.

27.4.1 Closed-Loop Equations

Substitute (27.11) and (27.12) into the block diagram Fig. 27.1, to obtain the following closed-loop equations (with e being the output):

$$\begin{bmatrix} x_d[k+1] \\ x_c[k+1] \end{bmatrix} = \begin{bmatrix} \overline{A_d} & -\overline{B_d}C_c \\ B(h)\overline{C_d} & e^{A_c h} - B(h)\overline{D_d}C_c \end{bmatrix} \begin{bmatrix} x_d[k] \\ x_c[k] \end{bmatrix} + \begin{bmatrix} \overline{B_d}\delta_0 \\ B(h)\overline{D_d}\delta_0 \end{bmatrix} r[k](\theta) \tag{27.13}$$

where δ_0 is the Dirac delta distribution, defined by $\delta_0 r[k](\theta) := r[k](0)$. It represents the sampler at each time k.

The following formula for $e[k](\theta)$ is readily obtained:

$$\begin{aligned} e[k](\theta) &= r[k](\theta) - y[k](\theta) \\ &= r[k](\theta) - C_c e^{A_c\theta} x_c[k] - C_c B(\theta) v[k] \\ &= r[k](\theta) - C_c e^{A_c\theta} x_c[k] \\ &\quad - C_c B(\theta)(\overline{C_d} x_d[k] + \overline{D_d} e[k](0)) \\ &= r[k](\theta) - C_c e^{A_c\theta} x_c[k] \\ &\quad - C_c B(\theta) \left(\overline{C_d} x_d[k] + \overline{D_d}(r[k](0) - C_c x_c[k])\right) \\ &= \begin{bmatrix} -C_c B(\theta)\overline{C_d} & -C_c e^{A_c\theta} + C_c B(\theta)\overline{D_d} C_c \end{bmatrix} \begin{bmatrix} x_d[k] \\ x_c[k] \end{bmatrix} \\ &\quad + (I - C_c B(\theta)\overline{D_d}\delta_0) r[k](\theta). \end{aligned} \tag{27.14}$$

27.5 Design Method

We briefly sketch the design method and show how the present objective can be achieved.

We first place a strictly proper anti-aliasing filter $F(s)$ for the reference signal in Fig. 27.1. This places a certain decay for the reference signal, and simultaneously make the total closed-loop system well-defined as a bounded operator on L^2 into itself. This $F(s)$ can also be used as a weighting on the input reference signals, and as a design parameter. In contrast to common practice where one places emphasis on low frequency in $F(s)$, we choose to *place more emphasis on the high-frequency range that we wish to track*. This somewhat nonclassical idea indeed works for the present tracking purposes.

As stated in the main Problem, we demand that the delayed error $\tilde{e}(t) := r(t-L) - y(t)$ be minimized for some positive L. This allows us more freedom in designing

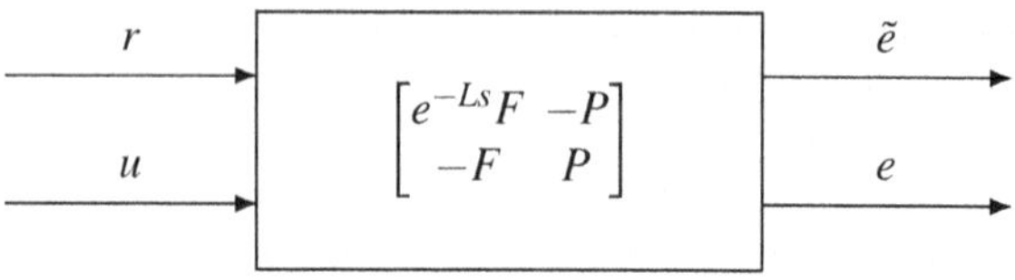

Fig. 27.2 Generalized plant. Reproduced with permission from © IEEE, Yamamoto et al. [17]

the controller and it improves the performance. See also [16] for a similar situation in signal reconstruction.

These considerations yield the generalized plant Fig. 27.2 for our design problem. The design parameter L is taken to be an integer multiple of h, with some small integer such as 4–10.

27.6 Numerical Example

Example 27.6.1 Take the plant

$$P(s) := \frac{2}{s^2 + 3s + 2} \tag{27.15}$$

with sampling period $h = 1$ in Fig. 27.1, having the Nyquist frequency π [rad/sec] $= 0.5$ [Hz]. Let $r = \sin \omega t$ be the tracking signal with $\omega = 1.25\pi$ [rad/sec]. This is equal to $0.625 > 0.5$ [Hz].

To accommodate more emphasis on this high-frequency tracking signal, take the following weighting function, which has a peak at the tracking frequency 1.25π [rad/sec], while deemphasizing low-frequency signals:

$$F(s) := \frac{s}{s^2 + 0.1s + (1.25\pi)^2}. \tag{27.16}$$

The figures here show the simulation results with $M = 8$ and $L = 4h$. Figure 27.3 shows the output response, and Fig. 27.4 its delayed error. It is seen that the output approximately tracks the reference input $\sin(1.25\pi)t$, with frequency higher than the Nyquist frequency π. Note also that the output is delayed approximately by 4 steps as required by the design specification. Figure 27.5 shows the output of the discrete-time controller. This clearly shows that the controller works as an approximate internal model for the reference sinusoid $\sin(1.25\pi)t$. It is naturally expected that it gives a more accurate internal model if the upsampling factor M is increased.

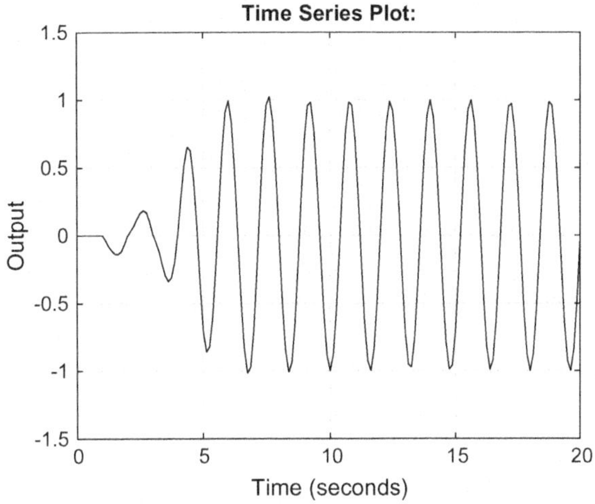

Fig. 27.3 System output tracking $\sin(1.25\pi)t$

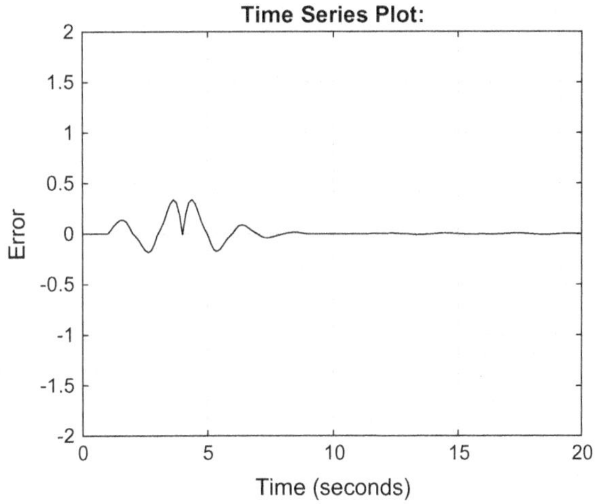

Fig. 27.4 Delayed error against $\sin(1.25\pi)t$

27.7 Discussion

It is shown that it is possible to track a signal containing higher frequency components than the Nyquist frequency. Instead of a continuous-time internal model, it is possible to construct an approximate internal model via upsampling, and this enables the desired high-frequency tracking. With a suitable choice of weighting function, H^∞ sampled-data control yields a suitable discrete-time controller leading to tracking in high frequency, even though the frequency may be beyond the Nyquist frequency. We note again that such a high-frequency error signal is captured as an alias component, and a proper design of a controller makes this tracking possible.

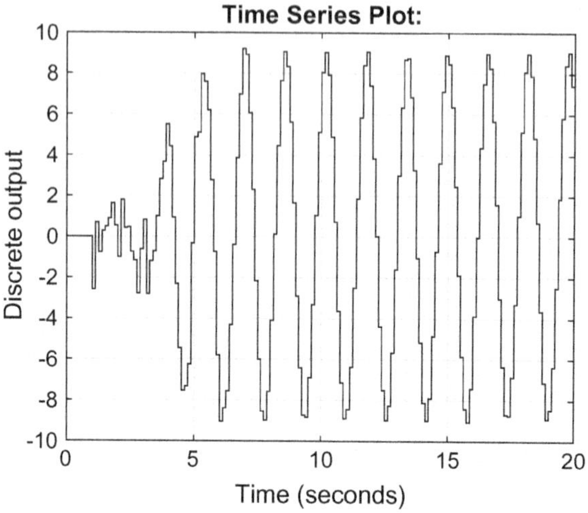

Fig. 27.5 Discrete-time controller output

We can employ the same idea and method for disturbance rejection with a suitable modification of the closed-loop configuration. This is particularly effective in disturbance rejection applications, e.g., [2, 19].

There is a challenge, when tracking or disturbance signals occur at multiple frequencies, for example, below and above the Nyquist frequency. Such a case requires an extra design technique; a design method is to be reported in [11].

References

1. Araki, M., Yamamoto, K.: Multivariable multirate sampled-data systems: state-space description, transfer characteristics, and nyquist criterion. IEEE Trans. Autom. Control **AC-31**, 145–154 (1986)
2. Atsumi, T.: Disturbance suppression beyond nyquist frequency in hard disk drives. Mechatronics **20**(1), 67–73 (2010)
3. Bamieh, B., Pearson, J.B.: A general framework for linear periodic systems with applications to H^∞ sampled-data control. IEEE Trans. Autom. Control **37**(4), 418–435 (1992)
4. Bamieh, B., Pearson, J.B., Francis, B.A., Tannenbaum, A.: A lifting technique for linear periodic systems with applications to sampled-data control. Syst. Control Lett. **17**(2), 79–88 (1991)
5. Baraniuk, R.G., Candes, E., Nowak, R., Vetterli, M.: Special issue on compressive sampling. In: IEEE Signal Processing Magazine, vol. 25, pp. 12–101. IEEE (2008)
6. Chen, T., Francis, B.A.: Optimal sampled-data control systems. Springer, London (1995)
7. Hagiwara, T., Araki, M.: Design of a stable state feedback controller based on the multirate sampling of the plant output. IEEE Trans. Autom. Control **33**(9), 812–819 (1988)
8. Mita, T., Chida, Y., Kaku, Y., Numasato, H.: Two-delay robust digital control and its applications-avoiding the problem on unstable limiting zeros. IEEE Trans. Autom. Control **35**(8), 962–970 (1990)
9. Shannon, C.E.: Communication in the presence of noise. Proc. IRE **38**, 10–21 (1949)
10. Vaidyanathan, P.P.: Multirate Systems and Filter Banks. Prentice Hall, Englewood Cliffs (1993)

11. Yamamoto, K., Yamamoto, Y., Nagahara, M.: Simultaneous rejection of signals below and above the Nyquist frequency. Proc. 1st IEEE Conf. Control Tech. and Appl. IEEE 1135–1139 (2017)
12. Yamamoto, Y.: A function space approach to sampled-data control systems and tracking problems. IEEE Trans. Autom. Control **AC-39**(4), 703–713 (1994)
13. Yamamoto, Y.: Digital control. In: Wiley Encyclopedia of Electrical and Electronics Engineering, vol. 5, pp. 445–457. Wiley (1999)
14. Yamamoto, Y.: Digital/analog converters and a design method for the pertinent filters. Japanese Patent No. 3820331 (2006)
15. Yamamoto, Y., Nagahara, M.: Sample-rate converters. Japanese patent No. 3851757 (2006)
16. Yamamoto, Y., Nagahara, M., Khargonekar, P.P.: Signal reconstruction via H^∞ sampled-data control theory—beyond the shannon paradigm. IEEE Trans. Signal Process. **SP-60**(2), 613–625 (2012)
17. Yamamoto, Y., Yamamoto, K., Nagahara, M.: Tracking of signals beyond the nyquist frequency. In: Proceedings of the 55th IEEE Conference Decision and Control, pp. 4003–4008. IEEE (2016)
18. Zayed, A.I.: Advances in Shannon's Sampling Theory. CRC Press, Boca Raton (1993)
19. Zheng, M., Sun, L., Tomizuka, M.: Multi-rate observer based sliding mode control with frequency shaping for vibration suppression beyond nyquist frequency. IFAC-PapersOnLine **49**, 13–18 (2016)

Chapter 28
Quadratic Hedging with Mixed State and Control Constraints

A. Heunis

Abstract We address a problem of stochastic optimal control in mathematical finance, namely quadratic hedging with constraints on both the portfolio invested and the wealth process. Quadratic hedging involves the minimization of a *quadratic loss* criterion. Constraints on the portfolio are essentially *control constraints* while constraints on the wealth process are *state constraints*, so the problem amounts to stochastic optimal control with the combination of control and state constraints. Few results are available on general problems of this kind. However, our particular problem has the nice properties of being convex, with simple linear dynamics and the state constraint in the form of a one-sided almost-sure inequality. These are key to the application of a powerful variational method of Rockafellar for abstract problems of convex programming. We construct an optimal portfolio by means of this approach.

28.1 Introduction and Motivation

A common problem in financial trading is the following: an agent begins at time $t = 0$ with some initial wealth $x_0 > 0$, and, with no further infusions of wealth, must trade in a market over a finite interval $0 \leq t \leq T$ such that the agent's wealth at close-of-trade $t = T$ is "near" to some stipulated random variable γ (usually referred to as a *contingent claim*).

To make this problem more precise suppose that the random variable γ is defined on a complete probability space $(\Omega, \mathcal{F}, P)$, and is bounded in the sense of being square-integrable, that is $\mathrm{E}[\gamma^2] < \infty$. Suppose that the agent can trade among a total of N different assets, the prices of which are modeled by random processes $\{S_n(t),\ 0 \leq t \leq T\}$ defined on the probability space $(\Omega, \mathcal{F}, P)$, $n = 1, 2, \ldots, N$. For each $t \in [0, T]$ and $n = 1, \ldots, N$ let $\pi_n(t)$ denote the monetary (i.e., dollar) amount invested in the n th stock (with price $S_n(t)$). The holdings of the agent in the N stocks at instant t is then given by the vector $\pi(t) := (\pi_1(t), \ldots, \pi_N(t)) \in \mathbb{R}^N$. A

A. Heunis
University of Waterloo, Waterloo, ON N2L 3G1, Canada
e-mail: heunis@uwaterloo.ca

R. Tempo et al. (eds.), *Emerging Applications of Control and Systems Theory*, Lecture Notes in Control and Information Sciences - Proceedings,
https://doi.org/10.1007/978-3-319-67068-3_28

standard result is that the wealth of the investor is the $\mathbb{R}$-valued process $\{X^\pi(t),\ t \in [0, T]\}$ on $(\Omega, \mathcal{F}, P)$ which satisfies the linear stochastic differential equation (SDE)

$$\mathrm{d}X^\pi(t) = r(t)X^\pi(t) + \pi'(t)\sigma(t)[\theta(t)\,\mathrm{d}t + \mathrm{d}W(t)], \qquad X^\pi(0) = x_0, \tag{28.1}$$

in which

(i) $\{r(t),\ t \in [0, T]\}$ is a given $\mathbb{R}$-valued *interest-rate* process;
(ii) $\{\theta(t),\ t \in [0, T]\}$ is a given $\mathbb{R}^N$-valued *market price of risk* process;
(iii) $\{\sigma(t),\ t \in [0, T]\}$ is a given N-by-N matrix-valued *volatility* process;
(iv) $\{W(t),\ t \in [0, T]\}$ is a given $\mathbb{R}^N$-valued *standard Brownian motion*.

These processes are defined on the same probability space $(\Omega, \mathcal{F}, P)$ as the contingent claim γ, with r, θ, and σ being uniformly bounded and progressively measurable with respect to the filtration $\{\mathcal{F}_t, t \in [0, T]\}$ of the Brownian motion W, that is

$$\mathcal{F}_t := \sigma\{W(\tau),\ \tau \in [0, t]\} \vee \{P - \text{ null sets of } \mathcal{F}\}. \tag{28.2}$$

The processes r, θ, and σ determine the asset prices S_n, and then (28.1) can be established on the basis of these prices. We regard (28.1) as a *stochastic dynamical system*, in which the $\mathbb{R}^N$-valued process $\{\pi(t), t \in [0, T]\}$ on $(\Omega, \mathcal{F}, P)$, which defines the monetary amounts allocated to the stocks by the investor (and henceforth called the *portfolio process*), is the *control input*, and the resulting process of investor wealth $\{X^\pi(t), t \in [0, T]\}$ is both the *state* and the *output*. For the SDE (28.1) to make sense, it must be stipulated that input process $\{\pi(t), t \in [0, T]\}$ is path-wise square-integrable, namely $\int_0^T \|\pi(t)\|^2\,\mathrm{d}t < \infty$ a.s., and $\mathcal{F}_t$-progressively measurable (this last really means that the agent is not "clairvoyant", that is the agent cannot anticipate the random evolution of the market beyond the instant t when investing in the stocks at instant t). Furthermore, since the contingent claim γ is assumed to be square-integrable, it is natural to limit attention to portfolio processes $\pi := \{\pi(t), t \in [0, T]\}$ which are square-integrable, so that from now on the input processes π are always members of the real vector space

$$\Pi := \left\{\pi : [0, T] \times \Omega \to \mathbb{R}^N \middle| \pi \text{ is } \mathcal{F}_t - \text{ prog. meas. } \& \text{ E} \int_0^T \|\pi(t)\|^2\,\mathrm{d}t < \infty\right\}. \tag{28.3}$$

With these formulations in place, we can loosely state the problem in the opening paragraph as one of determining the portfolio $\pi \in \Pi$ such that the investor wealth $X^\pi(T)$ at end-of-trade $t = T$ is as "close as possible" to the contingent claim γ. It remains to define the sense of "close as possible", and following Markowitz [7] this is customarily taken as the L_2-discrepancy between $X^\pi(T)$ and γ, that is $\mathrm{E}[(X^\pi(T) - \gamma)^2]$, so that the problem becomes one of minimizing this quantity over all $\pi \in \Pi$. If we define the quadratic criterion

$$J(x, \omega) := x^2 - 2\gamma(\omega)x, \qquad (x, \omega) \in \mathbb{R} \times \Omega, \tag{28.4}$$

then we have the following *stochastic optimal control problem*

$$\text{minimize } \mathrm{E}[J(X^{\pi}(T))] \quad \text{over} \quad \pi \in \Pi. \tag{28.5}$$

This is called the problem of *unconstrained quadratic hedging*, and the minimizing portfolio $\pi \in \Pi$ (assuming it exists) is an *unconstrained quadratic hedge* of the contingent claim γ. This problem was addressed and solved by Lim et al. [6], who construct an unconstrained quadratic hedge by means of stochastic linear-quadratic optimal control.

It is often that case that the agent cannot make a completely unconstrained allocation of capital to the N stocks, and must choose the vector of allocations $(\pi_1(t), \ldots, \pi_N(t))$ to satisfy some constraints (often imposed by regulatory agencies) such as a prohibition on short selling. This constraint is typically defined by some nonempty closed convex set $K \subset \mathbb{R}^N$, and the requirement that the portfolio $\pi(t)$ take values in the set K, more precisely that the process π be restricted to the convex set

$$\mathcal{A} := \{\pi \in \Pi \mid \pi(t, \omega) \in K \quad \lambda \otimes P - \text{a.e.} \quad \text{on } [0, T] \times \Omega\}. \tag{28.6}$$

For reasons of liquidity, the agent is allowed to not invest in any of the available stocks, that is, to allocate all wealth to a money market account, and for this reason it will always be assumed that $0 \in K$. We then have the problem of *constrained quadratic hedging*, namely

$$\text{minimize } \mathrm{E}[J(X^{\pi}(T))] \quad \text{subject to} \quad \pi \in \mathcal{A}. \tag{28.7}$$

Now (28.7) amounts to a stochastic optimal control problem with a control constraint, and as such is significantly more challenging than the unconstrained problem (28.5), and in particular must be addressed by an approach quite different from that used for (28.5). One possibility is dynamic programming, but this is precluded by the fact that the resulting Bellman equation is a *random* PDE, and therefore completely intractable. The appropriate method is in fact based on *convex duality*, and this was used by Labbé [5] to construct a constrained quadratic hedge for problem (28.7).

There is one significant drawback associated with quadratic hedging in the form of problems (28.5) and (28.7): even if π is the minimizing portfolio one could nevertheless still have $P\{X^{\pi}(T) < 0\} > 0$; that is one could have negative wealth at end-of-trade with positive probability, an obviously undesirable outcome for any agent. The only way to overcome this difficulty is to stipulate the condition $X^{\pi}(T) \geq 0$ as a further constraint on the portfolio π. In fact, more generally, we will stipulate a random variable B on the probability space $(\Omega, \mathcal{F}, P)$ and then require that $X^{\pi}(T) \geq B$ a.s., so that B defines a minimum "floor-level" of wealth at close of trade. Adding this constraint to problem (28.7) we then get the problem

$$\text{minimize } \mathrm{E}[J(X^{\pi}(T))] \quad \text{subject to} \quad \pi \in \mathcal{A} \quad \text{and} \quad X^{\pi}(T) \geq B \text{ a.s.} \tag{28.8}$$

The constraint $X^{\pi}(T) \geq B$ in this problem is an *indirect constraint* on the portfolio π defined through the SDE (28.1) which relates the input π to the output $X^{\pi}(T)$, and is usually known as a *state constraint*. That is, (28.8) amounts to a stochastic optimal control problem with the combination of a control constraint (on the input π) and a state constraint (on X^{π}), a challenging problem for which essentially no general results are available. Indeed, even for deterministic optimal control with state and control constraints there are few results available, and these are typically in the form of Pontryagin-type necessary conditions for optimality (see for example Dubovitskii et al. [1] and Makowski et al. [4]). For stochastic optimal control, there are basically no comparable necessary conditions for optimality, and, even if available, such results would be of limited value in addressing problem (28.8) since necessary conditions for optimality would not shed any light at all on the question of existence of optimal portfolios. Problem (28.8) nevertheless has the following very nice properties: (a) the state dynamics (28.1) are quite simple; (b) the problem is convex; (c) the state is $\mathbb{R}$-valued with the state constraint being the simple one-sided inequality $X^{\pi}(T) \geq B$ a.s. These properties are key to the application of a powerful *variational method* of Rockafellar [8] to problem (28.8). We outline this approach next.

28.2 The Rockafellar Variational Approach

Consider the following abstract convex optimization problem: one is given a real vector space $\mathbb{X}$, together with a convex "constraint set" $\mathbf{C} \subset \mathbb{X}$ and a convex function $f_0 : \mathbb{X} \to \mathbb{R}$. The *primal problem* is to minimize $f_0(x)$ subject to the constraint $x \in \mathbf{C}$. This is a convex optimization problem, and is best approached by the introduction of a *dual optimization problem* over a vector space of *dual variables*. The challenge in applying this approach is that the space of dual variables and dual functional are often not a priori clear. The Rockafellar variational approach gives a rational method to synthesize an appropriate vector space of dual variables, together with a dual functional on the space of dual variables. Define

$$f(x) := \begin{cases} f_0(x), & x \in \mathbf{C}, \\ +\infty, & x \notin \mathbf{C}. \end{cases} \tag{28.9}$$

The primal problem is, therefore, to minimize $f(x)$ over all $x \in \mathbb{X}$. We summarize the Rockafellar approach in the following three steps:

Step 1—Perturbation of the Primal Problem: Fix some normed vector space $\mathbb{U}$ (the "space of perturbations") and some convex "perturbation mapping"

$$F : \mathbb{X} \times \mathbb{U} \to (-\infty, \infty] \tag{28.10}$$

subject to the "consistency relation" (recall (28.9))

$$F(x, 0) = f(x), \qquad x \in \mathbb{X}. \tag{28.11}$$

Step 2—Space of dual variables and Lagrangian: Define the normed vector space of *dual variables*

$$\mathbb{Y} := \mathbb{U}^*, \tag{28.12}$$

i.e., $\mathbb{Y}$ is the space of norm-continuous linear functionals $y : \mathbb{U} \to \mathbb{R}$, and define the *Lagrangian function* $K : \mathbb{X} \times \mathbb{Y} \to [-\infty, \infty]$ as follows:

$$K(x, y) := \inf_{u \in \mathbb{U}} [y(u) + F(x, u)], \quad (x, y) \in \mathbb{X} \times \mathbb{Y}. \tag{28.13}$$

Step 3—Dual functional: Define the *dual functional* $g : \mathbb{Y} \to [-\infty, \infty]$ by

$$g(y) := \inf_{x \in \mathbb{X}} K(x, y) = \inf_{(x,u) \in \mathbb{X} \times \mathbb{U}} [y(u) + F(x, u)], \quad y \in \mathbb{Y}. \tag{28.14}$$

Remark 28.1 The preceding steps constitute the core of the Rockafellar variational approach. Once the space $\mathbb{U}$ and perturbation $F : \mathbb{X} \times \mathbb{U} \to [-\infty, \infty]$ are fixed in Step 1 then the space of dual variables $\mathbb{Y}$, Lagrangian $K(\cdot, \cdot)$, and dual function $g(\cdot)$ are completely determined. Notice that in Step 1 there is complete freedom in defining the space of perturbations $\mathbb{U}$ and perturbation function $F(\cdot, \cdot)$ provided that F is convex on $\mathbb{X} \times \mathbb{U}$ and (28.11) holds.

From (28.9) to (28.14), we see that $g(\cdot)$ is *concave* on $\mathbb{Y}$, and get the "sandwich relation"

$$f(x) \geq K(x, y) \geq g(y), \quad x \in \mathbb{X}, \quad y \in \mathbb{Y}, \tag{28.15}$$

from which one obtains

$$\inf_{x \in \mathbb{X}} f(x) - \sup_{y \in \mathbb{Y}} g(y) \geq 0, \tag{28.16}$$

(the quantity on the left side of (28.16) is known as the *duality gap*). It is immediate from (28.15) that, for arbitrary $(\bar{x}, \bar{y}) \in \mathbb{X} \times \mathbb{Y}$, one has

$$f(\bar{x}) = g(\bar{y}) \quad \Rightarrow \quad \bar{x} \text{ minimizes } f(\cdot) \text{on} \, \mathbb{X} \text{ (and } \bar{y} \text{ maximizes } g(\cdot) \text{ on } \mathbb{Y}), \tag{28.17}$$

that is, if we can somehow construct $(\bar{x}, \bar{y}) \in \mathbb{X} \times \mathbb{Y}$ such that $f(\bar{x}) = g(\bar{y})$ then $\bar{x}$ solves the primal problem (of minimizing f over $\mathbb{X}$) and $\bar{y}$ solves the so-called *dual problem* of maximizing g over the dual space $\mathbb{Y}$. This observation is, by itself, not especially useful. It is entirely possible that, having worked through Step 1–3, we end up with a strictly positive duality gap (so that there fails to exist any $(\bar{x}, \bar{y}) \in \mathbb{X} \times \mathbb{Y}$ such that $f(\bar{x}) = g(\bar{y})$); moreover, even if the duality gap is zero, there is no guarantee that $g(\cdot)$ attains its supremum on $\mathbb{Y}$, so that again there cannot exist $(\bar{x}, \bar{y}) \in \mathbb{X} \times \mathbb{Y}$ such that $f(\bar{x}) = g(\bar{y})$. This means that the choice of space of perturbations and/or perturbation function F in Step 1 is somehow inappropriate. The following result of Rockafellar and J.J. Moreau is, therefore, invaluable in suggesting appropriate choices of these entities:

Theorem 28.1 (Rockafellar–Moreau) *Suppose that*

$$\sup_{\substack{u \in \mathbb{U} \\ \|u\| < \epsilon}} F(\tilde{x}, u) < \infty, \qquad \textit{for some } \tilde{x} \in \mathbb{X} \textit{ and } \epsilon > 0. \tag{28.18}$$

Then

$$\inf_{x \in \mathbb{X}} f(x) = \sup_{y \in \mathbb{Y}} g(y) = g(\bar{y}) \quad \textit{for some } \bar{y} \in \mathbb{Y}. \tag{28.19}$$

Remark 28.2 From Theorem 28.1, one sees that the normed vector space $\mathbb{U}$ and perturbation function F in Step 1 must be chosen such that (28.18) holds. We then get *zero duality gap* and existence of a maximizer $\bar{y}$ in $\mathbb{Y}$ of the dual function g (which is usually called a *Lagrange multiplier*). Assuming we have obtained (28.19), we can contemplate the following strategy to construct an $\bar{x} \in \mathbb{X}$ which solves the primal problem:

(A) Establish *necessary conditions* which are a consequence of the fact that $\bar{y}$ is a maximizer of g in the dual space $\mathbb{Y}$;
(B) Use the necessary conditions from (A) to construct an $\bar{x} \in \mathbb{X}$ *in terms of* the maximizer $\bar{y}$ such that $f(\bar{x}) = g(\bar{y})$. It then follows from (28.17) that $\bar{x}$ solves the primal problem.

The feasibility of this approach depends on the dual functional $g(\cdot)$ being in reasonably tractable form in order to get useful necessary conditions from the optimality of $\bar{y}$. We must also be able to rewrite the statement $f(x) = g(y)$ (for *arbitrary* $(x, y) \in \mathbb{X} \times \mathbb{Y}$) in some tractable *equivalent form*, namely

$$f(x) = g(y) \quad \Longleftrightarrow \quad \left\{ \begin{array}{l} \text{some useful optimality relations} \\ \text{on } (x, y) \in \mathbb{X} \times \mathbb{Y} \text{ hold} \\ \text{e.g., complementary slackness conditions,} \\ \text{feasibility conditions, and transversality relations.} \end{array} \right. \tag{28.20}$$

With the equivalence (28.20) at hand we can use the necessary conditions obtained from **(A)** of Remark 28.2 to construct an $\bar{x} \in \mathbb{X}$ in terms of $\bar{y} \in \mathbb{Y}$ such that $(\bar{x}, \bar{y})$ satisfies the "useful optimality relations" on the right side of (28.20); we then conclude that $f(\bar{x}) = g(\bar{y})$ as required.

The variational approach outlined in this section has been used for static problems of abstract convex optimization, but seemingly not for stochastic optimal control problems with state and control constraints. In the following section, we shall see that the approach is ideally suited to problem (28.8).

28.3 Perturbation, Lagrangian, and Dual Functional

To focus on just the absolute essentials and avoid secondary technicalities, we shall simplify the dynamical relation (28.1) and suppose that the interest-rate process r is identically zero (not a bad approximation to current economic conditions!) and that the volatility process σ is the N-by-N unit matrix. Then (28.1) can be written in the integrated form

$$X^{\pi}(t) = x_0 + \int_0^t \pi'(\tau)\theta(\tau)\,\mathrm{d}\tau + \int_0^t \pi'(\tau)\,\mathrm{d}W(\tau), \qquad 0 \le t \le T, \tag{28.21}$$

for each $\pi \in \Pi$. We emphasize that nothing of central significance is lost by this simplification, and that all of the following considerations carry over to the general model (28.1), but the simplified model (28.21) results in a substantial gain in transparency. We therefore address problem (28.8) subject to the following conditions (some of which have already been stated):

Condition 28.2 *(i) For each $\pi \in \Pi$ the wealth process $\{X^{\pi}(t), t \in [0,T]\}$ is defined by* (28.21)*, in which the process W is an $\mathbb{R}^N$-valued standard Wiener process on $(\Omega, \mathcal{F}, P)$ and the market price of risk process θ is $\mathbb{R}^N$-valued, uniformly bounded, and $\mathcal{F}_t$-progressively measurable (recall* (28.2)*).*
(ii) The contingent claim γ (see (28.4)*) and random variable B (see* (28.8)*) are $\mathcal{F}_T$-measurable and square-integrable namely $E[\gamma^2] < \infty$ and $E[B^2] < \infty$.*
(iii) The constraint set $K \subset \mathbb{R}^N$ is closed and convex with $0 \in K$ (see (28.6)*).*

Remark 28.3 For problem (28.8) to even make sense it must of course be *assumed* that $X^{\pi}(T) \ge B$ for some $\pi \in \mathcal{A}$ (otherwise the constraints are incompatible). Assumptions of this kind are called "feasibility conditions" and are well known in finite-dimensional convex programming, from which it is also well known that a slight strengthening of the feasibility condition (called a "Slater condition") is essential for securing existence of Lagrange multipliers. In precisely the same spirit, we are going to slightly strengthen the feasibility condition for problem (28.8) to the following "Slater-type" condition:

Condition 28.3 *There is some nonrandom $\epsilon \in (0, \infty)$ and some $\tilde{\pi} \in \mathcal{A}$ such that $X^{\tilde{\pi}}(T) \ge B + \epsilon$ a.s.*

This is a mild condition which really just compels one to make a "non-greedy" stipulation for the "floor-level" wealth B when defining problem (28.8). It will be seen later that Condition 28.3 is *absolutely essential* for verifying the all-important condition (28.18) of Theorem 28.1.

We shall now express problem (28.8) as an abstract convex programming problem over the vector space of primal variables $\mathbb{X} := \Pi$ exactly along the lines of the reduction at (28.9), that is $f : \Pi \to (-\infty, \infty]$ is defined as

$$f(\pi) := \begin{cases} \mathrm{E}J(X^{\pi}(T)), & \text{when } \pi \in \mathcal{A} \text{ and} X^{\pi}(T) \geq B \text{ a.s.}, \\ +\infty, & \text{otherwise}, \end{cases} \tag{28.22}$$

for each $\pi \in \Pi$. Problem (28.8) then amounts to the minimization of f over the space of primal variables Π. We shall now "perturb" this problem in accordance with Step 1 of Sect. 28.2. To this end, define the normed vector space of perturbations

$$\mathbb{U} := L_2(\Omega, \mathcal{F}_T, P) \times L_\infty(\Omega, \mathcal{F}_T, P), \tag{28.23}$$

and define the perturbation function $F : \Pi \times \mathbb{U} \to (-\infty, \infty]$ by

$$F(\pi, u) := \begin{cases} \mathrm{E}J(X^{\pi}(T) - u_1), & \text{when } \pi \in \mathcal{A} \text{ and } X(T) \geq B + u_2 \text{ a.s.}, \\ +\infty, & \text{otherwise}, \end{cases} \tag{28.24}$$

for each $\pi \in \Pi$, and $u = (u_1, u_2) \in \mathbb{U} := L_2 \times L_\infty$. It is immediate that F is convex on $\Pi \times \mathbb{U}$, and from (28.22) and (28.24), we have the consistency relation

$$f(\pi) = F(\pi, 0), \qquad \pi \in \Pi, \tag{28.25}$$

(c.f. (28.11)). This completes Step 1.

The perturbation space $\mathbb{U}$ defined at (28.23) and perturbation function F defined at (28.24) are an adaptation to the dynamical problem (28.8) of the perturbations used for static problems of abstract convex programming (see Ekeland et al. [2], Rockafellar [8] and Rockafellar et al. [9]). We shall see later that the all-important condition (28.18) of Theorem 28.1 can be verified with these definitions of perturbation space and perturbation function.

We now proceed to Step 2 of Sect. 28.2 and define the space of dual variables as the norm-dual of the perturbation space at (28.23), namely

$$\mathbb{Y} := \mathbb{U}^* = L_2(\Omega, \mathcal{F}_T, P) \times L_\infty^*(\Omega, \mathcal{F}_T, P), \tag{28.26}$$

as well as the Lagrangian $K : \Pi \times \mathbb{Y} \to [-\infty, \infty]$ (c.f. (28.13)):

$$K(\pi, (Y, Z)) := \inf_{\substack{u_1 \in L_2 \\ u_2 \in L_\infty}} [\mathrm{E}(u_1 Y) + Z(u_2) + F(\pi, u)], \quad \pi \in \Pi, \quad Y \in L_2, \quad Z \in L_\infty^*. \tag{28.27}$$

Remark 28.4 For short, we write L_2 for $L_2(\Omega, \mathcal{F}_T, P)$, and similarly for L_∞. Also $L_\infty^*(\Omega, \mathcal{F}_T, P)$ (or just L_∞^*) denotes the usual normed vector space of norm-continuous linear functionals $Z : L_\infty(\Omega, \mathcal{F}_T, P) \to \mathbb{R}$. The notation $Z \leq 0$ for some $Z \in L_\infty^*$ denotes $Z(v) \leq 0$ for all a.s. nonnegative $v \in L_\infty$.

Upon substituting (28.24) into (28.27) and simplifying, we get the Lagrangian in explicit form as follows: for each $\pi \in \Pi$ and $(Y, Z) \in \mathbb{Y} := L_2 \times L_\infty^*$ we have

$$K(\pi, (Y, Z)) = \begin{cases} \mathrm{E}[X^{\pi}(T)Y] - \mathrm{E}[J^*(Y)] \\ \qquad + \inf\limits_{\substack{u_2 \in L_\infty \\ u_2 \leq X^{\pi}(T) - B}} Z(u_2), & \text{if } \pi \in \mathcal{A}_1,\ Z \leq 0, \\ -\infty, & \text{if } \pi \in \mathcal{A}_1,\ Z \not\leq 0, \\ +\infty, & \text{if } \pi \in \Pi \backslash \mathcal{A}_1, \end{cases} \tag{28.28}$$

in which the set $\mathcal{A}_1$ (arising naturally in the calculation of the Lagrangian) is the subset of the set of admissible portfolios $\mathcal{A}$ (see (28.6)) defined by

$$\mathcal{A}_1 := \{\pi \in \mathcal{A} \mid \text{ there is some } \alpha \in \mathbb{R} \text{ s.t. } X^{\pi}(T) - B \geq \alpha \quad \text{a.s.}\}. \tag{28.29}$$

That is $\mathcal{A}_1$ comprises admissible portfolios $\pi \in \mathcal{A}$ such that $X^{\pi}(T) - B$ is a.s. lower bounded by a constant $\alpha \in \mathbb{R}$ (this is necessary for the infimum on the right of (28.28) to make sense). Furthermore, the function $J^* : \mathbb{R} \times \Omega \to \mathbb{R}$, also arising naturally in the calculation of the Lagrangian, is the *Fenchel conjugate* of the mapping $x \to J(x, \omega)$ (see (28.4)) defined as follows:

$$J^*(y, \omega) := \sup_{x \in \mathbb{R}}[xy - J(x, \omega)] = \frac{(y + 2\gamma(\omega))^2}{4}, \quad (y, \omega) \in \mathbb{R} \times \Omega. \tag{28.30}$$

Finally, in accordance with Step 3 of Sect. 28.2, we define the dual function $g : \mathbb{Y} \to [-\infty, \infty)$ as follows (c.f. (28.14)):

$$g(Y, Z) := \inf_{\pi \in \Pi} K(\pi, (Y, Z)), \quad (Y, Z) \in \mathbb{Y} := L_2 \times L_\infty^*. \tag{28.31}$$

Upon combining (28.28) and (28.31) we get the dual function g in the explicit form

$$g(Y, Z) = \begin{cases} -\varkappa(Y, Z) - \mathrm{E}J^*(Y), & \text{if } Z \leq 0, \\ -\infty, & \text{if } Z \not\leq 0, \end{cases} \tag{28.32}$$

for all $(Y, Z) \in \mathbb{Y} := L_2 \times L_\infty^*$, in which the mapping

$$\varkappa(Y, Z) := \sup_{\pi \in \mathcal{A}_1} \left\{ -\mathrm{E}[X^{\pi}(T)Y] - \inf_{\substack{u_2 \in L_\infty \\ u_2 \leq X^{\pi}(T) - B}} Z(u_2) \right\}, \quad (X, Z) \in \mathbb{Y}, \tag{28.33}$$

is the *support functional* of the set $\mathcal{A}_1$ (well known from finite-dimensional convex programming). From (28.25), (28.27) and (28.31) we get the relation

$$f(\pi) \geq K(\pi, (Y, Z)) \geq g(Y, Z), \qquad \pi \in \Pi, \qquad (Y, Z) \in \mathbb{Y}, \tag{28.34}$$

hence, in particular, we have the weak duality principle

$$\inf_{\pi \in \Pi} f(\pi) \geq \sup_{(Y,Z) \in \mathbb{Y}} g(Y, Z). \tag{28.35}$$

With reference to the abstract formulation at (28.20) we shall now establish optimality relations which are fully equivalent to equality of the primal function (28.22) and the dual function (28.32). After some calculation (not given here) we obtain the following equivalence: for arbitrary $\bar{\pi} \in \Pi$ and $(\bar{Y}, \bar{Z}) \in \mathbb{Y}$ one has

$$f(\bar{\pi}) = g(\bar{Y}, \bar{Z}) \iff \begin{cases} (1)\ X^{\bar{\pi}}(T) \geq B, \quad (2)\ \bar{\pi} \in \mathcal{A}, \quad (3)\ \bar{Z} \leq 0, \\ (4) \quad \inf\limits_{\substack{u_2 \in L_\infty \\ u_2 \leq X(T) - B}} \bar{Z}(u_2) = 0, \\ (5)\ \mathrm{E}[X^{\bar{\pi}}(T)\bar{Y}] + \varkappa(\bar{Y}, \bar{Z}) = 0, \\ (6)\ X^{\bar{\pi}}(T) = (\partial J^*)(\bar{Y}), \end{cases} \tag{28.36}$$

in which, from (28.30),

$$\partial J^*(y, \omega) = (y + 2\gamma(\omega))/2, \quad (y, \omega) \in \mathbb{R} \times \Omega. \tag{28.37}$$

Remark 28.5 The optimality relations (28.36)(1)–(6) are similar to those encountered in finite-dimensional convex optimization. In particular (28.36)(1), (2) are *feasibility conditions* on the primal variable, and insist that $\bar{\pi} \in \Pi$ must satisfy the constraints in the primal problem, while (28.36)(3) is a familiar non-positivity condition on the dual variable (i.e., Lagrange multiplier) $\bar{Z} \in L_\infty^*$ (which "enforces" the inequality constraint $X^{\bar{\pi}}(T) \geq B$). On the other hand (28.36)(4), (5) are *complementary slackness relations* and (28.36)(6) is a *transversality relation*. It remains to construct some $(\bar{\pi}, (\bar{Y}, \bar{Z})) \in \Pi \times \mathbb{Y}$ such that $f(\bar{\pi}) = g(\bar{Y}, \bar{Z})$, for it then follows from (28.35) that $\bar{\pi}$ minimizes f on the primal space Π and is therefore an optimal portfolio for problem (28.8). We address this in the next section.

28.4 Construction of the Optimal Portfolio

We shall use Theorem 28.1 to establish that the duality gap is zero and the dual functional g attains its supremum (over $\mathbb{Y}$) at some maximizer $(\bar{Y}, \bar{Z}) \in \mathbb{Y}$. To this end, observe from Condition 28.3 and (28.24) that

$$F(\tilde{\pi}, u) = \mathrm{E}J(X^{\tilde{\pi}}(T) - u_1) \ \text{ for all } (u_1, u_2) \in L_2 \times L_\infty \text{ s.t. } \|u_2\|_{L_\infty} < \epsilon, \tag{28.38}$$

and, since $u_1 \to \mathrm{E}J(X^{\tilde{\pi}}(T) - u_1) : L_2 \to \mathbb{R}$ is clearly norm-continuous, it follows from (28.38) that

$$\sup_{\substack{(u_1, u_2) \in \mathbb{U} = L_2 \times L_\infty \\ \|u_1\|_{L_2} < \epsilon,\ \|u_2\|_{L_\infty} < \epsilon}} F(\tilde{\pi}, u) < \infty, \tag{28.39}$$

which verifies (28.18), so that Theorem 28.1 gives

$$\inf_{\pi\in\Pi} f(\pi) = \sup_{(Y,Z)\in\mathbb{Y}} g(Y,Z) = g(\bar{Y},\bar{Z}), \quad \text{for some } (\bar{Y},\bar{Z}) \in \mathbb{Y} = L_2 \times L^*_\infty. \tag{28.40}$$

Having secured a maximizer $(\bar{Y},\bar{Z}) \in \mathbb{Y}$ of the dual function g, it remains to construct some $\bar{\pi} \in \Pi$ such that relations (28.36)(1)–(6) hold, for then we have $f(\bar{\pi}) = g(\bar{Y},\bar{Z})$ as required (recall Remark 28.5). To this end define the processes

$$H(t) := \exp\left\{-\int_0^t \theta'(\tau)\,\mathrm{d}W(\tau) - (1/2)\int_0^t \|\theta(\tau)\|^2\,\mathrm{d}\tau\right\}, \quad t \in [0,T], \tag{28.41}$$

$$\bar{X}(t) := H^{-1}(t)\mathrm{E}\left[\partial J^*(\bar{Y})H(T) \mid \mathcal{F}_t\right], \quad t \in [0,T], \tag{28.42}$$

(here $\theta(\cdot)$ is the market price of risk parameter, recall (28.21)). From (28.42) we see that $\bar{X}H$ is a $\mathcal{F}_t$-martingale, and therefore, in view of (28.2), it follows from the the Itô martingale representation theorem that there exists a unique $\mathbb{R}^N$-valued, $\mathcal{F}_t$-progressively measurable and path-wise square-integrable "integrand process" $\{\bar{\psi}(t), t \in [0,T]\}$ such that

$$\bar{X}(t)H(t) = \bar{X}(0) + \int_0^t \bar{\psi}'(\tau)\,\mathrm{d}W(\tau), \quad t \in [0,T]. \tag{28.43}$$

Upon expanding (28.43) by Itô's formula (using (28.41)) we find that

$$\bar{X}(t) = \bar{X}(0) + \int_0^t \bar{\pi}'(\tau)\theta(\tau)\,\mathrm{d}\tau + \int_0^t \bar{\pi}'(\tau)\,\mathrm{d}W(\tau), \quad 0 \le t \le T, \tag{28.44}$$

in which

$$\bar{\pi}(t) := [\sigma'(t)]^{-1}\left\{H^{-1}(t)\bar{\psi}(t) + \bar{X}(t)\theta(t)\right\}, \quad t \in [0,T]. \tag{28.45}$$

Moreover, by a calculation which is not given here (and which relies on the fact that θ is uniformly bounded, see Condition 28.2(i)) one finds

$$\mathrm{E}\int_0^T \|\bar{\pi}(t)\|^2\,\mathrm{d}t < \infty \quad \text{so that} \quad \bar{\pi} \in \Pi \text{ (see (28.3))}. \tag{28.46}$$

It remains to verify that $\bar{\pi} \in \Pi$ defined at (28.45) and the maximizer $(\bar{Y},\bar{Z}) \in \mathbb{Y}$ of the dual function g (see (28.40)) satisfy relations (28.36)(1)–(6). From (28.22) and feasibility (see Condition 28.3) it is immediate that $\inf_{\pi\in\Pi} f(\pi) \in \mathbb{R}$, so that (28.40) gives $g(\bar{Y},\bar{Z}) \in \mathbb{R}$. In view of (28.32) it then follows that

$$\bar{Z} \le 0. \tag{28.47}$$

We now use the optimality of $(\bar{Y}, \bar{Z})$: from (28.40) one has

$$g(\bar{Y}, \bar{Z}) \geq g(\bar{Y} + \epsilon Y, \bar{Z}), \qquad \text{for all } \epsilon \in (0, \infty) \text{ and } Y \in L_2. \tag{28.48}$$

A calculation based on (28.48) (which is not given here) then establishes that

$$\bar{X}(0) = x_0, \qquad \bar{\pi} \in \mathcal{A}. \tag{28.49}$$

In view of the first relation of (28.49), with (28.44) and (28.21), we get

$$\bar{X}(t) = X^{\bar{\pi}}(t), \qquad t \in [0, T], \tag{28.50}$$

and therefore

$$X^{\bar{\pi}}(T) = \bar{X}(T) = (\partial J^*)(\bar{Y}), \tag{28.51}$$

in which the second equality at (28.51) follows from (28.42) with $t := T$. We again use the optimality of $(\bar{Y}, \bar{Z})$: from (28.40) one has

$$g(\bar{Y}, \bar{Z}) \geq g(\bar{Y} + \epsilon Y, \bar{Z} - \epsilon Y), \qquad \text{for all } \epsilon \in (0, \infty) \text{ and } Y \in L_2. \tag{28.52}$$

Notice that in (28.52), we identify $Y \in L_2$ as an element of L_∞^*, since the mapping $v \to \mathrm{E}[vY] : L_\infty \to \mathbb{R}$ is norm-continuous, thus a member of L_∞^*, for each $Y \in L_2$. A calculation based on (28.52) (which is again not given here) then establishes that

$$X^{\bar{\pi}}(T) \geq B. \tag{28.53}$$

From (28.40) again one has

$$g(\bar{Y}, \bar{Z}) \geq g(\bar{Y} - \epsilon \bar{Y}, \bar{Z} - \epsilon \bar{Z}), \qquad \text{for all } \epsilon \in (0, \infty), \tag{28.54}$$

and a calculation based on (28.54), also not given here, then establishes

$$\mathrm{E}[X^{\bar{\pi}}(T)\bar{Y}] + \varkappa(\bar{Y}, \bar{Z}) = \inf_{\substack{u_2 \in L_\infty \\ u_2 \leq X(T) - B}} \bar{Z}(u_2) = 0. \tag{28.55}$$

Remark 28.6 In the preceding we have verified (28.36)(1) (see (28.53)), (28.36)(2) (see (28.49)), (28.36)(3) (see (28.47)), (28.36)(4), (5) (see (28.55)) and (28.36)(6) (see (28.51)). From the equivalence at (28.36) we get $f(\bar{\pi}) = g(\bar{Y}, \bar{Z})$, which establishes that $\bar{\pi}$ is an optimal portfolio for problem (28.8) (recall Remark 28.5).

28.5 Concluding Remarks

Problem (28.8) discussed above is addressed in [3] in full generality. Here, we focus on a special case with the simplified dynamics (28.21), instead of the generality of the relation (28.1), and we have tried to bring out just the essential ideas while avoiding many of the technicalities in [3] which obscure the basic simplicity of the approach.

In place of the end-of-trade constraint $X^{\pi}(T) \geq B$ in problem (28.8), one can stipulate a "floor- level" process $\{B(t), t \in [0, T]\}$ and address the problem

$$\text{minimize } \mathrm{E}[J(X^{\pi}(T))] \quad \text{over all } \pi \in \mathcal{A} \text{ s.t. } X^{\pi}(t) \geq B(t), t \in [0, T] \text{a.s.} \tag{28.56}$$

This amounts to quadratic hedging with the portfolio constraint $\pi \in \mathcal{A}$ and a state constraint $X^{\pi}(t) \geq B(t), t \in [0, T]$ over the *full trading interval.* This is substantially more challenging than problem (28.8) because the state constraint can "bind" anywhere over the trading interval $t \in [0, T]$ (not just at $t = T$ as in problem (28.8)). Nevertheless, the approach based on the Rockafellar variational method and used for (28.8) does carry over to problem (28.56), although with significant additional technical effort. This problem is addressed in [10].

References

1. Dubovitskii, A.Y., Mil'yutin, A.A.: Necessary conditions for a weak extremum in problems of optimal control with mixed inequality constraints. Zhur. Vychislitel. Mat. Mat-Fys. **8**, 725–779 (1968). (transl: USSR Comp. Math. Math. Phys. **8**, 24–98 (1968))
2. Ekeland, I., Témam, R.: Convex Analysis and Variational Problems. North-Holland, Amsterdam (1976)
3. Heunis, A.J.: Quadratic minimization with portfolio and terminal wealth constraints. Ann. Financ. **11**, 243–282 (2015)
4. Makowski, K., Neustadt, L.W.: Optimal control with mixed control-phase variable equality and inequality constraints. SIAM J. Control Optim. **12**, 184–228 (1974)
5. Labbé, C., Heunis, A.J.: Convex duality in constrained mean-variance portfolio optimization. Adv. Appl. Probab. **39**, 77–104 (2007)
6. Lim, A.E.B., Zhou, X.Y.: Mean-variance portfolio selection with random parameters in a complete market. Math. Oper. Res. **27**, 101–120 (2002)
7. Markowitz, H.: Portfolio selection. J. Financ. **7**, 77–91 (1952)
8. Rockafellar, R.T.: Conjugate Duality and Optimization, (CBMS-NSF Ser. No. 16). SIAM, Philadelphia (1974)
9. Rockafellar, R.T., Wets, J.B.: Stochastic convex programming: singular multipliers and extended duality. Pac. J. Math. **62**, 507–522 (1976)
10. Zhu, D., Heunis, A.J.: Quadratic minimization with portfolio and intertemporal wealth constraints. Ann. Financ. (to appear)

The manufacturer's authorised representative in the EU is Springer Nature Customer Service Centre GmbH, Europaplatz 3, 69115 Heidelberg, Germany. If you have any concerns regarding our products, please contact ProductSafety@springernature.com

Printed and bound by CPI Group (UK) Ltd, Croydon, CR0 4YY
05/10/2026
02228133-0009